Electroweak Physics and the Early Universe

NATO ASI Series

Advanced Science Institutes Series

A series presenting the results of activities sponsored by the NATO Science Committee, which aims at the dissemination of advanced scientific and technological knowledge, with a view to strengthening links between scientific communities.

The series is published by an international board of publishers in conjunction with the NATO Scientific Affairs Division

A	**Life Sciences**	Plenum Publishing Corporation
B	**Physics**	New York and London
C	**Mathematical and Physical Sciences**	Kluwer Academic Publishers
D	**Behavioral and Social Sciences**	Dordrecht, Boston, and London
E	**Applied Sciences**	
F	**Computer and Systems Sciences**	Springer-Verlag
G	**Ecological Sciences**	Berlin, Heidelberg, New York, London,
H	**Cell Biology**	Paris, Tokyo, Hong Kong, and Barcelona
I	**Global Environmental Change**	

Recent Volumes in this Series

Volume 333—Perspectives in the Structure of Hadronic Systems
edited by M. N. Harakeh, J. H. Kock, and O. Scholten

Volume 334—Frontier Topics in Nuclear Physics
edited by Werner Scheid and Aurel Sandulescu

Volume 335—Hot and Dense Nuclear Matter
edited by Walter Greiner, Horst Stöcker, and André Gallmann

Volume 336—From Newton to Chaos: Modern Techniques for Understanding and Coping with Chaos in N-Body Systems
edited by Archie E. Roy and Bonnie A. Stevens

Volume 337—Density Functional Theory
edited by Eberhard K. U. Gross and Reiner M. Dreizler

Volume 338—Electroweak Physics and the Early Universe
edited by Jorge C. Romão and Filipe Freire

Volume 339—Nonlinear Spectroscopy of Solids: Advances and Applications
edited by Baldassare Di Bartolo

Volume 340—Confined Electrons and Photons: New Physics and Applications
edited by Elias Burstein and Claude Weisbuch

Volume 341—Turbulence: A Tentative Dictionary
edited by P. Tabeling and O. Cardoso

Series B: Physics

Electroweak Physics and the Early Universe

Edited by

Jorge C. Romão

Instituto Superior Técnico
Lisbon, Portugal

and

Filipe Freire

Universität Heidelberg
Heidelberg, Germany

Plenum Press
New York and London
Published in cooperation with NATO Scientific Affairs Division

Proceedings of a NATO Advanced Research Workshop on
Electroweak Physics and the Early Universe,
held March 23–25, 1994,
in Sintra, Portugal

NATO-PCO-DATA BASE

The electronic index to the NATO ASI Series provides full bibliographical references (with keywords and/or abstracts) to more than 30,000 contributions from international scientists published in all sections of the NATO ASI Series. Access to the NATO-PCO-DATA BASE is possible in two ways:

—via online FILE 128 (NATO-PCO-DATA BASE) hosted by ESRIN, Via Galileo Galilei, I-00044 Frascati, Italy

—via CD-ROM "NATO Science and Technology Disk" with user-friendly retrieval software in English, French, and German (©WTV GmbH and DATAWARE Technologies, Inc. 1989). The CD-ROM also contains the AGARD Aerospace Database.

The CD-ROM can be ordered through any member of the Board of Publishers or through NATO-PCO, Overijse, Belgium.

Library of Congress Cataloging-in-Publication Data

On file

ISBN 0-306-44909-9

A Division of Plenum Publishing Corporation
233 Spring Street, New York, N. Y. 10013

Printed in the United States of America

PREFACE

Electroweak Phase Transition and the Early Universe, a NATO Advanced Research Workshop, was held March 23-25, 1994, at the Hotel Tivoli in Sintra, Portugal. The meeting was co-sponsored by three other Lisbon-based institutions: the Fundação Gulbenkian, J. N. I. C. T. (Junta Nacional para Investigação Cientifica e Tecnológica) and G. T. A. E. (Grupo Teórico de Alta Energias). The workshop brought together a large number of theoretical physicists who are actively researching topics relevant to the understanding of the standard model of electroweak interactions in the early universe. We were pleased and overwhelmed by the positive, and sometimes instantaneous response that our enterprise raised right from its inception.

The old town of Sintra provided a serene and pleasant environment for the participants. Some heated and controversial discussions on many unanswered questions in the standard model took place throughout the three days of the workshop. If one consensus emerged from the meeting, it was the imperative need for non-perturbative techniques for the understanding of the electroweak phase transition.

To our knowledge, this was the first meeting that specialized on electroweak phase transition-related topics, an event that takes us back to as early as 10^{-12} seconds after the Big Bang (if we are to accept the standard cosmological model) or to an estimated temperature of the universe of 200 GeV. This was surprising because it was first conjectured more than 20 years ago by D. Kirzhnits from the Lebedev Institute in Moscow. Kirzhnits reached his theory by drawing an analogy between a then recent standard model that unifies weak and electromagnetic interactions and the more established BCS model for superconductivity. By looking at the consequences of considering a large system of weakly interacting particles in thermodynamical equilibrium, he arrived at the conclusion that symmetry ought to be restored at temperatures higher than 10^3 GeV.

The field acquired a more quantitative ground after the works of S. Coleman and E. Weinberg, S. Weinberg, and L. Dolan and R. Jackiw, which up to the present provided the main paradigm for the study of the hot standard model. Most of what was learned about the electroweak phase transition during the 1970's is due to the work of A. Linde. His papers are standard reference on topics related to cosmological phase transitions.

Why didn't this exciting problem raise much interest within the community at the time? It was believed that no major outcome of cosmological relevance was likely to manifest itself as a consequence of the electroweak phase transition. Only for a very light Higgs boson could one expect some interesting consequences. In 1985, V. Kuzmin, V. Rubakov and M. Shaposhnikov, realised that any baryon asymmetry of the universe generated prior to the electroweak phase transition is washed out through thermal baryon number-violating transitions. This view is now widely accepted and numerical simulations have strongly supported this scenario. Of course, if we accept B+L to be conserved, this requires electroweak baryogenesis.

The possibility that a strong baryon number-violating process occurred during the electroweak phase transition had been pointed out in the 1970's but it was only after the work of V. Kuzmin, V. Rubakov and M. Shaposhnikov that its relevance became apparent. It is well known that the electroweak phase transition satisfies the three Sakharov conditions for baryogenesis provided the phase transition is first order. Hitherto, all proposed baryogenesis scenarios require a further understanding of the nature of the phase transition. In other words, it is necessary to know more about the details of the phase transition in order to make reliable quantitative tests of the different mechanisms. This could require further research in topics like vortex solutions, formation and evolution of critical bubbles, non-perturbative effects in hot non-abelian gauge theories, sphaleron transitions, extensions of the standard model, development of new non-perturbative techniques, amongst others.

This briefly outlines the motivation behind the current interest in the study of the standard model at high temperatures, in particular near the electroweak phase transition temperature, that preceded the Sintra workshop. We are still far from a complete description of the most prominent processes, and their respective relics, that occurred at high temperatures in the electroweak world. Consequently, we would be happy if more meetings specializing in the topics discussed in Sintra would follow elsewhere and that Sintra '94 was just the first of a series of stimulating encounters. Hopefully, these proceedings will provide a useful source of consultation for future research and incentive to new physicists in this difficult, but challenging, field.

We are indebted to Luís Bettencourt, Pedro Ferreira and Margaret Kernan for their most valuable support in the organization of this meeting. Some of the material used during this meeting was provided by the Instituto Superior Técnico and we are grateful to Ana Mourão for her help in this matter. To the staff of the Hotel Tivoli we express our appreciation for the optimal conditions and good service that they provided. We are also grateful to the Câmara Municipal de Sintra for hosting a reception for the participants in the pleasant venue of the Palácio Valenças. Finally, one of us (FF), acknowledges that this workshop was born out of an initial proposal by Rocky Kolb during the PASCOS '92. Acknowledgement is also due to Tom Kibble for his support and advice for this enterprise during its early stages.

Filipe Freire and Jorge Romão

July 1994

CONTENTS

3D PHYSICS AND THE ELECTROWEAK PHASE TRANSITION: PERTURBATION THEORY AND LATTICE MONTE CARLO ANALYSIS

K. Kajantie

Department of Theoretical Physics,
P.O.Box 9, 00014 University of Helsinki, Finland

M. Shaposhnikov

CERN/TH, CH-1211 Geneve 23, Switzerland

INTRODUCTION

We know the Lagrangian of the SU(3)×SU(2)×U(1) standard model (SM) and, therefore, also the equilibrium thermodynamics of the quantum system with the spectrum of the SM: it is given by the functional integral

$$e^{-F/T} = \int \mathcal{D}\Phi \exp\left[-\frac{1}{\hbar}\int_0^{\beta\hbar} d\tau \int d^3x\, \mathcal{L}(\Phi(\tau,\mathrm{x}))\right], \tag{1}$$

where the symbol Φ denotes generically all the fields of the SM. In a sense the problem of the electroweak (EW) phase transition thus is trivial: just calculate the above functional integral to give the free energy F and all the thermodynamics.

The technique which first suggests itself is a perturbative evaluation of the functional integral [1]. In fact, for the usual zero T SM perturbation theory works extremely well in the SU(2)×U(1) sector. This is due to the fact that we live in the broken phase of this symmetry so that various particles have masses and momentum integrals are more regular. Thus there are all reasons to expect that perturbation theory works at least for low T. Low here means T clearly below the symmetry restoration temperature T_c. Even the lowest order perturbation theory predicts that something physically quite natural happens: the symmetry is restored at high T [2]. But is perturbation theory correct for T close to T_c? If not, what else can be done?

There has been lots of progress in the application of perturbation theory to this problem [3]-[21]. Qualitatively the conclusion is that perturbation theory may work in the broken phase up to T_c and even up to the highest temperature T_+ at which the broken phase exists, provided the Higgs boson mass is small enough. However, it does not work in the symmetric phase and thus also not for any problem requiring a comparison of the symmetric and broken (disordered and ordered) phases, like the determination of T_c, the latent heat L or the interface tension σ. This requires the use of nonperturbative techniques, [19]-[24], lattice Monte Carlo methods, in particular.

One additional aspect of the progress is that replacing the hot 4d theory by an effective 3d theory offers definite advantages. These are both of conceptual and practical nature.

The purpose of these lectures is to review the work on this topic in refs. [20]-[24]. First the development of the perturbative treatment starting from the early work of [2] is outlined. Finally the Monte Carlo treatment of the 3d effective theory is discussed. The whole topic is technically rather complicated and a unified discussion can be found in the original references. Much of the material here is on purpose complementary in nature.

One obvious reservation is in order: of course neither the Higgs mass m_H nor the precise nature of symmetry breaking are known and thus all kinds of surprises might be waiting for us near T_c – and are even desirable. Here we limit ourselves to the standard picture and give the results parametrized by the unknown m_H. The values of m_H we are interested in are from $m_H = m_W = 80$ GeV upwards. For $T = 0$ parameters we choose

$$g = \frac{2}{3}, \qquad m_W = 80.6\,\text{GeV}, \qquad \lambda = \frac{g^2 m_H^2}{8 m_W^2}. \tag{2}$$

implying $v(T=0) = 241.6$ GeV. We shall also take $g' = 0$ and neglect fermions as inessential for the thermodynamics of symmetry restoration – their number density at large wave lengths is negligible. Actually the top quark will affect the numerical value of various temperature scales, but not the details of symmetry restoration.

We proceed now to describe how the computation of the parameters of the transition has developed.

1-LOOP PERTURBATION THEORY

Calculating the scalar self energy in the leading high T approximation along the lines in [2] gives an effective potential of the form

$$V(\phi, T) = \frac{1}{2}\gamma(T^2 - T_0^2)\phi^2 + \frac{1}{4}\lambda\phi^4, \tag{3}$$

where

$$\gamma = \frac{3}{16}g^2 + \frac{1}{2}\lambda + \frac{m_{\text{top}}^2}{2v^2} \tag{4}$$

and, for $m_H = 80$ GeV,

$$\begin{aligned} T_0 = \frac{m_H}{\sqrt{2\gamma}} &= 170.0 \text{ GeV}, \quad m_{\text{top}} = 0 \\ &= 94.6 \text{ GeV}, \quad m_{\text{top}} = 170 \text{ GeV}. \end{aligned} \tag{5}$$

Thus the transition is of second order with a critical temperature equal to T_0. It may seem ridiculous to give temperature values to such an accuracy. However, the most accurate perturbative value available at present from 2-loop optimized perturbation theory is $T_c = 173.3$ GeV [21] and the non-perturbative value is $T_c = 163\pm$ a few

GeV [23]. Although the simple formula totally underestimates the complexity of the problem, it nevertheless gives the overall scale rather correctly.

1-LOOP IMPROVED PERTURBATION THEORY

(i) The next stage of approximation involves still 1-loop perturbation theory but in next to leading high T approximation. To give the result, define first the masses

$$m_T = \frac{1}{2}g\phi, \qquad m_L^2 = \frac{1}{4}g^2\phi^2,$$
$$m_1^2 = -\frac{m_H^2}{2} + 3\lambda\phi^2, \quad m_2^2 = -\frac{m_H^2}{2} + \lambda\phi^2. \tag{6}$$

The 1-loop potential in Landau gauge is then

$$\begin{aligned} V_1(\phi) &= \frac{1}{2}\gamma(T^2 - T_0^2)\phi^2 + \frac{1}{4}\lambda\phi^4 - \\ &\quad -\frac{1}{12\pi}(6m_T^3 + 3m_L^3 + m_1^3 + 3m_2^3). \end{aligned} \tag{7}$$

As a further approximation to eq.(7), valid for sufficiently small Higgs masses, set there $m_1 = m_2 = 0$ [5] - [8]. The potential then is of the general form

$$V(\phi, T) = \frac{1}{2}\gamma(T^2 - T_0^2)\phi^2 - \frac{1}{3}\alpha T\phi^3 + \frac{1}{4}\lambda\phi^4 \tag{8}$$

with $\alpha = 9g^3/(32\pi)$. The potential predicts a first order transition with

$$T_c = \frac{T_0}{\sqrt{1 - \frac{2}{9}\frac{\alpha^2}{\lambda\gamma}}} \tag{9}$$

and

$$\frac{\phi(T_c)}{T_c} = \frac{2}{3}\frac{\alpha}{\lambda}. \tag{10}$$

(ii) The potential (7), however, does not take into account the Debye screening phenomenon of non-abelian electric fields in the plasma [9]-[12]. This effect reduces the strength of the first order phase transition [11] and for sufficiently small Higgs masses the effective potential is given by (7) with $m_1 = m_2 = 0$ and $m_L^2 = m_D^2 + \frac{1}{4}g^2\phi^2$, where the electric screening mass is $m_D^2 = \frac{5}{6}g^2T^2$.

(iii) The next step is to sum up the leading temperature corrections to the scalar masses [10],[13]-[15]. The result is still given by (7) but with

$$m_1^2 = \gamma(T^2 - T_0^2) + 3\lambda\phi^2, \quad m_2^2 = \gamma(T^2 - T_0^2) + \lambda\phi^2. \tag{11}$$

The improved 1-loop potential thus systematically predicts a first order transition, but the predicted parameters vary depending on approximation. To illustrate this, some relevant numbers are given in Table 1 for the three choices described above.

The strengths of the phase transition depends crucially on the Higgs mass m_H. If we take the vacuum expectation value of the Higgs field at $T = T_0$,

$$\frac{\phi(T_c)}{T_c} = \frac{g^3 + 4(\sqrt{3}+1)\lambda^{3/2}}{8\pi\lambda}, \tag{12}$$

Table 1. Numerical values of T_c and ϕ_c for $m_H = 80$ GeV and for various 1-loop potentials (see text). The last column gives the value of the potential barrier between the two degenerate minima at $\phi = 0, \phi_c$ in units of $g^6T^4/12$.

potential	T_c GeV	ϕ_c GeV	barrier
(i), $m_1 = m_2 = 0$	172.3	55	0.0012
(ii), $m_1 = m_2 = 0$	185.3	42	0.0003
(iii), full V_1	189.5	51	0.0005

as a measure of the strength of the phase transition, then it is large if $g^2 \gg \lambda$. Thus for small λ or small m_H the gauge degrees of freedom make the transition strong. Numerically, the two terms on the numerator of eq.(12) are equal if $m_H \approx 1.4 m_W \approx 110$ GeV. Above this value one expects the scalar degrees of freedom to dominate and the transition may become a second order one, specific for a pure scalar theory. However, this may not be the case, since in a region near $\phi = 0$ perturbation theory is not valid.

Even the qualitative conclusions derived from different approximations to the effective potential are different, so the question arises whether there is a stability against higher order corrections.

2-LOOP PERTURBATION THEORY

The 2-loop resumed potential in the 4d SU(2)-Higgs model has been given in the extensive computations of refs. [16]-[18]. The result can be made more transparent by first passing to a 3d effective theory, and only then computing the 2-loop potential [21].

The basis for the necessity to proceed via an effective 3d theory using MS scheme for renormalization is as follows [25, 21]. When one is using dimensional regularization and the minimal subtraction scheme, the counter terms are mass independent and the renormalization scheme introduces one single mass scale μ. In a 4d hot EW theory the mass scales are T, m_L, m_T, m_1, m_2 and one meets with logarithmic terms of the type

$$\log\frac{T}{\mu}, \quad \log\frac{m_L}{\mu}, \quad \log\frac{m_T}{\mu}, \ldots \tag{13}$$

To minimize the large logs one may choose $\mu = T$. This unavoidably leads to logarithmic terms of the type

$$\log\frac{m_L}{T}, \log\frac{m_T}{T}, \ldots, \tag{14}$$

which are large. The way to get around these large logs is to first integrate over the nonstatic field modes with mass $2\pi n T$, $n = 1, 2, \ldots$. This leads to a 3d effective theory with correction terms $\sim$ inverse powers of T, which are neglected.

In the EW case the corresponding effective 3d theory is an SU(2)+fundamental Higgs +adjoint Higgs theory with the couplings $g_3^2 = g^2T$, $\lambda_3 = \lambda T$ (the $\phi^2 A_0^2$ coupling is given in terms of g_3^2 and for the A_0^4 self coupling one approximately has $\lambda_A = 0$). The action of the theory is

$$S_{\text{eff}} = \int d^3x \Big\{ \frac{1}{4} F_{ij}^a F_{ij}^a + \frac{1}{2}(D_i A_0)^a (D_i A_0)^a + (D_i\phi)^\dagger (D_i\phi) +$$

$$\frac{1}{2}m_D^2 A_0^a A_0^a + \frac{1}{4}\lambda_A (A_0^a A_0^a)^2 + m_3^2 \phi^\dagger\phi + \lambda_3(\phi^\dagger\phi)^2 + h_3 A_0^a A_0^a \phi^\dagger\phi\Big\}. \tag{15}$$

For its total 2-loop effective potential in the Landau gauge one obtains [21]

$$V_{\rm tot}(\phi,\mu_3) = V_1(\phi,\mu_3) + V_2(\phi,\mu_3) \tag{16}$$

with the 1-loop part as in eq. (7):

$$\begin{aligned} V_1(\phi) &= \frac{1}{2}m_3^2(\mu_3)\phi^2 + \frac{1}{4}\lambda_3\phi^4 - \\ &\quad -\frac{1}{12\pi}(6m_T^3 + 3m_L^3 + m_1^3 + 3m_2^3), \end{aligned} \tag{17}$$

and the 2-loop part

$$\begin{aligned} V_2(\phi,\mu_3) = \frac{1}{16\pi^2}\Big\{ & \\ -\frac{3g_3^4}{16}\phi^2\Big[& 2\bar H(m_1,m_T,m_T) - \frac{1}{2}\bar H(m_1,m_T,0) + \bar H(m_1,m_L,m_L) \\ & +\frac{m_1^2}{m_T^2}[\bar H(m_1,m_T,0) - \bar H(m_1,m_T,m_T)] \\ & +\frac{m_1^4}{4m_T^4}[\bar H(m_1,0,0) + \bar H(m_1,m_T,m_T) - 2\bar H(m_1,m_T,0)] \\ & -\frac{m_1}{2m_T} - \frac{m_1^2}{4m_T^2}\Big] \\ -3\lambda_3^2\phi^2[& \bar H(m_1,m_1,m_1) + \bar H(m_1,m_2,m_2)] \\ +2g_3^2 m_T^2 & \Big[\frac{63}{16}\bar H(m_T,m_T,m_T) + \frac{3}{16}\bar H(m_T,0,0) - \frac{41}{16}\Big] \\ & +\frac{3}{2}g_3^2[(3m_T^2 + 4m_D^2)\bar H(m_L,m_L,m_T) + 2m_T m_L + m_T^2 + m_D^2] \\ +4g_3^2 m_T^2 & + \frac{3}{8}g_3^2(2m_T + m_L)(m_1 + 3m_2) + \frac{3}{4}\lambda_3(m_1^2 + 2m_1 m_2 + 5m_2^2) \\ -\frac{3}{8}g_3^2[& (m_T^2 - 2m_1^2 - 2m_2^2)\bar H(m_1,m_2,m_T) + (m_T^2 - 4m_2^2)\bar H(m_2,m_2,m_T) \\ & +\frac{(m_1^2 - m_2^2)^2}{m_T^2}[\bar H(m_1,m_2,m_T) - \bar H(m_1,m_2,0)] \\ & +(m_1^2 - m_2^2)(m_1 - m_2)/m_T + m_T(m_1 + 3m_2) - m_1 m_2 - m_2^2]\Big\}, \end{aligned} \tag{18}$$

where

$$\bar H(m_1,m_2,m_3) = \log\frac{\mu_3}{m_1 + m_2 + m_3} + \frac{1}{2}. \tag{19}$$

The masses of the 3d theory are

$$\begin{aligned} m_T &= \frac{1}{2}g_3\phi, \qquad m_L^2 = m_D^2 + \frac{1}{4}g_3^2\phi^2, \\ m_1^2 &= m_3^2(\mu_3) + 3\lambda_3\phi^2, \quad m_2^2 = m_3^2(\mu_3) + \lambda_3\phi^2, \end{aligned} \tag{20}$$

where

$$m_3^2(\mu_3) = \frac{1}{16\pi^2} f_{2m} \log\frac{\Lambda_m}{\mu_3} \tag{21}$$

and

$$
\begin{aligned}
f_{2m} &= \frac{81}{16}g_3^4 + 9\lambda_3 g_3^2 - 12\lambda_3^2 \\
&= \left[1-\left(\frac{m_H}{3m_W}\right)^2\right]\left[3\left(\frac{m_H}{3m_W}\right)^2+1\right]T^2.
\end{aligned} \tag{22}
$$

The constant Λ_m is the integration constant of a renormalization group equation and it can be obtained by comparing with the 4d computations in [16]:

$$
m_3^2(\mu_3) = \frac{3}{16}g_3^2 T + \frac{1}{2}\lambda_3 T + \frac{g_3^2}{16\pi^2}\left(\frac{149}{96}g_3^2 + \frac{3}{4}\lambda_3\right) + \frac{1}{16\pi^2}\left[f_{2m}\left(\log\frac{3T}{\mu_3}+c\right)\right] - \frac{1}{2}m_H^2, \tag{23}
$$

where $c = -0.348725$.

To this order $V_{\rm tot}(\phi,\mu_3)$ is μ_3-independent:

$$
\mu_3 \frac{\partial[V_{\rm tot}(\phi,\mu_3) - V_{\rm tot}(0,\mu_3)]}{\partial\mu_3} = O(3 - \text{loop order.}) \tag{24}
$$

MASSES OF THE THEORY

The derivation of the 3d effective theory is based on integrating away the nonstatic (in imaginary time) field modes with masses $2\pi nT$, $n = 1,2,\ldots$. The remaining masses are those in eq.(20) above. What are their relative magnitudes [27]?

Firstly (from now on we go back to 4d notation)

$$
\frac{m_T^2}{m_L^2} = \frac{\phi^2}{\frac{10}{3}T^2 + \phi^2}, \tag{25}
$$

and if the relevant field values satisfy $\phi < T$, the longitudinal mass is much larger than the transverse mass. Perturbatively $\phi(T_c)/T_c \approx 9g^3/(32\pi\lambda) < 1$ corresponds to $m_H^2/m_W^2 > 3/(2\pi)$ and for these larger Higgs masses one may explicitly integrate out also the A_0 field relative to A_i [21]. For the scalar masses, neglecting a few constants in eq.(23),

$$
\frac{m_1^2}{m_T^2} = \frac{\gamma(T^2 - T_0^2) + \frac{f_{2m}}{16\pi^2}\log\frac{3T}{\mu} + 3\lambda\phi^2}{\frac{1}{4}g^2\phi^2}, \tag{26}
$$

where a natural optimized scale choice would be μ = const $\times m_T \sim \phi$ (see below). Without the f_{2m}-term the numerator has a zero – implying that m_1 becomes imaginary – at a value of ϕ which grows when T decreases further and further below T_0. This is a spurious zero and has to do with the breakdown of perturbation theory for small ϕ. One sees this from the the appearance of the 2-loop f_{2m}-term. For $m_H < 3m_W$ (see eq.(22)) and with the natural scale $\mu \sim \phi$ this term is positive and thus counteracts the zero. For T not very far below T_0, within the expected range of validity of the effective theory, m_1 actually is always real and no spurious imaginary part appears. For reasonable parameter values the above ratio decreases monotonically with ϕ and for large ϕ approaches the limit $12\lambda/g^2 = \frac{3}{2}m_H^2/m_W^2$.

The Higgs mass range in which one in the first place is interested in is m_H close to and somewhat above m_W. This is where the present lower limit lies, both the experimental limit (60 GeV) and the limit [26] from vacuum stability and $m_{\rm top} = 170$ GeV (somewhat more than 100 GeV). For these Higgs masses the scales m_1, m_2, m_T are roughly equal and m_L is larger.

OPTIMIZING PERTURBATION THEORY

To be able make predictions one still has to choose the scale μ. In fact, the 2-loop effective potential is almost μ_3 independent (see (24)), so are the predictions made with it. In the contrary, the one-loop effective potential exhibits μ_3 dependence through Eq. (23). Thus we can use the advantage of the arbitrariness of the μ_3 choice and take it in such a way that higher order terms (in our case 2-loop) are small in comparison with the leading approximation (in our case 1-loop). This is chosen by minimizing large logarithms. Assume first that the scales are ordered as $m_1, m_2 < m_T < m_L$. Then from eq.(18) the coefficient of the $\frac{1}{2}g^4\phi^2$ term is

$$-\frac{3}{8}\log\frac{\mu}{m_L}+\frac{9}{4}\log\frac{\mu}{m_T+2m_L}+\frac{51}{16}\log\frac{\mu}{m_T}\approx\frac{81}{16}\log\frac{\mu}{m_L^{30/81}m_T^{51/81}}. \tag{27}$$

It is natural to choose μ so that this log vanishes. When also m_1, m_2 are included one similarly obtains

$$\mu=\kappa m_L(\phi)^{1-\zeta}m_T(\phi)^{\zeta}, \tag{28}$$

with

$$\zeta=\frac{f_{2m}^0}{f_{2m}}\approx\frac{\frac{17}{27}+4\lambda-12\lambda^2}{1+4\lambda-12\lambda^2},\qquad f_{2m}^0=\frac{51}{16}g^4+9\lambda g^2-12\lambda^2, \tag{29}$$

where f_{2m}^0 is the 2-loop running mass coefficient for a theory without the triplet Higgs field A_0^a and κ is a parameter, $\kappa \sim 1$. It is to be determined by minimizing, for example, the 2-loop correction to the 1-loop result for the derivative $dV/d\phi$. The derivative has to be used since V itself contains a scale dependent additive constant, which is not physical.

Results for $m_H = 80$ GeV are shown in Table 1 and Figs.1 and 2. Table 1 gives numbers down to very small values of T. The effective theory is not valid for the smallest values of T, but the constraint $v(0) = 241.6$ GeV fixes the behaviour of $v(T)$ at small T. Figs.1 and 2 show the potential close to the perturbative $T_c^{\text{pert}} = 173.3$ GeV. As an example of what happens at still larger masses, Fig.3 shows the potential for various temperatures for $m_H = 132$ GeV, for which $T_0 = 234.9$ GeV.

VALIDITY OF PERTURBATION THEORY

The expansion parameter of the perturbative expansion in the 3d theory with a dimensionful coupling g_3^2 of dimension GeV must be the ratio of this coupling to some relevant mass scale. When the gauge degrees of freedom dominate, a typical term in the 2-loop potential is $4g_3^2m_T^2/(16\pi^2)$ and in the 1-loop potential is $m_T^3/2\pi$. Their ratio gives the expansion parameter ($\phi_{4d} = \sqrt{T}\phi_{3d}$)

$$\text{expansion parameter}=\frac{g_3^2}{2\pi m_T}=\frac{g_3}{\pi\phi}=\frac{gT}{\pi\phi}. \tag{30}$$

Thus the typical 3-loop term is

$$\frac{1}{(2\pi)^3}g_3^4m_T(\phi). \tag{31}$$

To 3 and 4 loops there are also terms $\sim \log(\mu_3/m_T(\phi))$ which are known on the basis of the 1-loop and 2-loop potentials and RG invariance [21].

Table 2. Values of $v(T)$, the value of ϕ at the broken minimum of the potential, for $m_H = 80$ GeV. The optimization parameter κ in eq. (28) is determined so that the 1- and 2-loop results coincide at the minimum (they then approximately coincide at all ϕ). The last column gives the depth $V(0) - V(v(T))$ for the 2-loop potential in units of $g^6T^4/12$.

T	$v(T)$	$v(T)/T$	κ	ΔV
50	246.6	4.93	1.5	1017
100	227.0	2.27	1.5	40.7
130	199.6	1.54	1.55	7.27
150	168.0	1.12	1.6	1.62
159	147.2	0.93	1.64	0.683
162	138.5	0.85	1.64	0.466
164	132.0	0.80	1.66	0.343
170	106.2	0.62	1.7	0.0797
172	93.25	0.54	1.72	0.0255
173	84.25	0.49	1.741	0.0050
173.3	80.86	0.47	1.748	0.0002
174.1	65.8	0.38	1.775	-0.0115

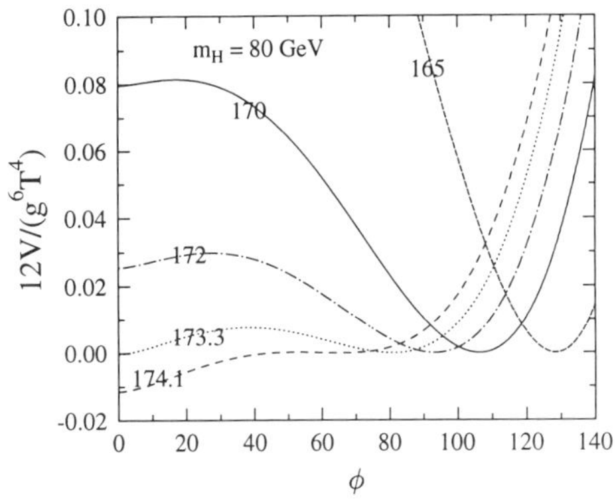

Figure 1. The perturbative 2-loop potential in units of $g^6T^4/12$ for m_H = 80 GeV and for various temperatures optimized so that the 1- and 2-loop potentials coincide at the broken minimum. They then essentially coincide at all ϕ. The perturbative T_c = 173.3 GeV and the end of the superheated broken phase is at T_+ = 174.1 GeV.

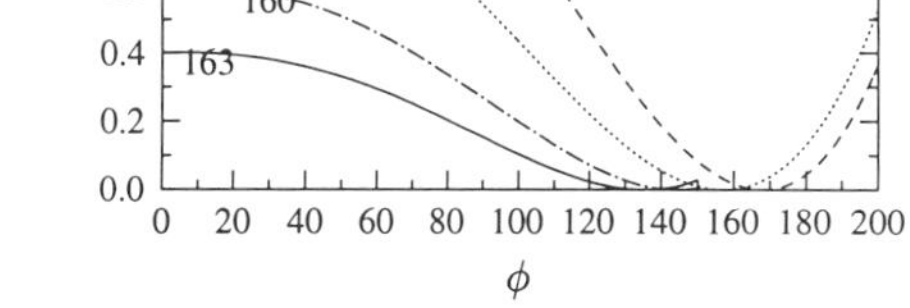

Figure 2. As Fig.1 but for lower temperatures. The nonperturbative T_c is about 163 GeV (see below).

On the basis of eq.(30) the perturbative computation is unreliable for

$$m_T < \frac{1}{2\pi} g_3^2, \quad \phi < \frac{g}{\pi} T. \tag{32}$$

Essentially the same estimate comes from the partial resummation of corrections with the help of exact evolution equations [28], which reveals that the effective gauge coupling becomes large at $Q^2 \sim (g^2 T/\pi)^2$, making the naive perturbative analysis at these momentum transfers misleading. For perturbative results to make any sense, at least $\phi(T_c)$ should be outside this nonperturbative domain. Using eq.(12) it is easy to see that this holds only for m_H less than about m_W.

How big could one expect the nonperturbative effects to be? One estimate could be to say that they are as large as a typical perturbative term at the limit of perturbation theory, i.e., from eq.(32) for $m_T = g_3^2/(2\pi)$. This gives an effect of expected magnitude

$$\Delta V_{\text{nonpert}} = \frac{1}{(2\pi)^4} g_3^6 = \frac{1}{(2\pi)^4} g^6 T^4. \tag{33}$$

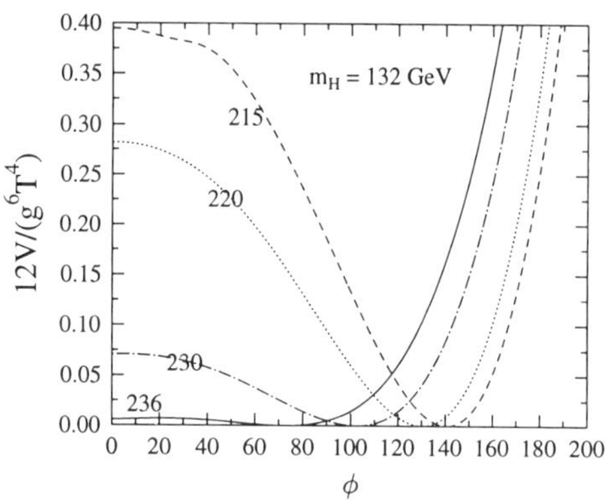

Figure 3. As Fig.1 but for $m_H = 132$ GeV. The perturbative $T_c = 237$ GeV. If nonperturbative effects at this mass are the same as at $m_H = 80$ GeV the true critical T_c will be about 215 GeV.

This is the same as a typical 4-loop term. It is also characteristic that it is $\mathcal{O}(g^6)$, at and beyond which 4d perturbative computation of the free energy anyway diverges. Its numerical value in eq.(33) is set by the $1/(2\pi)^4$ factor. In ref. [22] it has been argued, that non-perturbative effects can be much larger. The convincing analysis of the possible non-perturbative effects can be done at present only with lattice Monte-Carlo methods [21, 23].

LATTICE MONTE CARLO RESULTS

To begin with one has to give the lattice action corresponding to the continuum action in eq. (15). It is, in standard notation,

$$\begin{aligned}
S = {} & \beta_G \sum_x \sum_{i<j} (1 - \frac{1}{2}\mathrm{Tr}\, P_{ij}) + \\
& + \frac{1}{2}\beta_G \sum_x \sum_i [\mathrm{Tr}\, A_0(\mathbf{x}) U_i^{-1}(\mathbf{x}) A_0(\mathbf{x}+i) U_i(\mathbf{x}) - \mathrm{Tr}\, A_0^2(\mathbf{x})] + \\
& + \sum_x \Big\{ 10\Sigma - \frac{5}{3}\frac{4}{g^2\beta_G} \Big\} \frac{1}{2}\mathrm{Tr}\, A_0^2(\mathbf{x}) + \\
& + \sum_x \frac{17}{16}\frac{g^2\beta_G}{3\pi^2} \Big(\frac{1}{2}\mathrm{Tr}\, A_0^2(\mathbf{x})\Big)^2 + \\
& + \beta_H \sum_x \sum_i \Big[\frac{1}{2}\mathrm{Tr}\, \Phi^\dagger(\mathbf{x})\Phi(\mathbf{x}) - \frac{1}{2}\mathrm{Tr}\, \Phi^\dagger(\mathbf{x}) U_i(\mathbf{x}) \Phi(\mathbf{x}+i)\Big] + \\
& + \sum_x \Big[(1 - 2\beta_R - 3\beta_H)\frac{1}{2}\mathrm{Tr}\, \Phi^\dagger(\mathbf{x})\Phi(\mathbf{x}) + \beta_R\Big(\frac{1}{2}\mathrm{Tr}\, \Phi^\dagger(\mathbf{x})\Phi(\mathbf{x})\Big)^2\Big] + \\
& - \frac{1}{2}\beta_H \sum_x \Big[\frac{1}{2}\mathrm{Tr}\, A_0^2(\mathbf{x}) \frac{1}{2}\mathrm{Tr}\, \Phi^\dagger(\mathbf{x})\Phi(\mathbf{x})\Big].
\end{aligned} \tag{34}$$

where $\Sigma = 0.252731$.

All the three lattice coupling constants are related to g, T and m_H by the following equations:

$$\beta_G = \frac{4}{g^2}\frac{1}{Ta}, \tag{35}$$

$$\beta_R = \frac{1}{4}\lambda T a \beta_H^2 = \frac{m_H^2}{8m_W^2}\frac{\beta_H^2}{\beta_G}, \tag{36}$$

and the mass parameter m_3 is related to the lattice variables by

$$m_3^2 = \frac{2(1 - 2\beta_R - 3\beta_H)}{\beta_H a^2}. \tag{37}$$

The 1-loop counterterm $\sim 1/a$ removing divergences from the Higgs mass has been determined in [20], and the 2-loop counterterm $\sim \log(a)$ has been estimated in [21]. Given these counter terms, the constant physics curve, expressing the temperature in terms of $\beta_G, \beta_H, \beta_R$ then is

$$\begin{aligned}
\frac{m_H^2}{4T^2} = {} & \frac{1}{2}\gamma + \frac{1}{T^2a^2}\Big[3 - \frac{1}{\beta_H} - \frac{2\beta_R}{\beta_H} + \\
& + \frac{1}{2}Taf_{1m}\Sigma + \frac{f_{2m}}{32\pi^2}T^2a^2(\log\frac{Ta}{2} + \eta)\Big],
\end{aligned} \tag{38}$$

where Ta is given in eq.(35),

$$f_{1m} = -(\frac{9}{4}g^2 + 6\lambda), \quad f_{2m} = \frac{81}{16}g^4 + 9\lambda g^2 - 12\lambda^2, \tag{39}$$

and

$$\gamma = \frac{3}{16}g^2 + \frac{1}{2}\lambda + \frac{g^4}{16\pi^2}\Big(\frac{149}{96} + \frac{3}{4}\lambda\Big). \tag{40}$$

Finally, the number η is the g^2 and λ dependent scale factor in the logarithmic counter term. Its precise numerical value can be obtained by computing the 2-loop effective potential in lattice perturbation theory. This has been done for the scalar theory but not if gauge fields are included. However, its value can also be obtained by requiring that continuum and lattice results coincide for any measurable quantity, like $R^2 = \langle \phi^\dagger \phi \rangle$.

In setting up the lattice computations on an N^3 lattice, $N = 8, 12, 16, 24, 32$, one first fixes the lattice spacing in physical units by choosing the value of β_G:

$$a = \frac{4}{g^2 \beta_G T}; \tag{41}$$

the values $\beta_G = 12, 20, 32$ make a reasonably small but are not too large for the numerics of the Monte Carlo. Choosing $m_H = 35, 60, 80, 90$ GeV fixes β_R and leaves β_H as the variable with which one fixes the physical temperature T and searches for a two-state signal in some variable, as an indication of a first order transition.

Choosing $m_H = 80$ GeV, for which perturbative values are given in Table 2, the procedure thus ideally is as follows. One first chooses β_H so that T is deep in the broken phase. For example, when $\beta_G = 12$, one obtains $T = 113, 142$ GeV for $\beta_H \approx 0.352, 0.349$, respectively. Deep in broken phase one expects perturbation theory to work rather well. To estimate the expected accuracy in computing the value $v(T)$ of the field at the minimum, assume there is in the potential a 3-loop term of the form in eq.(31). Parametrizing its magnitude by a constant β, where this constant is expected to be of the order of one, the potential as approximately of the form

$$\frac{1}{2} m^2 \phi^2 + \frac{1}{4} \lambda \phi^4 + \frac{\beta}{(2\pi)^3} g_3^4 m_T(\phi). \tag{42}$$

From this it is simple to estimate that

$$\frac{\delta v}{v} = \frac{\beta}{(2\pi)^3} \frac{2m_W^2}{m_H^2} \left(\frac{gT}{v}\right)^3 \approx \frac{\beta}{400} \left(\frac{T}{v}\right)^3. \tag{43}$$

In the range $v/T = 1...4$ (see Table 2) one thus expects the 2-loop calculation to be accurate up to a relative error of $\beta/400....\beta/25000$. With $\beta = \mathcal{O}(1)$ the accuracy thus is better than 1%, especially at lower temperatures.

Requiring that lattice and perturbative results for a physical measurable quantity like $R^2 = \langle \phi^\dagger \phi \rangle$ coincide, one thus expects to be able to establish a reliable connection between lattice and physical parameters in the broken phase. In particular, the constant η in eq.(38) can be determined. Then one can proceed to higher temperatures and search for the value of β_H for which a two-state signal is observed. After a careful inclusion of finite volume V and finite lattice spacing a effects this can reliably be translated to a physical value of T_c. The details of such an analysis will be published elsewhere [24], we present here just some qualitative results of Monte-Carlo simulations.

The first question is the very existence of the first-order phase transition. To answer it, we searched for a two-peak signal in the distributions of the different order parameters (such as length of the Higgs field R^2, the average of the link operator $L = \mathrm{Tr}\, V^\dagger(x) U_i(x) V(x+i)$, where $\Phi = RV$, etc.) A typical picture at $\beta_H = 0.347710$, $\beta_G = 12$ and $V = 24^3$ is shown in Fig. 4. The two-peak structure is clearly seen. At the same time, due to the fact that the lattice volume is finite, the width of any of the peaks is comparable with the distance between them. For comparison, we present in Fig. 5 the distribution of R^2 for a Higgs mass $m_H = 35$ GeV, where finite-size effects are less important.

The determination of the critical value of β_H^c in the continuum limit requires the analysis of finite-size and finite spacing effects. In principle, a number of different methods can be used for its determination:

(i) Equal-area signal. The value of β_H^c is determined as the value of β_H for which the *areas* [29] under the two peaks in the distribution of some order parameter (actually, it does not matter which order parameter is chosen) are equal.

(ii) One looks for a maximum of the heat capacity (action susceptibility), considered as a function of β_H [29, 30].

(iii) The same as in (ii), but for L-susceptibility [30, 31].

All three methods must give the same results in the continuum limit. For finite lattices and for finite β_G the results are different, and the continuum limit should be found by extrapolation.

The evolution of R^2 distributions in the broken phase with β_H, used for the determination of parameter η is shown in Fig. 6. With the value of η determined, the observed values of β_H^c can be converted to results for the critical temperature of the electroweak phase transition using eq. ((38)). From the $\beta_G = 12$ and 20 data we get $T_c = 163$ GeV with estimate of error about 3 GeV. The T_c determined by lattice methods is smaller than the value $T_c = 173.3$ GeV determined by 2-loop effective potential.

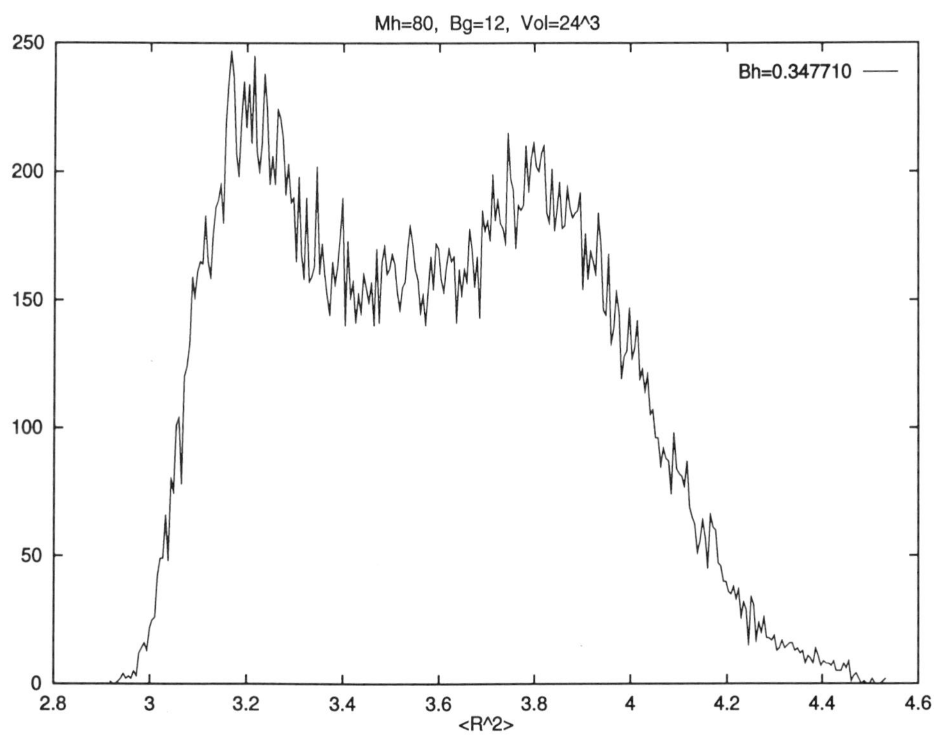

Figure 4. Distribution of the order parameter $R^2 = \frac{1}{2}\mathrm{Tr}\,\Phi^\dagger\Phi$ at the critical temperature.

The convergence of perturbation theory in the broken phase is better at the lattice T_c, since the vacuum expectation value substantially grows with the temperature drop.

At present we cannot extract from available lattice data the parameters relevant for cosmological applications. In particular, the region of metastability of the symmetric phase, together with the bubble nucleation temperature T^* cannot be determined. The direct lattice determination of T^* is hardly possible at all, so that some indirect methods are necessary. One of them is related to the measurement of the interface tension at the critical temperature, which allows one to compute the bubble nucleation rate at least in the vicinity of the critical temperature. Another obvious option is to study details of the phase transition for the heavy Higgs boson, since we are approaching the non-perturbative region from the side of the broken phase, increasing m_H.

As has been discussed in [22], non-perturbative effects are likely to decrease the value of the potential in the vicinity of the origin. It seems that our lattice results are in a qualitative agreement with this picture. Indeed, consider the plot of the 2-loop effective potentials at the lattice critical temperature $T_c \simeq 163$ GeV shown in Fig. 2. Clearly, this potential is completely wrong at small fields ϕ, since we know that *we are* at the critical temperature. Therefore, the effective potential must have a contribution making the broken and symmetric phases degenerate. According to Fig. 2 this contribution at $\phi = 0$ is of order

$$V_{\rm pert}(v(T_c)) - V_{\rm pert}(0) \approx 0.03 g_3^6 \approx 60(\alpha_W T)^3. \tag{44}$$

What is most striking is the large magnitude of the nonperturbative effects. The value of T_c is changed only mildly, but the relevant scale to estimate the change with is the

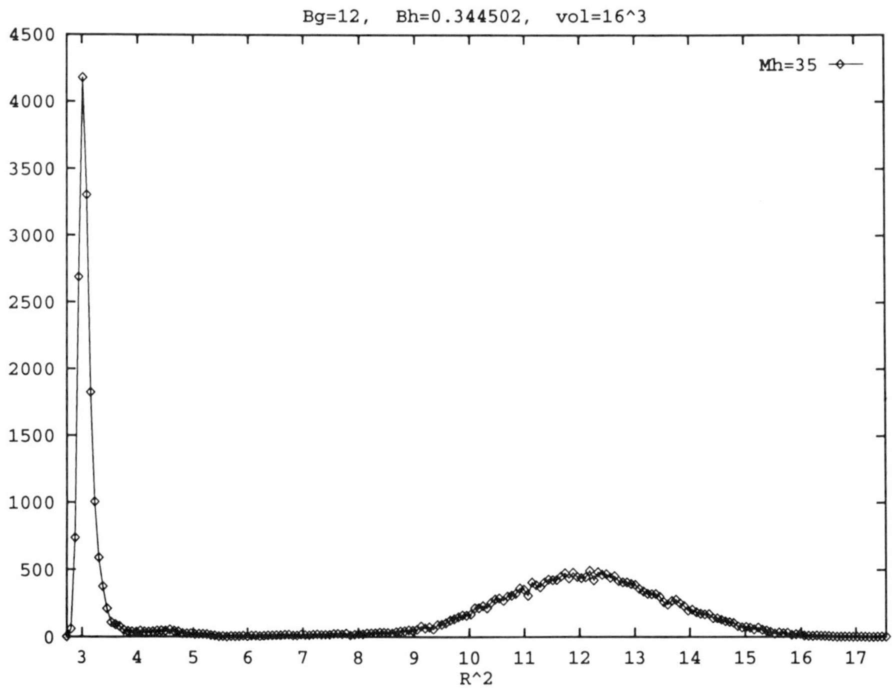

Figure 5. Distribution of the order parameter R^2 at the critical temperature, with $m_H = 35$ GeV.

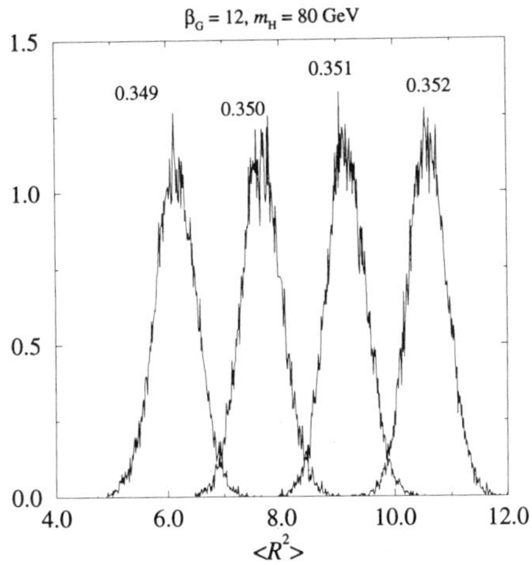

Figure 6. Evolution of the distribution of R^2 in the broken phase with β_H for $\beta_G = 12$, with $m_H = 80$ GeV.

magnitude of the different perturbative terms at $T_c^{\rm pert}$ (which is roughly the magnitude of the potential barrier between the minima) and the value of $V(\phi = 0)$ at $T_c^{\rm latt}$. Measured like this nonperturbative effects are almost two orders of magnitude larger than the estimate (33).

CONCLUSIONS

We can conclude the following about the perturbative and non-perturbative treatment of finite T field theories, hot EW theory in particular.

1. To optimize perturbation theory by avoiding large logarithmic terms one should first dimensionally reduce the full 4d theory to an effective 3d theory by integrating over the nonstatic modes with large "masses" $2\pi nT$.

2. For reliable perturbative results one needs both the 1- and 2-loop potentials. Only with both of them can one determine the unknown renormalization scale by some optimisation criterion.

3. The 3-loop potential can and should be computed. Evaluating that with, say, the scale μ given by the 1- and 2-loop analysis will be a quantitatively meaningful statement of the magnitude of higher order corrections.

4. Perturbation theory is valid only for $m_T > g_3^2/(2\pi)$ or $\phi/T > g/\pi$ and the value of $V(\phi = 0)$ is inaccurate at least by the amount $g_3^6/(2\pi)^4 = g^6T^4/(2\pi)^4$. Perturbation theory is not valid in the symmetric phase, hence not for the determination of T_c, latent heat, interface tension or correlation lengths in the symmetric phase.

5. Perturbation theory works in the broken phase if $\phi(T)/T > 1$. At the perturbative T_c, $\phi(T_c^{\rm pert})/T_c^{\rm pert} \sim (g^3 + \lambda^{3/2})/\lambda$ and this can be > 1 for $m_H < m_W$.

6. The lattice simulations of 3d gauge-Higgs theories can clarify many non-perturbative aspects of the electroweak phase transition, especially for sufficiently heavy Higgs boson. In particular, they indicate that the EW phase transition is stronger first order than follows from 2-loop effective potential.

7. The following combination of the lattice and perturbative methods may appear to work at $m_H \simeq m_W$. The quantities requiring information on the unbroken phase (such as critical temperature, latent heat, interface tension etc.) can be found by the lattice methods. Hopefully, at the true nonperturbative $T_c^{\rm latt}$, measured numerically with the lattice regulated theory, perturbation theory in the broken phase works. Then it can be used for the determination of parameters relevant for the broken phase only, such as sphaleron rate.

8.If the specific model for the description of non-perturbative effects considered in [22] is correct, then with the value of the non-perturbative energy shift defined by (44) electroweak baryogenesis may be possible up to a Higgs mass of about 100 GeV. The exact determination of this bound, however, requires a much better understanding of the symmetric phase of the electroweak theory at high temperatures.

REFERENCES

[1] R. Jackiw, Phys. Rev. D9 (1974) 1686

[2] D. A. Kirzhnitz, JETP Lett. 15 (1972) 529;
D. A. Kirzhnitz and A. D. Linde, Phys. Lett. 72B (1972) 471

[3] L. Dolan and R. Jackiw, Phys.Rev. D9 (1974)3320

[4] S. Weinberg, Phys.Rev. D9 (1974) 3357

[5] D.A. Kirzhnitz and A.D. Linde, Ann. Phys. 101 (1976) 195;
A.D. Linde, Nucl. Phys. B216 (1983) 421, Rep. Prog. Phys. 47 (1984)925

[6] M.E. Shaposhnikov. Nucl. Phys. B287 (1987) 757; A.I. Bochkarev and M.E. Shaposhnikov, Mod. Phys. Lett. 2A (1987) 417

[7] G.W. Anderson and L.J. Hall, Phys. Rev. D45 (1992) 2685

[8] K. Enqvist, J. Ignatius, K. Kajantie and K. Rummukainen, Phys. Rev. D45 (1992) 3415

[9] M.E. Shaposhnikov, Phys. Lett. B277 (1992)324; B282 (1992) 483(E);

[10] M. Carrington, Phys. Rev. D45 (1992) 2933

[11] M. Dine, R. G. Leigh, P. Huet, A. Linde and D. Linde, Phys. Rev. D46 (1992) 550

[12] C. G. Boyd, D. E. Brahm and S. Hsu, Phys. Rev. D48 (1993) 4963

[13] M. Quiros, J.R. Espinosa and F. Zwirner, Phys. Lett. B314 (1993) 206

[14] W. Buchmüller, T. Helbig and D. Walliser, Nucl. Phys. B407 (1993) 387

[15] W. Buchmüller, Z. Fodor, T. Helbig and D. Walliser, DESY Preprint DESY-93-021 (1993), Ann. Phys., to be published

[16] P. Arnold and O. Espinosa, Phys. Rev. D47 (1993) 3546

[17] A. Hebecker, Z. Phys. C60 (1993) 271

[18] Z. Fodor and A. Hebecker, DESY preprint DESY 94-025, 1994

[19] B. Bunk, E.-M. Ilgenfritz, J. Kripfganz, A. Schiller, Phys. Lett. B284 (1992) 371; Nucl. Phys. B403 (1993) 453

[20] K. Kajantie, K. Rummukainen and M. Shaposhnikov, Nucl. Phys. B407 (1993) 356

[21] K. Farakos, K. Kajantie, K. Rummukainen and M. Shaposhnikov, CERN Preprint CERN-TH.6973/94, hep-ph 9404201

[22] M. Shaposhnikov, Phys. Lett. B316 (1993) 112

[23] K. Farakos, K. Kajantie, K. Rummukainen and M. Shaposhnikov, CERN Preprint CERN-TH.7244/94, hep-ph 9404234

[24] K. Farakos, K. Kajantie, K. Rummukainen and M. Shaposhnikov, CERN Preprint CERN-TH.7220/94, in preparation

[25] S. Weinberg, Phys. Lett. 91B (1980) 51

[26] M. Lindner, M. Sher and H. Zaglauer, Phys. Lett. B228 (1989) 139

[27] A. Jakovác, K. Kajantie and A. Patkós, Helsinki preprint HU-TFT-94-01, hep-ph-9312355

[28] M. Reuter and C. Wetterich, Nucl. Phys. B408 (1993) 91

[29] H.J. Herrmann, W. Janke and F. Karsch (eds.), Proc. Workshop on Dynamics of First Order Phase Transitions, Julich , June 1992 (World Scientific, Singapore, 1992)

[30] C. Borgs, R. Kotecký and S. Miracle-Sole, J. Stat. Phys. 62 (1991) 529

[31] A. M. Ferrenberg and R. H. Swendsen, Phys. Rev. Lett. 61 (1988) 2635

DYNAMICAL EVOLUTION OF THE ELECTROWEAK PHASE TRANSITION*

Joseph I. Kapusta

School of Physics and Astronomy
University of Minnesota
Minneapolis, MN 55455

INTRODUCTION

The dynamics of first-order phase transitions has fascinated scientists at least since the time of Maxwell and van der Waals. Much work on the classical theory of nucleation of gases and liquids was carried out in the early part of this century, culminating in the theory of Becker and Döring[4]. There were and still are many important applications, such as cloud and bubble chambers, freezing of liquids, and precipitation in the atmosphere. The modern theory of nucleation was pioneered by Langer[5]. Langer's theory is based in a more fundamental way on the microscopic interactions of atoms and molecules. It can also be applied close to a critical point where in fact most of the current interest in the condensed matter community has been. Finally nucleation theory has been extended to relativistic quantum field theory by Callan and Coleman[6] for zero temperature and by Affleck[7] and Linde[8] for finite temperature. Applications here are to elementary particle phase transitions in the early universe[9,10] and even to the nucleation rate for black holes[11].

The goal of nucleation theory is to compute the probability that a bubble or droplet of one phase appears in a system initially in the other phase near the critical temperature. Homogeneous nucleation theory applies when the system is pure; inhomogeneous nucleation theory applies when impurities cause the formation of bubbles or droplets. For the application we have in mind, namely the early universe, it seems that homogeneous nucleation theory is appropriate. In the everyday world it is usually the opposite; dust or ions in the atmosphere are much more efficient in producing precipitation. Nucleation theory is applicable for first-order phase transitions when the matter is not dramatically supercooled or superheated. If substantial supercooling or superheating is present, or if the phase transition is second-order, then the relevant dynamics is

*Work done originally in collaboration with Laszlo P. Csernai[1,2] and Margaret E. Carrington[3].

spinodal decomposition. In this paper we concern ourselves only with homogeneous nucleation theory. For an excellent overview of all these topics see Gunton, San Miguel and Sahni[12].

Suppose that a system is cooled below its critical temperature. Then there exists a critical sized droplet (or bubble, depending on whether the energy density in the lower temperature phase is greater or less than the higher temperature phase). If a droplet that forms because of statistical fluctuations is too small, its surface free energy is relatively large and the cost in total free energy is positive. The droplet will evaporate. If the droplet is large, its surface free energy is unimportant, and the droplet will accrete molecules and grow. A droplet of critical size is metastable, it is balanced between evaporation and accretion. The classical theory of Becker and Döring[4], which is nicely reviewed by McDonald[13], says that the probability per unit time per unit volume to nucleate the dense liquid phase from a dilute gas is given by

$$I = a(i_*) \left(\frac{\epsilon''(i_*)}{2\pi T} \right)^{1/2} \bar{n}(1) e^{-\epsilon(i_*)/T} , \tag{1}$$

where $\epsilon(i_*)$ is the formation energy of a critical sized droplet consisting of i_* molecules, prime denotes differentiation with respect to the number of molecules i, T is the temperature, $\bar{n}(1)$ is the density of single molecules and $a(i_*)$ is the accretion rate of single molecules on a critical droplet. Usually the accretion rate is taken to be

$$a(i_*) = \frac{1}{2} \bar{n}(1) \bar{v} 4\pi R_*^2 s , \tag{2}$$

which is the flux of particles ($\bar{v}$ is the mean speed of gas molecules) striking the surface of the critical droplet times a 'sticking fraction' s less than one. The first term in eq. (1) is a dynamical factor influencing the growth rate, the second term characterizes fluctuations about the critical droplet, and the product of the third and fourth terms gives the quasi-equilibrium number density of critical sized droplets.

Langer's modern theory of nucleation yields the following formula for the rate:

$$I = \frac{\kappa}{2\pi} \Omega_0 e^{-\Delta F/T} \tag{3}$$

where ΔF is the change in the free energy of the system due to the formation of the critical droplet. Ω_0 is a statistical prefactor which measures the available phase space volume. κ is a dynamical prefactor which determines the exponential growth rate of critical droplets which are perturbed from their quasi-equilibrium radius R_*.

$$\kappa = \frac{d}{dt} \ln[R(t) - R_*] \tag{4}$$

The basic structure is the same as in the classical theory, but the prefactors are different. The dynamical prefactor has been calculated by Langer and Turski[14,15] and by Kawasaki[16] for a liquid-gas phase transition near the critical point, where the gas is not dilute, to be

$$\kappa = \frac{2\lambda\sigma T}{\ell^2 n_\ell^2 R_*^3} . \tag{5}$$

This involves the thermal conductivity λ, the surface free energy σ, the latent heat per molecule ℓ and the density of molecules in the liquid phase n_ℓ. The interesting physics in this expression is the appearance of the thermal conductivity. In order for the droplet to grow beyond the critical size, latent heat must be conducted away from the surface into the gas. For a relativistic system of particles or quantum fields which has no net conserved charge, such as baryon number, the thermal conductivity vanishes. The reason is that there is no rest frame defined by the baryon density to refer to heat transport. Hence this formula obviously cannot be applied to such systems.

A relativistic quantum field theory approach has been worked out by Callan and Coleman[6] for nucleation from one vacuum to another and extended by Affleck[7] and Linde[8] to finite temperature. In the limit that thermal fluctuations dominate quantum fluctuations the rate is

$$I = \frac{\omega_-}{\pi}\left(\frac{S_3}{2\pi T}\right)^{3/2}\left(\frac{det'[-\nabla^2 + V''(\phi,T)]}{det[-\nabla^2 + V''(0,T)]}\right)^{-1/2} e^{-S_3/T}\ , \tag{6}$$

where S_3 is the three-dimensional action associated with the formation of a critical sized bubble or droplet. The ω_- is the frequency of the unstable mode. The ratio of determinants contains some interesting dynamics. The V'' is the second derivative of the potential with respect to the 'order parameter' ϕ which is actually a field describing the shape of the critical bubble or droplet. This ratio of determinants is almost never evaluated because it must be done numerically. A notable example of such an evaluation is presented by Carson, Li, McLerran and Wang[17] for the sphaleron of electroweak theory. Usually dimensional analysis is invoked to approximate the pre-exponential factor by T^4 or by T_c^4, so that

$$I = T^4\, e^{-\Delta F/T} \quad \text{or} \quad I = T_c^4\, e^{-\Delta F/T}. \tag{7}$$

It is clear that the nonrelativistic expressions are inadequate when we deal with the electroweak phase transitions since the baryon density is negligible. The classical expression is probably not accurate either because the two phases of electroweak matter are expected to have comparable densities near the critical temperature, and neither may be considered dilute with respect to the other. We want to do better than estimate the prefactor by dimensional analysis as quoted above. To evaluate the ratio of determinants is extremely cumbersome and must be done numerically. Therefore I will describe some work we have done to estimate the pre-exponential factor using the basic formalism of Langer as applied to a course-grained effective field theory with zero net baryon number.

Following Langer, we first construct a free energy functional for a coarse-grained effective field theory. This is expressed in terms of physical quantities, such as the equation of state of the two phases, the latent heat, the critical temperature, the surface free energy, the correlation lengths, and the viscosities. Then we determine the growth rate κ and the statistical prefactor Ω_0. We also will need the asymptotic growth velocity of bubbles. All of the physical quantities were computed to the best of our abilities or taken from other works assuming the validity of the standard model without extra Higgs. As better computations or measurements of physical observables become available, as reported at this workshop, they should be used in place of the ones we use here.

With this input we solve for the time evolution of the universe as it passes through the electroweak phase mixture. In particular we compute, as functions of time, the temperature, the fraction of space which has been converted from the symmetric phase to the asymmetric phase, the average bubble size, and the average bubble separation. The time dependence of these quantities has been discussed before[18]. In this paper we calculate these quantities to the best accuracy available given our present understanding of equilibrium and nonequilibrium finite temperature field theory.

REVIEW OF LANGER'S FORMALISM

The general theory of nucleation developed by Langer[5] starts with the introduction of a set of variables η_i, $i = 1, ..., N$, which describe N degrees of freedom of the system of interest. Oftentimes it is convenient to take these to be collective coordinates. In the system with which we are concerned, for example, the η_i will be the energy density $e(\vec{r})$ and flow momentum $\vec{M}(\vec{r})$ at positions $\vec{r}$ in the system. Thus a sum over the index i represents an integration over $\vec{r}$ and a sum over each of the density and flow fields.

We next introduce a distribution function $\rho(\{\eta\}, t)$, which is a probability density over configurations $\{\eta\}$ and is also a function of time t. The equilibrium configurations, for which $\partial_t \rho = 0$, have the probability distribution of the form

$$\rho_{eq}\{\eta\} \quad \propto \quad \exp[-F\{\eta\}/T] \; , \tag{8}$$

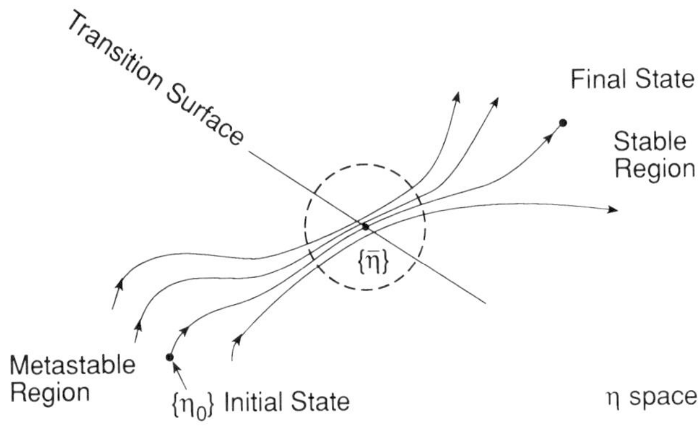

Figure 1. Paths for nucleating a stable phase from a metastable phase in η space.

where $F\{\eta\}$ is a coarse-grained free energy. Such configurations are the initial metastable point in the η-space denoted by $\{\eta_0\}$, or the final state. In this space the phase transition starts from a metastable point $\{\eta_0\}$ and moves to the vicinity of a stable point, a point where F has its minimum. In this process the system most likely has to pass a saddle point $\{\bar{\eta}\}$. See figure 1. This configuration, $\{\bar{\eta}\}$, describes a situation close to $\{\eta_0\}$ except for the presence of one critical size droplet of the new phase. At the saddle point we can assume stationary flow $\partial_t \rho = 0$, and we can calculate the current across this saddle. The rate of probability flow across this saddle, $\{\bar{\eta}\}$, determines the droplet formation rate in the system. This rate

$$I = I_0 \exp[-\Delta F/T] \; , \tag{9}$$

telling us the number of critical size droplets created in unit volume in unit time, was evaluated by Langer[5]. The activation energy ΔF is given by

$$\Delta F = F\{\bar{\eta}\} - F\{\eta_0\} \; . \tag{10}$$

The prefactor I_0 is the product of two terms

$$I_0 = \frac{\kappa}{2\pi}\Omega_0 \; , \tag{11}$$

the dynamical prefactor κ (of dimension $[c/fm]$) and the statistical prefactor Ω_0 (of dimension $[fm^{-3}]$). In terms of the eigenvalues $\bar{\lambda}_\alpha$ and $\lambda_\alpha^{(0)}$ of the matrix

$$\frac{\partial^2 F\{\eta\}}{\partial\eta_i \partial\eta_j} \; ,$$

evaluated at points $\{\bar{\eta}\}$ and $\{\eta_0\}$ respectively, the statistical prefactor can be written as

$$\Omega_0 = \mathcal{V} \left(\frac{2\pi T}{|\bar{\lambda}_1|}\right)^{1/2} \prod_{\alpha=\alpha_0+2}^{N} \left(\frac{2\pi T}{\bar{\lambda}_\alpha}\right)^{1/2} \prod_{\alpha=1}^{N} \left(\frac{\lambda_\alpha^{(0)}}{2\pi T}\right)^{1/2} \; , \tag{12}$$

where $\mathcal{V}$ is the volume of the η-space available for the flux of probability flow[5]. We will evaluate Ω_0 in a later section.

Since $\{\eta_0\}$ is a minimum of F, all the $\lambda_\alpha^{(0)}$'s must be positive. Similarly, the $\bar{\lambda}_\alpha$ are eigenvalues of the above second-derivative matrix, in this case evaluated at the saddle point $\{\bar{\eta}\}$. Because $\{\bar{\eta}\}$ locates the highest point along the path of lowest energy leading away from $\{\eta_0\}$, there must be one and only one eigenvalue $\bar{\lambda}_\alpha$, which is negative. This is the eigenvalue denoted by $\bar{\lambda}_1$ in eq. (12). If $F\{\eta\}$ has translational symmetry, then there must be at least three other eigenvalues $\bar{\lambda}_\alpha$ which are identically zero, corresponding to the three independent translations of the position of the symmetry-breaking fluctuation (bubble or droplet) described by $\{\bar{\eta}\}$. The product of $\bar{\lambda}$'s appearing in eq. (12) starts with $\alpha = \alpha_0 + 2$, where α_0 is the total number of symmetries of F which are broken by $\{\bar{\eta}\}$. The integration over these α_0 degrees of freedom defines the factor $\mathcal{V}$ in eq. (12). That is, $\mathcal{V}$ is the volume of the α_0-dimensional region of η-space spanned by the set of configurations $\{\bar{\eta}\}$ which leave $F\{\bar{\eta}\}$ invariant.

The dynamical factor κ is the exponential growth rate of the unstable mode of deformation $\{\bar{\eta}\}$. To compute κ, we linearize about $\eta_i = \bar{\eta}_i$. That is, we write $\nu_i = \eta_i - \bar{\eta}_i$ and set $\nu \propto e^{\kappa t}$. Then κ must be determined by solving some set of linearized dynamical equations of motion. In the nucleation problem the instability described by κ is the initial growth rate of a bubble or droplet which has just exceeded the critical size.

COARSE-GRAINED EFFECTIVE FIELD THEORY

The model of nucleation which we propose here will be defined by the choice of the statistical variables, η_i, and the corresponding coarse-grained free energy $F\{\eta\}$. The conventional formulation of classical many-body statistical mechanics in terms of particle positions and momenta is not very convenient for the present purpose. Nucleation is characterized by semi-macroscopic fluctuations involving large numbers of particles. Therefore hydrodynamic-type of collective variables are more appropriate to describe the formation of bubbles or droplets.

Hydrodynamics can be derived from microscopic kinetic theory by a coarse - graining or cellular method[19,20]. That is, one divides up the macroscopic system into semi-macroscopic cells of given volume and assigns specific densities and flows to each of these cells. The free energy computed by performing a partition sum subject to the cellular constraints is the coarse-grained F that we are talking about. There is no problem, in principle, in summing over the cellular densities and flows to obtain the true equilibrium free energy. Moreover, as long as each cell comes to local thermal equilibrium rapidly compared to the times required for the hydrodynamic processes that one wants to consider, then one can use the coarse-grained F for computing non-equilibrium properties of the system.

The question which arises at this point is: What is a suitable size for the coarse-graining cells? In order for the hydrodynamic description to make sense, the cell volume must be much larger than the average volume per molecule. For our purposes, however, the cells cannot have linear dimensions appreciably larger than a correlation length. If the cells are chosen to be too large, phase separation will occur within single cells, and the interesting details of the condensation mechanism will be lost in the process of taking cellular averages. To put this another way, we expect F as a function of the average energy density ϵ to be a nonconvex function with distinct minima corresponding to the two phases. But, if the cell size is large enough for well-defined phase separation to occur within a cell, then F must approach its convex envelope and cannot possibly have the above property. Therefore the cell size can be neither much larger nor much smaller than a correlation length.

Relativistic Hydrodynamics

The equations of motion of relativistic fluid dynamics, $\partial_\mu T^{\nu\mu} = 0$, can be given in terms of $E \equiv T^{00}$ and $M^i = T^{0i}$, that is, $E = (e + pv^2)\gamma^2$ and $\vec{M} = (e + p)\gamma^2\vec{v}$, where e is the energy density and p is the pressure[21]. The low speed limit of relativistic fluid dynamics ($\gamma^2 \approx 1$ and $pv^2 \ll e$, but p not assumed small compared to e) is given by

$$\partial_t e = -\nabla \cdot \vec{M} \tag{13}$$

and

$$\partial_t \vec{M} = -\nabla \cdot \left(\frac{1}{w}\vec{M} \otimes \vec{M}\right) - \nabla p \,. \tag{14}$$

Here $w = e + p$ is the enthalpy density, we assumed that the relativistic energy density is $E = (e + pv^2)\gamma^2 \approx e$, and that the relativistic momentum density is $\vec{M} = w\gamma^2\vec{v} \approx w\vec{v}$.

With the above mentioned restrictions in mind we will try to find a suitable form for the coarse-grained free energy F. This is not a trivial problem. We choose as our basic variables the local energy density and momentum density fields, $e(\vec{r}, t)$ and $\vec{M}(\vec{r}, t)$. The free energy F must consist of a kinetic energy F_K and an interaction term F_I. The kinetic term is simply

$$F_K(e, \vec{M}) = \frac{1}{2}\int d^3r \; w\vec{v}^{\,2} = \int d^3r \; \frac{1}{2w}\vec{M}^2 \,. \tag{15}$$

We shall assume that F_I is a functional of e only, and that it can be written in the form[22,23]

$$F_I\{e(\vec{r})\} = \int d^3r \; \left[\frac{1}{2}K(\nabla e)^2 + f(e)\right] \tag{16}$$

where $f(e)$ is the Helmholtz free energy density and $\frac{1}{2}K(\nabla e)^2$ is the usual gradient energy. K is a constant to be determined. Note that in this discussion we assume that the temperature T is constant.

Parametrization of the Free Energy

Imagine having two phases in equilibrium with each other at temperature T, and furthermore consider an interface separating them. This interface cannot be perfectly sharp. The interface must have a finite thickness on the order of a correlation length. In a local density picture the energy density e should vary smoothly from one phase to the other. Since first-order phase transitions have a latent heat, this means that we should know the free energy density $f(e)$ for the energy density ranging between one phase and the other. The low temperature, low energy density asymmetric phase of electroweak theory will be denoted by a subscript A and the high temperature, high energy density symmetric phase will be denoted by a subscript S. In addition to the need to know $f(e)$ for $e_A < e < e_S$ we would also encounter situations where we would need to know $f(e)$ for a range of values about e_A and e_S. Statistical fluctuations about local thermal equilibrium would require such knowledge, for example.

For a range of temperatures about T_c, $f(e)$ should have minima located at $e_A(T)$ and $e_S(T)$. There should also be a barrier between these two minima located at some $e_0(T)$. We require that

$$\begin{aligned} f(e_A(T)) &= -p_A(T)\,, \\ f(e_S(T)) &= -p_S(T)\,. \end{aligned} \tag{17}$$

Therefore, at fixed T we shall parametrize $f(e)$ by a fourth order polynomial in e. Due to the pinning of the two local minima shown above, $f(e)$ will have its global minimum at $e_S(T)$ when $T > T_c$, and will have its global minimum at $e_A(T)$ when $T < T_c$. At the critical temperature the two minima of $f(e)$ are equal. Our parametrization is

$$\begin{aligned} f(e) &= f_0 + \frac{f_0''(e-e_0)^2}{2} - \frac{(e_A + e_S - 2e_0)f_0''}{3(e_A - e_0)(e_S - e_0)}(e-e_0)^3 \\ &+ \frac{f_0''}{4(e_A - e_0)(e_S - e_0)}(e-e_0)^4\,, \end{aligned} \tag{18}$$

where $e_A(T)$, $e_S(T)$, $p_A(T)$ and $p_S(T)$ are specified functions of T, and f_0'' is the curvature of f at the top of the barrier located at e_0 ($f_0'' < 0$). Let us define $\Delta e \equiv e_S - e_A$ and $\Delta p \equiv p_A - p_S$. In terms of these variables

$$e_0 = \frac{e_A + e_S}{2} + \frac{f_0''(\Delta e)^3}{12\Delta p} \pm \left[\left(\frac{f_0''(\Delta e)^3}{12\Delta p}\right)^2 + \frac{(\Delta e)^2}{4}\right]^{1/2} \tag{19}$$

where the $+$ ($-$) is chosen when $\Delta p > 0$ ($\Delta p < 0$) and

$$f_0 = -p_S + \frac{f_0''}{12}\frac{(e_S - e_0)^2(e_S - 2e_A + e_0)}{(e_A - e_0)}\,. \tag{20}$$

Thus, if the location of the two minima and their depths are given for fixed T, then only one free parameter, f_0'', remains. In particular, this parameter determines the barrier height, position, and curvature at all energy densities. Unless we can extract this free energy function from the Lagrangian in a more fundamental way we shall be content to use this parametrization in the following analyses.

Surface Profile

For the remainder of this paper we restrict ourselves to the case of idealized bubbles or droplets. That is, we consider only the limit in which the nucleating fluctuation described by $\{\bar{\eta}\}$ is, indeed, a well-defined sphere of the asymmetric A phase with radius R large compared to the interface thickness or the correlation length ξ (to be defined below). In principle we need not make this restriction in the present theory. As we shall see, however, this limit is the appropriate one in most cases of interest here.

The stationary point $\{\bar{\eta}\}$ is given in our model by $\vec{v}(\vec{r}) = 0$ and $e(\vec{r}) = \bar{e}(\vec{r})$, where $\bar{e}$ satisfies

$$\frac{\delta F_I}{\delta \bar{e}(r)} = -K\nabla^2 \bar{e} + \frac{\partial f}{\partial \bar{e}} = 0 \,. \tag{21}$$

For a spherical bubble of asymmetric A phase surrounded by the symmetric S phase at $T < T_c$ the energy density $\bar{e}$ depends only on the distance r from the center of the bubble. The energy density profile $\bar{e}(r)$ then describes a smooth transition from one phase to the other[24,25]. Using our parametrization of $f(e)$, and introducing a correlation length defined at the top of the barrier by $\xi_0^2 \equiv -K/f_0''$, the static profile equation becomes

$$\frac{d^2\bar{e}}{dr^2} + \frac{2}{r}\frac{d\bar{e}}{dr} + \frac{(\bar{e}-e_0)(\bar{e}-e_A)(\bar{e}-e_S)}{\xi_0^2(e_A-e_0)(e_S-e_0)} = 0. \tag{22}$$

At the critical temperature $f(e_A) = f(e_S)$. Then the free energy, eq. (18), becomes symmetric, $e_0 = (e_A + e_S)/2$ and $\xi_S^2 = \xi_A^2 = \xi_0^2/2$. In this case the interfacial profile has a nice analytical solution in the planar $(R \to \infty)$ limit:

$$\bar{e}(x) = \frac{1}{2}\left[e_A + e_S + \Delta e \tanh\left(\frac{x}{2\xi_A}\right)\right] \,. \tag{23}$$

Here the surface is located at $x = 0$ with A phase on the left and S phase on the right.

Suppose that a large bubble has formed at $T < T_c$ because of statistical fluctuations. The change in free energy of the system is:

$$\Delta F = \frac{4\pi}{3}(f_A - f_S)R^3 + 4\pi R^2\sigma, \tag{24}$$

where σ is the surface free energy. In our case for baryon free matter:

$$\Delta F = \frac{4\pi}{3}[p_S(T) - p_A(T)]R^3 + 4\pi R^2\sigma. \tag{25}$$

The bubble is stationary if $\partial_R \Delta F = 0$, which leads to the Young-Laplace equation:

$$p_A(T) - p_S(T) = \frac{2\sigma}{R_*(T)} \,. \tag{26}$$

Thus the activation energy is

$$\Delta F = \frac{4}{3}\pi\sigma R_*^2 \,. \tag{27}$$

The surface free energy can be calculated from our parametrization of F_I. For a planar interface or for a sphere whose radius is much greater than its surface thickness the formula is[22]:

$$\sigma = K \int_{-\infty}^{\infty} dx \left(\frac{d\bar{e}}{dx}\right)^2 \,. \tag{28}$$

Inserting the solution for the planar interface at T_c from eq. (23) this integral takes the form

$$\sigma = K\left(\frac{\Delta e}{2}\right)^2 \frac{1}{2\xi_A}\int_{-\infty}^{\infty} dz \frac{1}{\cosh^4(z)} = \frac{K(\Delta e)^2}{6\xi_A} \,. \tag{29}$$

The correlation length and the surface free energy determine the parameters $-f_0''$ and K in the course-grained free energy. In principle these parameters are temperature dependent. Their temperature dependence is, however, unknown to us. When numerical calculations are made later in this paper we shall use these values independent of T.

The free energy of critical size bubbles is accurately described by eq. (27) only when the radius is much greater than the correlation length, or equivalently stated, when the temperature approaches T_c. Deviations appear when R_* becomes less than about ten times the correlation length. This is known both in molecular physics[26] and in field theory[8,27,28,29]. This finite size effect needs to be taken into account in rate calculations for the early universe, because it has been estimated that the argument of the exponential in eq. (9) is on the order of -160 when the transition occurs[18], and a 10% change in the free energy has a big effect when exponentiated.

Here we use the notation of Liu, McLerran and Turok[29]. The authors express the free energy in terms of a dimensionless integral

$$\Delta F/T \propto \int dr\, r^2 \left[\frac{1}{2}\left(\frac{dg}{dr}\right)^2 + \frac{\zeta}{2}g^2 - g^3 + \frac{1}{4}g^4\right] \tag{30}$$

where g is a dimensionless field, r is a dimensionless coordinate, and we shall not be concerned with the constant of proportionality. The parameter ζ is a function of the temperature. It has the value 2 at T_c and decreases towards zero as the temperature decreases. The potential has local minima at $g_S = 0$ and at $g_A = (3+\sqrt{9-4\zeta})/2$. The profile $g(r)$ has been computed numerically for different values of ζ and the integral evaluated to yield the free energy; the results are presented graphically in figure 1 of Liu, McLerran and Turok[29]. The ratio $\Delta F(\text{finite size})/\Delta F(\text{volume + surface})$ can be read off of that figure. For $\zeta > 1.45$ an accurate (better than 1%) parametrization of this ratio is

$$\Delta F(\text{finite size}) = \Delta F(\text{volume} + \text{surface}) \times (g_A/2)^{9/2}. \tag{31}$$

Note that this correction factor extrapolates to unity at $\zeta = 2$.

The parameter ζ still needs to be related to other physical quantities to be useful in our rate equations. This can be done in one of two ways. One way is to take the scalar condensate ϕ as the independent field and to rescale it and the radial coordinate to put $\frac{1}{2}(\nabla\phi)^2 + V_{eff}(\phi, T)$ in the dimensionless form used by Liu, McLerran and Turok[29], thereby extracting ζ (assuming the effective potential can be represented by a fourth-order polynomial). The other way is to take the local energy density e as the independent field, as we are doing here, and to rescale it and the radial coordinate to put the effective coarse-grained free energy functional in the dimensionless form. The second way is actually the more convenient one. After some algebra, one can infer from the expressions given by Csernai and Kapusta[1] that

$$\begin{aligned}\zeta &= \frac{18\chi}{(2+\chi)^2},\\ \chi &= 1 + \frac{R_*}{4\xi} - \left[\left(\frac{R_*}{4\xi}\right)^2 + 1\right]^{1/2}\end{aligned} \tag{32}$$

where R_* is given by the Young-Laplace equation (26). We shall apply this correction factor to the free energy, eq. (27), when computing the thermal nucleation rate. We shall not be concerned with any modifications to the pre-exponential factor due to finite size effects.

THE PREFACTOR

In this section we evaluate the exponential prefactor in the nucleation rate. This prefactor is a product of two terms: the statistical prefactor and the dynamical prefactor. The statistical prefactor, Ω_0, is a measure of both the available phase space as the system goes over the saddle and of statistical fluctuations at the saddle relative to the equilibrium states. The dynamical prefactor, κ, is the exponential growth rate of the bubble or droplet sitting on the saddle.

Statistical Prefactor

The general expression for the statistical prefactor was given in eq. (12). To evaluate it, we first consider the eigenvalues of the matrix of second derivatives of F, the λ_α. The $\lambda_\alpha^{(0)}$ are eigenvalues of the operator

$$\left.\frac{\delta^2 F_I}{\delta e(\vec{r})\delta e(\vec{r}\,')}\right|_{e=e_S} = \left(-K\nabla^2 + \frac{\partial^2 f}{\partial e_S^2}\right)\delta(\vec{r}-\vec{r}\,') \,. \tag{33}$$

Here by $\partial^2 f/\partial e_S^2$ we mean the second derivative of f with respect to e at fixed temperature evaluated in the equilibrium S phase. This is a measure of fluctuations in the system and cannot be determined from knowledge of the equation of state alone. Since the right side of eq. (33) depends on $\vec{r}$ only through ∇^2, its eigenfunctions are plane waves, with wave vectors $\vec{q}$ and eigenvalues

$$\lambda_{\vec{q}}^{(0)} = K\vec{q}^{\,2} + \frac{\partial^2 f}{\partial e_S^2}. \tag{34}$$

At the saddle point, $e(\vec{r}) = \bar{e}(r)$, the operator

$$\left.\frac{\delta^2 F_I}{\delta e(\vec{r})\delta e(\vec{r}\,')}\right|_{e=\bar{e}(r)} = \left(-K\nabla^2 + \frac{\partial^2 f}{\partial \bar{e}^2}\right)\delta(\vec{r}-\vec{r}\,') \,. \tag{35}$$

is no longer translationally invariant because of the r dependence of $\bar{e}$. As has been discussed previously by Langer[24], the resulting spherically symmetric Schrödinger-like eigenvalue equation has an s-wave ground state with a radial eigenfunction proportional to $d\bar{e}/dr$ and a negative eigenvalue

$$\bar{\lambda}_1 \approx -2K/R^2 \,. \tag{36}$$

This eigenstate is associated with the instability of the critical bubble against uniform expansions or contractions. The next states are the three p-waves, with eigenvalues $\bar{\lambda} = 0$, which occur because of the broken translational symmetry. Then there are higher-order partial waves with positive $\bar{\lambda}$ corresponding to volume-conserving deformations of the shape of the droplet. Finally, there is a continuum of nonlocalized eigenfunctions of eq. (35) starting at $\bar{\lambda} = \partial^2 f/\partial e_S^2$. These eigenfunctions are similar to the states associated with the $\lambda^{(0)}$ in that they describe fluctuations in the bulk plasma, but here these fluctuations are perturbed by the presence of the bubble. As before, the eigenvalues associated with the kinetic part of F are spurious and can be disregarded.

We can recognize the products over α in eq. (12) as representing fluctuation corrections to the mean-field excess free energy of the bubble. If we were to evaluate ΔF using measured values of the surface energy and thermodynamic potential, it would be inconsistent to include fluctuation corrections to ΔF in the prefactor Ω_0. Strictly speaking, the nucleation formula used here requires that ΔF first be evaluated at the stationary point obtained from eq. (21), and then be corrected by the fluctuation terms in Ω_0. But this procedure would imply that the radius of the critical droplet be determined by the σ given in eq. (28), which is not necessarily the same as the experimental surface free energy because of the fluctuation corrections. What we shall do, instead, is delete the explicit fluctuation terms in Ω_0, and interpret σ everywhere as the true surface energy; we shall make a similar assumption concerning other thermodynamic quantities that appear. This procedure possibly can be justified by going beyond the Gaussian approximations for η-space integrations which were used in deriving eq. (12); that is, by constructing a renormalized perturbation expansion in the neighborhood of $\{\bar{\eta}\}$. If this program can be carried out, we might also be able to compute systematically curvature corrections to the surface energy. These corrections will be omitted here, and we shall focus our attention on other ingredients of the nucleation formula, particularly the dynamical prefactor. See also Langer and Turski[15] for additional comments on this aspect of the analysis.

Note, now, that there are $\alpha_0 + 1 = 4$ more terms in the product over the $\lambda_\beta^{(0)}$ than in that over the $\bar{\lambda}_\alpha$ in eq. (12). This means that the logarithm of the combined products is not precisely a free-energy difference. To see what is happening here, it is useful to think in terms of a one-to-one pairing between the $\lambda_\beta^{(0)}$ and the $\bar{\lambda}_\alpha$. At the top of the spectra (large positive $\lambda^{(0)}$ and $\bar{\lambda}$) both kinds of eigenvalues correspond to short-wavelength fluctuations which extend throughout the volume of the system V. We can pair these eigenvalues so that their contributions cancel each other to the extent that the droplet volume is negligible compared to the total volume of the system. At the bottom of the continuum a finite set of $\bar{\lambda}$, which correspond to localized deformations of the bubble, fall appreciably below their associated $\lambda^{(0)}$. Thus, by pairing the λ's as described, the correction to ΔF remains of order R^3 in the limit $V \to \infty$, as it must. This procedure leaves four unpaired $\lambda^{(0)}$'s at the bottom of the spectrum which are not accounted for by the revised ΔF. Specifically, we have

$$\lim_{V\to\infty} \prod_{\beta=1}^{4} \left(\frac{\lambda_\beta^{(0)}}{2\pi T} \right)^{1/2} = \left(\frac{1}{2\pi T} \frac{\partial^2 f}{\partial e_S^2} \right)^2 \tag{37}$$

remaining as the sole explicit contribution from the complicated products over the α eq. (12).

Having written down the value for $\bar{\lambda}_1$ in eq. (36), we need only evaluate the factor $\mathcal{V}$ to complete the calculation of Ω_0. The formula for $\mathcal{V}$ has been given by Langer[24,5]. It is

$$\mathcal{V} = V \left[\frac{1}{3} \int dr (\nabla \bar{\epsilon})^2 \right]^{3/2} = V \left[\frac{4\pi R^2 \sigma}{3\ K} \right]^{3/2} . \tag{38}$$

Here we have used eq. (28) for σ and that $d\bar{\epsilon}/dr$ is appreciable only in a narrow region near $r = R$ for the bubble.

The resulting expression for Ω_0 is

$$\Omega_0 = V \left(\frac{4\pi R^2 \sigma}{3\ K} \right)^{3/2} \left(\frac{\pi T R^2}{K} \right)^{1/2} \left(\frac{1}{2\pi T} \frac{\partial^2 f}{\partial e_S^2} \right)^2 . \tag{39}$$

Identifying the correlation length ξ_S in the S phase by

$$\frac{1}{K} \frac{\partial^2 f}{\partial e_S^2} = \frac{1}{\xi_S^2} , \tag{40}$$

we can write eq. (39) in the form

$$\Omega_0 = \frac{2}{3\sqrt{3}} \left(\frac{\sigma}{T}\right)^{3/2} \left(\frac{R}{\xi_S}\right)^4 V. \tag{41}$$

If one considers the nucleation rate to be per unit volume then the volume V should be divided out of the above expression. Usually in this paper we do mean the rate per unit volume and so Ω_0 will not include the factor V in subsequent discussion.

Dynamical Prefactor

In order for a bubble (or droplet) to grow, latent heat must be transported away from the surface region as discussed by Langer and Turski[15]. For the nonrelativistic systems they were considering, heat conduction was necessary to allow for growth. This eventually led to eq. (5), which says that κ is proportional to the thermal conductivity λ. It is clear that to get our bubble to grow we must include the effects of dissipation on the dynamics.

In relativistic dissipative fluid dynamics one adds new terms to the stress-energy-momentum tensor and to the baryon current as follows[21,30]:

$$T^{\mu\nu} = -pg^{\mu\nu} + (p+e)u^\mu u^\nu + \Delta T^{\mu\nu}, \tag{42}$$

$$N^\mu = nu^\mu + \Delta N^\mu. \tag{43}$$

In the absence of dissipation $\Delta T^{\mu\nu}$ and ΔN^μ are both zero, and u^μ is the flow four-velocity of the matter. When dissipation is present one has a choice of defining u^μ to be the velocity of baryon flow or the velocity of energy flow. These are known as the Eckart and Landau-Lifshitz approaches, respectively. In the Eckart approach ΔN^μ is zero by definition. Then $\Delta T^{\mu\nu}$ is a linear combination of a shear tensor, a projection tensor on the hyperplane normal to u^μ, and a heat flow vector, with coefficients being the shear viscosity η, the bulk viscosity ζ, and heat conductivity λ, respectively. In the Landau-Lifshitz approach $\Delta T^{\mu\nu}$ does not have the term involving the thermal conductivity; rather, ΔN^μ is nonzero and is in fact proportional to λ. These two approaches are completely equivalent and describe the same physics. However, the systems we are dealing with in this paper have no net baryon number, no baryon chemical potential, and so the Eckart approach is indeterminate. We must define u^μ to be the velocity of energy flow. The thermal conductivity simply is not defined because there is no net baryon density to define a frame of reference with respect to which energy can be conducted.

The contribution to the stress-energy-momentum tensor due to dissipation is

$$\Delta T_{\mu\nu} = -\eta\left(\partial_\mu u_\nu + \partial_\nu u_\mu - u_\mu u^\iota \partial_\iota u_\nu - u_\nu u^\iota \partial_\iota u_\mu\right) - (\zeta - \frac{2}{3}\eta)(\partial_\iota u^\iota)\left(g_{\mu\nu} - u_\mu u_\nu\right) . \tag{44}$$

The coefficients of shear and bulk viscosity must be positive from the requirement that the entropy should not decrease. In fact, the divergence of the entropy current is

$$\partial_\mu(su^\mu) = -\frac{1}{T}\Delta T^\mu_\nu \partial_\mu u^\nu , \tag{45}$$

from which one can calculate the total entropy change of the system. This equation is a direct consequence of the equations of motion, which are just the usual ones of conservation of energy and momentum, $\partial_\mu T^{\mu\nu} = 0$. In general in our applications the flow three-velocity is small compared to the speed of light. The relativistic treatment is required mainly by the fact that the pressure is comparable to the energy density, and by the absence of a net baryon number to define the motion of the matter.

We now want to determine the equations of motion of dissipative fluid dynamics for small deviations about the stationary configuration $e(\vec{r},t) = \bar{e}(r), \vec{v}(\vec{r},t) = 0$. To that end we write $e = \bar{e}(r) + \nu(\vec{r},t)$ and $\vec{v} = \vec{v}(\vec{r},t)$ and linearize the full equations of motion, including the gradient term F_K, in terms of ν and $\vec{v}$.

To determine κ we look for radial perturbations of the form

$$\nu(\vec{r},t) = \nu(r)e^{\kappa t} \; , \tag{46}$$

$$\vec{v}(\vec{r},t) = v(r)\hat{r}e^{\kappa t} \; . \tag{47}$$

These radial deviations are governed by the equations of motion. Eliminating $\nu(r)$ using one of them we obtain a linear, third-order differential equation with r-dependent coefficients for the velocity profile:

$$\begin{aligned} \kappa^2 \bar{w} v(r) &= -\frac{d\bar{e}}{dr}\left[K\left(\frac{d^2}{dr^2} + \frac{2}{r}\frac{d}{dr}\right) - f''\right]\left[\frac{1}{r^2}\frac{d}{dr}\left(r^2 \bar{w} v(r)\right)\right] \\ &\quad + \frac{d}{dr}\left[\kappa(\zeta + 4\eta/3)\frac{1}{r^2}\frac{d}{dr}\left(r^2 v(r)\right)\right] . \end{aligned} \tag{48}$$

Self-consistent solutions of this equation, together with the boundary conditions, should provide us with the allowed values of κ. Unfortunately, this is not a trivial equation to solve. Fortunately, there is a shortcut to determine κ which does not require knowledge of the complete solution for $v(r)$ and $\nu(r)$.

It is necessary to distinguish between the actual radius of the bubble, R, and the radius of the bubble in the stationary or metastable configuration, R_*, determined by the Young-Laplace equation (26). If the stationary bubble is perturbed only slightly then the energy density profile will only change by a minute amount. Transport of heat away from the surface will be a very slow process because of the assumed smallness of the viscosity. As the bubble slowly begins to expand, the energy density profile will not change much, only the profile needs to be moved out a small distance dR in a time dt. The energy flux density (energy per unit area per unit time) which must be transported outwards is $\Delta w \, dR/dt$. Here we do not distinguish between the difference of energy densities and the difference in enthalpy densities of the two bulk phases because the pressure difference is small compared to the energy density differences; we shall refer to them interchangeably as the latent heat. This energy flux must be balanced by that due to dissipation, which is[21] $-(\zeta + 4\eta/3)v \, dv/dr$. We evaluate the flow velocity just outside the surface of the bubble; it can be shown that the derivative is $dv/dr \approx -2v/R$. Therefore energy balance gives us the relation

$$\Delta w \frac{dR}{dt} = 2(\zeta + 4\eta/3)\frac{v^2}{R} \; . \tag{49}$$

The outward momentum flux density (momentum per unit area per unit time) is $\Delta w \, v^2$. (This neglects a small contribution from viscous terms which can be considered a higher order effect.) The momentum flux density must be equated to the force per unit area which comes from the Laplace formula,

$$\Delta w \, v^2 = 2\sigma\left(\frac{1}{R_*} - \frac{1}{R}\right) . \tag{50}$$

Again, the velocity is to be evaluated just outside the surface.

Using both energy and momentum conservation we can eliminate the velocity and solve for dR/dt,

$$\frac{dR}{dt} = \frac{4(\zeta + 4\eta/3)\sigma(R - R_*)}{(\Delta w)^2\, R^2\, R_*} \, . \tag{51}$$

This is a differential equation for $R(t)$ from which we can read off the value of κ. It is

$$\kappa = \frac{4\sigma(\zeta + 4\eta/3)}{(\Delta w)^2\, R_*^3} \, . \tag{52}$$

Note the similarity to the result of Langer and Turski, eq. (5). In the relativistic case the linear combination of viscosity coefficients replaces the thermal conductivity times the temperature. Actually, it is possible to do a fully relativistic treatment including heat conduction as well as viscosity. This has been done very recently by Venugopalan and Vischer[31].

Inserting eqs. (41) and (52) into eq. (3) gives us the nucleation rate

$$I = \frac{4}{\pi}\left(\frac{\sigma}{3T}\right)^{3/2} \frac{\sigma(\zeta_S + 4\eta_S/3)R_*}{\xi_S^4(\Delta w)^2} e^{-\Delta F/T} \, . \tag{53}$$

This is the probability per unit volume per unit time to nucleate a bubble of A phase out of the S phase. At the critical temperature, $R_* \to \infty$, and the rate vanishes because of the exponential. The system must supercool at least a minute amount in order that the rate attain a finite value. Note that at the critical temperature the pre-exponential factor is linearly divergent in R_* which is qualitatively unlike the simple dimensionless estimates of eqs. (7).

EFFECTIVE POTENTIAL AND EQUATION OF STATE

Carrington has computed the effective potential in the standard model by evaluating both the one-loop diagrams and the sum of ring diagrams[32]. The ring diagrams are necessary in this context because it turns out that the phase transition is very weakly first order and several of the scalar and vector boson masses are very small near the critical temperature. Infrared contributions to the effective potential become important as the boson masses go to zero, and it is well-known that these contributions can be accurately taken into account by summing the ring diagrams[33,34].

Let us denote the effective potential by $V_{eff}(\phi, T)$ where ϕ is the value of the scalar condensate field. It turns out that with inclusion of the ring diagrams there is no simple analytic expression for this potential, it must be evaluated numerically. To a good approximation it looks very much like a fourth-order polynomial in ϕ with temperature-dependent coefficients. In the neighborhood of T_c it has two minima, one corresponding to the S phase at $\phi_S = 0$ and the other corresponding to the A phase at $\phi_A > 0$. Above T_c the global minimum is at the origin, below T_c the global minimum is at ϕ_A, and at T_c the two minima are degenerate.

The thermodynamic pressure is the negative of the effective potential at the minimum. Thus

$$\begin{aligned} p_S(T) &= -V_{eff}(0, T) \\ p_A(T) &= -V_{eff}(\phi_A(T), T) \, . \end{aligned} \tag{54}$$

The effective potential is normalized so that the pressure of the vacuum is zero. The pressure difference defined as $\Delta p = p_A - p_S$ is equal to zero at T_c and is positive below T_c. We use a vacuum Higgs mass of 60 GeV and a top quark mass of 120 GeV, all other parameters are standard. With these parameters $T_c = 97.27$ GeV. For supercooling of up to 2% ($0.98T_c < T < T_c$) we have found an accurate parametrization of the pressure difference to be

$$\Delta p = (5.977x + 498.7x^2 - 7621x^3) \times 10^5 \text{ TeV/fm}^3 \tag{55}$$

where $x = 1 - T/T_c$.

It is not necessary to know the absolute value of the pressure as accurately as the pressure difference. The pressure difference determines the size and free energy of critical-size bubbles as well as the bubble growth velocity. The magnitude of the pressure influences the expansion rate of the universe, but since the universe is expanding so slowly at this epoch, a few percent change in the absolute magnitude of the pressure has no real consequences for the phase transition. Close inspection of the expressions given by Carrington[32] leads one to the approximate form of the pressure in the high temperature phase as

$$p_S(T) = \frac{427}{360}\pi^2 T^4 - \frac{m_H^2}{8\sqrt{2}G_F} \tag{56}$$

where m_H is the zero temperature Higgs mass and G_F is the Fermi constant. The pressure in the low temperature phase is $p_A = p_S + \Delta p$.

The entropy, energy and enthalpy densities in each phase can be obtained via the canonical thermodynamic relationships. For the parameters chosen in this paper the ratio of the latent heat to the energy density in the S phase at T_c is very small: $\Delta e/e_S = \Delta w/e_S = (5.98 \times 10^5 \text{ TeV fm}^{-3})/(4.13 \times 10^8 \text{ TeV fm}^{-3}) = 1.45 \times 10^{-3} = 0.145\%$. This is a very weak first order phase transition even though in a terrestrial sense the latent heat is huge.

The static correlation length ξ for the scalar field is expressed in terms of the second derivative of the effective potential at a minimum.

$$\xi^{-2} = \frac{d^2 V_{eff}}{d\phi^2}. \tag{57}$$

Static, spatial inhomogeneities in the scalar field fall off exponentially with this decay length, $\exp(-z/\xi)$. In general the correlation lengths in the two phases are different. For an effective potential that is quartic in the field they are the same at T_c. Even though the ring-improved effective potential is not exactly of the quartic form, we have found numerically that the curvatures at the minima are equal at T_c. The value is

$$\xi_A = \xi_S = 17.94/T_c = 0.0364 \text{ fm}. \tag{58}$$

As the temperature decreases, the correlation length increases in the S phase and decreases in the A phase. Since we expect relatively little supercooling we shall neglect this temperature variation. This approximation could be relaxed if necessary.

The surface free energy at T_c is given by the well-known expression[22]

$$\sigma = \int_{-\infty}^{\infty} dz \left(\frac{d\bar{\phi}}{dz}\right)^2 \tag{59}$$

where $\bar{\phi}(z)$ satisfies the classical equation of motion with the potential $V_{eff}(\bar{\phi}, T)$. Since the potential is well-approximated by a fourth-order polynomial, we use the solution

$$\bar{\phi}(z) = \frac{\phi_A}{2}[1 + \tanh(z/2\xi)]. \tag{60}$$

The surface free energy then is

$$\sigma = \frac{\phi_A^2(T_c)}{6\xi}\,. \tag{61}$$

We find numerically that $\phi_A(T_c) = 0.5227\,T_c$. Therefore $\sigma = 60$ TeV/fm^2. It is problematical to rigorously define a surface free energy at any temperature other than T_c.

VISCOSITY OF THE HIGH TEMPERATURE PHASE

The dynamical prefactor κ involves the shear viscosity in the high temperature S phase. We have computed it to leading-log order in the coupling constants within the relaxation time approximation. This is sufficient for our purposes, since a factor of two error in the shear viscosity introduces a factor of two error in the exponential prefactor in the nucleation rate, and this will have a negligible influence on the numerical results which are dominated by the exponential factor. In the S phase all particles are massless to zero'th order. The bulk viscosity is very small for a system of point particles with no internal rotational degrees of freedom. The shear viscosity must, on dimensional grounds, be of the form cT^3, where c is a number. The whole thrust of the calculation is to compute c. Let me outline how it goes.

For the purpose of illustration, consider a weakly interacting system of massless particles. Denote the one-particle Wigner distribution by $n_p(\vec{x}, t)$. In the local rest frame, and in local thermal equilibrium, the distribution function has the form $n_p^{(0)} = 1/(e^{E_p/T} \pm 1)$ where the upper sign is for fermions and the lower for bosons. For small departures from equilibrium one can compute $\delta n_p = n_p - n_p^{(0)} \ll n_p^{(0)}$ from the relativistic Boltzmann equation. In the relaxation time approximation

$$\frac{\partial}{\partial t}\delta n_p + \vec{v}_p \cdot \nabla \delta n_p = -\delta n_p/\tau \tag{62}$$

where τ is the collision time and $\vec{v}_p$ is the single-particle velocity. The viscosity is given in terms of the collision time as[35]

$$\eta = \frac{\tau}{15}\int \frac{d^3p}{(2\pi)^3}\frac{p^4}{E_p^2} n_p^{(0)}(1 \pm n_p^{(0)}). \tag{63}$$

We obtain the collision times from the cross-sections with the expression

$$\tau^{-1}(p) = \int \frac{d^3p'}{(2\pi)^3} n_{p'} v_{rel}\, \sigma(p, p') \tag{64}$$

where $s = (p + p')^2$ and $v_{rel} = s/(2EE')$. We assume that the momentum dependence is weak so that we can make the replacement $\tau^{-1}(p) = \langle \tau^{-1}(p)\rangle \equiv \tau^{-1}$. Finally, we replace the averaged product by the product of the averages, and use the fact that $\langle v_{rel}\rangle = 1$, to obtain

$$\tau^{-1} = n\langle \sigma(p, p')\rangle. \tag{65}$$

In a calculation of viscosities the expression that should be used for the cross-section is actually the transport cross-section, given by

$$\sigma = \int d(\cos\theta)\,(1 - \cos\theta)\frac{d\sigma}{d(\cos\theta)}. \tag{66}$$

This can be interpreted as follows. The factor $(1 - \cos\theta)$ gives a suppression of the contribution of forward scattering to the coefficient of viscosity. Physically this is a result of the fact that, when calculating viscosity, the relevant part of the dissipative energy-momentum tensor is the part that gives the energy-momentum transferred in a direction perpendicular to the forward scattering direction. The origin of this cutoff is the finite temperature screening of the long-ranged Coulomb force[36]. This is especially important for the exchange of massless quanta. In this case, the small angle limit of the two-particle scattering cross-section is dominated by the Coulomb divergence

$$\left(\frac{d\sigma}{d(\cos\theta)}\right)_{\text{Coulomb}} \sim \sin^{-4}(\theta/2). \tag{67}$$

The angular integral still has a divergence at $\theta = 0$, although it has been softened by the nature of the transport cross-section. A small angle cut-off is imposed on the integral: $\theta_{min} = \alpha^{3/2}$.

These equations are modified in a straightforward way when the system contains internal degrees of freedom and/or more than one particle species. We need to calculate all two-particle scattering cross-sections in the standard $SU(3) \times SU(2) \times U(1)$ model. This is quite tedious; details may be found in Carrington and Kapusta[3]. Actually, only the forward (t-channel) contributions or, in the case of identical particles, also the backward (u-channel) contributions, are important because these dominate the transport cross-section. The final answer is

$$\eta = 82.5\,T^3\,, \tag{68}$$

where we have used $\alpha_s = 0.1$ and standard values for the electroweak couplings α and α'.

BUBBLE GROWTH VELOCITY

The asymptotic growth velocity of bubbles was considered in detail by Liu, McLerran and Turok[29]. They used relativistic transport theory. Such a microscopic calculation is necessary in order to compute the velocity of the interface; energy and momentum conservation in perfect fluid dynamics is not sufficient.

We start from eq. (111) of Liu, McLerran and Turok[29]. It is

$$\frac{9\pi}{4N_F}\,\frac{u}{T\ln(AT^2/M^2)}\int_{-\infty}^{\infty} dz\,\phi^2\left(\frac{d\phi}{dz}\right)^2 = \Delta p\,. \tag{69}$$

Here A is an as yet undetermined constant, M is the vector meson mass in the A phase, $N_F = 48$ is the number of left-handed fermion degrees of freedom, Δp is the pressure difference between the two phases, $\phi(z)$ gives the variation in the Higgs field across the interface, and $u = \gamma v = v/\sqrt{1-v^2}$ with v the velocity. Since the constant A is unknown we set the logarithm to one, as done by Liu, McLerran and Turok[29].

For temperatures below but very close to T_c the Higgs field follows the usual hyperbolic arctangent behavior as given in eq. (60). Direct integration gives

$$\frac{9\pi}{80N_F}\,\frac{\phi_A^4}{\xi T_c}u = \Delta p\,. \tag{70}$$

With the numerical values for ϕ_A, T_c, ξ as given earlier we get

$$u = 3.26 \times 10^4\,\frac{\Delta p}{T_c^4}\,. \tag{71}$$

Thus the growth velocity vanishes as the temperature approaches T_c, as it should.

Given the nucleation rate one would like to know the (volume) fraction of space $f(t)$ which has been converted from the S phase to the A phase at the proper time t as measured in the local comoving frame of an expanding system. This requires kinetic equations which use the nucleation rate I as an input. Langer and Schwartz[37] have discussed such kinetic equations and compared predictions of their theory to cloud-point data in near-critical fluids. Guth and Weinberg[38] proposed a formula for $f(t)$ and applied it to cosmological first-order phase transitions. One may find other kinetic equations in the literature. It does not seem possible to derive such kinetic equations from first principles. Here we use a rate equation first proposed by Csernai and Kapusta[2]. The nucleation rate I is the probability to form a bubble of critical size per unit time per unit volume. If the system cools to T_c at time t_c then at some later time t the fraction of space which has been converted to the A phase is

$$f(t) = \int_{t_c}^{t} dt' I(T(t'))[1 - f(t')]V(t', t). \tag{72}$$

$V(t', t)$ is the volume of a bubble at time t which was nucleated at the earlier time t'; this takes into account bubble growth. The factor $1 - f(t')$ takes into account the fact that new bubbles can only be nucleated in the fraction of space not already occupied by the A phase. This conservative approach does not take into account collisions and fusion of bubbles, which would tend to decrease somewhat the time needed to complete the transition[39]. It also neglects any spatial variation in the temperature.

Next we need a dynamical equation which couples the time evolution of the temperature to the fraction of space converted to the A phase. We use Einstein's equations as applied to the early universe, neglecting curvature. The evolution of the energy density ϵ is

$$\frac{d\epsilon}{dR} = -\frac{3w}{R}, \tag{73}$$

where R is the scale factor at time t. This assumes kinetic but not phase equilibrium, and is basically a statement of energy conservation. We express the energy density as

$$\epsilon = f\epsilon_A(T) + [1 - f]\epsilon_S(T), \tag{74}$$

where ϵ_A and ϵ_S are the energy densities in the two phases at the temperature T. There is a similar equation for w. The time dependence of the scale factor is determined by the equation of motion

$$\frac{1}{R}\frac{dR}{dt} = \sqrt{\frac{8\pi G\epsilon}{3}}. \tag{75}$$

This expression can be used to rewrite the rate equation in the alternate form

$$f(R) = \sqrt{\frac{3}{8\pi G}} \int_1^R \frac{dR'}{R'\sqrt{\epsilon(R')}} I(T(R'))\,[1 - f(R')]\,V(R', R). \tag{76}$$

We have chosen the normalization $R(t_c) = 1$. Because the universe is expanding very slowly on the electroweak timescale ($t_c \approx 10^{-11}$ sec) and the phase transition is only weakly first order ($\Delta\epsilon \ll \epsilon$), it is a very good approximation to neglect the variation of $R'\sqrt{(\epsilon(R')}$ in the denominator of the integrand of the rate equation above. Similarly, during the phase transition it is a very good approximation to integrate the equation of motion of R to get

$$t - t_c = \sqrt{\frac{3}{8\pi G\epsilon(t_c)}}(R - 1). \tag{77}$$

This is useful in relating the time into the transition to the amount of expansion of the universe.

We also need to know how fast a bubble expands once it is created. We shall only attempt a crude description of this growth process here. After applying a perturbation, a critical size bubble begins to grow. As the radius increases, the surface curvature decreases, and an asymptotic interfacial velocity is approached. The asymptotic radial growth velocity will be referred to as $v(T)$. The expected qualitative behavior of $v(T)$ is that the closer T is to T_c the slower the bubbles grow. At T_c there is no motivation for bubbles to grow at all since one phase is as good as the other. Our simple illustrative model for bubble growth is

$$V(t',t) = \frac{4\pi}{3}\left(r_*(T(t')) + \int_{t'}^{t} dt'' v(T(t''))\right)^3 . \tag{78}$$

This expression can also be written in terms of R, R', R'' instead of t, t', t''. As pointed out by Linde[8] one could argue for the insertion of the dilution factor $R(t')/R(t'')$ behind the integration symbol above. However, the universe expands so terribly little during this very weak first order transition that this is a negligible effect.

To get a first impression as to how much the universe must supercool before nucleation begins, we plot the nucleation time in figure 2 as a function of temperature. Nucleation time is defined by

$$\tau_{\rm nucleation} = \frac{4\pi r_*^3}{3} I . \tag{79}$$

This is the characteristic time scale to make the transition in the absence of bubble growth; that is, the transition would complete only because every point in space had been nucleated. The solid curve represents our best estimate of the nucleation rate, as discussed in previous sections. The dashed curve shows what happens when the finite

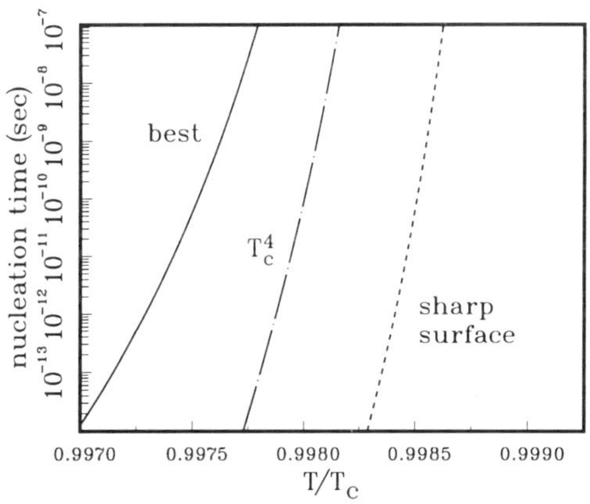

Figure 2. The nucleation time as a function of temperature. The solid curve represents our best estimate. The dashed curve approximates the bubble free energy as a sum of volume and surface terms, and neglects the finite-size correction. The dot-dashed curve approximates the pre-exponential factor by T_c^4.

size correction to the critical bubbles is neglected; see eq. (31). Although this curve is labeled *sharp surface*, this is not an entirely correct designation. The reason is that it does make use of the surface tension σ, and this quantity is computed with a smooth surface of finite thickness, albeit planar. The dot-dashed curve shows what happens when the pre-exponential factor in the rate is taken to be T_c^4 rather than the expression given by eq. (53). Both approximations differ from the best estimate of the nucleation time by many orders of magnitude. On the other hand, one may say that use of the more sophisticated estimates for the bubble free energy and the pre-exponential factor simply results in a delay of the onset of the phase transition.

To relate the expansion scale of the universe to the local time we recall eq. (77). Defining $\Delta t = t - t_c$ and $\Delta R = R - 1$, and using the equation of state from eqs. (55) and (56), we obtain

$$\Delta t = 4.97 \times 10^{-11}\, \Delta R \text{ sec.} \tag{80}$$

Comparison with figure 2 suggests that we will see supercooling of about 0.25% in the temperature.

Figure 3 shows the temperature as a function of the scale factor. The solid curve represents the Maxwell construction for phase coexistence at T_c and generates no entropy. The dashed curve is a result of solving numerically the coupled equations discussed earlier in the paper. The difference between these two curves is remarkable. Nucleation does not even begin until long after an idealized adiabatic Maxwell phase transition would have completed. However, once nucleation begins, the transition proceeds and completes in a much shorter time interval. For practical purposes, nucleation begins at the bottom of the cooling line. Thereafter, nucleation and growth of bubbles releases latent heat, which causes the temperature to rise. Once the transition is completed, the universe again cools. A very small amount of entropy is generated during the transition because the universe is out of equilibrium. The amount generated can be inferred by comparing T^3 for the dashed curve to T^3 for the solid curve. An earlier treatment of the electroweak phase transition gave a trajectory of temperature versus time very similar in shape to this one[18].

Figure 4 shows the average bubble density

$$n(t) = \int_{t_c}^{t} dt' I(T(t'))[1 - f(t')] \tag{81}$$

as a function of time. Bubble nucleation is not noticeable until about 6.5×10^{-14} sec after the universe has cooled to T_c. At about 6.87×10^{-14} sec, bubble nucleation has already turned off because of reheating to higher temperatures. The transition is not completed until about 7.05×10^{-14} sec, as indicated by the termination point on the curve. Figure 5 shows the fraction of space $f(t)$ which has made the conversion to the A phase. One could say, from the point of view of this quantity, that the transition requires only about 3×10^{-15} sec from start to completion. This is much shorter than the characteristic expansion time scale of the universe, which is $dt/dR = 5 \times 10^{-11}$ sec. Figure 6 shows the average bubble radius as a function of time. There is a linear relation between the average radius and time, indicating constant radial growth velocity, after about 6.87×10^{-14} sec. This is the time when nucleation has just turned off. At this time only about 10% of the universe has been converted to the new phase. The remaining 90% is converted because of bubble growth. From the figure it may be seen that this growth velocity is about 0.83 times the speed of light.†

†Figures 4-6 differ slightly from those in the original paper[3] because of a typographical error in a computer program.

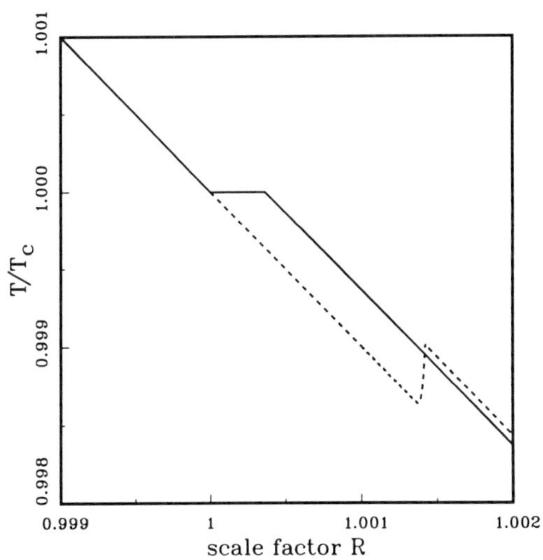

Figure 3. Temperature vs scale factor. The solid curve represents the adiabatic Maxwell construction for phase coexistence at T_c. The dashed curve represents the results of numerically integrating the dynamical equations of motion.

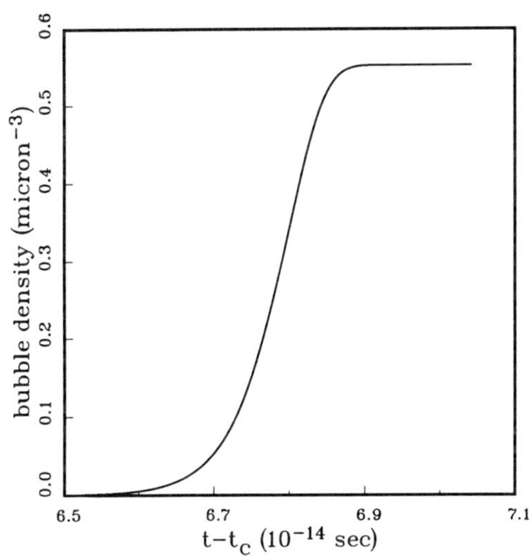

Figure 4. The density of A phase bubbles as a function of time.

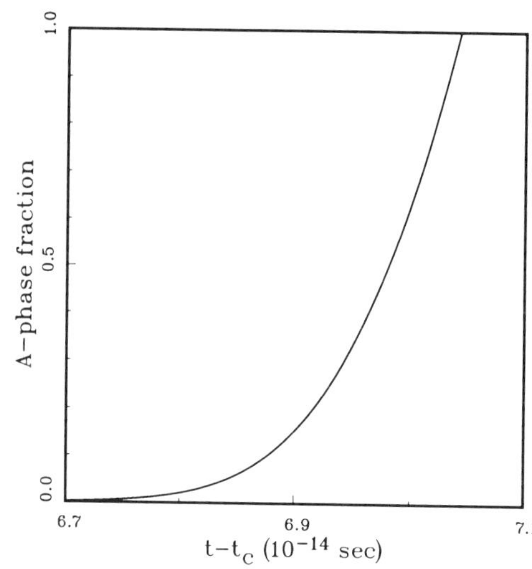

Figure 5. The fraction of space which has been converted to the A phase as a function of time.

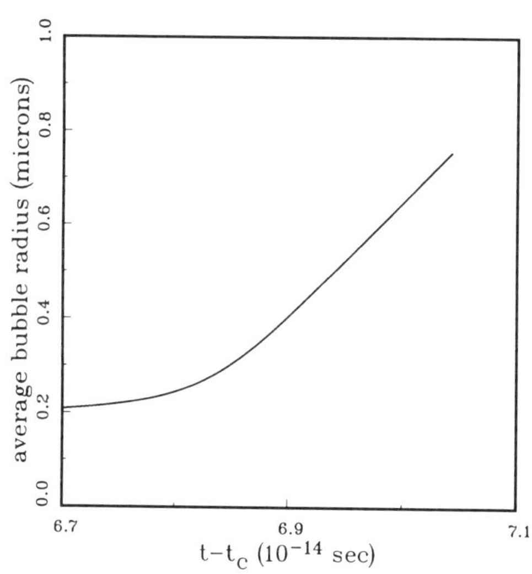

Figure 6. The average bubble radius as a function of time.

A useful quantity for baryogenesis is the value of the Higgs condensate inside the bubble. For the actual range of temperatures seen here during the transition, we know that $\phi_A/T_c \approx 0.54 - 0.58$. At zero temperature the condensate has the known value $\phi_A = 246$ GeV $= 2.53\ T_c$. A larger value than that seen during the transition would have been more favorable for baryogenesis. However, this is intimately tied to the very small latent heat in the standard model. Extended models of the electroweak interactions could have larger latent heats and larger values of the Higgs condensate inside the bubbles.

We must add a cautionary note. When the fraction of space occupied by bubbles exceeds about 50%, interactions among the bubbles probably cannot be neglected. It is unlikely, though, that further improvements in the dynamics would qualitatively change the current picture of the transition. Indeed, crude estimates of the effects of bubble fusion on the dynamics of the QCD transition indicate that the transition completes only a little faster, and that the average bubble size is greater[39]. Further developments on this topic would be welcome.

CONCLUSIONS

We have analyzed in detail the dynamics of the electroweak phase transition during the early universe. We used the ring-improved effective potential for the standard model. From this potential we extracted the equation of state, the surface free energy and the correlation lengths. We calculated the viscosity in the high temperature S phase, which is an important factor in the nucleation rate, and which may be useful in other contexts as well. We used a recently computed expression for the pre-exponential factor in the nucleation rate, and took into account the finite size effect for critical bubbles. We used a recent expression for the growth velocity of bubbles. All of these ingredients were joined together with a rate equation which allows for completion of the phase transition. Coupling this rate equation with Einstein's equations allowed us to solve for the dynamical evolution of the universe through the electroweak phase transition.

We computed the temperature, fraction of space converted, bubble density and average bubble radius as functions of time. The temporal evolution of the universe does not resemble an adiabatic Maxwell construction of coexisting phases. This is remarkable because the characteristic expansion time scale of the universe is 10^{-11} sec, very short compared to the characteristic time scale 10^{-26} sec of electroweak theory. The explanation is that the electroweak phase transition, at least as studied here, is first order but very weak. The pressure difference between the two phases near T_c is very small. This leads to times and structures which are mesoscopic during the phase transition.

The universe is out of equilibrium during the phase transition. This is one of the necessary ingredients for baryogenesis. Baryon number changing processes are expected to take place in the bubble surface; therefore, our results should be useful for such studies. They can also be easily repeated for extensions of the standard model.

Improvements to many aspects of our study may be considered, especially for the late stage of the phase transition when bubbles begin to overlap. However, it is becoming clear that a quantitative analysis of the electroweak phase transition during the early universe can be put on firm footing, and this will help us to contemplate our origins via baryogenesis.

Acknowledgments

I am happy to point out that the material presented here is based on collaborative work with my colleagues Laszlo Csernai and Meg Carrington. I would like to thank the organizers for arranging an exciting workshop in such a wonderful setting. This work was supported by the U. S. Department of Energy under Grant Number DOE/DE-FG02-87ER40328. Partial travel support to and at the workshop was provided by NATO.

References

[1] L. P. Csernai and J. I. Kapusta, *Phys. Rev. D* 46:1379 (1992).

[2] L. P. Csernai and J. I. Kapusta, *Phys. Rev. Lett.* 69:737 (1992).

[3] M. E. Carrington and J. I. Kapusta, *Phys. Rev. D* 47:5304 (1993).

[4] R. Becker and W. Döring, *Ann. Phys.* 24:719 (1935).

[5] J. S. Langer, *Ann. Phys.* 54:258 (1969).

[6] S. Coleman, *Phys. Rev. D* 15:2929 (1977); 16:1248(E) (1977);
C. G. Callan and S. Coleman, *Phys. Rev. D* 16:1762 (1977).

[7] I. Affleck, *Phys. Rev. Lett.* 46:388 (1981).

[8] A. D. Linde, *Nucl. Phys. B* 216:421 (1983). There was an overall factor of $\omega_-/\pi T$ missing in the expression for I in this paper.

[9] T. DeGrand and K. Kajantie, *Phys. Lett. B* 133:172 (1983).

[10] K. Enqvist, J. Ignatius, K. Kajantie and K. Rummukainen, *Phys. Rev. D* 45:3415 (1992).

[11] D. J. Gross, M. J. Perry and L. G. Yaffe, *Phys. Rev. D* 25:330 (1982).

[12] J. D. Gunton, M. San Miguel and Paramdeep S. Sahni, *in*: "Phase Transitions and Critical Phenomena," C. Domb and J. L. Lebowitz, eds., Academic Press, London (1983).

[13] J. E. McDonald, *Amer. J. Phys.* 30:870 (1962); 31:31 (1963).

[14] L. A. Turski and J. S. Langer, *Phys. Rev. A* 22:2189 (1980).

[15] J. S. Langer and L. A. Turski, *Phys. Rev. A* 8:3230 (1973).

[16] K. Kawasaki, *J. Stat. Phys.* 12:365 (1975).

[17] L. Carson, Xu Li, L. McLerran and R. T. Wang, *Phys. Rev. D* 42:2127 (1990).

[18] K. Enqvist, J. Ignatius, K. Kajantie and K. Rummukainen, *Phys. Rev. D* 45:3415 (1992).

[19] N. G. Van Kampen, *Phys. Rev.* 135:A362 (1964).

[20] J. L. Jackson, *Phys. Fluids* 3:928 (1960).

[21] L. D. Landau and E. M. Lifshitz, "Fluid Mechanics," Pergamon, London (1959).

[22] J. W. Cahn and J. E. Hilliard, *J. Chem. Phys.* 28:258 (1958); 31:688 (1959).

[23] B. Widom, *in*: "Phase Transitions and Critical Phenomena," C. Domb and M. S. Green, eds., Academic Press, NY (1972).

[24] J. S. Langer, *Ann. Phys.* 41:108 (1967).

[25] J. S. Langer, *Ann. Phys.* 65:53 (1971).

[26] A. H. Falls, L. E. Scriven and H. T. Davis, *J. Chem. Phys.* 75:3986 (1981).

[27] N. Turok, *Phys. Rev. Lett.* 68:1803 (1992).

[28] M. Dine, R. G. Leigh, P. Huet, A. D. Linde and D. Linde, *Phys. Rev. D* 46:550 (1992).

[29] B. H. Liu, L. McLerran and N. Turok, *Phys. Rev. D* 46:2668 (1992).

[30] S. Weinberg, "Gravitation and Cosmology," Wiley, New York (1972).

[31] R. Venugopalan and A. Vischer, *Phys. Rev. E* 49:5849 (1994).

[32] M. E. Carrington, *Phys. Rev. D* 45:2933 (1992).

[33] A. L. Fetter and J. D. Walecka, "Quantum Theory of Many-Particle Systems," McGraw-Hill, New York (1971).

[34] J. I. Kapusta, "Finite Temperature Field Theory," Cambridge University Press, Cambridge, England (1989).

[35] A. Hosoya and K. Kajantie, *Nucl. Phys. B* 250:666 (1985).

[36] F. Reif, "Fundamentals of Statistical and Thermal Physics," McGraw-Hill, New York (1965).

[37] J. S. Langer and A. J. Schwartz, *Phys. Rev. A* 21:948 (1980).

[38] A. H. Guth and E. J. Weinberg, *Phys. Rev. D* 23:876 (1981).

[39] L. P. Csernai, J. I. Kapusta, Gy. Kluge and E. E. Zabrodin, *Z. Phys. C* 58:453 (1993).

GAUGE INVARIANT EFFECTIVE POTENTIAL AND 4-d LATTICE SIMULATIONS OF THE ELECTROWEAK PHASE TRANSITION

Zoltán Fodor*

Deutsches Elektronen-Synchrotron DESY
22603 Hamburg, Germany

1. INTRODUCTION

At high temperatures the spontaneously broken symmetry of electroweak interactions is restored[1]. This is a direct consequence of the Higgs mechanism of symmetry breaking, and the corresponding phase transition is an interesting aspect of electroweak interactions. This has important cosmological implications, since the baryon-number violating processes are unsuppressed at high temperatures. In particular it is conceivable that the observed baryon asymmetry can be understood within the standard model (SM)[2]. In order to provide the necessary departure from thermal equilibrium a sufficiently strong first order phase transition via bubble nucleation is needed.

The descriptions of the phase transition usually start from the finite-temperature effective potential. Considerable effort has been devoted to evaluate the effective potential in perturbation theory[3–11] where, due to infrared problems, a useful expansion is only obtained after a resummation. Alternatively results have been obtained by means of lattice Monte Carlo simulations[12–15], reduced three-dimensional actions[16], average action at finite temperature[17], renormalization group analysis and ϵ-expansion[18–20].

In Section 2 the finite-temperature wave function corrections of the SU(2)-Higgs model to one-loop order and the effective potential to order g^4, λ^2 will be studied. This allows one to give a range of Higgs boson masses for which the derivative expansion of the effective action is reliable. Section 3 contains the gauge-invariant treatment of the finite temperature electroweak effective potential. This is of particular importance for comparison with future lattice simulations, where the expectation value of $\Phi\Phi^\dagger$ is well suited to characterize the broken phase, and the corresponding effective potential has been evaluated[21]. First results of a lattice study of the electroweak phase transition will be given in Section 4. The conclusions can be found in Section 5.

2. THE EFFECTIVE ACTION AT FINITE TEMPERATURE

The decay of a metastable homogeneous state involves "critical droplets" and there-

*On leave from Institute for Theoretical Physics, Eötvös University, Budapest, Hungary fore inhomogeneous field configurations. The free energy of these configurations is estimated based on an effective action which is the sum of the finite-temperature effective potential and the canonical kinetic term of the Higgs field. Thus, there is a need to calculate the wave function correction terms. Higher order corrections to the effective potential are also essential to describe the electroweak phase transition.

2.1 The wave function correction term

Consider the SU(2)-Higgs model at finite temperature which is described by the action

$$S_\beta[\Phi, W] = \int_\beta dx \left(-\frac{1}{4}W^a_{\mu\nu}W^{a\mu\nu} + (D_\mu\Phi)^\dagger(D^\mu\Phi) - V_0(\varphi^2)\right) \quad , \tag{1}$$

where

$$V_0(\varphi^2) = \frac{1}{2}\mu^2\varphi^2 + \frac{1}{4}\lambda\varphi^4 \,, \quad \frac{1}{2}\varphi^2 \equiv \Phi^\dagger\Phi \quad ,$$
$$\int_\beta dx \equiv \int_0^\beta d\tau \int d^3x \quad , \quad \beta \equiv \frac{1}{T}. \tag{2}$$

D_μ and $W^a_{\mu\nu}$ are the covariant derivative and the Yang-Mills field strength, respectively. The gauge fixing part of the lagrangian has been omitted, and in this section Landau gauge is used. The SU(2)-Higgs model contains the essential features of the standard model of electroweak interactions. The effects of the three generations of quarks and leptons with the usual gauge and the top Yukawa coupling ($m_t = f_t v/\sqrt{2}$) have been included.

In order to find the thermal equilibrium states and to compute properties of the first-order phase transition we have to know the effective action $\Gamma_\beta[\Phi]$ at finite temperature as functional of the Higgs field Φ which, after gauge fixing, plays the role of an order parameter. The effective action can be given as a functional integral

$$e^{-\Gamma_\beta[\Phi]} = \int_\beta [D\hat{\Phi}][D\hat{W}_\mu] \exp\left(-S_\beta[\Phi + \hat{\Phi}, \hat{W}] + \int_\beta dx \frac{\delta\Gamma_\beta[\Phi]}{\delta\Phi(x)}\hat{\Phi}(x)\right) \quad , \tag{3}$$

where the integration extends over fields with periodic boundary conditions in τ. Again, gauge fixing and ghost terms have not been written explicitly. For stationary fields $\Phi(\vec{x})$ the effective action corresponds to the free energy density of the system,

$$\Gamma_\beta[\Phi] = \beta F[\Phi, T] \quad . \tag{4}$$

A systematic expansion is obtained by means of a resummation, where in all propagators the tree-level masses are replaced by one-loop plasma masses to order g^2 and λ (cf. Ref. 6). The mass differences are then treated as counter terms,

$$\delta S_\beta = -\frac{1}{2}\int_\beta dx \left[W^{a\mu}\left(\delta m_L^2 P_{L\mu\nu} + \delta m_T^2 P_{T\mu\nu}\right)W^{a\nu} - \delta m_\varphi^2\varphi^2 - \delta m_\chi^2\chi^2\right] \quad . \tag{5}$$

Here P_L and P_T are the usual projection operators on longitudinal and transverse degrees of freedom of the vector field. The plasma mass terms δm_i^2 are the differences between the masses

$$m_L^2 = \frac{11}{6}g^2T^2 + \frac{g^2\varphi^2}{4} \quad , \tag{6}$$

$$m_T^2 = \frac{g^2\varphi^2}{4} \quad , \tag{7}$$

$$m_\varphi^2 = \left(\frac{3}{16}g^2 + \frac{\lambda}{2} + \frac{1}{4}f_t^2\right)(T^2 - T_b^2) + 3\lambda\varphi^2 \quad , \tag{8}$$

$$m_\chi^2 = \left(\frac{3}{16}g^2 + \frac{\lambda}{2} + \frac{1}{4}f_t^2\right)(T^2 - T_b^2) + \lambda\varphi^2 \quad , \tag{9}$$

and the tree-level masses

$$m^2 = \frac{1}{4}g^2\varphi^2 \quad , \quad \bar{m}_\varphi^2 = \lambda(3\varphi^2 - v^2) \quad , \quad \bar{m}_\chi^2 = \lambda(\varphi^2 - v^2) \quad . \tag{10}$$

The temperature T_b, at which the barrier between the symmetric and the broken phase vanishes, reads in leading order of the couplings

$$T_b^2 = \frac{16\lambda v^2}{3g^2 + 8\lambda + 4f_t^2} \quad . \tag{11}$$

Note, that the tree-level Higgs- and Goldstone-boson masses are obtained from the scalar potential by ($\varphi^2 = \sum_{I=1}^4 \varphi_I\varphi_I$)

$$\frac{\partial^2 V_0}{\partial\varphi_I\partial\varphi_J} = \bar{m}_\varphi^2 P_{IJ}^\varphi + \bar{m}_\chi^2 P_{IJ}^\chi \quad , \tag{12}$$

where

$$P_{IJ}^\varphi = \frac{\varphi_I\varphi_J}{\varphi^2} P_{IJ}^\chi = \delta_{IJ} - \frac{\varphi_I\varphi_J}{\varphi^2} \tag{13}$$

are the projection operators on the Higgs and the Goldstone fields.

At one-loop order the improved perturbation theory yields the effective potential to order g^3, $\lambda^{3/2}$,

$$\begin{aligned} V_{eff}(\varphi^2, T) &= \frac{1}{2}\left(\frac{3g^2}{16} + \frac{\lambda}{2} + \frac{1}{4}f_t^2\right)(T^2 - T_b^2)\varphi^2 + \frac{\lambda}{4}\varphi^4 \\ &\quad -(3m_L^3 + 6m_T^3 + m_\varphi^3 + 3m_\chi^3)\frac{T}{12\pi} + \mathcal{O}(g^4, \lambda^2, f_t^4), \end{aligned} \tag{14}$$

which is equivalent to the result of the ring summation[3]. Two-loop contributions are also known[4,5,10].

The strength of the electroweak phase transition is rather sensitive to the magnetic mass of the gauge bosons, whose calculation requires non-perturbative techniques. In Landau gauge the one-loop gap equations[6,7,22] yield $m_T = (g^2/3\pi)T$ at $\varphi = 0$. In order to estimate its effect on parameters of the phase transition we will replace eq. (7) by

$$m_T^2 = \gamma^2\frac{g^4}{9\pi^2}T^2 + \frac{g^2}{4}\varphi^2 \quad , \tag{15}$$

and compute sensitive quantities for different values of γ.

The effective potential (14) has degenerate local minima at $\varphi = 0$ and $\varphi = \varphi_c > 0$ at a critical temperature T_c and therefore predicts a first-order phase transition. The evaluation of the transition rate requires knowledge of a stationary point of the free energy which interpolates between the two local minima. The effective action (3) can be expanded in powers of derivatives, and for time-independent fields one has

$$\Gamma_\beta[\Phi] = \beta\int d^3x \left(V_{eff}(\varphi^2, T) + \frac{1}{2}(\delta_{IJ} + Z_{IJ}(\Phi, T))\vec{\nabla}\varphi_I\vec{\nabla}\varphi_J + \ldots\right) \quad . \tag{16}$$

From eqs. (3) and (16) it is obvious that the functions Z_{IJ} can be obtained from the inverse scalar propagator in the homogeneous scalar background field Φ with spacial momentum $\vec{k}$,

$$
\begin{aligned}
D^{-1}(\vec{k}^2, \Phi, T)_{IJ} &= \vec{k}^2 \delta_{IJ} + \frac{\partial^2 V_0(\varphi^2)}{\partial \varphi_I \partial \varphi_J} + i\Sigma_{IJ}(\vec{k}^2, \Phi, T) \\
&= m_\varphi^2(\varphi^2, T) P^\varphi_{IJ} + m_\chi^2(\varphi^2, T) P^\chi_{IJ} \\
&\quad + (\delta_{IJ} + Z_{IJ}(\Phi, T))\vec{k}^2 + \mathcal{O}(\vec{k}^4) \quad . \qquad (17)
\end{aligned}
$$

The one-loop contributions to Z_{IJ} are shown in Fig 1. It is a well-known problem in finite temperature field theory that the order of a given Feynman graph can be smaller than the order coming from the vertices. This is due to the fact that loop diagrams accumulate couplings in the denominators which decrease the power of the usual couplings coming from the Feynman rules. However, it can be shown that only graphs of the type shown in Fig. 1 give contributions to the wave function correction in leading order, where one has to include not only the tree level but also the plasma masses for internal lines.

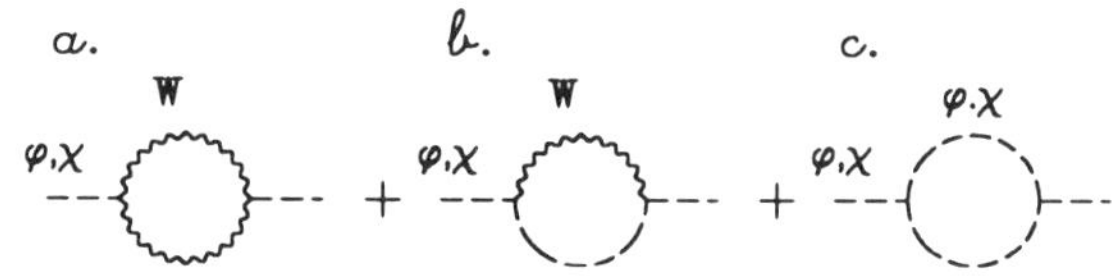

Figure 1. One loop contributions to the wave function correction factors

Eq. (16) is an expansion in powers of $(\vec{\nabla}/m_i)^2$, where m_i denotes all masses which appear in propagators of internal lines. The terms with zero Matsubara frequencies for the internal lines dominate, since they have the smallest denominators. For the same reason the top-quark contribution is suppressed, since all Matsubara frequencies are non-zero for fermions. It is straightforward to evaluate the contributions of Fig. 1

$$
Z_{IJ}(\Phi, T) = Z_\varphi(\varphi^2, T) P^\varphi_{IJ} + Z_\chi(\varphi^2, T) P^\chi_{IJ} \quad , \qquad (18)
$$

where

$$
\begin{aligned}
Z_\varphi(\varphi^2, T) &= \frac{T}{4\pi}\left[\frac{1}{4}\lambda\bar{m}^2\left(\frac{3}{m_\varphi^3} + \frac{1}{m_\chi^3}\right) - 2g^2\frac{1}{m_\chi + m_T}\right. \\
&\qquad \left. + \frac{1}{8}g^2 m^2\left(\frac{1}{2m_L^3} + \frac{5}{m_T^3}\right)\right] \qquad (19)
\end{aligned}
$$

and

$$
Z_\chi(\varphi^2, T) = \frac{T}{6\pi}\left[\frac{2\lambda\bar{m}^2}{(m_\varphi + m_\chi)^3} - g^2\left(\frac{2}{m_\chi + m_T} + \frac{1}{m_\varphi + m_T}\right)\right] \quad , \qquad (20)
$$

with $\bar{m}^2 = \lambda\varphi^2$ and $m^2 = g^2\varphi^2/4$. In terms of the complex field Φ this result can be written as

$$
\frac{1}{2} Z_{IJ} \vec{\nabla}\varphi_I \vec{\nabla}\varphi_J = \frac{1}{4}(\Phi^\dagger\Phi)^{-1}[\vec{\nabla}(\Phi^\dagger\Phi)]^2 Z_\varphi + \left(\vec{\nabla}\Phi^\dagger\vec{\nabla}\Phi - \frac{1}{4}(\Phi^\dagger\Phi)^{-1}[\vec{\nabla}(\Phi^\dagger\Phi)]^2\right) Z_\chi \quad . \qquad (21)
$$

Wave function correction terms in the 't Hooft-Feynman gauge have recently been evaluated in Ref. 11. Note, that despite the divergence of Z_φ at $\varphi \sim 0$ the correction to the surface tension $\sigma = \int_0^{\varphi_c} d\varphi \sqrt{2(1 + Z_\varphi(\varphi^2, T_c))V_{eff}(\varphi^2, T_c)}$ is finite.

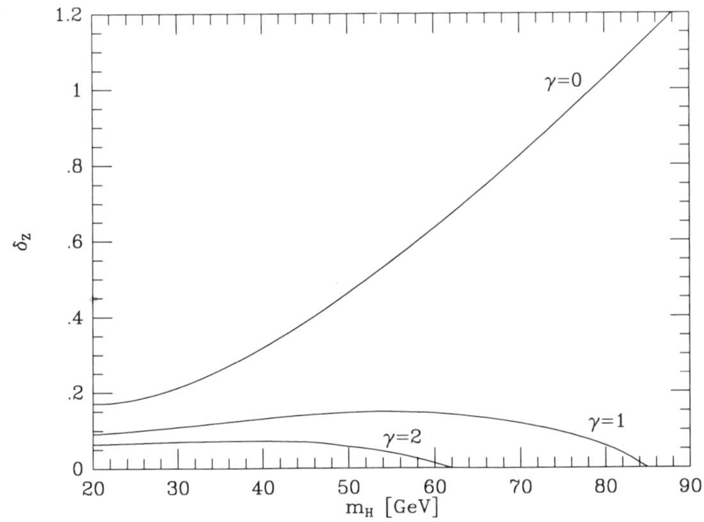

Figure 2. One loop wave function correction as a function of the Higgs mass for different values of γ.

A measure for the size of the one-loop correction to the Z-factor is the ratio

$$\delta_Z = \frac{\int d^3x Z_{\bar{\varphi}} \left(\vec{\nabla}\bar{\varphi}\right)^2}{\int d^3x \left(\vec{\nabla}\bar{\varphi}\right)^2}, \qquad (22)$$

where the $\bar{\varphi}$ is the saddle point solution at the nucleation temperature calculated from the one loop potential. This correction is shown in Fig.2 as function of the Higgs mass for different values of γ. For $\gamma = 0$ the perturbative expansion becomes unreliable at $m_H \sim 80$ GeV. For $\gamma = 1, 2$ the correction never exceeds 10% which demonstrates the importance of the infrared cutoff provided by the magnetic mass.

The results presented above are based on the considerations of Ref. 8, where additional details can also be found.

2.2 The effective potential to order g^4, λ^2

The principal method[9] of the calculation is based on the Dyson-Schwinger equation for the derivative of the potential $\partial V/\partial\varphi$. Essentially a summation of tadpole diagrams is performed. However, the vertex need not be corrected to obtain the full result in this order.

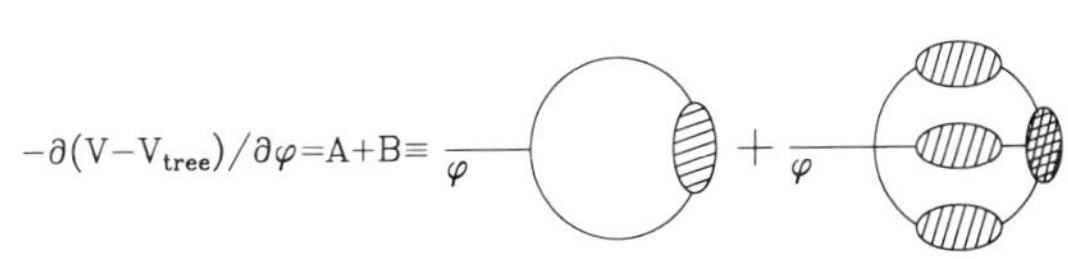

The calculation has been done in an alternative way too, using the resummation[5] method of P. Arnold and O. Espinosa, who calculated the effective potential to order

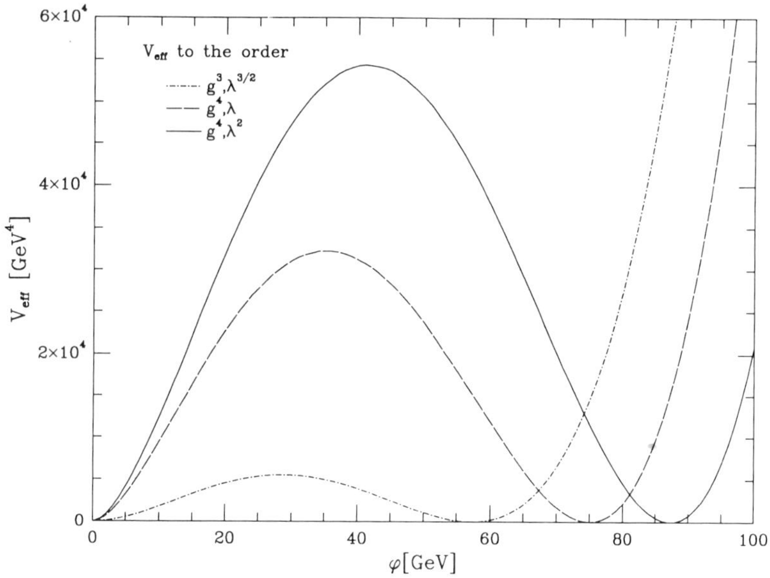

Figure 3. Different approximations of the effective potential plotted at their respective critical temperatures at a Higgs mass of 70 GeV.

g^4, λ. The two methods give identical results. Furthermore, dropping the appropriate terms of V the lower order g^4, λ-result can be derived.

The full result of order g^4, λ^2 predicts a stronger first order phase transition than the lower order results. It is instructive to plot the potential using different approximations for the $SU(2)$ Higgs-model (without fermions) at their respective critical temperatures. As it can be seen on Fig. 3 the expectation value of the Higgs field does not change dramatically, but there is an order of magnitude difference between the heights of the barrier. No convergence of the perturbation series can be claimed for these parameters. A more detailed picture is provided by the surface tension

$$\sigma = \int_0^{\varphi_+} d\varphi \sqrt{2V(\varphi, T_c)}, \tag{23}$$

which may be seen as a measure of the strength of the phase transition. It can be used conveniently to discuss the properties of the potential as a function of the Higgs mass. The results are shown in Fig. 4. For very small Higgs masses the third order potential gives a much larger value for the surface tension than the more complete calculations. The reason for that is the $g^4\varphi^4$ contribution, which takes over the role of the tree level $\lambda\varphi^4$ term for small scalar coupling. This radiatively induced quartic term ensures that σ does not increase for small Higgs masses. As expected, corrections of higher order in λ do not change the g^4, λ result if the scalar mass is small.

This picture changes drastically if larger Higgs masses are considered. In this region higher order corrections produce an enormous increase in the surface tension.

It is instructive to compare first the results of order $g^3, \lambda^{3/2}$ and g^4, λ^2. The increase in the strength of the phase transition, studied already[4], can be traced back to the infrared features of a non-Abelian gauge theory. The crucial contribution is the one coming from the non-Abelian setting sun diagram. It produces contributions to the potential of type $\varphi^2 \ln(\beta m_T)$ with negative sign. The huge effect of the $\ln \beta m_T$-contribution to the coefficient of $g^4\varphi^2$ can be understood by recalling that at the critical temperature the leading order φ^2-terms essentially cancel. However, the $\varphi^2 \ln \varphi$-type behaviour can not be absorbed in a correction of T_c, these terms increase the strength of the phase transition.

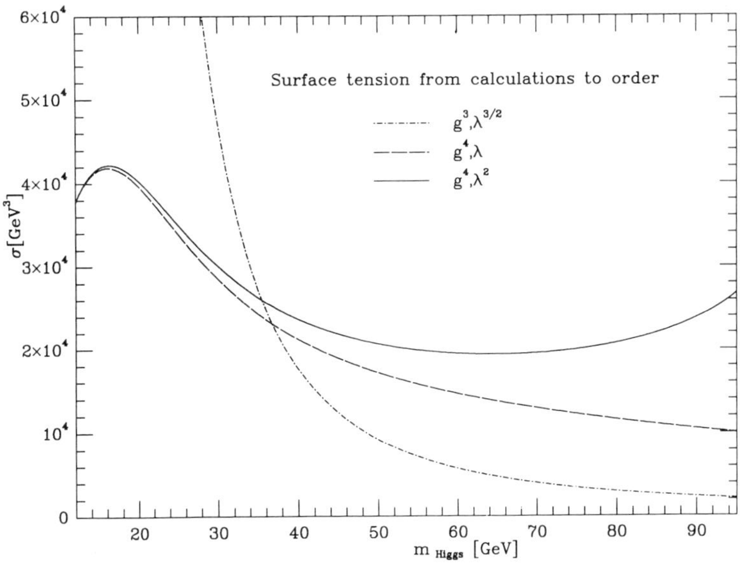

Figure 4. The surface tensions calculated from the different potentials as functions of the zero temperature Higgs mass.

Lets compare now the g^4, λ of Ref.5 and the g^4, λ^2 results. The complete calculation produces contributions of type $\varphi^2 \ln \beta(m_T + m_{\varphi,\chi})$ with positive sign. These terms are coming from the scalar-vector setting sun diagrams. In Ref. 5 scalar masses have been neglected, resulting in spurious $\varphi^2 \ln \beta m_T$-terms with positive sign, which reduces the surface tension. Another important contribution is the one proportional to $g^2(m_\varphi + 3m_\chi)m_L$. On the relevant scale ($\varphi < T$) it produces a very steep behaviour of the potential, again increasing the surface tension. The observed difference between the result of Ref. 5 and the complete g^4, λ^2 calculation presented here is mostly due to these two effects, together with the well known influence of the cubic scalar mass.

The application to the full standard model has been also carried out and the renormalization at zero temperature has been done. In the case of the full standard model the qualitative behaviour of the potential is essentially the same as for the SU(2)-Higgs model. The main difference is a decrease of the surface tension by a factor ~ 4. This can be traced back to the large top mass.

As it has been shown in the previous subsection wave function correction terms suggest that the perturbative approach becomes unreliable for large Higgs masses. In this subsection another analysis of the higher order terms in the effective action has been presented (namely, the g^4, λ^2 contributions to the effective potential), which leads to a very similar conclusion.

The results of this subsection are from Ref 10., where the details of the calculation and the full standard model analysis can be also found.

3. GAUGE INVARIANT TREATMENT OF THE ELECTROWEAK PHASE TRANSITION

In this section the effective potential for the composite field $\sigma = 2\Phi^\dagger\Phi$ will be calculated[23]. For simplicity, the Higgs model in three dimensions is studied.

The SU(2)-Higgs model in three dimensions is described by the lagrangian

$$\mathcal{L} = \frac{1}{4}W^a_{\mu\nu}W^a_{\mu\nu} + (D_\mu\Phi)^\dagger D_\mu\Phi + V_0(\varphi^2) \quad , \tag{24}$$

where

$$V_0(\varphi^2) = \frac{1}{2}m^2\varphi^2 + \frac{1}{4}\mu\lambda\varphi^4 \ , \ \varphi^2 = 2\Phi^\dagger\Phi \tag{25}$$

Here $W^a_{\mu\nu}$ is the ordinary field strength tensor and $D_\mu = \partial_\mu - i\mu^{1/2}\frac{g}{2}W^a_\mu\tau^a$ is the covariant derivative; g and λ are gauge and scalar self-coupling, respectively, and μ is the mass scale used to define dimensionless couplings in three dimensions.

In order to obtain the effective potential for the field $\sigma = 2\Phi^\dagger\Phi$, one has to evaluate first the "free energy" in the presence of an external source J,

$$e^{-\Omega W(J)} = \int DWD\Phi D\Phi^\dagger e^{-S(\Phi,\Phi^\dagger,W) - \int d^3x 2\Phi^\dagger\Phi J} \quad , \tag{26}$$

where Ω is the total volume. For constant J the effective potential is obtained as the Legendre transform,

$$\begin{aligned} \frac{\partial}{\partial J}W(J) &= \sigma \ , \\ V(\sigma) &= W(J(\sigma)) - \sigma J \ . \end{aligned} \tag{27}$$

The free energy $W(J)$ can be calculated in the standard semiclassical or loopwise expansion. The equation for a spatially constant stationary point,

$$(m^2 + 2\mu\lambda\Phi_c^\dagger\Phi_c + 2J)\Phi_c = 0 \ , \tag{28}$$

has two solutions, $\Phi_c = \Phi_s$ and $\Phi_c = \Phi_b$, which correspond to the symmetric and the broken phase, respectively,

$$\Phi_s = 0 \ , \tag{29}$$

$$\Phi_b = \left(-\frac{1}{2\mu\lambda}(m^2 + 2J)\right)^{1/2}\Phi_0 \ , \tag{30}$$

where $|\Phi_0| = 1$. The determinants of fluctuations around the two stationary points depend on the masses of vector bosons, Higgs (φ) and Goldstone (χ) bosons. In the broken phase ($\Phi_c = \Phi_b$) one has, in any covariant gauge,

$$m_W^2 = -\frac{g^2(m^2 + 2J)}{4\lambda} \ , \ m_\varphi^2 = -2(m^2 + 2J) \ , \ m_\chi^2 = 0 \ , \tag{31}$$

whereas in the symmetric phase ($\Phi_c = \Phi_s$) the masses are given by

$$m_W^2 = 0 \ , \ m_\varphi^2 = m_\chi^2 = m^2 + 2J \ . \tag{32}$$

Φ_s and Φ_b correspond to the global minima of the classical action for $m^2 + 2J > 0$ and $m^2 + 2J < 0$, respectively.

The one-loop contribution to the free energy in covariant gauge is given by

$$W_1(J) = \frac{1}{2}\int \frac{d^3k}{(2\pi)^3}\Big(6\ln(k^2+m_W^2)+\ln(k^2+m_\varphi^2)$$
$$+3\ln(k^4+k^2m_\chi^2+\alpha m_W^2 m_\chi^2)-6\ln k^2\Big)\,, \qquad (33)$$

where α is the gauge parameter. This expression is gauge independent since the product $m_W m_\chi$ vanishes in the symmetric and broken phase. One easily verifies that the same result for $W_1(J)$ is obtained in R_ξ-gauge.

Subtracting linear divergencies by means of dimensional regularization and performing the Legendre transformation according to eq. (27) gives the effective potential.

$$V(\sigma) = V_b(\sigma)\Theta(\sigma) + V_s(\sigma)\Theta(-\sigma)\,, \qquad (34)$$

where

$$V_b(\sigma) = \frac{1}{2}m^2\sigma + \frac{1}{4}\mu\lambda\sigma^2 - \frac{1}{2\pi}\left(\frac{1}{4}\mu g^2\sigma\right)^{3/2} - \frac{1}{12\pi}(2\mu\lambda\sigma)^{3/2}\,, \qquad (35)$$

$$V_s(\sigma) = \frac{1}{2}m^2\sigma - \frac{\pi^2}{6}\sigma^3\,. \qquad (36)$$

Here the couplings depend on the renormalization scale, i.e., $g = g(\mu)$, $\lambda = \lambda(\mu)$, $m^2 = m^2(\mu)$.

It is instructive to compare this result with the familiar effective potential for the field Φ in Landau gauge. The one-loop correction is given by the integral in eq. (33). Shifting the scalar field Φ in the usual way by the background field $\varphi/\sqrt{2}$ yields the masses,

$$m_W^2 = \frac{1}{4}\mu g^2\varphi^2\,,\ m_\varphi^2 = m^2 + 3\mu\lambda\varphi^2\,,\ m_\chi^2 = m^2 + \mu\lambda\varphi^2\,. \qquad (37)$$

For the potential in Landau gauge one then obtains

$$V_{LG}(\varphi^2) = \frac{1}{2}m^2\varphi^2 + \frac{1}{4}\mu\lambda\varphi^4 - \frac{1}{12\pi}\left(6m_W^3 + m_\varphi^3 + 3m_\chi^3\right)\,. \qquad (38)$$

Comparing the two potentials (34) and (38) the first striking difference is the range of the fields. For the potential (38) one has $0 \leq \varphi^2 < \infty$, whereas for the potential (34) the field varies in the range $-\infty < \sigma < \infty$. In the first case the symmetric phase is represented by the point $\varphi = 0$, whereas in the second case it corresponds to the half-axis $\sigma \leq 0$. This difference is a consequence of the different source terms for which the "free energy" W is calculated. Note, that for the gauge invariant potential at one-loop order only the four scalar degrees of freedom contribute in the symmetric phase. At small values of σ the potential increases very steeply.

The second important difference between the potentials (34) and (38) concerns the contribution of scalar loops in the broken phase. Contrary to the ordinary potential, the non-analytic terms of the gauge invariant potential do not depend on m^2. Hence, this potential can also be used for $m^2 < 0$, where the symmetric phase is unstable.

We are interested in the SU(2)-Higgs model at finite temperature, for which the ordinary, ring-improved one-loop potential is given by (cf. Ref. 6)

$$V_{ring}(\varphi^2, T) = \frac{1}{2}\left(\frac{3}{16}g^2 + \frac{1}{2}\lambda\right)\left(T^2 - T_b^2\right)\varphi^2 + \frac{1}{4}\lambda\varphi^4 \qquad (39)$$
$$-\frac{T}{12\pi}\left(3m_L^3 + 6m_T^3 + m_\varphi^3 + 3m_\chi^3\right) + \mathcal{O}(g^4, \lambda^2). \qquad (40)$$

The masses of longitudinal and transverse vector bosons, Higgs and Goldstone bosons are

$$\begin{aligned} m_L^2 &= \frac{5}{6}g^2T^2 + \frac{1}{4}g^2\varphi^2 \ , \ m_T^2 = \frac{1}{4}g^2\varphi^2 \ , && (41) \\ m_\varphi^2 &= \left(\frac{3}{16}g^2 + \frac{1}{2}\lambda\right)(T^2 - T_b^2) + 3\lambda\varphi^2 \ , && (42) \\ m_\chi^2 &= \left(\frac{3}{16}g^2 + \frac{1}{2}\lambda\right)(T^2 - T_b^2) + \lambda\varphi^2 \ . && (43) \end{aligned}$$

For values of the Higgs field φ small compared to the temperature T, one can expand m_L in powers of φ^2/T^2. Up to terms of order $\mathcal{O}(g^3\varphi^6/T^2)$ the result is then identical with the effective potential of the three-dimensional theory in Landau gauge (cf. Ref. 16),

$$V_{ring}(\varphi^2, T) = TV_{LG}\left(\frac{\varphi^2}{T}\right) \ , \quad (44)$$

if parameters are identified as follows,

$$\mu = T \ , \ \lambda(\mu) = \lambda - \frac{3}{128\pi}\sqrt{\frac{6}{5}}g^3 \ , \ g(\mu) = g \ , \quad (45)$$

$$m^2(\mu) = \left(\frac{3}{16}g^2 + \frac{1}{2}\lambda - \frac{3}{16\pi}\sqrt{\frac{5}{6}}g^3\right)(T^2 - \bar{T}_b^2) \ , \ \bar{T}_b^2/T_b^2 = \frac{3g^2 + 8\lambda}{3g^2 + 8\lambda - \frac{3}{\pi}\sqrt{\frac{5}{6}}g^3} \ . \quad (46)$$

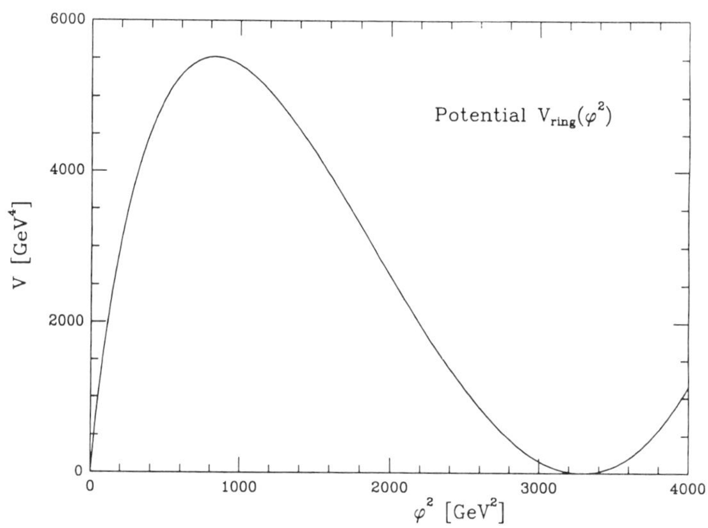

Figure 5. Effective potential in Landau gauge, $m_{Higgs} = 70\ GeV$.

Inserting these parameters into eqs. (34) - (36) we finally obtain for the gauge invariant finite-temperature potential,

$$V(\sigma, T) = V_s(\sigma, T)\Theta(\sigma) + V_b(\sigma, T)\Theta(-\sigma) \ , \quad (47)$$

where

$$V_b(\sigma, T) = \frac{1}{2}m^2(T)\sigma + \frac{1}{4}\lambda\sigma^2 - \frac{T}{12\pi}\left(6\left(\frac{1}{4}g^2\sigma\right)^{3/2} + (2\lambda\sigma)^{3/2}\right) \ , \quad (48)$$

$$V_s(\sigma, T) = \frac{1}{2} m^2(T)\sigma - \frac{\pi^2}{6} \frac{\sigma^3}{T^2} \,. \tag{49}$$

Contrary to the conventional potential $V_{ring}(\varphi^2, T)$, the gauge invariant potential $V(\sigma, T)$ is valid at temperatures above and below the barrier temperature $\bar{T}_b$.

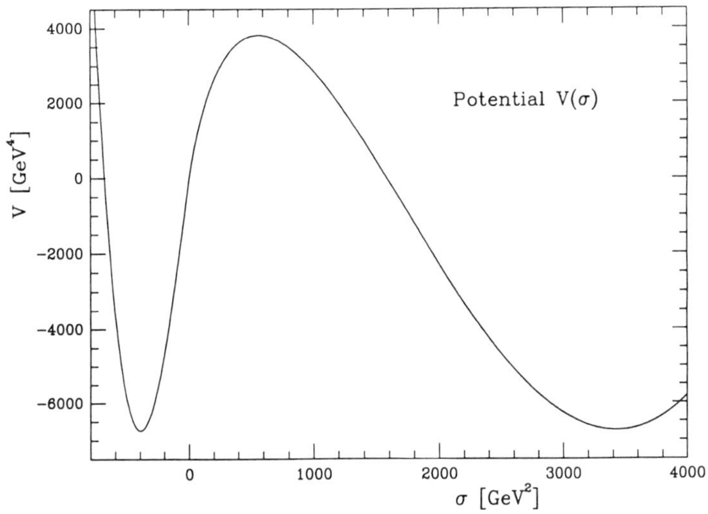

Figure 6. The gauge invariant effective potential, $m_{Higgs} = 70\ GeV$.

In Fig. 5 and 6 the ordinary potential $V_{ring}(\varphi^2, T)$ and the gauge invariant potential $V(\sigma, T)$ are shown for a Higgs mass of $m_H = 70$ GeV. Each potential is shown at its critical temperature, defined by the degeneracy of the two minima. In the broken phase the two potentials are very similar. The main difference concerns the symmetric phase where the gauge invariant potential shows a strong increase at small values of σ. Note, that the barrier is higher by about a factor of two for the gauge invariant potential.

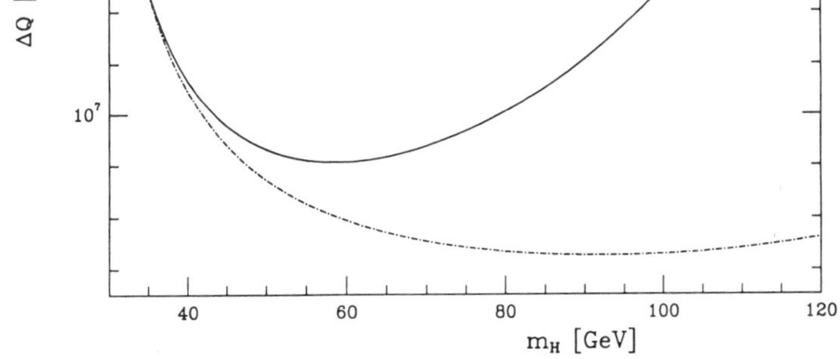

Figure 7. Latent heat as a function of the Higgs mass, predicted by the potential in Landau gauge (dashed line) and the gauge invariant potential (solid line).

We have evaluated several observables for the two potentials. The critical temperatures are different, but very similar. For Higgs masses between 30 GeV and 120 GeV the ratio $(T_c - \bar{T}_b)/\bar{T}_b$ differs by at most 40%. The latent heat, shown in Fig. 7, differs by about 70% at $m_H = 120$ GeV.

4. 4-d LATTICE SIMULATION OF THE ELECTROWEAK PHASE TRANSITION

A brief description of the used machine, the generalization of the overrelaxation algorithm to gauge-Higgs systems and preliminary results of the 4-d lattice simulation for the electroweak phase transition are presented in this section.

4.1 The machine

The simulations are carried out at DESY using an APE quadrics computer. This machine is a massive parallel supercomputer designed for lattice gauge theories. The hardware architecture is based on the fact that the updating algorithms are local. For instance, having a Wilson-type gauge action, the update of a single link with Metropolis, heatbath or overrelaxation algorithms requires to know only the links belonging to the incomplete plaquettes around it.

The performance of one node is ≈ 50 Megaflops, and the memory capacity is 4 Megabytes. The available hardware configuration at DESY is one Q1 machine (eight nodes, 0.4 Gigaflops and 32 Megabytes, used for developing programs) and two Q16 machines (128 nodes, 6.4 Gigafolps and 0.5 Gigabytes each, for long runs). Roughly half of the CPU-time has been used for the study of the electroweak phase transition. The hardware architecture can be extended to a 2048 nodes, 100 Gigaflops configuration.

The programing language of the machine is TAO, produces a highly optimized code. The language has a very strong FORTRAN flavour, but it is more flexible. For instance, defining a, b, c as $SU(2)$ variables, the matrix multiplication can be given as $c = a * b$. There are a few commands necessary for communication between nodes (e.g. indices like *left*, *right*, *up* etc.). Using this high level language one can reach ≈ 25 $\mu sec/node$ for an overrelaxation update in the pure $SU(2)$ case. The typical length of an $SU(2)$-Higgs program is a few 1000 lines.

4.2 Overrelaxation algorithm for coupled gauge-Higgs system

In the study of the electroweak phase transition it is sufficient to concentrate on the SU(2) group alone and to neglect all fermion fields and the additional U(1) group. This simplified model is supposed to contain all the essential features of the full model. As it has been shown in the previous sections the perturbative approach of the electroweak phase transition suffers from large uncertainties. This phenomenon is mainly due to the bad infrared behaviour of the problem. A systematic way to handle this infrared features is based on lattice Monte-Carlo simulations. Therefore, it seems necessary to confront the perturbative approach with numerical "experiments" to test the reliability of the perturbative results and to understand this infrared behaviour.

The Monte-Carlo simulations are necessarily done on a discretized (euclidean) space time lattice. To obtain results which are not distorted by the lattice one would like to go close to the continuum and work in the "scaling region". To approach this limit some of the correlation lengths ξ_i have to diverge. With growing correlation lengths the problem of critical slowing down arises. In order to obtain independent configurations one has to take into account the autocorrelation time which is the number of iterations with a given algorithm to reach a new independent configuration. The autocorrelation time grows with the correlation length as $\tau \propto \xi^z$ with z the so called dynamical critical exponent. For local algorithms like Metropolis or heatbath z is known to be 2. This implies that for ξ's of the order of 5-10, which are realistic values of todays simulations of the SU(2)-Higgs model, the autocorrelation time can be of the order of 25-100.

For gauge theories an interesting approach, the overrelaxation method, as initiated by Adler[24] as a generalization of the heatbath algorithm seems to be the most promising way to solve this problem. In its popular limiting case the field evolution is deterministic and conserves energy (microcanonical). In order to restore the necessary ergodicity one uses standard, ergodic updates (e.g. Metropolis or heatbath) and microcanonical updates alternately. These so called hybrid algorithms turned out to be extremely efficient in pure gauge theories[25]. In this case the overrelaxation step is a kind of reflection of the local field with the following features:
a) the energy is unchanged
b) since the reflection is an element of the gauge group, the measure in the defining functional integral is invariant

An extension of this approach to coupled gauge-Higgs systems will be presented[14]. Consider an SU(2) gauge-Higgs system with the following lattice action:

$$S = S_g + S_h, \tag{50}$$

where

$$S_g = -\frac{\beta}{2}\mathrm{Tr}(\sum_p U_p) \tag{51}$$

is the usual Wilson plaquette action and

$$S_h = -\kappa \mathrm{Tr}(\sum_{x,\mu} \Phi_x^\dagger U_{x,\mu}\Phi_{x+\mu}) + \sum_x \left[\Phi_x^\dagger\Phi_x + \lambda(\Phi_x^\dagger\Phi_x - 1)^2\right] \tag{52}$$

describes the self interaction of the four component Higgs field Φ_x and its coupling to the gauge field. Here one can introduce the standard 2×2 matrix notation, $\Phi_x = \rho_x\alpha_x$, where $\rho_x \in R$, $\rho_x \geq 0$, and $\alpha_x \in SU(2)$.

First the link-updating will be studied. The action can be written as a sum of two terms: a term which contains the local link variable and a constant term.

$$S = \mathrm{Tr}(U_{x,\mu}V) + const, \tag{53}$$

where

$$V = -\frac{\beta}{2}\sum_{\nu\neq\mu,\nu\neq-\mu} U_{x+\mu,\nu}U^\dagger_{x+\nu,\mu}U^\dagger_{x,\nu} - \kappa\sum_\mu \rho_x\rho_{x+\mu}\alpha_{x+\mu}\alpha_x^\dagger. \tag{54}$$

Consider an update as a "reflection" of U

$$U'_{x,\mu} = V_0^\dagger U^\dagger_{x,\mu}V_0^\dagger, \tag{55}$$

where V_0 is the normalized V, thus $qV_0 = V$, where $q \in R$ and $V_0 \in SU(2)$. As in the case of pure gauge theory it is easy to see that this updating satisfies the above mentioned a) and b) conditions.

The microcanonical updating of the angular part of the scalar field can be done completely analogously.

$$S = Tr(\alpha_x^{\dagger} V) + const, \tag{56}$$

where

$$V = -\kappa\rho_x \sum_{\mu} \rho_{x+\mu} U_{x,\mu} \alpha_{x+\mu}, \tag{57}$$

thus the updated angle

$$\alpha_x' = V_0 \alpha_x^{\dagger} V_0, \tag{58}$$

where V_0 is, as above, the normalized V. Again, one can show that the above a) and b) conditions are satisfied.

The most interesting case is the microcanonical updating of the radial mode of the scalar field. Separating the angular and the radial modes as $\Phi = \rho\alpha$, the $\mathcal{D}\Phi\mathcal{D}\Phi^{\dagger}$ measure will get a form of $d\alpha d\rho\rho^3$. Therefore one gets a potential

$$V(\rho_x) = -C\rho_x + \rho_x^2 + \lambda(\rho_x^2 - 1)^2 - 3log(\rho_x), \tag{59}$$

where

$$C = \kappa \mathrm{Tr}(\sum_{\mu} \rho_{x+\mu} \alpha_x^{\dagger} U_{x,\mu} \alpha_{x+\mu}) \ . \tag{60}$$

In our case it is easy to show that $V(\rho)$ has only one minimum.

The procedure to perform the overrelaxation step in ρ is now clear. For each ρ_x calculate the value of $\rho_x' \neq \rho_x$ such that $V(\rho_x) = V(\rho_x')$. Note that this is not just a reflection with respect to the minimum of the potential as $V(\rho)$ is not symmetric.

The non-trivial part of the ρ updating is the observation that ρ' is a nonlinear function of ρ. Therefore, the transformation of ρ to ρ' does not leave the measure invariant. As a consequence, one has to correct for this by a reject/accept step.

The probability of finding the system in a state with local energy between V and $V + \Delta V$ is a sum of two probabilities. The probability to find it between ρ and $\rho + \Delta\rho$ plus the probability to find it between $M(\rho)$ and $M(\rho) + \Delta M(\rho)$, where $V(\rho) = V(M(\rho)) = V$ and $V(\rho + \Delta\rho) = V(M(\rho) + \Delta M(\rho)) = V + \Delta V$. Since we want the algorithm to be microcanonical, $dV = V'(\rho)d\rho - V'(\rho')d\rho' = 0$ $(\rho' = M(\rho))$.

The Monte-Carlo realization of the above idea can be done in the following way. One determines $M(\rho)$ "mirror image" of ρ and calculates the $V'(\rho)$ and $V'(M(\rho))$ derivatives. The $\rho \to M(\rho)$ updating is accepted if

$$\left|\frac{V'(\rho)}{V'(M(\rho))}\right| > r, \tag{61}$$

where r is a random number between 0 and 1. Otherwise the updating is rejected. In practice one finds the acceptance always to be larger than 80%.

The algorithm described above has been tested in the SU(2)-Higgs model. For parameter values $\lambda = 0.0001$, $\beta = 8$ and $\kappa = 0.129$ the system is in the Higgs region of the model. The normalized autocorrelation function Γ_O of some operator O which is defined as

$$\Gamma_O(t) = \frac{< O(0)O(t) > - < O(0) >^2}{< O(0) >^2} \tag{62}$$

can be measured, where t indicates a fictitious Monte Carlo time corresponding to the number of sweeps. Γ_O falls off exponentially

$$\Gamma_O(t) \propto \exp(-t/\tau) \ \text{for} \ t \to \infty \tag{63}$$

which defines the exponential autocorrelation time τ. In Fig. 8 the autocorrelation function Γ_ρ for the length of the Higgs field ρ has been plotted. This operator in general shows the longest autocorrelation time in the SU(2)-Higgs model. The solid line corresponds to a pure Metropolis simulation and the dotted line shows the result of a hybrid overrelaxation with a mixing ratio of Metropolis to overrelaxation 1:1. The measurements have been performed after each sweep of a Metropolis or an overrelaxation update through the lattice. Although the lattice used is quite small (8^4) the autocorrelation function for the Metropolis algorithm shows a very slow fall off, indicating an autocorrelation time of about $\tau_{Met} \approx 130$. On the other hand the hybrid overrelaxation algorithm gives a considerable improvement, $\tau_{OR} \approx 25$, over the Metropolis algorithm. The results for the exponential autocorrelation times τ eq.(19), given above, have been checked against the integrated autocorrelation times and a complete agreement has been found.

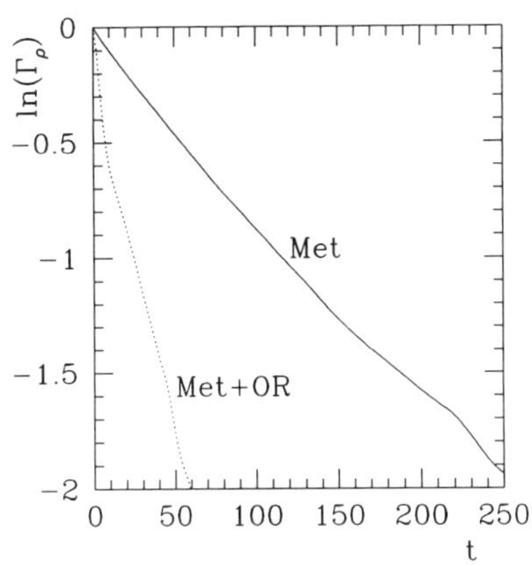

Figure 8. The logarithm of the autocorrelation function for the length of the Higgs field ρ as a function of the Monte Carlo "time" t. The lattice is 8^4. The solid line is the Metropolis algorithm alone. The dotted line is a hybrid of one Metropolis to one overrelaxation step.

It is expected that the hybrid overrelaxation algorithm has a dynamical critical exponent of $z \approx 1$ as compared to the Metropolis or heatbath with $z \approx 2$. Therefore one can expect even better improvements on larger lattices.

4.3 Results of the Monte-Carlo simulation

In this subsection some results of the runs of the first two weeks are presented. The detailed results of the long run and the description of the methods can be found in Ref. 15.

The bare parameters used in the action of the previous subsection are $\beta \equiv 4/g^2 = 8$ for the gauge coupling, (corresponds approximately to the real physical value), $\lambda = 0.0001$ and $\lambda = 0.0005$ for the scalar quartic coupling and κ between 0.1283 and 0.1289 for the scalar hopping parameter related to the bare mass square μ_0^2 by $\mu_0^2 = (1 - 2\lambda)/\kappa - 8$.

The above parameters resulted in Higgs masses of $\approx 19\ GeV$ and $\approx 48\ GeV$. These Higgs masses are excluded experimentally by LEP, therefore this study is merely exploratory.

The temperature in lattice unit is $T = 1/2$, since lattices with temporal extension $L_t = 2$ have been used. ($L_t = 3, 4, ...$ are planned in near future.) $V = 2 \cdot 32 \cdot 32 \cdot 256$ was the largest lattice size we have used.

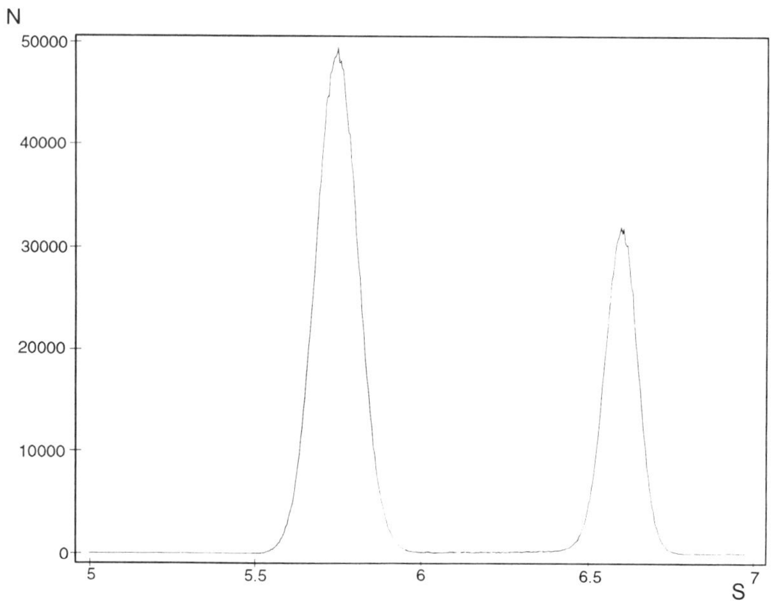

Figure 9. The action (used as an order parameter) distribution obtained on a $V = 2 \cdot 8 \cdot 8 \cdot 32$ lattice, $\lambda = 0.0001$, the Higgs mass is $\approx 19\ GeV$.

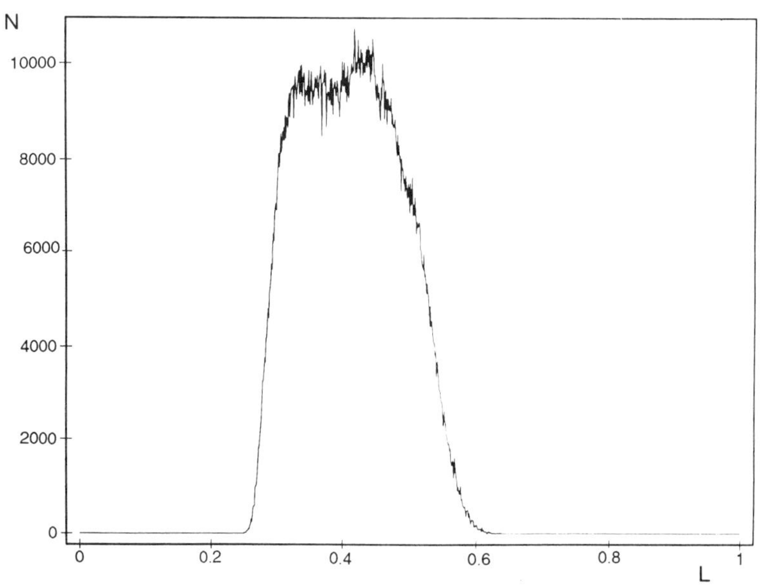

Figure 10. The α-link, $Tr(\alpha_x^\dagger U_{x,\mu} \alpha_{x+\mu})$ as an order parameter, distribution obtained on a $V = 2 \cdot 12 \cdot 12 \cdot 36$ lattice, $\lambda = 0.0005$, the Higgs mass is $\approx 48\ GeV$.

A two-peek structure of some order parameter is the clear signal of a first order phase transition. It is quite easy to obtain it for $m_{Higgs} \approx 19\ GeV$ (see Fig. 9). For $m_{Higgs} \approx 48\ GeV$ it is more difficult, one needs fairly large lattices. The two peaks start to separate on a lattice of size $V = 2 \cdot 12 \cdot 12 \cdot 32$ (see Fig. 10).

The obtained critical temperatures are smaller then the critical temperatures obtained from the perturbative approach. This difference is quite large ($\approx 20\%$) for $m_{Higgs} \approx 19\ GeV$ and not so large (a few %) for $m_{Higgs} \approx 48\ GeV$. This discrepancy is mainly due to the fact that the high temperature expansion (used in the perturbative approach) not reliable for small Higgs masses.

The gauge invariant effective potential ($V_{eff}(\Phi^\dagger\Phi)$ as in Section 3) has been also determined for $m_{Higgs} \approx 19\ GeV$ based on the lattice simulation (see Fig. 11). The similarities with Fig. 5 are remarkable, however a quantitative comparison needs further investigations.

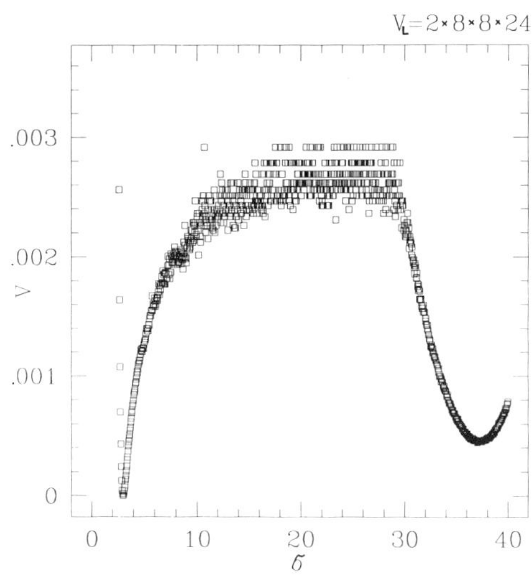

Figure 11. The gauge invariant effective potential as a function of $\Phi^\dagger\Phi$ obtained on a $V = 2 \cdot 8 \cdot 8 \cdot 32$ lattice, $\lambda = 0.0001$, the Higgs mass is $\approx 19\ GeV$.

Note that our long run[15] showed that the lattice results for the physical quantities as latent heat and surface tension are in qualitative agreement with the results of the two-loop perturbative results. The phase transition is quite strong for small Higgs masses and rather weak for larger masses. We plan to analyse the physically relevant mass region in the near future.

5. CONCLUSION, OUTLOOK

In the previous sections the electroweak phase transition has been analysed. Several physical quantities (e.g. nucleation temperature, surface tension, latent heat) have been determined as a function of the Higgs mass. However, the perturbative results does not permit a definitive statement about the features of the phase transition for realistic Higgs masses. This is mainly due to the unknown low momentum behaviour of the problem[6,8,10,20]. It is not clear whether these effects decrease or increase the strength of the phase transition. The only way out of this problem seems to be the use of Monte-

Carlo simulations on a lattice. Understanding the infrared behaviour of the problem on the lattice and building it into the perturbative model gives a possibility to go further in describing the electroweak phase transition.

6. REFERENCES

1. D. A. Kirzhnits and A. D. Linde, Phys. Lett. B72 (1972) 471; S. Weinberg, Phys. Rev. D9 (1974) 3357; L. Dolan and R. Jackiw, Phys. Rev. D9 (1974) 3320.
2. V. A. Kuzmin, V. A. Rubakov and M. E. Shaposhnikov, Phys. Lett. B155 (1985) 36.
3. M. E. Carrington, Phys. Rev. D45 (1992) 2933.
4. J. E. Bagnasco and M. Dine, Phys. Lett. B303 (1993) 308.
5. P. Arnold and O. Espinosa, Phys. Rev. D47 (1993) 3546.
6. W. Buchmüller, Z. Fodor, T. Helbig and D. Walliser, preprint DESY 93-21 (1993), Ann. Phys. (NY) in press.
7. J.R. Espinosa, M. Quirós and F. Zwirner, Phys. Lett. B314 (1993) 206.
8. D. Bödeker, W. Buchmüller, Z. Fodor and T. Helbig, preprint DESY-93-147 (1993), Nucl. Phys., in press.
9. C.G. Boyd, D.E. Brahm and S.D.H. Hsu, Phys. Rev. D48 (1993) 4963; A. Hebecker, Z. Phys. C60 (1993) 271.
10. Z. Fodor, A. Hebecker, preprint DESY-94-025 (1994).
11. M. Hellmund, J. Kripfganz and M. G. Schmidt, preprint HD-THEP-93-23 (1993).
12. B. Bunk, E.-M. Ilgenfritz, J. Kripfganz, A. Schiller, Nucl. Phys. B403 (1993) 453.
13. K. Kajantie, K. Rummukainen, M.E. Shaposhnikov, Nucl. Phys. B407 (1993) 356; K. Farakos, K. Kajantie, K. Rummukainen, M.E. Shaposhnikov, preprint CERN-TH.7244/94 (1994); M.E. Shaposhnikov these proceedings.
14. Z. Fodor, K. Jansen, preprint DESY-94-046, Phys. Lett., in press.
15. F. Csikor, Z. Fodor, J. Hein, K. Jansen, A. Jaster, I. Montvay, preprint DESY 94-088 (1994).
16. A. Jakovác, K. Kajantie, A. Patkós, preprint HU-TFT-94-01 (1994); K. Farakos, K. Kajantie, K. Rummukainen, M.E. Shaposhnikov, preprint CERN-TH.6973/94 (1994); A. Jakovác these proceedings; K. Kajantie these proceedings.
17. M. Reuter and C. Wetterich, preprint DESY-94-017 (1994); N. Tetradis and C. Wetterich, preprint DESY 93-128 (1993); M. Reuter these proceedings; N. Tetradis these proceedings; C. Wetterich these proceedings.
18. J. March-Russel, Phys. Lett. B296 (1992) 364; M. Gleiser and E. W. Kolb, Phys. Rev. D48 (1993) 1560.
19. P. Arnold and L. G. Yaffe, Phys. Rev. D49 (1994) 3003; P. Arnold these proceedings.
20. W. Buchmüller and Z. Fodor, preprint DESY 94-043 (1994), Phys. Lett. B, in press; W. Buchmüller these proceedings.
21. M. Lüscher, unpublished notes (1988)
22. O. Philipsen, these proceedings; A. Rebhan, these proceedings.
23. W. Buchmüller, Z. Fodor and A. Hebecker, preprint DESY-94-043 (1994), Phys. Lett. B, in press.
24. S.L. Adler, Phys. Rev. D23 (1981) 2901; Phys. Rev. D37 (1988) 458.
25. K. Decker and Ph. de Forcrand, Nucl. Phys. (Proc.Suppl.) B17 (1990) 567.

ON THE STRENGTH OF FIRST ORDER PHASE TRANSITIONS

Marcelo Gleiser

Department of Physics and Astronomy

Dartmouth College

Hanover, NH 03755, USA

Electroweak baryogenesis may solve one of the most fundamental questions we can ask about the universe, that of the origin of matter. It has become clear in the past few years that it also poses a multi-faceted challenge. In order to compute the tiny primordial baryonic excess, we probably must invoke physics beyond the standard model (an exciting prospect for most people), we must push perturbation theory to its "limits" (or beyond), and we must deal with nonequilibrium aspects of the phase transition. In this talk, I focus mainly on the latter issue, that of nonequilibrium aspects of first order transitions. In particular, I discuss the elusive question of "weakness". What does it mean to have a weak first order transition, and how can we distinguish between weak and strong? I argue that weak and strong transitions have very different dynamics; while strong transitions proceed by the usual bubble nucleation mechanism, weak transitions are characterized by a mixing of phases as the system reaches the critical temperature from above. I show that it is possible to clearly distinguish between the two, and discuss consequences for studies of first order transitions in general.

I. INTRODUCTION

Since the pioneering work of Becker and Döring on the nucleation of droplets in fluids [1], the theoretical investigation of nonequilibrium aspects of first order phase transitions has been of interest to researchers in subjects ranging from metereology and materials science to quantum field theory and cosmology. Phenomenological field theories were developed by Cahn and Hilliard and by Langer [2] in the context of coarse-grained time-dependent Ginsburg-Landau models, in which an expression for the decay rate of the metastable stable state was obtained within a steady-state formulation [3]. The study of metastable decay in zero-temperature quantum field theory was initiated by Voloshin, Kobzarev and Okun [4] and soon after put onto firmer theoretical ground by Coleman and Callan [5]. Finite temperature vacuum decay was first analysed by Linde [6] and has received considerable attention in the recent literature [7].

The main motivation for studying metastable decay for quantum fields comes from the possibility that the Universe underwent a series of phase transitions as it expanded from its initial hot, and presumably highly symmetric, state [8]. These cosmological phase transitions have the potential of not only answering many questions left open by

Electroweak Physics and the Early Universe, Edited by
Romao and Freire, Plenum Press, New York, 1994

the standard big-bang model, but also of serving as windows to high energy physics unnaccessible to current (and future) terrestrial accelerators. However, as it has become increasingly clear during the past few years, in order to reliably compute quantities of interest, a deeper understanding of the nonequilibrium aspects of the transitions is needed. And we all know that nonequilibrium statistical mechanics is a hard subject; most of the nice universal properties that appear in equilibrium statistical mechanics are lost, and we are forced to study how nonlinearities will influence the approach to equilibrium of this or that particular system. But things are not all that bad. The very fact that we cannot but embark into the study of nonequilibrium dynamics in the context of finite temperature relativistic quantum field theories is also a blessing; as decades of research on the dynamics of phase transitions in condensed matter systems have shown, the subject is extremely rich, offering a wealth of interesting possibilities. It is tempting to speculate that this will also be the case for cosmological phase transitions and that new, unsuspected phenomena are lurking behind our present level of understanding.

This talk is organized as follows. In the next Section I briefly discuss some of the problems related with the weakness of the electroweak potential, such as its evaluation and possible scenarios for the dynamics of the transition. In Section 3 I present the results of a numerical experiment in which a very clear criterion to distinguish between "weak" and strong first order transitions emerges. In Section 4 I briefly review some of the work on subcritical bubbles emphasizing the qualitative agreement between the subcritical bubbles picture and the results of the numerical experiment of Section 3. In Section 5 I present some concluding remarks.

II. The Issue: The Weakness of the Electroweak Transition

Of the many interesting possibilities raised by primordial phase transitions, the generation of the baryon number of the Universe during the electroweak phase transition has been explored extensively since the seminal work of Kuzmin, Rubakov, and Shaposhnikov [9]. For the purpose of this talk, the important aspect of the electroweak phase transition is that it is, in most scenarios proposed so far (see, e.g. [10] for an alternative approach), a first order phase transition. And, at least within the context of the standard model of particle physics, the transition is very possibly a weak one; the standard computation for nucleation of critical bubbles shows that the thin-wall approximation fails and that the bubbles are rather thick [11]. In fact, this seems to be the case even for extensions of the standard model, which according to some authors offer better hope of producing the correct baryon asymmetry. (Contrast, for example, the contributions of Shaposhnikov and Turok in these Proceedings.)

The weakness of the transition poses a tremendous challenge even to the study of equilibrium properties of the system; improving the perturbative evaluation of the effective potential proves to be a very demanding and ungrateful task, as technical difficulties are compounded by the fact that nonperturbative effects must be called for to regulate the perturbative expansion near the symmetric phase [12]. Using alternative methods such as the ϵ-expansion offer an interesting posibility, which, nevertheless relies on the success that these methods have on different systems [13]. Another alternative is to go to the computer and study the equilibrium properties of the standard model on the lattice [14]. Recent results are encouraging inasmuch as they seem to be consistent with perturbative results in the broken phase for fairly small Higgs masses. The transition also seems to be stronger than the perturbative estimates would predict. For larger Higgs masses (for this author, smaller than the 80 GeV claimed in Ref. 14) the interpretation of the results is not very straight forward, as finite-size effects become more important, and distinguishing the two phases by the double-peak structure of the distribution function becomes trickier.

It is clear from the above paragraph that much work must be done before we can claim we understand the equilibrium properties of the electroweak phase transition in the context of the standard model. Given that we do not know the exact shape of the effective potential for realistic Higgs masses, quantities such as the curvature of the potential around the symmetric minimum, the critical temperature of the transition, and maybe even the order of the transition are still unknown. Even if one goes beyond the standard model, as most people prefer, some of these problems persist.

However, one thing seems to be certain; that the transition is weakly first order. What does this mean exactly? "A weak first order phase transition..." In the context of electroweak baryogenesis, the usage of the term *weak* to characterize the transition is usually identified with the wall thickness of the nucleating bubbles. A weak transition would have typically thick bubbles, in that their radius is not much larger than the wall's thickness. (Of course, in this case the definition of radius is somewhat blurred.) On the other hand, the transition cannot be too weak or not enough baryon number is generated. This is equated with the discontinuity in the magnitude of the Higgs field at the critical temperature, $\langle\phi(T_c)\rangle/T \lesssim 1$. As I hope will be clear later on, this definition of weak, or not too weak, is not enough. Looking at the effective potential (assuming you know it), identifying a barrier between the symmetric and broken-symmetric phase, and proceeding to use the nucleation rate formula for critical bubbles is not necessarily the right thing to do. The reason for this is simple; the vacuum decay formula is obtained by assuming that there is a nicely behaved, nearly-homogeneous background about which we expand the partition function in order to obtain the semi-classical approximation. The reader is referred to Ref. 15 for details. However, for small enough barriers between the two phases, large amplitude fluctuations will be present, invalidating the assumption of a near-homogeneous background. Homogeneous nucleation breaks down for these situations. Two questions immediately come to mind. When does homogeneous nucleation break down, and what should one do when it does? (Getting drunk on Port is not a very good long term solution, posing problems immediately after you wake up the next day, *if* you wake up the next day.)

With these questions in mind, and motivated by the "weakness" of the electroweak transition, a few years ago Gleiser and Kolb (GK) proposed a novel mechanism by which such transitions evolve [16]. Rather than having nucleation of thick bubbles below the critical temperature of the transition, the transition would be characterized by a substantial phase mixing as the critical temperature is approached from above. GK modelled the dynamics of this phase mixing by estimating the fraction of the volume occupied by subcritical (correlation volume) thermal fluctuations of each phase at the critical temperature. In their initial approach, they neglected the fact that these subcritical fluctuations were unstable and thus were criticized on the grounds that they overestimated the equilibrium fraction of the volume in the broken-symmetric phase [17]. In their simple analytical approach the equilibrium fraction of both phases should be an identical 0.5 each when the two phases are degenerate (at T_c). Their method was recently extended by the authors of Ref. [18] to include the shrinking of the bubbles. The results of Ref. [18] indicate that GK are at least qualitatively correct; there **will** be a regime in which the transition is weak enough that considerable phase mixing occurs above T_c. (It is, of course, possible that this interesting regime lies beyond the validity of the perturbative evaluation of the electroweak effective potential. Presently this does not appear to be so [18].) I will briefly review the subcritical bubbles method later on. Now I would rather proceed with the discussion of the strength of the transition.

III. Distinguishing the Weak from the Strong: A Simple Model

In order to sharpen the distinction between weak and strong first order transitions, I decided to investigate this question within the context of a simple toy model in 2+1 dimensions which could be studied numerically [19]. Due to the complex nonequilibrium nature of the system, any analytical approach (at least those proposed so far) is bound to be severely limited. One is justified in regarding these simple models with suspicion. The need for a numerical investigation of this question is clear. This need is even more justified by noting that several of the *gross* features of the electroweak transition may appear in other unrelated physical systems, such as nematic liquid crystals and certain magnetic materials below their critical temperature. Moreover, numerical simulations of first-order transitions in the context of field theories (as opposed to discrete Ising models [20]) are scarse. Recent work has shown that the effective nucleation barrier is accurately predicted by homogeneous nucleation theory in the context of 2+1-dimensional classical field theory [21]. These results were obtained for strong transitions, in which the nucleation barrier B was large. Nucleation was made possible due to the fairly large temperatures used in the simulations. (Recall that the decay time is proportional to $\exp(B/T)$.)

In order to study how the weakness of the transition will affect its dynamics, the homogeneous part of the free-energy density (the effective potential to some order in perturbation theory) is written as follows

$$U(\phi,T) = \frac{a}{2}\left(T^2 - T_2^2\right)\phi^2 - \frac{\alpha}{3}T\phi^3 + \frac{\lambda}{4}\phi^4 \,. \tag{1}$$

With this parameterization, the free-energy resembles the finite-temperature effective potential used in the description of the electroweak transition, where α is determined by the masses of the gauge bosons and T_2, the spinodal instability temperature, is related to the zero-temperature mass of the Higgs boson [9]. Here, we will not be concerned with the limits of validity of this effective potential in the context of electroweak transitions. The goal is to explore the possible dynamics of a transition with free energy given by Eq. 1. This free-energy is also remarkably similar to the de Gennes-Ginsburg-Landau free energy used in the study of the nematic-isotropic transition of liquid crystals [22]. This transition is known to be weakly first-order; there is a discontinuity in the order parameter, even though there is no release of latent heat [23]. In fact, departures from mean field estimates for the correlation length were detected as the degeneracy temperature is approached from above, signalling the presence of "pseudo-universal phenomena", characterized by long-wavelength fluctuations observed by light-scattering experiments.

In 2+1 dimensions, it proves convenient to introduce dimensionless variables, $\tilde{x} = x\sqrt{a}T_2$, $\tilde{t} = t\sqrt{a}T_2$, $X = \phi/\sqrt{T_2}$, $\theta = T/T_2$, so that we can write the Hamiltonian (free energy) as

$$\frac{H[X]}{\theta} = \frac{1}{\theta}\int d^2\tilde{x}\left[\frac{1}{2}\mid\nabla X\mid^2 + \frac{1}{2}\left(\theta^2 - 1\right)X^2 - \frac{\tilde{\alpha}}{3}\theta X^3 + \frac{\tilde{\lambda}}{4}X^4\right] , \tag{2}$$

where $\tilde{\alpha} = \alpha/(a\sqrt{T_2})$, and $\tilde{\lambda} = \lambda/(aT_2)$. (From now on the tildes will be dropped.) For temperatures above $\theta_1 = (1-\alpha^2/4\lambda)^{-1/2}$ there is only one minimum at $X = 0$. At $\theta = \theta_1$ an inflection point appears at $X_{\text{inf}} = \alpha\theta_1/2\lambda$. Below θ_1 the inflection point separates into a maximum and a minimum given by $X_{\pm} = \frac{\alpha\theta}{2\lambda}\left[1 \pm \sqrt{1 - 4\lambda\left(1 - 1/\theta^2\right)/\alpha^2}\right]$. At the critical temperature $\theta_c = (1 - 2\alpha^2/9\lambda)^{-1/2}$ the two minima, at $X_0 = 0$ and X_+ are degenerate. Below θ_c the minimum at X_+ becomes the global minimum and the X_0-phase becomes metastable. Finally, at $\theta = 1$ the barrier between the two phases disappears. Note that $\alpha^2 < 4\lambda$ for a solution as described above.

In order to study numerically the approach to equilibrium at a given temperature θ, the coupling of the order parameter X with the thermal bath will be modelled by a Markovian Langevin equation,

$$\frac{\partial^2 X}{\partial t^2} = \nabla^2 X - \tilde{\eta}\frac{\partial X}{\partial t} - \frac{\partial U(X,\theta)}{\partial X} + \tilde{\xi}(x,t) , \tag{3}$$

where $\tilde{\eta} = \eta/\sqrt{a}T_2$ is the dimensionless viscosity coefficient, and $\tilde{\xi} = \xi/aT_2^{5/2}$ is the dimensionless stochastic noise with vanishing mean, related to η by the fluctuation-dissipation theorem,

$$\langle\xi(\vec{x},t)\xi(\vec{x}',t')\rangle = 2\eta T\delta(t-t')\delta^2(\vec{x}-\vec{x}') . \tag{4}$$

The viscosity coefficient was set to unity in all simulations. Two important comments are in order. In principle it should be possible to obtain a Langevin-like equation for the slower long-wavelength modes from a microscopic approach by integrating out the short-wavelength modes which have faster relaxation time-scales. This programme is quite complicated in the context of relativistic field theories since dissipation is a two-loop effect and progress has been slow [24]. Recent results indicate that one should expect corrections to the above equation, as noise will in general be colored and multiplicative as opposed to additive as above [25]. It is possible that these corrections will change time-dependent quantities, such as equilibration time-scales and nucleation rates, although they should not affect final equilibrium properties of the system, such as the fractional volume in each phase, or its critical properties, which are most important here. In lack of a better understanding of the microscopic dynamics of such systems, the above equation will be adopted here as a starting point. A second important point is to note that as this equation will be solved on a lattice, the lattice spacing works as an effective hard-momentum cut-off. Therefore, the lattice formulation is an effective coarse-grained formulation of the continuum theory and one should be careful when mapping from the lattice to the continuum theory. If one is to probe physics at short-wavelengths, renormalization countertems should be included in the lattice formulation so that the results are cut-off independent and a proper continuum limit is obtained. This point was emphasized and renormalization counterterms obtained for a temperature-independent potential in the nucleation study of Alford and Gleiser in Ref. [21]. Here, due to the temperature dependence of the potential, the renormalization prescription of Alford and Gleiser does not work. Instead, the lattice spacing will be fixed at $\ell = 1$. It turns out that in all cases of interest here the mean-field correlation length $\xi^{-2} = U''(X_0, \theta)$ (not to be confused with the random noise) will be sufficiently larger than unity to justify this choice.

The Langevin equation was integrated using the fifth-degree Nordsiek-Geer algorithm which allows for fast integration with high numerical accuracy [26]. The time step used was $\delta t = 0.2$, and the results discussed here were obtained with a square lattice with $L = 64$. (Comparison with $L = 40$ and $L = 128$ produced negligible differences for our present purposes.) No dependence of the results was found on the lattice length, time-step, random noise generator, and the random noise seed.

The strategy adopted was to study the behavior of the system given by Eq. [2] at the critical temperature when the two minima are degenerate. The reason for this choice of temperature is simple. If at θ_c most of the system is found in the $X = 0$ phase then as the temperature drops below θ_c one expects homogeneous nucleation to work; the system is well-localized in its metastable phase. This is what happens when a system is rapidly cooled below its critical temperature (rapid quench) so that it finds itself trapped in the metastable state. The large amplitude fluctuations which will eventually appear are the nucleating bubbles. If at θ_c one finds a large probability for the system to be in the X_+-phase, then considerable phase-mixing is occuring and homogeneous nucleation should not be accurate in describing the transition. Large amplitude fluctuations are present initially in the system, before it is quenched to temperatures below θ_c. For definiteness call the two phases the 0-phase and the +-phase. The phase distribution of the system can be measured if the idea of fractional area (volume in 3 dimensions) is introduced. As the field evolves according to Eq. [3], one counts how much of the total area of the lattice belongs to the 0-phase with $X \leq X_-$ (*i.e.* to the left of the maximum), and how much belongs to the +-phase with $X > X_-$ (*i.e.* to the right of the maximum). Dividing by the total area one obtains the fractional area in each phase, so that $f_0(t) + f_+(t) = 1$, independently of L^2.

The system is prepared initially in the 0-phase, $f_0(0) = 1$ and $f_+(0) = 0$. Thus, the area-averaged value of the order parameter, $\langle X\rangle(t) = A^{-1}\int X dA$ is initially zero. The coupling with the thermal bath will induce fluctuations about $X = 0$. By keeping $\lambda = 0.1$ fixed, the dependence of $f_0(t)$, $f_+(t)$, and $\langle X\rangle(t)$ on the value of α can be measured. Larger values of α imply stronger transitions. This is clear from the expression for θ_c which approaches unity as $\alpha \to 0$. That is, for small α the critical temperature approaches the spinodal temperature. (In the electroweak case, the same argument applies, as what is relevant is the ratio α^2/λ; α is fixed but λ increases as the Higgs mass increases.) The results are shown in Figure 1 for several values of α between $\alpha = 0.3$ and $\alpha = 0.4$. Each one of these curves is the result of an ensemble average over 200 runs. The two important features here are the final equilibrium fraction in each phase and the equilibration time-scale.

The approach to equilibrium can be fitted at all times to a slow exponential,

$$f_0(t) = \left(1 - f_0^{\rm EQ}\right) \exp\left[-\left(t/\tau_{\rm EQ}\right)^{\sigma}\right] + f_0^{\rm EQ}\,, \tag{5}$$

where $f_0^{\rm EQ}$ is the final equilibrium fraction and $\tau_{\rm EQ}$ is the equilibration time-scale. In Table 1 the values of $\tau_{\rm EQ}$ and σ are listed for different values of α.
Note that the slot for $\alpha = 0.36$ is empty. For this value of α the approach to equilibrium cannot be fitted at all times to a slow exponential; at large times it must be fitted to a power law,

$$f_0(t)\,|_{\alpha=0.36} \;\propto\; t^{-k}\,, \tag{6}$$

where k is the critical exponent controlling the approach to equilibrium. An excellent fit is obtained for $k = 0.25\ (\pm 0.05)$, as shown in Figure 2.

The fact that there is a slowing down of the dynamics for $\alpha = 0.36$ is indicative of the presence of a phase transition near $\alpha \simeq 0.36$. This transition reveals itself in a striking way if we define as an order parameter the equilibrium fractional difference $\Delta F_{\rm EQ}$,

$$\Delta F_{\rm EQ} = f_0^{\rm EQ} - f_+^{\rm EQ}\,. \tag{7}$$

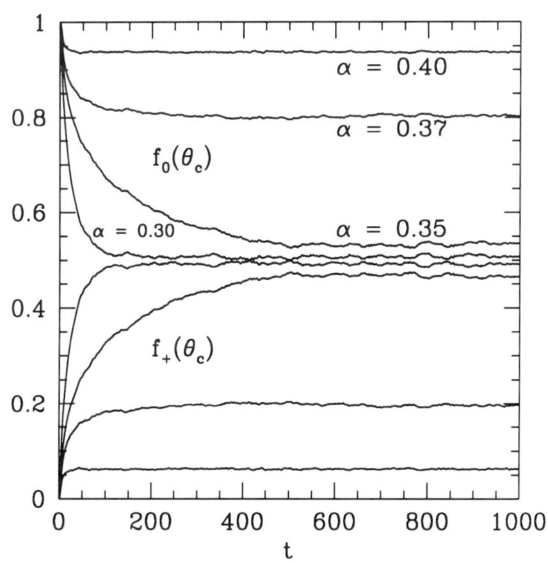

Figure 1. The approach to equilibrium for several values of α.

Table 1. The values of the equilibration time-scales and the exponents for the exponential fit of Eq. 5 for several values of α. Also shown are the equilibrium fractions $f_0(\theta_c)$ and $f_+(\theta_c)$. Uncertainties are in the last digit.

α	τ_{EQ}	σ	$f_0(\theta_c)$	$f_+(\theta_c)$
0.30	21.0	0.80	0.505	0.495
0.33	40.0	0.80	0.514	0.486
0.35	75.0	0.60	0.525	0.475
0.36	–	–	0.580	0.420
0.37	25.0	0.65	0.800	0.200
0.38	15.0	0.80	0.870	0.130
0.40	5.0	1.0	0.937	0.063

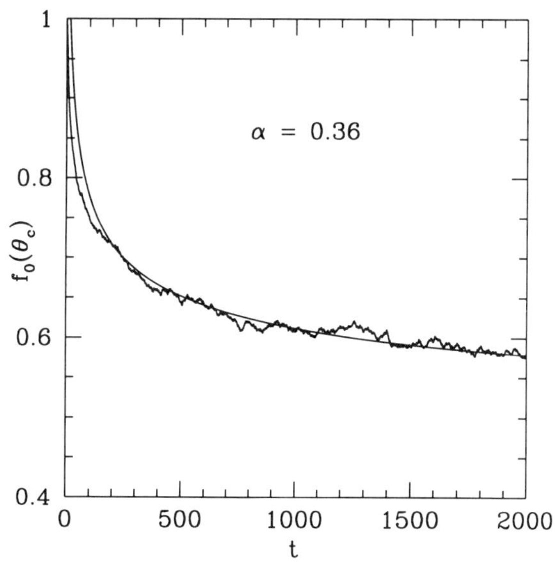

Figure 2. Fitting $f_0(\theta_c)$ to a power law at large times for $\alpha = 0.36$.

In Figure 3 $\Delta F_{EQ}(\theta_c)$ is plotted as a function of α. Clearly, there is a sharp transition in the behavior of the system around $\alpha = \alpha_c \simeq 0.36$. For $\alpha < \alpha_c$ the fractional area occupied by both phases is practically the same at 0.5. There is considerable mixing of the two phases, with the system unable to distinguish between them. One may call this phase the symmetric phase with respect to the order parameter $\Delta F_{\rm EQ}$. For $\alpha > \alpha_c$ there is a clear distinction between the two phases, with the +-phase being sharply suppressed. This may be called the broken-symmetric phase. It is clear from the behavior of $\Delta F_{\rm EQ}$ with α that this is a second order phase transition. The rounding of the curve about α_c is due to the finite size of the lattice. This curve is to be compared with the behavior of the magnetization as a function of the temperature in Ising models. α plays the rôle of "inverse" temperature; small α means high temperature, when symmetry is restored.

As a consequence of this behavior a very clear distinction between a strong and weak transition is possible. A strong transition has $\alpha > \alpha_c$ so that the system is dominated by the 0-phase at θ_c. For a weak transition neither phase clearly dominates the system, and as argued above, the dynamics should be quite different from the usual nucleation mechanism. A mixing of the two phases occurs as the system approaches the critical temperature from above.

It should be possible to parameterize the behavior of $\Delta F_{\rm EQ}$ in the neighborhood of α_c with a power law,

$$\Delta F_{\rm EQ} \propto (\alpha - \alpha_c)^{\beta} , \tag{8}$$

where β is the critical exponent controlling the behavior of $\Delta F_{\rm EQ}$ near α_c. The determination of this critical exponent and of a more precise value of α_c will be left for a future publication as it involves a detailed analysis of finite-size scaling [20].

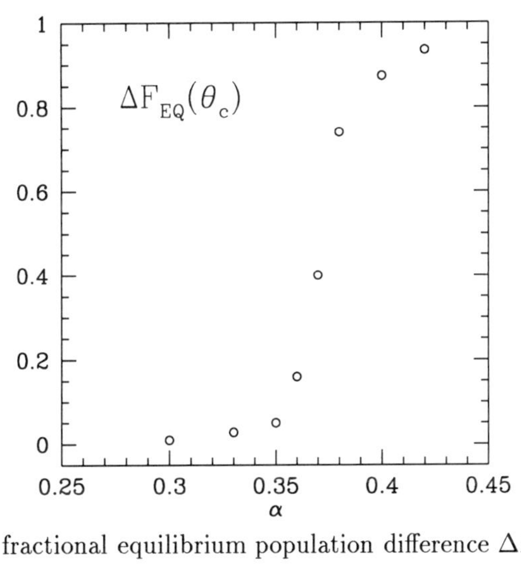

Figure 3. The fractional equilibrium population difference $\Delta F_{\rm EQ}$ as a function of α.

In order to understand the reason for the sharp change of behavior of the system near α_c, in Figure 4 the equilibrium area-averaged order paramemeter $\langle X \rangle_{\rm EQ}$ and the inflection point $X_{\rm inf} = \frac{\alpha\theta}{3\lambda}\left[1 - \sqrt{1 - 3\lambda(1 - 1/\theta^2)/\alpha^2}\right]$, are shown as a function of α.

Also shown is the rms amplitude of correlation-size fluctuations $X^2_{\rm rms} = \langle X^2 \rangle_T - \langle X \rangle^2_T$, where $\langle \cdots \rangle_T$ is the normalized thermal average with probability distribution $P[X_{sc}] = \exp[-F[X_{sc}]/\theta]$. $F[X_{sc}]$ is the free energy of a gaussian-shape subcritical fluctuation. For details see Ref. [27]. It is clear from this Figure that the transition from weak to strong occurs as $\langle X \rangle_{\rm EQ}$ drops below $X_{\rm inf}$. This can be interpreted as an effective Ginzburg criterion for the weak-to-strong transition. It matches quite well the fact that the critical slowing down occurs for $\alpha \simeq 0.36$. This result is in qualitative agreement with the study of Langer *et al.* contrasting the onset of nucleation vs. spinodal decomposition for binary fluid and solid solutions [28], where it was found that the transition between the two regimes occurs roughly at the spinodal (*i.e.* at the inflection point). Note, however, that here we are dealing with relativistic field theories, while the work of Langer and collaborators had to do with phenomenological models of phase transitions for systems with conserved order parameter. These systems will typically have slower dynamics than the field theories of interest in cosmology. Also, for Langer and his collaborators, as in most applications in condensed matter physics, the dynamics is studied *after* the system is quenched (rapidly cooled) to below its critical temperature. In cosmology, the cooling is provided by the expansion rate of the universe.

Even though $X_{\rm rms}$ drops below $X_{\rm inf}$ for a smaller value of α, being a much less computer intensive quantitity to obtain, it should serve as a rough indicator of the weak-to-strong transition.

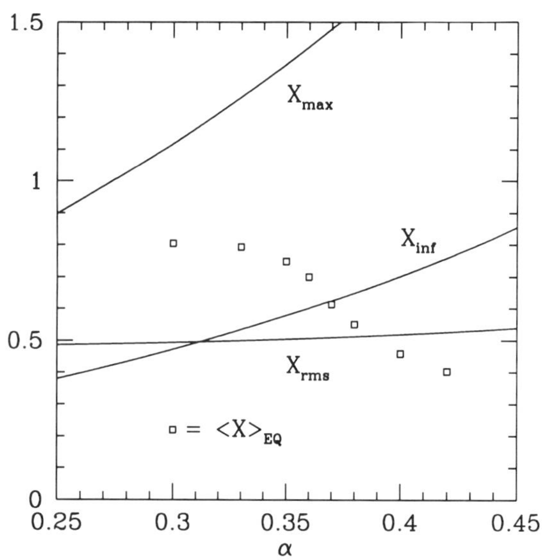

Figure 4. Comparison between area-averaged field and location of the inflection point as a function of α. Also shown are the location of the barrier, $X_{\rm MAX}$, and the rms fluctuation $X_{\rm rms}$.

To summarize the results of this Section, the distinction between a weak and a strong first order transition can be made quantitative by studying the behavior of the effective free energy *at the critical temperature.* A clear change in behavior occurs at a critical value of the parameter used to define the "strength" of the barrier separating the two phases. Here, the cubic coupling was chosen as an example. For values of the parameter larger than the critical value, the system is mostly in the symmetric phase and the transition should be well described by homogeneous nucleation theory. For values of the parameter below its critical value, the system is in a mixed phase and no phase is preferred. (Of course this assumes that the system had time to reach its equilibrium distribution. If we rapidly quench the system to below its critical temperature, it will not reach the mixed-phase state. In cosmology this would correspond to a *very* early transition, when the universe's expansion rate is relatively fast compared to typical fluctuation rates in the system.) In this case, the transition will proceeed by a mechanism closer to spinodal decomposition, although more studies are needed to obtain a clear picture of the approach to equilibrium in this case. The transition between weak and strong is itself a (second order) phase transition, with the equilibrium fractional population difference playing the rôle of the order parameter.

Finally let me stress that these results are not particular to 2+1 dimensions. Indeed, recent work on 3+1 dimensions produced qualitatively identical results. There is a second order transition for a critical value of α which for a given value of λ, will be lower than the 2+1 dimensional value. (Systems fluctuate more in 2 dimensions.) In order to make (some) contact with the electroweak transition, we should investigate how λ (related to the Higgs mass) varies for fixed α (related to the mass of the gauge bosons). Even though we investigated a model with a real scalar field, this should give us an indication of the transition from weak to strong in the electroweak case as well [29].

IV. Subcritical Bubbles: A Simple Approach to Nonequilibrium Dynamics

Now I finally, and briefly, turn to the subject of subcritical bubbles, as a possible method to study the approach to equilibrium in cosmology (and in the laboratory). Consider a system described by an effective (coarse-grained) free-energy density as discussed above, for example. If we prepare the system in the symmetric phase (or, in the example above –with no symmetry– at $\langle\phi\rangle = 0$), at any temperature there will be fluctuations which will probe the other phase. These fluctuations will be suppressed by a Boltzmann factor. The larger the amplitude of the field and the larger the volume of the fluctuation, the larger the suppression. Also, once they appear, they will disappear, unless the system is below the critical temperature and they happen to be larger than the critical fluctuations for nucleation. However, at a given temperature, *they will always be there.* The whole discussion and results of the previous Section offer convincing evidence that systems fluctuate about their equilibrium values, sometimes quite dramatically so. (The nice thing about numerical –and most– experiments is that they are reproducible. You can, if so disposed, always convince yourself that there is a clear distinction between "weak" and strong first order transitions.) The subcritical bubbles method was proposed in order to obtain a semi-analytic description of these fluctuations so that we can examine their importance. In the light of the previous results, for strong transitions they should be irrelevant, while for weak transitions they are crucial. In passing, I note that subcritical fluctuations of the broken-symmetric phase within the symmetric phase have been observed *above the critical temperature* in the isotropic-nematic liquid crystal transition [30]. Nematic fluctuations within the isotropic phase were identified and their relaxation time measured, in order to study departures from mean-field theory predictions. As expected, only in the neighborhood of the critical temperature substantial departures from mean-field were observed.

Expanding on the GK approach, Gelmini and Gleiser (GG) obtained a kinetic equation that incorporates both the shrinking of the subcritical bubbles, and their possible "destruction" by thermal noise [18]. If $n(R,t)$ is the number density of subcritical bubbles of radius R of the broken-symmetric phase within the symmetric phase, the rate equation is

$$\frac{\partial n(R,t)}{\partial t} = -\frac{\partial n(R,t)}{\partial R}\left(\frac{dR}{dt}\right) + \left(\frac{V_0}{V}\right)\Gamma_{0\to+}(R) - \left(\frac{V_+}{V}\right)\Gamma_{+\to 0}(R) - \left(\frac{V_+}{V}\right)\Gamma_{TN}(R) \tag{9}$$

Here, $\Gamma_{0\to+}(R)$ ($\Gamma_{+\to0}(R)$) is the rate per unit volume for the thermal nucleation of a bubble of radius R of phase $\phi = \phi_+$ within the phase $\phi = 0$ (phase $\phi = 0$ within the phase ϕ_+). $\Gamma_{TN}(R) \simeq aT/\frac{4}{3}\pi R^3$ is the (somewhat ad hoc) expression used for the thermal destruction rate, with a a constant. Also, V_+ must be understood as the volume of the (+)-phase in bubbles of radius R *only*, since we are following the evolution of $n(R,t)$.

In order to obtain analytical solutions of this equation, GG solved it only for temperatures just below the temperature when the broken-symmetric minimum appears which (θ_1 in the model above), of course, is above the critical temperature. For this temperature, most of the system will always be in the symmetric phase and we can write $V_0/V \simeq 1$. Another important assumption is that most of the subcritical bubbles will be of correlation volume. This is due to the fact that larger fluctuations will be suppressed, while smaller fluctuations are inconsistent with the coarse-graining procedure. With these assumptions, it is possible to solve the kinetic equation and obtain two crucial quantities; the equilibration time-scale typical for each of the processes that suppress subcritical bubbles (shrinking, thermal nucleation of a subcritical bubble of the symmetric phase inside a region of broken symmetric phase, and thermal destruction), and the equilibrium number density of these bubbles. This way we can distinguish which process is the dominant process for the suppression of subcritical bubbles for different parameters of the free energy density. Applying this formalism to the 1-loop electroweak potential, GG showed that for Higgs masses above 55 GeV or so, considerable phase mixing is occurring even for temperatures above the critical temperature. Thus, the subcritical bubble picture is in excellent qualitative agreement with the numerical results described in the previous Section; for weak enough transitions we should expect substantial departures from the usual vacuum decay mechanism.

V. Concluding Remarks

In this talk I discussed some of the issues related to the dynamics of weak vs. strong first order phase transitions. As I hope was made clear, if indeed the electroweak phase transition is "weak" in the sense defined here, novel aspects of nonequilibrum dynamics will have to be taken into account when dealing with the computation of the net baryon number generated in a given model. Taken at face value, the results here may be bad news for electroweak baryogenesis in the context of the standard model. Even if lattice computations show that for Higgs masses of 80 GeV or so the transition is still strong, the Higgs may weigh much more than that. This being the case, there will always be a regime in which phase mixing will occur and nucleation theory will fail. On the other hand, apart from hand-waiving arguments, it is not clear that the domain coarsening dynamics that will take place in a weak transition will not produce a net baryon number. At present, we simply do not know enough about the nature of the approach to equilibrium to decide on this issue, or set it aside. If baryon number is to be generated in extensions of the standard model, then the results here will provide a useful constraint in the usually large parameter space of these models. In order to have a strong enough transition, the Ginzburg-like criterion discussed here must be satisfied, and at least one parameter of the model may be eliminated this way. If the transition in these models is weak enough, one should again expect departures from the standard vacuum decay estimates for bubble nucleation rates. This is due to the inescapable conclusion that hot systems fluctuate, and if they can these fluctuations will produce some dramatic effects. Critical opalescence in nematic liquid crystals is but one example of these "pre-transitional" phenomena in nature. It may well be that they will also be of crucial importance in cosmological phase transitions.

The possibility of generating the baryon number of the universe during the electroweak phase transition is a tremendous challenge to present research in the cosmology/high energy physics interface. We most probably must invoke physics beyond the standard model to get enough CP violation, we must tackle hard problems related to infrared divergences in gauge theories, and, last but not least, we must understand the nonequilibrium aspects of phase transitions in the context of field theories in a cosmological background. Judging from what happened during the past 5 years or so, and by the number of people working in this topic, progress will keep coming fast. Maybe in another 5 years, we should all get together again (hopefully in Sintra) to see how far we managed to get.

Acknowledgements

I am grateful to my collaborators Rocky Kolb and Graciela Gelmini, as well as M. Alford and R. Ramos for the many long discussions on bubbles and phase transitions. This work is partially supported by a National Science Foundation grant No. PHYS-9204726.

References

1. E. Becker and W. Döring, Ann. Phys. (Leipzig) **24**, 719 (1935).

2. J. W. Cahn and J. E. Hilliard, J. Chem. Phys. **31**, 688 (1959); J. S. Langer, Ann. Phys. (NY) **41**, 108 (1967); *ibid.* **54**, 258 (1969)

3. J. D. Gunton, M. San Miguel, and P. S. Sahni, in *Phase Transitions and Critical Phenomena*, edited by C. Domb and J. L. Lebowitz (Academic, London, 1983), Vol. 8.

4. M. B. Voloshin, I. Yu. Kobzarev, and L. B. Okun', Yad. Fiz. **20**, 1229 (1974) [Sov. J. Nucl. Phys. **20**, 644 (1975)].

5. S. Coleman, Phys. Rev. **D15**, 2929 (1977); C. Callan and S. Coleman, Phys. Rev. **D16**, 1762 (1977).

6. A. D. Linde, Nucl. Phys. **B216**, 421 (1983); [Erratum: **B223**, 544 (1983)].

7. L. P. Csernai and J. I. Kapusta, Phys. Rev. **D46**, 1379 (1992); M. Gleiser, G. Marques, and R. Ramos, Phys. Rev. **D48**, 1571 (1993); D. Brahm and C. Lee, Phys. Rev. **D49**, 4094 (1994).

8. E. W. Kolb and M. S. Turner, *The Early Universe*, Addison-Wesley (1990).

9. V. A. Kuzmin, V. A. Rubakov, and M. E. Shaposhnikov, Phys. Lett. **B155**, 36 (1985). For a recent review see, A. Cohen, D. Kaplan, and A. Nelson, Ann. Rev. Nucl. Part. Sci. **43**, 27 (1993).

10. R.H. Brandenberger and A.-C. Davis, Phys. Lett. **B308**, 79 (1993).

11. B. Liu, L. McLerran, and N. Turok, Phys. Rev. **D46**, 2668 (1992); M. Dine and S. Thomas, Santa Cruz preprint No. SCIPP 94/01.

12. P. Arnold and O. Espinosa, Phys. Rev. **D47**, 3546 (1993); M. Dine, P. Huet, R.G. Leigh, A. Linde, and D. Linde, Phys. Rev. **D46**, 550 (1992); C.G. Boyd, D.E. Brahm, and S. Hsu, Phys. Rev. **D48**, 4963 (1993); M. Quiros, J.R. Spinosa, and F. Zwirner, Phys. Lett. **B314**, 206 (1993); W. Buchmüller, T. Helbig, and D. Walliser, Nucl. Phys. **B407**, 387 (1993); M. Carrington, Phys. Rev. **D45**, 2933 (1992).

13. P. Arnold and L. Yaffe, Phys. Rev. **D49**, 3003 (1994); M. Gleiser and E.W. Kolb, Phys. Rev. **D48**, 1560 (1993).

14. K. Farakos, K. Kajantie, K. Rummukainen, and M. Shaposhnikov, CERN preprint No. TH.7244/94, May 1994; K. Kajantie, K. Rummukainen, and M. Shaposhnikov, Nucl. Phys. **B407**, 356 (1993); B. Bunk, E.-M. Ilgenfritz, J. Kripfganz, and A. Schiller, Phys. Lett. **B284**, 372 (1992); Nucl. Phys. **B403**, 453 (1993)

15. See, e.g. the work of Gleiser, Marques, and Ramos in Ref. 7; P. Arnold and L. McLerran, Phys. Rev. **D36**, 581 (1987); S. Dodelson and B. Gradwohl, Nucl. Phys. **B400**, 435 (1993).

16. M. Gleiser and E. W. Kolb, Phys. Rev. Lett. **69**, 1304 (1992); Phys. Rev. **D48**, 1560 (1993); M. Gleiser, E. W. Kolb, and R. Watkins, Nucl. Phys. **B364**, 411 (1991).

17. M. Dine, P. Huet, R.G. Leigh, A. Linde, and D. Linde, Phys. Rev. **D46**, 550 (1992); G. Anderson, Phys. Lett. **B295**, 32 (1992).

18. G. Gelmini and M. Gleiser, in press, Nucl. Phys. B.

19. M. Gleiser, Dartmouth preprint No. DART-HEP-94/01.

20. D. P. Landau in *Finite Size Scaling and Numerical Simulations of Statistical Systems*, edited by V. Privman (World Scientific, Singapore, 1990).

21. M. Alford and M. Gleiser, Phys. Rev. **D48**, 2838 (1993); O. T. Valls and G. F. Mazenko, Phys. Rev. **B42**, 6614 (1990).

22. S. Chandrasekhar, *Liquid Crystals*, (Cambridge University Press, Cambridge [Second edition, 1992]).

23. T. W. Stinson and J. D. Litster, *Phys. Rev. Lett.* **25**, 503 (1970); H. Zink and W. H. de Jeu, *Mol. Cryst. Liq. Cryst.* **124**, 287 (1985).

24. K. Chou, Z. Su, B. Hao and L. Yu, Phys. Rep. **118**, 1 (1985); R. R. Parwani, Phys. Rev. **D45**, 4695 (1992); S. Jeon, Phys. Rev. **D47**, 4586 (1993).

25. M. Gleiser and R. Ramos, Dartmouth preprint No. DART-HEP-93/06; B. L. Hu, J. P. Paz and Y. Zhang, in *The Origin of Structure in the Universe*, Ed. E. Gunzig and P. Nardone (Kluwer Acad. Publ. 1993); D. Lee and D. Boyanovsky, Nucl. Phys. **B406**, 631 (1993); S. Habib, in *Stochastic Processes in Astrophysics*, Proc. Eighth Annual Workshop in Nonlinear Astronomy (1993).

26. A. Nordsiek, Math. Comp. **16**, 22 (1962).

27. M. Gleiser and R. Ramos, Phys. Lett. **B300**, 271 (1993).

28. J. Langer, Physica **73**, 61 (1974); J. Langer, M. Bar-on, and H. Miller, Phys. Rev. **A11**, 1417 (1975).

29. J. Borrill and M. Gleiser, in progress.

30. F. W. Deeg and M. D. Fayer, Chem. Phys. Lett. **167**, 527 (1990); J. D. Litster and T. W. Stinson, J. App. Phys., **41**, 996 (1970).

AN OVERVIEW OF THE ϵ EXPANSION AND THE ELECTROWEAK PHASE TRANSITION

Peter Arnold

Department of Physics, FM-15
University of Washington
Seattle, WA 98195
U.S.A.

In this talk I'm going to summarize work done with Larry Yaffe [1] on applying ϵ expansion techniques to the electroweak phase transition. As I shall review, analysis based on the perturbative expansion of the effective potential, which has already been discussed in this conference, is only valid (in the minimal standard model) when the zero-temperature Higgs mass is small compared to the zero-temperature W mass. The goal of my talk will be to discuss what can be done in the experimentally more realistic case when this condition fails. I shall discuss how one might compute various parameters of the transitions such as the correlation length at the critical temperature, the latent heat of the transition, the bubble nucleation rate, and the baryon violation rate after the transition. Note that these are all quasi-equilibrium quantities. In particular, I shall modestly not attempt to tackle the much harder problems directly associated with the generation of baryon asymmetry on the expanding bubble walls.

Throughout this talk I will work in a simple toy model: the minimal standard model with a single Higgs doublet. I do this because the model has relatively few unknown parameters, and so it is relatively easy to analyze and results are not obscured by complicated parameter dependence. I call it a toy model because I don't care to step into the debate of whether the minimal model has enough CP violation[2, 3]. For simplicity, I will also be ignoring the Weinberg angle and focusing on a pure SU(2) Higgs theory.

Why life is simple when $\mathrm{m(Higgs)}_{T=0} \ll \mathrm{M(W)}_{T=0}$

Consider the classical Mexican-hat potential for the Higgs:

$$V_0 \sim -\mu^2\phi^2 + \lambda\phi^4 \,. \tag{1}$$

Electroweak Physics and the Early Universe, Edited by
Romao and Freire, Plenum Press, New York, 1994

And now consider the free energy of the system at finite temperature in the background of a Higgs field ϕ. One contribution will be from the classical energy V_0 above for that ϕ. But there will also be contributions from the free energy of the real particles present in the plasma. Focus on the contribution from W bosons. This contribution (at one-loop) is nothing more than a formula you can look up in your old graduate statistical mechanics book; it is the free energy of an ideal Bose gas:

$$\Delta F \sim T \int d^3k \ln\left(1 - e^{-\beta E_k}\right), \tag{2}$$

where the energy of a relativistic particle is

$$E_k = \sqrt{\vec{k}^2 + M_W^2} \sim \sqrt{\vec{k}^2 + g^2\phi^2}. \tag{3}$$

Note that the W gets its mass from the background Higgs field, so that $M_W \sim g\phi$. Thus ΔF above depends on the background value of ϕ. At high temperatures, one can expand ΔF in powers of M_W/T, and one gets a high-temperature expansion of the form

$$\Delta F \sim \#T^4 + \#M_W^2T^2 - \#M_W^3T + \cdots, \tag{4}$$

where # indicates numerical constants that I'm not going to bother writing down.

Putting the classical potential V_0 together with the W contribution to ΔF gives*

$$F \sim V_0 + \Delta F \sim \text{const} + (-\mu^2 + g^2T^2)\phi^2 - g^3\phi^3T + \lambda\phi^4 + \cdots \tag{5}$$

(where I'm no longer going to bother even putting in the # signs). The $g^2T^2\phi^2$ term comes from the M^2T^2 term of the high-temperature expansion of ΔF. This term is responsible for the phase transition by turning the curvature of the potential at the origin from negative at low temperature to positive at high temperature. The $g^3\phi^3T$ term, which comes from the M^3T^2 term, is responsible for making the phase transition first-order as shown in fig. 1. Specifically, there exists a temperature T_c at which there are two degenerate minima of the potential, and the expectation of ϕ changes discontinuously as one cools through the transition. I shall ignore the generally small effects of subsequent terms in the high-temperature expansion (4).

Now consider the $T = T_c$ curve in fig. 1. The shape comes from the interplay of the ϕ^2, ϕ^3 and ϕ^4 terms in the free energy (5): the ϕ^2 term turns it up at the origin, the ϕ^3 term turns it down again, and the ϕ^4 term turns it up at large ϕ. In particular, in the region of ϕ near the symmetry-breaking minimum (where $\phi \neq 0$) or near the hump in the potential, these three terms must be the same order of magnitude, so that

$$(-\mu^2 + g^2T^2)\phi^2 \sim g^3\phi^3T \sim \lambda\phi^4. \tag{6}$$

From the last relation in particular, one finds that this region of ϕ is characterized by

$$\phi \sim \frac{g^3}{\lambda}T. \tag{7}$$

Now consider the order of the loop expansion parameter if we perturbatively analyze the physics of the phase transition. Each loop costs a factor of g^2. However, at high temperatures, each loop is also associated with a factor of the temperature T, so that the cost of a loop is actually g^2T. (One way to see this factor of T is to remember that, at finite temperature, Euclidean time becomes periodic with period

*Of course, there are also scalar and other contributions to ΔF. I'm just focusing on the W contributions here for the sake of brevity and pedagogy.

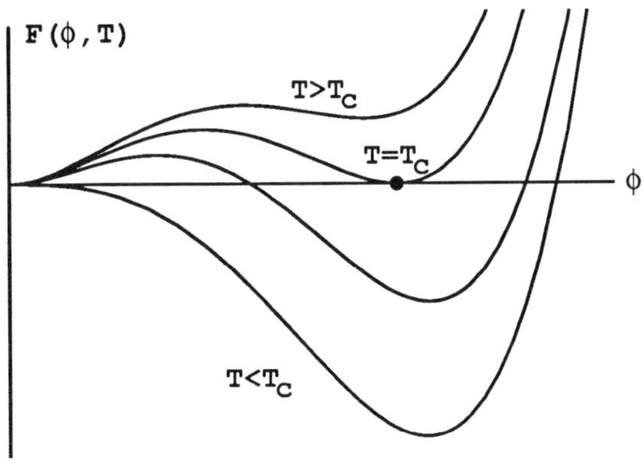

Figure 1. The form of the free energy, as a function of ϕ, for different temperatures.

$\beta = 1/T$. As a result, zero-temperature loop momentum integrals $\int d^4p$ are replaced by finite Fourier sums $T\sum_{p_0}\int d^3p$ over the frequencies, and a factor of T then appears explicitly.) Finally, by dimensional analysis, g^2T must be divided by the mass scale M_W of the problem to determine the final, dimensionless cost of adding a loop. The loop expansion parameter (after resummation) is then

$$\frac{g^2T}{M_W} \sim \frac{gT}{\phi} \sim \frac{\lambda}{g^2}, \tag{8}$$

where I have used $M_W \sim g\phi$ and (7). λ and g^2, however, are related to the zero-temperature Higgs and W masses, so that the loop expansion parameter can be rewritten as

$$\frac{\lambda}{g^2} \sim \frac{m^2(\text{Higgs})_{T=0}}{M^2(\text{W})_{T=0}}. \tag{9}$$

So we see that the perturbative loop expansion is only valid when the zero-temperature Higgs mass is small compared to the zero-temperature W mass. All of this power counting is reviewed in ref. [4].

Why life is not simple

One of the constraints of electroweak baryogenesis is that there must not be any significant amount of baryon number violation after the completion of the phase transition. Otherwise, any baryon asymmetry generated during the transition would be washed away to zero. The rate of baryon number violation is exponentially sensitive to the sphaleron mass:

$$\text{rate} \sim e^{-\beta E(\text{sphaleron})} \sim e^{-\#M_W/g^2T} \sim e^{-\#g^2/\lambda}. \tag{10}$$

The second equality above comes because the sphaleron mass is of order M_W/α_w, and the final equality because the exponent M_W/g^2T is (not coincidentally) nothing other than the inverse of the loop expansion parameter (8), which we previously determined for the broken-symmetry phase. You can now see that the rate will be small after the transition only if the zero-temperature Higgs mass (and hence λ) is small. The inverse

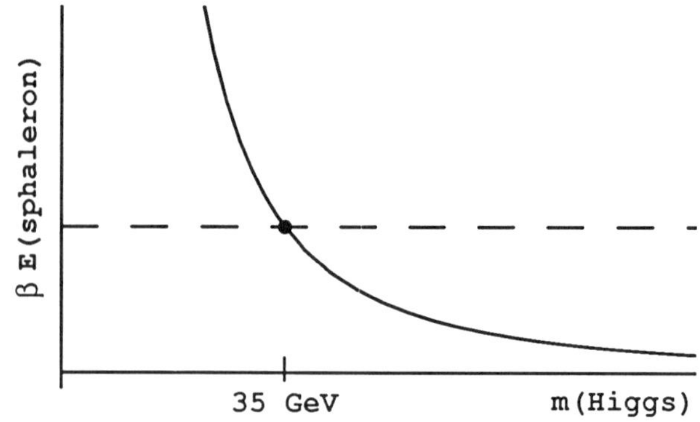

Figure 2. The Boltzmann exponent for baryon number violation vs. the zero-temperature Higgs mass.

relationship between the rate exponent and the zero-temperature Higgs mass is shown schematically in fig. 2. Requiring that the rate be small compared to the expansion rate of the universe puts a lower bound on βE(sphaleron) and hence an upper bound on the Higgs mass. A careful analysis, based on the one-loop effective potential, yields an upper bound on the Higgs mass of 35–40 GeV due to this constraint[5]. This seems to rule out electroweak baryogenesis in the minimal standard model.

But now it's time to wonder whether a one-loop analysis can be trusted. In particular, is a Higgs mass of 35 GeV in fact small compared to the W mass of 80 GeV? That depends on all the factors of 2 left out of the order-of-magnitude analysis of loop corrections that I presented above. To settle this issue, the two-loop potential was computed in ref. [4] (and the most dominant pieces independently in ref. [6]). Fig. 3 shows the difference between the one-loop and two-loop potentials at the critical temperature. Whereas the VEV changed only by about 20%,† the height of the hump changed by a factor of three! The moral is that, for some quantities, (35 GeV)/(80 GeV) is *not* a small number. And so it is not clear, in particular, whether higher-loop corrections to the baryon number violation rate will be small.

But what can we do when mean field theory (*i.e.* perturbation theory) fails—that is, when fluctuations around the mean field ϕ are large and cannot be treated perturbatively. To phrase it another way, what do we do when the correlation length $1/M_W$ becomes very large at T_c so that the loop expansion parameter (8) is large? Condensed matter physicists have long dealt with this problem in the context of second-order phase transitions, where the correlation length is infinite at T_c. Our strategy is to borrow one of their techniques—the ϵ expansion—which has been very successful for some systems.‡

The ϵ expansion

The ϵ expansion consists of the following steps.

†Beyond leading order the VEV is an unphysical and gauge-dependent quantity. The result shown here is for Landau gauge.

‡The idea of applying the ϵ expansion to the electroweak phase transition has been suggested previously by Gleisser and Kolb [7], March-Russel [8], and probably by others that I'm not familiar with.

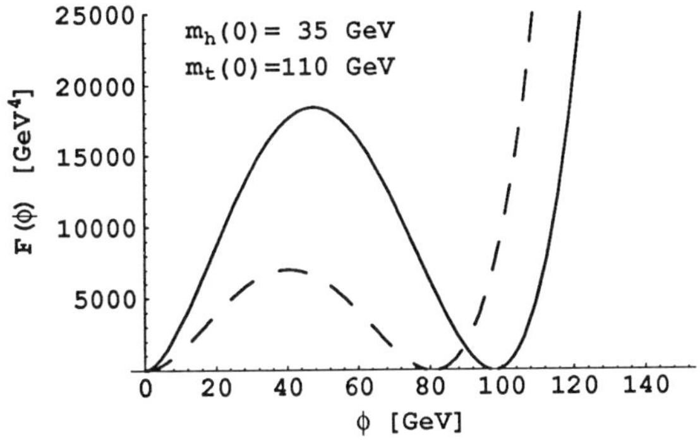

Figure 3. The (resummed) two-loop vs. one-loop free energy, as functions of ϕ, for $\mathrm{m(Higgs)}_{T=0} = 35$ GeV at the critical temperature (as determined individually for each potential).

Step 1. Recall that equilibrium quantities can generally be expressed in terms of the partition function $Z = \mathrm{tr}\, e^{-\beta H}$. The exponential $e^{-\beta H}$ looks just like an evolution operator e^{iHt} when the time $t = i\beta$ is imaginary. Just like the evolution operator, Z has a path-integral representation, and the only difference with standard zero-temperature path-integrals is that time only extends for Euclidean time β:

$$Z = \int [\mathcal{D}\phi] \exp\left(- \int_0^\beta d\tau \, L_{\mathrm{E}} \right) . \tag{11}$$

Also, the trace is implemented by integrating over all configurations where the initial state is the same as the final state, which translates into the periodic boundary condition $\phi(0, \vec{x}) = \phi(\beta, \vec{x})$ on the path integral. (Fermions have anti-periodic boundary conditions.)

Step 2. Since the Euclidean time dimension only has extent β, it disappears in the high temperature limit $\beta \to 0$. That is, if one is studying (equilibrium) physics at distance scales l large compared to $1/T$, then the Euclidean time dimension decouples and the four-dimensional theory is reduced to an effective three-dimensional theory of the static modes ($p_0 = 0$). The effects of fluctuations ($p_0 \neq 0$) in the Euclidean time direction decouple, as powers of β/l, except for renormalizations of relevant operators. For example, the three dimensional mass and couplings are related to the original four-dimensional ones by

$$m_3^2 = m_4^2 + g^2 T^2 + \text{higher-order} , \tag{12}$$
$$g_3^2 = g_4^2(T) + \text{higher-order} . \tag{13}$$

The g^2T^2 term in the mass is just the term I discussed earlier that turns the curvature of the potential at $\phi = 0$ from concave down at low temperature to concave up at high temperature, and so drives the phase transition. In four-dimensional language, the phase transition is achieved by varying T. In three-dimensional language, that corresponds to varying the mass m_3^2 from negative values up through positive values.

Step 3. As we shall see, the three-dimensional theory is difficult to directly solve in a well-defined systematic fashion. The trick of the ϵ expansion is to solve instead a related class of theories by generalizing the 3 spatial dimensions to $4 - \epsilon$ spatial

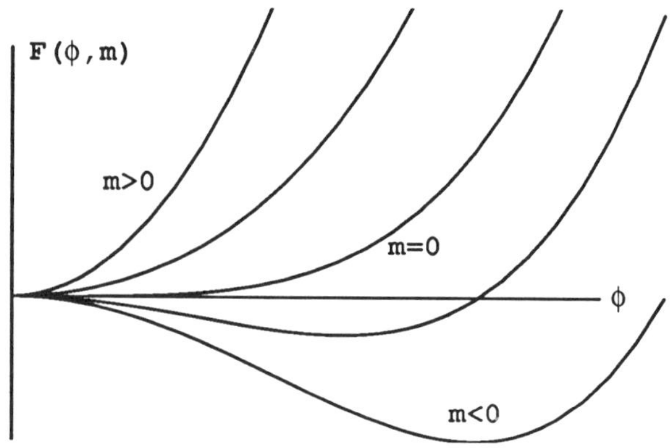

Figure 4. The classical potential for scalar theory, as a function of the mass of the effective 3 dimensional theory.

dimensions, replacing the action in the usual way by

$$S(d{=}3) \rightarrow \int d^{4-\epsilon}x \left[(D\phi)^2 + m^2\phi^2 + \mu^\epsilon\lambda\phi^4 + \cdots\right] . \tag{14}$$

(Note that I am defining my couplings, such as λ, to be dimensionless.)

Step 4. For $\epsilon \ll 1$, the theory turns out to be solvable perturbatively in ϵ by using renormalization-group (RG) improved perturbation theory. I'll discuss this is more detail in a moment.

Step 5. After computing to a given order in ϵ, return to the world of three spatial dimensions by taking $\epsilon \rightarrow 1$.

Review of a success story: pure scalar theory

Forget about gauge theories for a moment and consider a theory of a single, real scalar field:

$$\mathcal{L}_{\rm E} \sim (\partial\phi)^2 + m^2\phi^2 + \mu^\epsilon\lambda\phi^4 . \tag{15}$$

The phase transition of the tree-level potential is depicted in Fig. 4, which shows a 2nd-order transition. The 1-loop potential (appropriately resummed) for this model turns out to predict a 1st-order phase transition, but the loop expansion parameter turns out to be order 1 and so mean field theory results cannot be trusted.[§] In fact, this scalar theory goes by a familiar name near its critical point: it is the Ising model, and the Ising model is known to have a 2nd-order phase transition. (More accurately, the scalar theory is in the same universality class as the Ising model. The field ϕ roughly corresponds to the Ising spin averaged over a volume of size $1/\mu^3$.)

In a 2nd-order phase transition, the correlation length becomes infinite. To study such a transition, one should therefore examine the system on larger and larger distance scales and try to determine its scaling behavior. But, as particle physicists, we're all quite familiar with how to change scale in $4-\epsilon$ dimensions. We use the renormalization group:

$$\mu\partial_\mu\lambda = -\epsilon\lambda + c\lambda^2 + O(\lambda^2) . \tag{16}$$

[§]For a review, see section II.A of ref. [4].

Figure 5. The renormalization group flow, into the infrared, of the pure scalar theory.

If one set ϵ to zero, this would just be the usual four-dimensional renormalization group for a pure scalar theory, with the one-loop β-functions depicted explicitly as $c\lambda^2$. c is a numerical constant that is easily computed. Slightly away from four dimensions, the additional term $-\epsilon\lambda$ on the right-hand side simply represents the classical scaling of the interaction $\lambda\phi^4$ due to its dimension. For example, at tree level, a change in μ by a factor of 2 can be compensated in $\mu^\epsilon\lambda\phi^4$ by changing λ by a factor of $2^{-\epsilon}$. This trivial scaling is the sole content of the $-\epsilon\lambda$ term in (16). It is the other terms which contain the physics of integrating out degrees of freedom as the scale μ is changed.

The flow of the RG equation (16) is shown in Fig. 5. The arrows depict flow into the infrared ($\mu \to 0$). At small λ, the $-\epsilon\lambda$ term dominates and drives λ bigger; at large λ, the $+c\lambda^2$ term dominates and drives it smaller. There is an infrared-stable fixed point when the two terms balance at $\lambda_* = \epsilon/c$. The long-distance behavior of the system will be described by this fixed point.

When ϵ is small, the fixed point $\lambda_* = \epsilon/c$ will be at small coupling, and therefore our neglect of two-loop and higher-order contributions to the renormalization group in (16) will be justified. More generally, the loop expansion in λ at the fixed point is equivalent to an expansion in ϵ. We know, however, that the loop expansion does not converge but is asymptotic, blowing up at order λ_*^n where $n \sim 1/\lambda_*$. That means that the expansion in ϵ will also be asymptotic, blowing up at order ϵ^n where $n \sim 1/\epsilon$. If we're lucky, $n \sim 1/\epsilon$ will mean 3 or 4 when $\epsilon \to 1$ and the first few terms of the expansion will be useful in three dimensions, giving a reasonable approximation for quantities of interest. If we're unlucky, $n \sim 1/\epsilon$ will mean zero when $\epsilon \to 1$, and the asymptotic series will start blowing up immediately after the first term and be useless in three dimensions.

Are we lucky or unlucky? Let's look at some examples. The first is the susceptibility exponent γ defined by $\chi \sim |T - T_c|^\gamma$ where χ measures the response of $\langle\phi\rangle$ to an external source $J\phi$. The result is

$$\gamma = 1 + 0.167\epsilon + 0.077\epsilon^2 - 0.049\epsilon^3 + \cdots , \tag{17}$$

which, for $\epsilon = 1$, gives $1.195 + \cdots$. This compares favorably to the result 1.2405 ± 0.0015 which comes from more sophisticated techniques (the ϵ expansion combined with Pade resummation, or renormalization group studies directly in three dimensions). It is also consistent with lattice results, which have larger error bars. It is also consistent with experiments, which have yet larger error bars. I should also give a much more marginal example, which is the correlation exponent η defined by $\langle\phi(r)\phi(0)\rangle \sim r^{2-d-\eta}$ as $r \to \infty$. η is simply twice the anomalous dimension of ϕ and is given by

$$\eta = 0.0185\epsilon^2 + 0.0187\epsilon^3 - 0.0083\epsilon^4 + 0.0359\epsilon^5 + \cdots . \tag{18}$$

If one adds the first three terms for $\epsilon = 1$, and ignores the fourth where the series is clearly blowing up, one gets 0.0289. The actual result is believed to be about 0.035. So the naive application of the ϵ expansion gives an approximation to within about 20%.

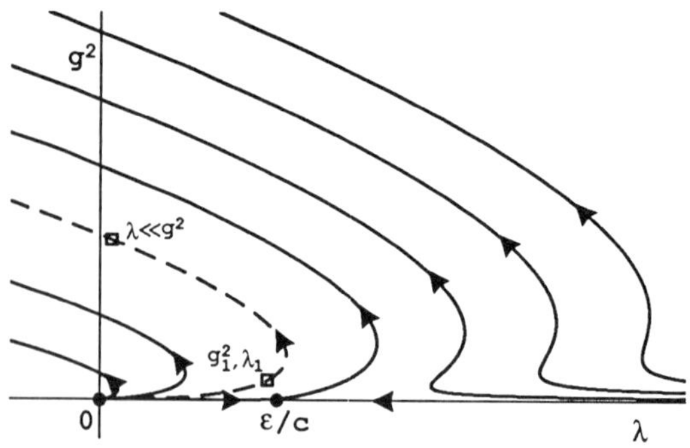

Figure 6. The renormalization group flow, into the infrared, of electroweak theory.

Electroweak theory

The one-loop renormalization group equations for the couplings of electroweak theory are similar to the case of scalar theory:

$$\mu\partial_\mu g^2 = -\epsilon g^2 + \beta_0 g^4 , \tag{19}$$

$$\mu\partial_\mu \lambda = -\epsilon g^2 + (a g^4 + b g^2 \lambda + c\lambda^2) . \tag{20}$$

The scalar β-function is a little more complicated now because there are terms involving g^2 as well as λ.

When one computes the numerical coefficients in these equations and plots the renormalization group flow, one gets Fig. 6. The $g^2 = 0$ line is similar to our previous result for pure scalar theory. Howvere, because of the flow in g^2, the fixed point is no longer infrared stable. As observed by Ginsparg [9], there are in fact no infrared stable fixed points seen in the ϵ expansion. This suggests that the phase transition might always be 1st-order. Another, extremely hand-waving, suggestion of this comes from the observation that all theories flow into the region $\lambda < 0$ and, at tree level, $\lambda < 0$ would be an unstable theory. This suggests that there is something suspect about flowing to arbitrarily large distances, as one would desire if the transition were 2nd-order.

The fact that the transition is 1st-order can be made rigorous, for small ϵ, by remembering that we know how to solve the theory when $\lambda \ll g^2$. When $\lambda \ll g^2$, straightforward perturbation theory is under control and predicts a 1st-order transition. So now consider a theory initially defined by the couplings g_1^2 and λ_1 as depicted in Fig. 6. The renormalization group allows us to flow to larger distance scales and fins a completely equivalent theory where $\lambda \ll g^2$, also depicted in the figure. This equivalent theory can then be solved with straightforward perturbation theory and predicts a 1st-order transition. In summary, our method is as follows.

Step 1. Start with initial couplings g_1^2, λ_1 at scale $\mu_1 \sim T$ (which is the scale where the Euclidean time fluctuations first decouple).

Step 2. Use the renormalization group to flow to an equivalent theory with $\lambda \ll g^2$.

Step 3. We can then perturbatively compute the effective potential or whatever other quantities are of interest.

Step 4. Expand all result in ϵ.

This procedure is closely related to the work of Rudnick [10] and Chen, Lubensky and Nelson [11].¶

In the first step, we started with initial couplings g_1^2 and λ_1. We know what the initial couplings should be in the three dimensional theory; they are perturbatively related to the couplings in the original 3+1 dimensional theory as in (13). But in the ϵ expansion we have generalized the 3-dimensional theory to a $4-\epsilon$ dimensional theory. What couplings g_1^2 and λ_1 in the $4-\epsilon$ dimensional theory most naturally extrapolate to those of the 3-dimensional theory as $\epsilon \to 1$?

This question is easily answered if one observes that the one-loop renormalization group equations are independent of ϵ if appropriately rescaled. Specifically, take

$$g^2 \to \epsilon \bar{g}^2, \qquad \lambda \to \epsilon \bar{\lambda}, \qquad \mu \to \bar{\mu}^{-1/\epsilon}. \tag{21}$$

The one-loop RG equations (20) then become

$$\mu \partial_\mu \bar{g}^2 = -\bar{g}^2 + \beta_0 \bar{g}^4, \tag{22}$$
$$\mu \partial_\mu \bar{\lambda} = -\bar{g}^2 + (a\bar{g}^4 + b\bar{g}^2\bar{\lambda} + c\bar{\lambda}^2) \tag{23}$$

and are independent of ϵ. Thus, if I had simply plotted g^2/ϵ vs. λ/ϵ in Fig. 6 rather than g^2 vs. λ, the trajectories would have been *independent* of ϵ. The natural generalization of any particular theory from 3 to $4-\epsilon$ dimensions is then clear: just keep the same place on the same trajectory. So the relationship between the couplings is

$$(g^2_{4-\epsilon}, \lambda_{4-\epsilon}) = (\epsilon g_3^2, \epsilon\lambda_3). \tag{24}$$

Note that, even though the trajectories are independent of ϵ once the couplings are scaled by ϵ, the rate at which those trajectories are traversed is exponentially sensitive to ϵ as a result of the scaling of μ in (21). In particular, the trajectories are traversed *exponentially* slowly as $\epsilon \to 0$.

Example

I'll now very roughly and schematically outline a sample calculation. Consider the classical Higgs potential

$$V_{\rm cl} \sim m^2\phi^2 + \mu^\epsilon \lambda \phi^4. \tag{25}$$

Suppose I've used the renormalization group to run to $\lambda \ll g^2$. In particular, it's computationally convenient to run to $\lambda = 0$. I can now compute the 1-loop potential, and the result for $\epsilon \to 0$ is the familiar Coleman-Weinberg potential:

$$V^{(1-\text{loop})} \sim m^2\phi^2 + g^4\phi^4 \ln\left(\frac{g\phi}{\mu}\right). \tag{26}$$

The dependence of this potential on the effective mass m gives a 1st-order phase transition, as shown in Fig. 7. To study the system at the critical temperature, I should find the critical value of m where the two minima are degenerate. I can then compute, for example, the scalar correlation length ξ in the asymmetric phase at the critical temperature, which at this order is related to the curvature of the potential. One finds

$$\xi^2 \sim \frac{1}{V''(\phi_{\min})} \sim \frac{1}{g^2\mu^2}. \tag{27}$$

¶There is a long story about why the ϵ-expansion prediction that the phase transition is always 1st-order has, historically, been believed to be incorrect in the U(1) case, where the theory models the phase transition of superconductors. For a discussion of problems with and loopholes to this belief, see section I.E of ref. [1].

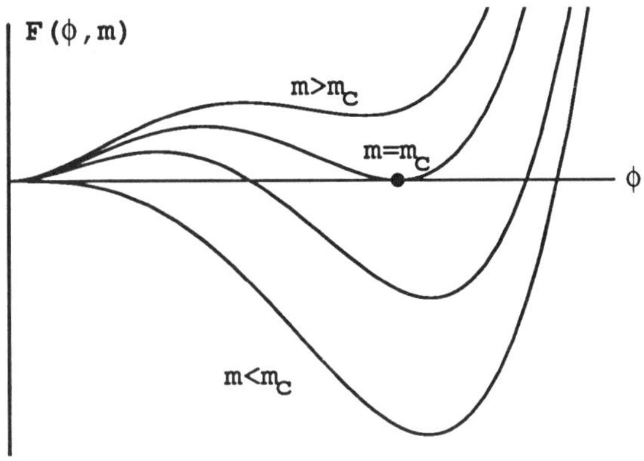

Figure 7. The Coleman-Weinberg potential as a function of m.

[The dependence $1/g^2\mu^2$ is easy to obtain from the form of (26) by simple scaling arguments.] Now recall from the previous section that g^2 is $O(\epsilon)$ and that the amount of running required to run to $\lambda = 0$ is exponentially long as $\epsilon \to 0$, so that

$$\mu \sim e^{O(1/\epsilon)} T\,. \tag{28}$$

As a result, the dependence of (27) on ϵ is

$$\xi^2 \sim \frac{1}{\epsilon} e^{\#/\epsilon} \frac{1}{T^2}\,, \tag{29}$$

where # is computed from the renormalization group equations. If one works to higher orders in ϵ, one finds that the prefactor of the exponential is a series in ϵ:

$$\xi^2 T^2 \sim \frac{1}{\epsilon} e^{\#/\epsilon} (1 + \#\epsilon + \#\epsilon^2 + \cdots)\,. \tag{30}$$

When I later refer to orders of the ϵ expansion, I shall be referring to the expansion of the prefactor. So "leading-order" will refer to a calculation that includes the first term of the prefactor, "next-to-leading order" includes the second term, and so forth.

Tests

Now that I've described how to go about computing things with the ϵ expansion, it would be nice to find some cases where we can check to see if the ϵ expansion is working. One case where we *already* know how to compute results is in the perturbative regime where the initial λ_1/g_1^2 is $\ll 1$. As a first check, we can compute the results of the ϵ expansion in the same limit. Table 1 shows the ratio of ϵ expansion results to the correct answer in the limit $\lambda_1/g_1^2 \to 0$. Note that leading-order in ϵ is generally within roughly 50% and next-to-leading order within roughly 5-10%. (The exception is the last quantity, which is the difference of free energies of the two ground states at the temperature where the $\phi = 0$ ground state is just loosing its meta-stability.)

We have also tested the ϵ expansion against large N results where N is the number of scalar fields in the theory. I didn't have time to discuss this in Sintra. The result is

Table 1. The ratio of the ϵ-expansion results, computing prefactors through leading order (LO) and next-to-leading order (NLO) in ϵ, to the corresponding three-dimensional result when $\lambda_1 \ll g_1^2$.

observable ratio	LO	NLO
asymmetric phase correlation length	1.14	0.94
symmetric phase correlation length	1.62	0.92
latent heat	0.77	1.04
surface tension of a domain wall	0.60	0.98
$\Delta F(T_0)$	0.24	0.56

that the ϵ expansion works qualitatively but fails quantitatively at leading-order in ϵ. However, the ϵ-expansion self-diagnoses its own quantitative failure because a computation of next-to-leading order corrections in ϵ shows that they are large compared to the leading-order results.

The perturbative test discussed above is only of limited interest because, after all, we already knew how to solve the theory in the perturbative regime. The large N tests are of limited interest because we want to know what happens when N is small. (Indeed, the large N theory is very different. For example, it is not asymptotically free.) What we really want is a test of the ϵ expansion when $\lambda_1 \gtrsim g_1^2$. A natural thing to do is to check whether the ϵ expansion diagnoses its own failure—that is, whether next-to-leading order corrections are small or large compared to leading-order results. Computing results at next-to-leading order in this regime is involved, and we only had the stamina to make one test. Our test quantity is related to the specific heat. More specifically, for technical reasons, it is

$$(\text{asymmetric phase correlation length})^2 \times (\text{latent heat})\,. \tag{31}$$

The calculation to next-to-leading order requires the 2-loop renormalization group and a 2-loop computation of the potential.[||] Fig. 8 shows the results.

From this figure, we conclude that the ϵ expansion may be quantitatively useful for m_h below 150 GeV or so, where the correction is within $\pm 30\%$. The expansion may be useful qualitatively, but probably not quantitatively, for larger Higgs mass. You should be cautioned, however, that we have only tested one quantity in this way, and perhaps it's better behaved than most.

Qualitative results for B violation

The original motivation for this study was to examine the bounds on electroweak baryogenesis, and so I should say a few words about the implications of the ϵ expansion for the rate of B violation at the completion of the phase transition. In particular, one can compare the results of renormalization-group improved perturbation theory in $4-\epsilon$ dimensions to the results one would get with naive, unimproved perturbation theory in $4-\epsilon$ dimensions. At one loop, we find that RG improvement predicts a *stronger* phase transition than the unimproved one-loop result, consistent with the nature of the 2-loop correction of Fig. 3 discussed earlier. One might assume that a stronger phase transition implies a smaller rate of B violation after the transition. Instead, we

[||] If we had just computed the latent heat by itself, we would have needed the three-loop renormalization group, which is not currently known. See ref. [1] for details.

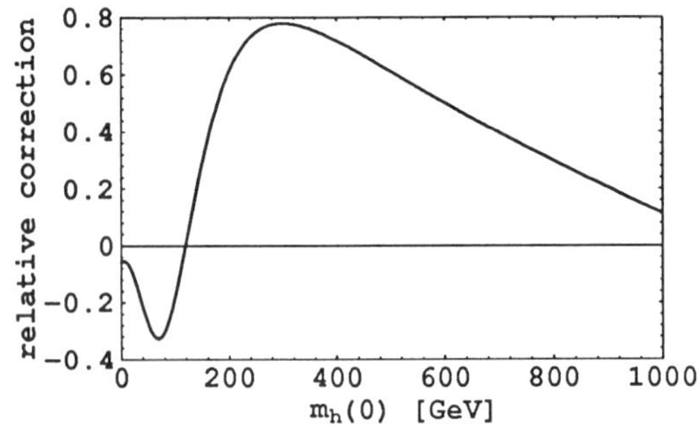

Figure 8. The relative size of the next-to-leading order correction compared to the leading-order result, as a function of zero-temperature Higgs mass.

find a *larger* rate of B violation in $4-\epsilon$ dimensions. This conclusion is, as always, only rigorous for $\epsilon \ll 1$ and may or may not hold qualitatively when $\epsilon = 1$. (Ideally, one should compute next-to-leading order corrections and see if they are under control.) However, our result offers an important warning. If one studies the transition on a lattice, and finds a transition much stronger than expected by perturbation theory, that does not necessarily mean that baryogenesis is more viable in the model being studied.

I should conclude with a final caveat, however. The calculation of the B violation rate in the ϵ expansion is beset by additional technical perils arising from the intrinsically 3-dimensional nature of the sphaleron. And so the extrapolation of $\epsilon \to 1$ may be even more suspect than usual. See ref. [1] for details.

REFERENCES

[1] P. Arnold and L. Yaffe, *Phys. Rev.* D49, 3003 (1994).

[2] G. Farrar and M. Shaposhnikov, CERN preprint CERN-TH-6732-93 (1993); *Phys. Rev. Lett.* 70, 2833 (1993); *ibid.* 71, 210(E) (1993); M. Shaposhnikov, *Phys. Lett.* B277, 324 (1992); *ibid.* B282, 483(E) (1992).

[3] M. Gavela, P. Hernadez, J. Orloff and O. Penne, CERN preprint CERN-TH-7081-93 (1993); P. Huet and E. Sather, SLAC preprint SLAC-PUB-6479 (April 1994).

[4] P. Arnold and O. Espinosa, *Phys. Rev.* D47, 3546 (1993).

[5] M. Dine, R. Leigh, P. Huet, A. Linde and D. Linde, *Phys. Lett.* B238, 319 (1992); *Phys. Rev.* D46, 550 (1992).

[6] J. Bagnasco and M. Dine, *Phys. Lett.* B303, 308 (1993).

[7] M. Gleisser and E. Kolb, *Phys. Rev.* D48, 1560 (1993).

[8] J. March-Russel, *Phys. Lett.* B296, 364 (1992); M. Alford and J. March-Russel, *Phys. Rev.* D48, 2838 (1993).

[9] P. Ginsparg, *Nucl. Phys.* B170 [FS1], 388 (1980).

[10] J. Rudnick, *Phys. Rev.* B11, 3397 (1975).

[11] J. Chen, T. Lubensky, and D. Nelson, *Phys. Rev.* B17, 4274 (1978).

THE ELECTROWEAK PHASE TRANSITION IN EXTENDED MODELS

Jose Ramón Espinosa

Instituto de Estructura de la Materia, CSIC
Serrano 123, 28006-Madrid, Spain

ABSTRACT

We study the possibility of relaxing the cosmological bound on the Higgs mass coming from the requirement of non-erasure of the baryon asymmetry by sphalerons. After reviewing the Standard Model case we obtain this bound in two extensions of it: 1) The Standard Model with an additional gauge singlet, and 2) The Minimal Supersymmetric Standard Model. Taking fully into account all experimental constraints and thermal screening effects we found that the situation can be slightly improved with respect to the Standard Model but only in case 1) a non negligible region in parameter space exists where the baryon washout is avoided and the experimental bounds evaded.

INTRODUCTION

The attractive possibility that the observed baryon asymmetry of the universe was created at the electroweak phase transition has deserved a lot of of attention in the last years[1]. It was realized[2] that all the conditions formulated by Sakharov[3] (for generating the baryon asymmetry) can be met in the Standard Model (SM). In particular, the sphaleron transitions were recognized as unsupressed at high temperatures and so playing a central role in the mechanisms generating the amount of ΔB we finally observe today.

These sphaleron processes, if unsupressed after the electroweak phase transition, will drive the previously created baryon asymmetry to zero, and so its rate Γ at that time[2,4] ($\Gamma \sim exp(-E_{sph}/T)$) should be less than the Hubble expansion rate H to go out of equilibrium. Due to the exponential sensitivity of Γ to the ratio $E_{sph}(T)/T$, imposing

$\Gamma < H$ gives the model independent condition[5]:

$$\frac{E_{sph}(T_c)}{T_c} \geq 45. \tag{1}$$

After relating the sphaleron energy[6] E_{sph} to the vacuum expectation value of the Higgs field at the critical temperature T_c ($E_{sph}(T_c) \sim \langle\phi(T_c)\rangle/g$) the central relation Eq. (1) translates into

$$\frac{\langle\phi(T_c)\rangle}{T_c} \gtrsim 1, \tag{2}$$

which tells that the transition should be strongly enough first order so as to avoid the erasure of the baryon asymmetry.

When the dependence of $\langle\phi(T_c)\rangle$ on the parameters of the Higgs potential is taken into account, condition (2) gives an upper bound on the mass of the Higgs boson (in the Standard Model where this calculation was originally performed and also in more general cases). The study of this bound is the central aim of the work here resumed.

All along this paper we will say that a phase transition is strongly first order if condition (2) is satisfied and weakly first order (or even second order) if it is not.

THE STANDARD MODEL

The effective potential at non zero temperature is the central tool in studying the electroweak phase transition. In one-loop approximation the temperature-dependent potential describes a first order phase transition between the symmetric phase (at high T) and the $SU(2) \times U(1)$ breaking one. Using a high temperature expansion the form of the potential is:

$$V(\phi, T) = D(T^2 - T_0^2)\phi^2 - ET\phi^3 + \frac{1}{4}\lambda(T)\phi^4, \tag{3}$$

where D, T_0, E and $\lambda(T)$ are easily calculable.

Approximating the critical temperature T_c as the temperature T_D at which two degenerate minima coexist (T_c will be somewhere in the narrow range between T_D and the temperature T_0 at which the minimum at $\phi = 0$ is destabilized) one gets

$$\frac{\langle\phi(T_c)\rangle}{T_c} = \frac{2E}{\lambda(T_c)}. \tag{4}$$

We see from this formula that the strenght of the transition is governed by E (which gives the cubic term in the potential). Now, E comes from bosonic loops only (from $n = 0$ Matsubara frecuencies). Each bosonic degree of freedom, with field dependent mass $M_i(\phi)$ contributes to the potential a term

$$\Delta_i V = -\frac{T}{12\pi}\left[M_i^2(\phi)\right]^{3/2}. \tag{5}$$

In the SM, the main contribution is the one coming from gauge boson loops:

$$\Delta V = -\frac{T}{12\pi}\left[6\left(\frac{1}{4}g^2\phi^2\right)^{3/2} + 3\left(\frac{1}{4}(g^2+g'^2)\phi^2\right)^{3/2}\right]. \tag{6}$$

Using the value of E obtained from Eq. (6) and relating λ to the Higgs mass one gets the bound[5,7]

$$M_h \lesssim (45 - 50)\ GeV, \tag{7}$$

region which is already ruled out by LEP.

This bound is obtained within the one-loop approximation to the effective potential and as is well known this will only be reliable for $\phi > T$ (the expansion parameter is $g^2T^2/M^2(\phi)$). For small ϕ the perturbative expansion is plagued with infrared divergences. To cure this, one has to reorganize the loop expansion resumming the most important infrared divergent diagrams[8] (the so called daisy diagrams). The effect of this daisy resummation is to shift the mass of the $n = 0$ modes contributing to the cubic term by a thermal mass (Debye screening effect). In the SM this thermal mass is of order g^2T^2 to leading order for the longitudinal gauge bosons, and zero (to this order) for the transverse ones. The effect of this thermal mass is to shield the cubic term of the longitudinal modes while leaving the transverse modes unaffected and so to effectively reduce E by a factor 2/3. The improved bound on the Higgs mass is then even lower[9]:

$$M_h \lesssim \sqrt{2/3}(45 - 50)\ GeV \sim (35 - 40)\ GeV. \tag{8}$$

There exist further refinements of this calculation in the literature. If the screening of the transverse modes at subleading order (g^4T^2) is taken into account[10] the strenght of the transition is further reduced. On the other hand, some non-perturbative effects may give a lower value for the critical temperature[11] translating into a larger ratio $\langle\phi(T_c)\rangle/T_c$. Also there are some two-loop (resummed) calculations[12] that give larger $\langle\phi(T_c)\rangle$.

Even with these uncertainties the SM has another problems to be able to generate the observed baryon asymmetry. In fact the CP violation effects are by far too small to give the correct amount of ΔB and some recent atempts[13] to solve this problem have been shown[14] to be flawn. So it seems necessary to move to some extensions of the SM to explain the generation of the baryon asymmetry at the electroweak phase transition.

From what we have learned by studying the Standard Model case we can work out an estrategy[15] for the extensions of it to give a strong first order phase transition. A look at Eq. (4) shows that we need a large cubic term in our potential. So, adding more bosonic degrees of freedom may help for only bosons contribute to this cubic term. The screened mass of such bosons will have the general form

$$\overline{M}^2(\phi) = M^2 + \lambda\phi^2 + \Pi(T), \tag{9}$$

where M is an invariant mass, λ is a generic coupling of the new bosons to the Higgs (here ϕ stands generically for the fields driving the transition) and Π is the thermal mass. For this to give a non-negligible contribution to the cubic term, small values of M and Π and large values of λ are required. Usually, there will exist an experimental lower bound for M so that one cannot take it to be arbitrarily small. Concerning the coupling λ the requirement of perturbativity up to some large scale will set an upper bound on it. In a given model one has to take all this conditions into account when looking for a region in the parameter space where the phase transition is strong enough.

Another way of increasing the strenght of the transition has been considered in the literature[15,16,17], namely the possibility of decreasing the value of the coupling in the denominator of Eq. (4) due to one loop corrections (even at zero temperature). But usually for this mechanism to give a sizeable effect large couplings and/or many bosonic degrees of freedom are needed.

Before moving to the discussion of extended models some words about the validity of the perturbative calculations are needed. Roughly speaking, one can trust the (resummed) perturbative calculations in the SM if $m_h^2 \lesssim m_W^2$ (or equivalently $\lambda/g^2 \lesssim 1$). When trying to improve the "sphaleron bound" $m_h \lesssim m_{h,crit}$, one can face cases where $m_{h,crit} \sim m_W$ rising some doubts about the reliability of the calculation of $m_{h,crit}$. In the cases we are going to consider the relevant parameter will no longer be λ/g^2. In fact g^2 will be substituted by ζ^2 in the SM + singlet case and by h_t^2 in the MSSM . As we will be interested in large values of these couplings, for a given Higgs mass the expansion parameter will be smaller than in the SM, and the perturbative calculation more reliable.

THE STANDARD MODEL WITH A GAUGE SINGLET

This is the simplest extension of the Standard Model that can in principle improve the situation concerning the electroweak baryogenesis problem, as already noted in Ref. [15].

The lagrangian of the model is defined as:

$$\mathcal{L} = \mathcal{L}_{SM} + \partial^\mu S^* \partial_\mu S - M^2 S^* S - \lambda_S (S^* S)^2 - 2\zeta^2 S^* S H^* H, \tag{10}$$

where H is the SM doublet with $\langle H \rangle = \phi/\sqrt{2}$, ϕ is the classical field, and $M^2, \lambda_S, \zeta^2 \geq 0$, to guarantee that $\langle S \rangle = 0$ at all temperatures.

The temperature dependent effective potential in the one loop approximation was studied in Ref. [15]. The daisy improvement was performed in Refs. [18,16]. The potential is a function of the classical field ϕ as in the SM, but with a new contribution coming from the S bosonic loops of the form (imposing renormalization conditions preserving the tree level value of the *vev* v)

$$\Delta V_S = g_S \left\{ \frac{m_S^2(\phi) T^2}{24} - \frac{\overline{M}_S^3(\phi) T}{12\pi} - \frac{m_S^4(\phi)}{64\pi^2} \left[\log \frac{m_S^2(v)}{c_B T^2} - 2 \frac{m_S^2(v)}{m_S^2(\phi)} \right] \right\}, \tag{11}$$

where $g_S = 2$ is the number of degrees of freedom of the (complex) singlet. $m_S(\phi)$ is the field dependent mass of S and $\overline{M}_S(\phi)$ its Debye screened mass. They are given by

$$\begin{aligned} m_S^2(\phi) &= M^2 + \zeta^2 \phi^2, \\ \overline{M}_S^2 &= m_S^2(\phi) + \Pi_S(T). \end{aligned} \tag{12}$$

Here Π_S is, to leading order

$$\Pi_S(T) = \frac{\lambda_S + \zeta^2}{3} T^2. \tag{13}$$

The thermal polarizations for the rest of SM particles can receive extra contributions coming from S loops (see Refs. [18,16]).

To understand the numerical results presented at the end of this section it is convenient to study analytically the effective potential assuming as will be the case that the bosonic contribution is dominated by the singlet (and this is the reason for the relevant expansion parameter to be λ/ζ^2 rather than λ/g^2). The high temperature expansion of the effective potential takes then the form

$$V(\phi) = A(T)\phi^2 + B(T)\phi^4 + C(T)\left(\phi^2 + K^2(T)\right)^{3/2}, \tag{14}$$

where

$$C(T) = -\frac{\zeta^3 T}{6\pi}, \tag{15}$$

$$K^2(T) = \frac{(\zeta^2 + \lambda_S)T^2 + 3M^2}{3\zeta^2}. \tag{16}$$

From these expressions and the discussion at the end of the previous section is clear that the best case for the phase transition to be stongly first order is to have low values of M and λ_S and large values of ζ^2. This is what was found numerically in Ref. [18].

For large values of M the transition is weaker and eventually, when $M \gg T$, the singlet decouples and the SM result is recovered. The dependence on ζ^2 is the most interesting one. Imposing the condition that all the couplings in the theory remain perturbative up to some high scale Λ an upper bound for ζ^2 at the electroweak scale is derived (as a function of course of the scale Λ and the unknown value of the top yukawa coupling h_t). This renormalization group study was also performed in Ref. [18].

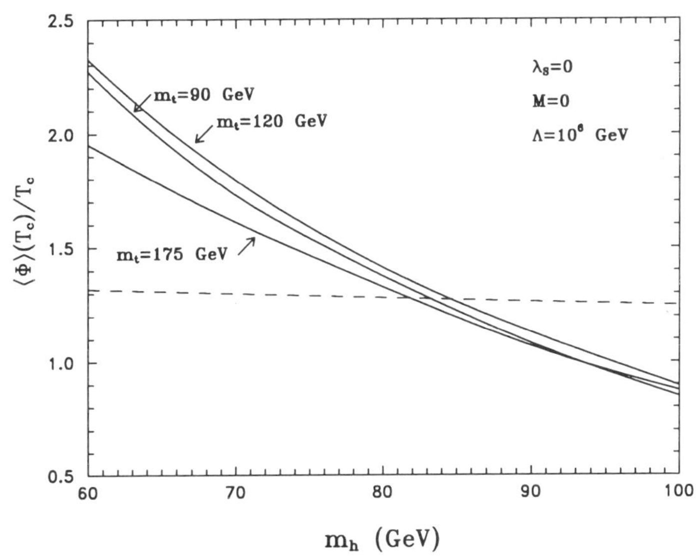

Figure 1. Plot of $\langle\phi(T_c)\rangle/T_c$ versus m_h for $\lambda_S = 0$, $M = 0$, $m_t = 90,\ 150$ and $175\ GeV$ and ζ corresponding to $\Lambda = 10^6\ GeV$.

Translating the requirement of a strong phase transition ($E_{sph}(T_c)/T_c > 45$ or $\langle\phi(T_c)\rangle \gtrsim 1.3$ with our definition of T_c) into a bound on the Higgs mass M_H^{crit}, one finds that in this model one can evade the stringent bound of the SM and go to larger values of M_H^{crit}. The numerical value of this critical mass depends on how large the coupling ζ^2 is, or equivalently, on the mass of the top and the scale of new physics Λ. Assuming that this scale is not lower than $10^6\ GeV$ the value of M_H^{crit} can be of order $80\ GeV$ (see Fig. 1) which is higher than the experimental bound. Larger values of this critical mass can only be obtained at the price of introducing new physics at lower scales.

THE MINIMAL SUPERSYMMETRIC STANDARD MODEL

The Minimal Supersymmetric Standard Model (MSSM) is the physically most motivated and phenomenologically most acceptable among the extensions of the SM. Concerning the generation of the baryon asymmetry the MSSM allows for extra CP-violating phases besides the Kobayashi-Maskawa one, which could help in generating the observed baryon asymmetry[19]. On the other hand the MSSM has a lot of new degrees of freedom and the nature of the phase transition could be significantly modified with respect to the SM.

The main tool for our study is the one-loop, daisy-improved (for previous studies in the one-loop approximation see Refs. [7,20]) finite-temperature effective potential of the MSSM, $V_{\rm eff}(\phi,T)$. Now the potential is actually a function of two fields: $\phi_1 \equiv {\rm Re}\, H_1^0$ and $\phi_2 \equiv {\rm Re}\, H_2^0$, where H_1^0 and H_2^0 are the neutral components of the Higgs doublets H_1 and H_2, thus ϕ will stand for (ϕ_1,ϕ_2). Working in the 't Hooft-Landau gauge and in the $\overline{DR}$-scheme, we can write

$$V_{\rm eff}(\phi,T) = V_0(\phi) + V_1(\phi,0) + \Delta V_1(\phi,T) + \Delta V_{\rm daisy}(\phi,T)\,, \tag{17}$$

where

$$V_0(\phi) = m_1^2\phi_1^2 + m_2^2\phi_2^2 + 2m_3^2\phi_1\phi_2 + \frac{g^2+g'^2}{8}(\phi_1^2-\phi_2^2)^2\,, \tag{18}$$

$$V_1(\phi,0) = \sum_i \frac{n_i}{64\pi^2} m_i^4(\phi)\left[\log\frac{m_i^2(\phi)}{Q^2} - \frac{3}{2}\right], \tag{19}$$

$$\Delta V_1(\phi,T) = \frac{T^4}{2\pi^2}\left\{\sum_i n_i\, J_i\left[\frac{m_i^2(\phi)}{T^2}\right]\right\}, \tag{20}$$

$$\Delta V_{\rm daisy}(\phi,T) = -\frac{T}{12\pi}\sum_i n_i\left[\overline{m}_i^3(\phi,T) - m_i^3(\phi)\right]. \tag{21}$$

The first term, Eq. (18), is the tree-level potential. The second term, Eq. (19), is the one-loop contribution at $T=0$: Q is the renormalization scale, where we choose for definiteness $Q^2=m_Z^2$, $m_i^2(\phi)$ is the field-dependent mass of the ith particle, and n_i is the corresponding number of degrees of freedom, taken negative for fermions. Since $V_1(\phi,0)$ is dominated by top (t) and stop ($\tilde t_1,\tilde t_2$) contributions, only these will be included in the following. The third term, Eq. (20), is the additional one-loop contribution due to temperature effects. Here $J_i = J_+(J_-)$ if the ith particle is a boson (fermion), and

$$J_\pm(y^2) \equiv \int_0^\infty dx\, x^2 \log\left(1 \mp e^{-\sqrt{x^2+y^2}}\right). \tag{22}$$

Since the relevant contributions to $\Delta V_1(\phi,T)$ are due to top (t), stops ($\tilde t_1,\tilde t_2$) and gauge bosons (W,Z), only these will be considered in the following. Finally, the last term, Eq. (21), is the correction coming from daisy diagrams. The sum runs over bosons only. As usual, the masses $\overline{m}_i^2(\phi,T)$ are obtained from the $m_i^2(\phi)$ by adding the leading T-dependent self-energy contributions, which are proportional to T^2.

The relevant degrees of freedom for our calculation are:

$$n_t=-12\,,\ n_{\tilde t_1}=n_{\tilde t_2}=6\,,\ n_W=6\,,\ n_Z=3\,,\ n_{W_L}=2\,,\ n_{Z_L}=n_{\gamma_L}=1\,. \tag{23}$$

The field-dependent top mass is

$$m_t^2(\phi) = h_t^2\phi_2^2\,. \tag{24}$$

The entries of the field-dependent stop mass matrix are

$$m_{\tilde{t}_L}^2(\phi) = m_{Q_3}^2 + m_t^2(\phi) + D_{\tilde{t}_L}^2(\phi)\,, \tag{25}$$
$$m_{\tilde{t}_R}^2(\phi) = m_{U_3}^2 + m_t^2(\phi) + D_{\tilde{t}_R}^2(\phi)\,, \tag{26}$$
$$m_X^2(\phi) = h_t(A_t\phi_2 + \mu\phi_1)\,, \tag{27}$$

where m_{Q_3}, m_{U_3} and A_t are soft supersymmetry-breaking mass parameters, μ is a superpotential Higgs mass term, and

$$D_{\tilde{t}_L}^2(\phi) = \left(\frac{1}{2} - \frac{2}{3}\sin^2\theta_W\right)\frac{g^2 + g'^2}{2}(\phi_1^2 - \phi_2^2), \tag{28}$$
$$D_{\tilde{t}_R}^2(\phi) = \left(\frac{2}{3}\sin^2\theta_W\right)\frac{g^2 + g'^2}{2}(\phi_1^2 - \phi_2^2) \tag{29}$$

are the D-term contributions. The field-dependent stop masses are then

$$m_{\tilde{t}_{1,2}}^2(\phi) = \frac{m_{\tilde{t}_L}^2(\phi) + m_{\tilde{t}_R}^2(\phi)}{2} \pm \sqrt{\left[\frac{m_{\tilde{t}_L}^2(\phi) - m_{\tilde{t}_R}^2(\phi)}{2}\right]^2 + \left[m_X^2(\phi)\right]^2}\,. \tag{30}$$

The corresponding effective T-dependent masses, $\overline{m}_{\tilde{t}_{1,2}}^2(\phi, T)$, are given by expressions identical to (30), apart from the replacement

$$m_{\tilde{t}_{L,R}}^2(\phi) \to \overline{m}_{\tilde{t}_{L,R}}^2(\phi, T) \equiv m_{\tilde{t}_{L,R}}^2(\phi) + \Pi_{\tilde{t}_{L,R}}(T)\,. \tag{31}$$

The $\Pi_{\tilde{t}_{L,R}}(T)$ are the leading parts of the T-dependent self-energies of $\tilde{t}_{L,R}$,

$$\Pi_{\tilde{t}_L}(T) = \frac{4}{9}g_s^2T^2 + \frac{1}{4}g^2T^2 + \frac{1}{108}g'^2T^2 + \frac{1}{6}h_t^2T^2\,, \tag{32}$$
$$\Pi_{\tilde{t}_R}(T) = \frac{4}{9}g_s^2T^2 + \frac{4}{27}g'^2T^2 + \frac{1}{3}h_t^2T^2\,, \tag{33}$$

where g_s is the strong gauge coupling constant. Only loops of gauge bosons, Higgs bosons and third generation squarks have been included, implicitly assuming that all remaining supersymmetric particles are heavy and decouple. If some of these are also light, the plasma masses for the stops will be even larger, further suppressing the effects of the associated cubic terms, and therefore weakening the first-order nature of the phase transition.

We expect the stops to play an important role in making the transition stronger as they have a large number of degrees of freedom and couple to the Higgses with strength h_t. On the other hand as they are coloured scalars their thermal mass is of order $g_s^2T^2$ as shown above and also they are not protected by any symmetry to have an invariant mass. The final importance they have in the strenght of the transition will result from the interplay between these opposite properties.

Finally, the field-dependent gauge boson masses are

$$m_W^2(\phi) = \frac{g^2}{2}(\phi_1^2 + \phi_2^2)\,, \qquad m_Z^2(\phi) = \frac{g^2 + g'^2}{2}(\phi_1^2 + \phi_2^2)\,, \tag{34}$$

and the effective T-dependent masses of the longitudinal gauge bosons are

$$\overline{m}_{W_L}^2(\phi, T) = m_W^2(\phi) + \Pi_{W_L}(T)\,, \tag{35}$$

$$\overline{m}_{Z_L,\gamma_L}^2(\phi, T) = \frac{1}{2}\left[m_Z^2(\phi) + \Pi_{W_L}(T) + \Pi_{B_L}(T)\right]$$
$$\pm \sqrt{\frac{1}{4}\left[\frac{g^2 - g'^2}{2}(\phi_1^2 + \phi_2^2) + \Pi_{W_L}(T) - \Pi_{B_L}(T)\right]^2 + \left[\frac{gg'}{2}(\phi_1^2 + \phi_2^2)\right]^2}\,. \tag{36}$$

In eqs. (35) and (36), $\Pi_{W_L}(T)$ and $\Pi_{B_L}(T)$ are the leading parts of the T-dependent self-energies of W_L and B_L, given by

$$\Pi_{W_L}(T) = \frac{5}{2} g^2 T^2 \,, \qquad \Pi_{B_L}(T) = \frac{47}{18} g'^2 T^2 \,, \tag{37}$$

where only loops of Higgs bosons, gauge bosons, Standard Model fermions and third-generation squarks have been included.

We need to analyse the effective potential (17) as a function of ϕ and T. Before doing this, however, we trade the parameters m_1^2, m_2^2, m_3^2 appearing in the tree-level potential (18) for more convenient parameters. To this purpose, we first minimize the zero-temperature effective potential, i.e. we impose the vanishing of the first derivatives of $V_0(\phi)+V_1(\phi,0)$ at $(\phi_1,\phi_2)=(v_1,v_2)$, where (v_1,v_2) are the one-loop vacuum expectation values at $T=0$. This allows us to eliminate m_1^2 and m_2^2 in favour of m_Z^2 and $\tan\beta \equiv v_2/v_1$ in the standard way. Moreover, m_3^2 can be traded for the one-loop-corrected mass m_A^2 of the CP-odd neutral Higgs boson[21] Therefore the whole effective potential (17) is completely determined, in our approximation, by the parameters $(m_A, \tan\beta)$ of the Higgs sector, and by the parameters $(m_t,\ m_{Q_3},\ m_{U_3},\ \mu,\ A_t)$ of the top/stop sector. The same set of parameters also determines the one-loop-corrected masses and couplings of the MSSM Higgs bosons.

The next steps are the computation of the critical temperature and of the location of the minimum of the effective potential at the critical temperature. We define here T_0 as the temperature at which the determinant of the second derivatives of $V_{\rm eff}(\phi,T)$ at $\phi=0$ vanishes:

$$\det \left[\frac{\partial^2 V_{\rm eff}(\phi,T_0)}{\partial\phi_i \partial\phi_j} \right]_{\phi_{1,2}=0} = 0 \,. \tag{38}$$

It is straightforward to compute the derivatives in eq. (38) from the previous formulae; the explicit expressions are given in Ref. [22].

Once eq. (38) is solved (numerically) and T_0 is found, one can minimize (numerically) the potential $V_{\rm eff}(\phi,T_0)$ and find the minimum $[v_1(T_0), v_2(T_0)]$. The quantity of interest is indeed, as will be discussed later, the ratio $v(T_0)/T_0$, where $v(T_0) \equiv \sqrt{v_1^2(T_0)+v_2^2(T_0)}$.

Before presenting the numerical results we discuss the experimental constraints on the parameters of the top/stop sector and of the Higgs sector. We treat m_{Q_3}, m_{U_3} and the other soft mass terms as independent parameters, even if they can be related in specific SUGRA models. We want to be as general as possible and in this way the region of parameters we are able to exclude will be excluded in any particular model.

Direct and indirect searches at LEP imply[23] that $m_{\tilde{b}_L} \gtrsim 45$ GeV, which in turn translates into a bound in the $(m_{Q_3}, \tan\beta)$ plane. Electroweak precision measurements[24] put stringent constraints on a light stop-sbottom sector: in first approximation, and taking into account possible effects[25] of other light particles of the MSSM, we conservatively summarize the constraints by $\Delta\rho(t,b)+\Delta\rho(\tilde{t},\tilde{b}) < 0.01$.

We finally need to consider the constraints coming from LEP searches for supersymmetric Higgs bosons[23]. Experimentalists put limits on the processes $Z \to hZ^*$ and $Z \to hA$, where h is the lighter neutral CP-even boson. We need to translate these limits into exclusion contours in the $(m_A, \tan\beta)$ plane, for given values of the top/stop parameters. In order to do this, we identify the value of $BR(Z \to hZ^*)$, which corresponds to the limit $m_\phi > 63.5$ GeV on the SM Higgs, and the value of $BR(Z \to hA)$, which best fits the published limits for the representative parameter choice $m_t = 140$ GeV, $m_{Q_3} = m_{U_3} \equiv \tilde{m} = 1$ TeV, $A_t = \mu = 0$. We then compare those values of $BR(Z \to hZ^*)$

and $BR(Z \to hA)$ with the theoretical predictions of the MSSM, for any desired parameter choice and after including the radiative corrections associated to top/stop loops[26,21]. Of course, this procedure is not entirely correct, since it ignores the variations of the efficiencies with the Higgs masses and branching ratios, as well as the possible presence of candidate events at some mass values, but it is adequate for our purposes.

We now present our numerical results, based on the effective potential of eq. (17), concerning the strength of the electroweak phase transition and the condition for preserving the baryon asymmetry.

Particularizing to the MSSM the studies of sphalerons in general two-Higgs models[27], we obtain that

$$E_{sph}^{MSSM}(T) \leq E_{sph}^{SM}(T)\,, \tag{39}$$

where, in our conventions,

$$\frac{E_{sph}^{SM}(T)}{T} = \frac{4\sqrt{2}\pi}{g} B\left\{\frac{\lambda_{\text{eff}}(T)}{4g^2}\right\}\frac{v(T)}{T}\,, \tag{40}$$

and B is a smoothly varying function whose values can be found in Ref. [6]. For example, $B(10^{-2}) = 1.67$, $B(10^{-1}) = 1.83$, $B(1) = 2.10$. It can also be shown that

$$\frac{v(T_D)}{T_D} < \frac{v(T_C)}{T_C} < \frac{v(T_0)}{T_0}\,, \tag{41}$$

where T_C is the actual temperature at which the phase transition occurs, satisfying the inequalities

$$T_0 < T_C < T_D\,, \tag{42}$$

if T_0 is defined by (38) and T_D is the temperature at which there are two degenerate minima.

Finally, the corrections in E_{sph}^{SM} due to $g' \neq 0$ have been estimated and shown to be small[28]. Therefore, a conservative bound to be imposed is

$$R \equiv \frac{v(T_0)}{T_0}\frac{4\sqrt{2}\pi B\left\{\frac{\lambda_{\text{eff}}(T_0)}{4g^2}\right\}}{45g} > 1\,. \tag{43}$$

The last point to be discussed is the determination of the value of $\lambda_{\text{eff}}(T_0)$ to be plugged into eq. (43). The B-function we use is taken from Ref. [6], where the sphaleron energy was computed using the zero-temperature 'Mexican-hat' potential, $V = \frac{\lambda}{4}(\phi^2 - v^2)^2$. The sphaleron energy at finite temperature was computed in Ref. [29], where it was proven that it scales like $v(T)$, i.e.

$$E_{sph}^{SM}(T) = E_{sph}^{SM}(0)\,\frac{v(T)}{v}\,, \tag{44}$$

with great accuracy. Therefore, to determine the value of $\lambda_{\text{eff}}(T_0)$ we have fitted $V_{\text{eff}}(\phi, T_0)$, in the direction of the minimum, to the one-dimensional approximate potential,

$$V_{\text{eff}}(\phi, T_0) \simeq \frac{1}{4}\lambda_{\text{eff}}(T_0)[\phi^2 - v^2(T_0)]^2\,, \tag{45}$$

The value of λ_{eff} obtained from (45),

$$\lambda_{\text{eff}}(T_0) = 4\,\frac{V_{\text{eff}}(0, T_0) - V_{\text{eff}}[v(T_0), T_0]}{v^4(T_0)}\,, \tag{46}$$

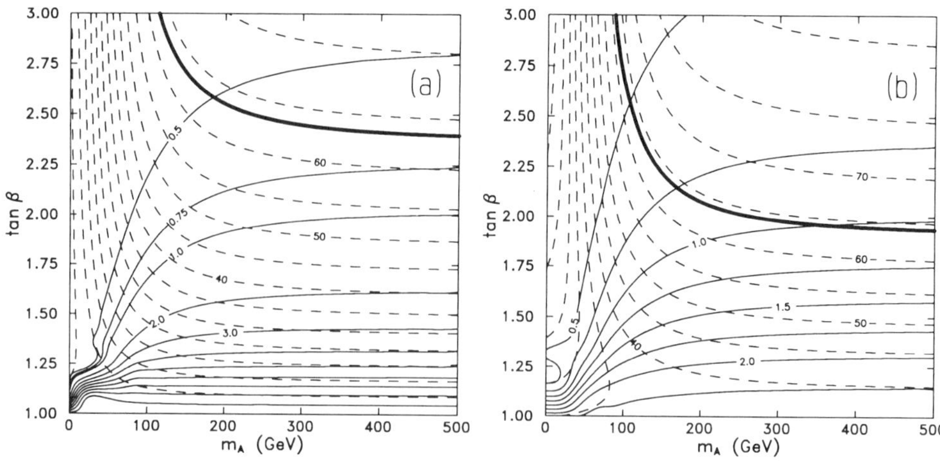

Figure 2. Contours of R in the $(m_A, \tan\beta)$ plane, for the parameter choices: a) $m_t = 130$ GeV, $m_{Q_3} = 50$ GeV, $m_{U_3} = 0$ ($m_{\tilde{t}} \sim 130$ GeV, $m_{\tilde{b}_L} \sim 50$ GeV), $A_t = \mu = 0$; b) $m_t = 170$ GeV, $m_{Q_3} = 280$ GeV, $m_{U_3} = 0$ ($m_{\tilde{t}_L} \sim 330$ GeV, $m_{\tilde{t}_R} \sim 170$ GeV, $m_{\tilde{b}_L} \sim 280$ GeV), $A_t = \mu = 0$. The region excluded by Higgs searches at LEP is delimited by the thick solid line. For reference, contours of constant m_h (in GeV) are also represented as dashed lines.

where all quantities on the right-hand side are calculated numerically from the potential of eq. (17), is then plugged into eq. (43) to obtain our bounds. Of course, the quality of the fit is good only for values of $\phi \lesssim v(T_0)$ but this is precisely the region of interest to determine the sphaleron energy.

Our numerical results are summarized in fig. 2, in the $(m_A, \tan\beta)$ plane and for two representative values of the top quark mass: $m_t = 130$ GeV (fig. 2a) and $m_t = 170$ GeV (fig. 2b). In each case, the values of the remaining free parameters have been chosen in order to maximize the strength of the phase transition, given the experimental constraints on the top-stop sector. Notice that arbitrarily small values of m_{U_3} cannot be excluded on general grounds, even if they are disfavoured by model calculations. Also, we have explicitly checked that, as in Ref. [30], mixing effects in the stop mass matrix always worsen the case. In fig. 2, solid lines correspond to contours of constant R: one can see that the requirement of large values of R favours small $\tan\beta$ and $m_A \gg m_Z$. The thick solid line corresponds to the limits coming from Higgs searches at LEP: for our parameter choices, the allowed regions correspond to large $\tan\beta$ and/or $m_A \gg m_Z$. For reference, contours of constant m_h (in GeV) have also been plotted as dashed lines. One can see that, even for third-generation squarks as light as allowed by all phenomenological constraints, only a very small globally allowed region can exist in the $(m_A, \tan\beta)$ plane, and that the most favourable situation is the one already discussed in Ref. [30]. More precisely, the region that is still marginally allowed corresponds to $m_A \gg m_Z$, $\tan\beta \sim 2$, stop and sbottom sectors as light as otherwise allowed, a heavy top, and a light Higgs boson with SM-like properties and mass $m_h \sim 65$ GeV, just above the present experimental limit. A less conservative interpretation of the limits from precision measurements, the inclusion of some theoretically motivated constraints on the model parameters, or a few GeV improvement in the SM Higgs mass limit, would each be enough to fully exclude electroweak baryogenesis in the MSSM.

ACKNOWLEDGMENTS

The results presented here are based on joint work done in collaboration with A. Brignole, M. Quirós and F. Zwirner. I must say it is always a pleasure to work with them.

This research was supported by a Comunidad de Madrid grant.

REFERENCES

1. For reviews and references see, e.g.:
   A.D. Dolgov, *Phys. Rep.* **222** (1992) 309; A.G. Cohen, D.B. Kaplan and A.E. Nelson, *Ann. Rev. Nucl. Part. Sci.* **43** (1993) 27; M. Dine, preprint SCIPP 94/03.
2. V.A. Kuzmin, V.A. Rubakov and M.E. Shaposhnikov, *Phys. Lett.* **B155** (1985) 36.
3. A.D. Sakharov, *JETP Lett.* **5** (1967) 24.
4. P. Arnold and L. McLerran, *Phys. Rev.* **D36** (1987) 581; *Phys. Rev.* **D37** (1988) 1020.
5. M.E. Shaposhnikov, *JETP Lett.* **44** (1986) 465; *Nucl. Phys.* **B287** (1987) 757 and **B299** (1988) 797; A.I. Bochkarev and M.E. Shaposhnikov, *Mod. Phys. Lett.* **A2** (1987) 417.
6. F.R. Klinkhamer and N.S. Manton, *Phys. Rev.* **D30** (1984) 2212.
7. M. Dine, P. Huet and R. Singleton Jr., *Nucl. Phys.* **B375** (1992) 625.
8. M.E. Carrington, *Phys. Rev.* **D45** (1992) 2933; M.E. Shaposhnikov, *Phys. Lett.* **B277** (1992) 324 and (E) **B282** (1992) 483; C.G. Boyd, D.E. Brahm and S.D.H. Hsu, *Phys. Rev.* **D48** (1993) 4952.
9. M. Dine, R.G. Leigh, P. Huet, A. Linde and D. Linde, *Phys. Lett.* **B283** (1992) 319 and *Phys. Rev.* **D46** (1992) 550.
10. J.R. Espinosa, M. Quirós and F. Zwirner, *Phys. Lett.* **B314** (1993) 206; W. Buchmüller, Z. Fodor, T. Helbig and D. Walliser, preprint DESY 93-021.
11. K. Kajantie, K. Rummukainen and M.E. Shaposhnikov, *Nucl. Phys.* **B407** (1993) 356; M.E. Shaposhnikov, *Phys. Lett.* **B316** (1993) 112.
12. P. Arnold and O. Espinosa, *Phys. Rev.* **D47** (1993) 3546; J. Bagnasco and M. Dine, *Phys. Lett.* **B303** (1993) 308.
13. G. Farrar and M.E. Shaposhnikov, *Phys. Rev. Lett.* **70** (1993) 2833 + (E) **71** (1993) 210 and preprint CERN-TH.6734/93.

14. M.B. Gavela, P. Hernández, J. Orloff and O. Pène, preprint CERN-TH.7081/93; P. Huet and E. Sather, preprint SLAC-PUB-6479.

15. G.W. Anderson and L.J. Hall, *Phys. Rev.* **D45** (1992) 2685.

16. K.E.C. Benson, *Phys. Rev.* **D48** (1993) 2456.

17. X. Zhang, preprint UMDHEP 93-074.

18. J.R. Espinosa and M. Quirós, *Phys. Lett.* **B305** (1993) 98.

19. A.G. Cohen and A.E. Nelson, *Phys. Lett.* **B297** (1992) 111.

20. G.F. Giudice, *Phys. Rev.* **D45** (1992) 3177; S. Myint, *Phys. Lett.* **B287** (1992) 325.

21. J. Ellis, G. Ridolfi and F. Zwirner, *Phys. Lett.* **B262** (1991) 477.

22. A. Brignole, J.R. Espinosa, M. Quirós and F. Zwirner, *Phys. Lett.* **B324** (1994) 181.

23. G. Coignet, Plenary talk at the XVI International Symposium on Lepton-Photon Interactions, Cornell University, Ithaca, New York, 10–15 August 1993.

24. G. Altarelli, Plenary talk given at the International Europhysics Conference on High Energy Physics, Marseille, 22–28 July 1993, preprint CERN-TH.7045/93.

25. R. Barbieri, M. Frigeni and F. Caravaglios, *Phys. Lett.* **B279** (1992) 169; G. Altarelli, R. Barbieri and F. Caravaglios, *Nucl. Phys.* **B405** (1993) 3, and *Phys. Lett.* **B314** (1993) 357; J. Ellis, G.L. Fogli and E. Lisi, *Phys. Lett.* **B285** (1992) 238, **B286** (1992) 85 and *Nucl. Phys.* **B393** (1993) 3.

26. Y. Okada, M. Yamaguchi and T. Yanagida, *Prog. Theor. Phys. Lett.* **85** (1991) 1 and *Phys. Lett.* **B262** (1991) 54; J. Ellis, G. Ridolfi and F. Zwirner, *Phys. Lett.* **B257** (1991) 83; H.E. Haber and R. Hempfling, *Phys. Rev. Lett.* **66** (1991) 1815; R. Barbieri and M. Frigeni, *Phys. Lett.* **B258** (1991) 395.

27. A. Bochkarev, S. Kuzmin and M. Shaposhnikov, *Phys. Rev.* **D43** (1991) 369 and *Phys. Lett.* **B244** (1990) 275; B. Kastening, R.D. Peccei and X. Zhang, *Phys. Lett.* **B266** (1991) 413.

28. J. Kunz, B. Kleihaus and Y. Brihaye, *Phys. Rev.* **D46** (1992) 3587.

29. S. Braibant, Y. Brihaye and J. Kunz, *Int. J. Mod. Phys.* **A8** (1993) 5563.

30. J.R. Espinosa, M. Quirós and F. Zwirner, *Phys. Lett.* **B307** (1993) 106.

SOME NON-PERTURBATIVE ASPECTS OF THE ELECTROWEAK PHASE TRANSITION

W. Buchmüller

Deutsches Elektronen-Synchrotron DESY
22603 Hamburg, Germany

INTRODUCTION

The theoretical description of the electroweak phase transition is based on the finite-temperature effective potential of the standard model, which can be approximated by the SU(2)-Higgs model for our purposes. Considerable effort has been devoted to evaluate the effective potential in perturbation theory where, due to infrared divergencies, a useful expansion is only obtained after a resummation of finite-temperature mass corrections[1].

Perturbation theory predicts a first-order phase transition, at least for Higgs boson masses below a critical mass $m_H^c \sim m_W$. The corresponding barrier, which separates the symmetric from the broken phase, is generated by radiative corrections involving fields with zero Matsubara frequency. This means that the quanta of the three-dimensional theory are responsible for the dynamics of the transition. In fact, the entire infrared behaviour of the finite-temperature theory is determined by the three-dimensional theory[2]. Hence, one expects a breakdown of perturbation theory for small values of the Higgs field due to the infrared Landau singularity of the three-dimensional gauge theory[3,4].

In the following we shall study the effect of the Landau singularity in more detail[5] by means of the ϵ-expansion[6]. The comparison between a renormalization group improved effective potential and the familiar one-loop potential will then show where pertubation theory becomes inadequate.

RENORMALISATION GROUP IMPROVED POTENTIAL

The SU(2)-Higgs model in $d = 4 - \epsilon$ dimensions is described by the lagrangian

$$\mathcal{L} = \frac{1}{4} W_{\mu\nu}^a W_{\mu\nu}^a + (D_\mu \Phi)^\dagger D_\mu \Phi + V_0(\varphi^2) \quad , \tag{1}$$

where

$$V_0(\varphi^2) = \frac{1}{2} m^2 \varphi^2 + \frac{1}{4} \mu^\epsilon \lambda \varphi^4 \ , \ \varphi^2 = 2\Phi^\dagger \Phi \ . \tag{2}$$

Here $W_{\mu\nu}^a$ is the ordinary field strength tensor and $D_\mu = \partial_\mu - i\mu^{\epsilon/2} \frac{g}{2} W_\mu^a \tau^a$ is the covariant

Electroweak Physics and the Early Universe, Edited by
Romao and Freire, Plenum Press, New York, 1994

derivative; g and λ are gauge and scalar self-coupling, respectively, and μ is the mass scale used to define dimensionless couplings in $4-\epsilon$ dimensions. We will consider the gauge invariant effective potential $V^{\epsilon}(\sigma)$, where $\sigma = 2\Phi^{\dagger}\Phi^{1}$.

Neglecting scalar loops, the one-loop potential in three dimensions is given by

$$V_1(\sigma) \;\equiv\; V_1^{\epsilon}(\sigma)\,|_{\epsilon=1} = \frac{1}{2}m^2\sigma + \frac{1}{4}\mu\lambda\sigma^2 - \frac{g^3}{16\pi}(\mu\sigma)^{3/2}\,. \tag{3}$$

This potential agrees with the finite-temperature potential $V(\Sigma, T)$ in the broken phase up to terms of order $\mathcal{O}(g^3\Sigma^3/T^2)$,

$$V(\Sigma,T) = TV_1(\sigma = \frac{\Sigma}{T})\,, \quad \Sigma = 2\Phi^{\dagger}\Phi\,|_{\epsilon=0}\,, \tag{4}$$

where

$$\begin{aligned} \mu &= T\,,\ \lambda(\mu) = \lambda - \frac{3}{128\pi}\sqrt{\frac{6}{5}}g^3\,,\ g(\mu) = g\,, \\ m^2(\mu) &= \left(\frac{3g^2}{16} + \frac{\lambda}{2} - \frac{3}{16\pi}\sqrt{\frac{5}{6}}g^3\right)(T^2 - T_b^2)\,, \end{aligned} \tag{5}$$

and T_b is the barrier temperature.

The one-loop approximation becomes unreliable as $\sigma \to 0$ due to non-perturbative effects in the symmetric phase. In $4-\epsilon$ dimensions one can improve the one-loop potential by summing the large logarithms by means of the renormalization group. Extrapolating this result to $\epsilon = 1$, one may hope to obtain an improvement of the one-loop result also in three dimensions. The same extrapolation is usually performed for the renormalization group equations of the couplings in connection with critical phenomena.

In $d = 4-\epsilon$ dimensions one obtains for the one-loop potential, after subtracting terms $\mathcal{O}(1/\epsilon)$ and again neglecting scalar loops,

$$\begin{aligned} V_1^{\epsilon}(\sigma) &= \frac{1}{2}m^2\sigma + \frac{1}{4}\mu^{\epsilon}\lambda\sigma^2 + \frac{1}{64\pi^2}\mu^{\epsilon}\sigma^2\left[a_1 g^4 \ln\left(\frac{\mu^{\epsilon}g^2\sigma e^{2/3}}{4\hat{\mu}^2}\right)\right. \\ &\left.-\frac{\epsilon}{2}\left(b_1 g^4 + b_2 g^4 \ln^2\left(\frac{\mu^{\epsilon}g^2\sigma e^{2/3}}{4\hat{\mu}^2}\right)\right)\right] + \mathcal{O}(\epsilon^2)\,, \end{aligned} \tag{6}$$

where $\hat{\mu}^2 = 4\pi\mu^2\exp(\gamma - 3/2)$, and a_1, b_1 and b_2 are constants[5]. From the invariance of the effective potential under a change of the scale μ, i.e., $dV_1^{\epsilon}(\sigma)/d\mu = 0$, one easily derives the one-loop renormalization group equations for couplings and mass.

The solutions of the evolution equations for $g(\mu)$ and $\lambda(\mu)$ are known exactly. The running gauge coupling

$$g^2(\mu) = \frac{x^{\epsilon}g_1^2}{1 + \frac{\beta_0}{\epsilon}(x^{\epsilon}-1)\,g_1^2}\,, \quad x = \frac{\mu_1}{\mu}\,, \tag{7}$$

and also the scalar coupling $\lambda(\mu)$ exhibit the well known Landau pole at

$$\mu_L = \left(-\frac{\epsilon}{\beta_0 g_1^2} + 1\right)^{-1/\epsilon}\mu_1\,. \tag{8}$$

For the gauge group SU(2) the one-loop β-function is $\beta_g = -(44-n)g^4/48\pi^2 \equiv \beta_0 g^4$, where n is the number of Higgs doublets.

In the finite-temperature case the critical temperature T_c is very close to the barrier temperature T_b. Hence, the effective mass m in the three-dimensional theory is very small, and its effect on the position of the global minimum σ_c can be neglected. For simplicity, we will therefore set $m = 0$ in the following, i.e., we consider the Coleman-Weinberg model in three dimensions. The renormalization group improved potential then takes the very simple form

$$V^{\epsilon}_{RGI}(\sigma) = \frac{1}{4}\bar{\mu}^{\epsilon}\bar{\lambda}(\bar{\mu})\sigma^2 , \tag{9}$$

where the mass scale $\bar{\mu}$ is given by $\bar{\mu}^2 = g^2\mu\sigma \exp{(\gamma - 5/6)}\zeta/16\pi$. We have included a factor $\zeta = \mathcal{O}(1)$, since a change of scale does not modify the leading logarithms in eq. (6). The coupling $\bar{\lambda}(\bar{\mu})$ takes into account the scale dependence of the field σ [5].Expanding the renormalization group improved potential (9) in powers of g^2, one easily verifies that the leading logarithms in eq. (6) are summed.

COMPARISON WITH THE ONE-LOOP POTENTIAL

In three dimensions the potential $V_{RGI} \equiv V^{\epsilon}_{RGI}\mid_{\epsilon=1}$ can now be compared with the one-loop potential (3). A formal expansion in powers of g^2 yields an expansion in powers of $\sqrt{\mu/g^2\sigma}$,

$$V_{RGI}(\sigma) \;=\; \frac{1}{4}\bar{\mu}\bar{\lambda}(\bar{\mu})\sigma^2 \;=\; \frac{1}{4}\mu\sigma^2\left(\lambda - \frac{9}{128\pi^2}g^4\frac{\mu}{\bar{\mu}} + \ldots\right) , \tag{10}$$

since $\bar{\mu} = \mathcal{O}(\sqrt{g^2\mu\sigma})$. The term $\mathcal{O}(\sigma^{3/2})$ is identical with corresponding term of the one-loop potential (3) if one chooses $\zeta = 81\exp{(-\gamma + 5/6)}/64\pi$.

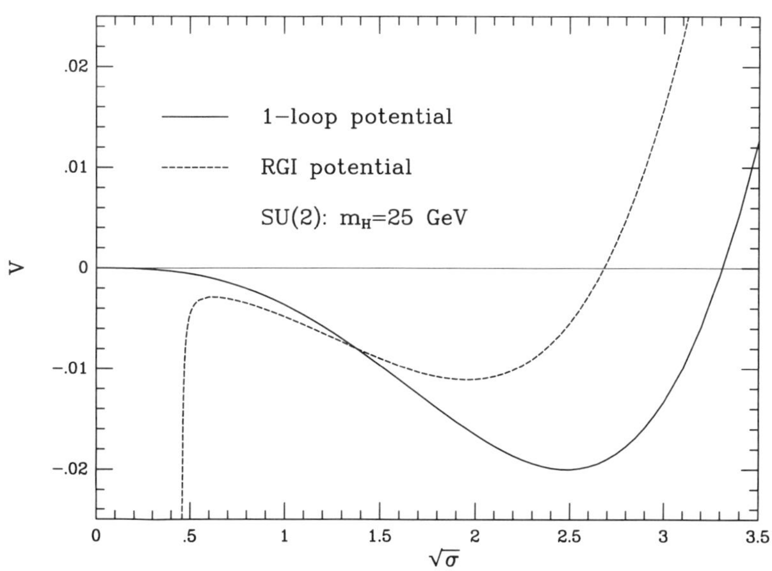

Figure 1. One-loop potential and renormalization group improved potential for the SU(2)-Higgs model in units of $\mu = T$. $m_H = 25$ GeV.

For $m^2 = 0$, the one-loop potential (3) has its global minimum at $\sigma_c > 0$. As one increases the Higgs mass, the position of the minimum moves closer to the origin. Clearly, perturbation theory breaks down as one approaches the Landau pole. The singularity is reached for $\bar{\mu}(\sigma_c) = \mu_L$, where μ_L is given by eq. (8) with $\mu_1 = \mu$. This condition for the breakdown of perturbation theory defines a critical Higgs mass, which is given by

$$\frac{{m_H^c}^2}{m_W^2} = -\frac{3}{64\pi\sqrt{\pi}} e^{\frac{\gamma}{2}-\frac{5}{12}} \frac{1}{\beta_0} \,. \tag{11}$$

For the SU(2)-Higgs model one obtains $m_H^c \simeq 0.86\ m_W$ which, with $m_W = 80$ GeV yields a critical Higgs mass of about 70 GeV. It is remarkable that at the same Higgs mass perturbation theory breaks down if a magnetic mass, determined from a one-loop gap equation, is included[1]. Also considerations based on "subcritical bubbles"[7], which are totally unrelated to non-perturbative effects in the gauge sector, suggest that for Higgs masses above 70 GeV the phase transition may no longer be first-order.

In fig. (1) the one-loop potential (3) and the renormalization group improved potential (9) are compared for a Higgs mass $m_H = 25$ GeV and the standard model gauge coupling $g^2 = 0.41$. At $\sqrt{\sigma_L} = 0.2\ \sqrt{\sigma_c}$ the potential shows the Landau singularity. At larger values of σ both potentials agree reasonably well.

SUMMARY

What is the effect of the strong interactions in the symmetric phase on the electroweak phase transition? It is conceivable that the non-perturbative effects increase the strength of the transition, for instance via the formation of condensates[4]. It could also be that the main non-perturbative effects can be parametrized by means of a "magnetic mass", which would decrease the strength of the transition[1]. At present, only lattice simulations appear to be able to settle this issue. However, in both cases we are led to the conclusion that the perturbative approach becomes inadequate at Higgs masses above $m_H^c \sim 70$ GeV.

REFERENCES

1. Z. Fodor, these proceedings
2. K. Kajantie, these proceedings
3. C. Wetterich, these proceedings
4. M. Shaposhnikov, these proceedings
5. W. Buchmüller and Z. Fodor, preprint DESY 94-043 (1994), Phys. Lett. B, in press
6. P. Arnold, these proceedings
7. M. Gleiser, these proceedings

ELECTROWEAK BARYOGENESIS: PROBLEMS AND PROSPECTS

Neil Turok

Joseph Henry Laboratories
Princeton University
Princeton, NJ 08544

In this talk I comment on some of the most recent work on electroweak baryogenesis. The simplest 'working' scenarios are those employing extended Higgs sectors to boost the level of CP violation in the standard model. Both 'local' and 'nonlocal' mechanisms seem viable, although accurate methods of calculating the asymmetry are still being developed. The 'nonlocal' τ lepton scenario appears the most tractable in this regard. I then discuss a new scenario for baryogenesis in the minimal standard model, in which CP is spontaneously violated by a Z condensate on bubble walls.

Electroweak baryon number violation and baryogenesis lays fair claim to being the most interesting development in standard model high energy physics in the last decade. Why? For at least three reasons: i) the proponderance of matter over antimatter in the universe is a fundamental mystery, ii) the physics involved is rich, involving topology, anomalies and fermion level crossing at the limits of (and just beyond) current field theory techniques, and iii) it is accessible to experiment within the foreseeable future [1], [2] It is no coincidence that many participants at this workshop felt it was the most interesting they had been to for some time. The field is unique in bringing together cosmologists, field theorists, lattice gauge theorists and phenomenologists, to focus on a specific set of challenging but well-defined problems where we can expect significant progress over the next few years (For recent reviews see [3], [4]).

The baryon asymmetry is a basic ingredient in the standard cosmology: in nucleosynthesis calculations the baryon-to-photon ratio has to be set by hand to $\eta \sim 4 - 6 \times 10^{-11}$ to fit the observed light element abundances. The possibility of instead being able to calculate η from first principles (as first suggested by Sakharov [5]) is attractive, and offers the prospect of extending the standard cosmology back to 10^{-10} seconds after the big bang. The value of η has many other effects - for example in fixing the epoch of matter-radiation decoupling. And if we could calculate the baryon abundance, this might help with another puzzle, namely why the abundance ratio of the baryons to dark matter is apparently of order unity $(.01 - .1)$.

Electroweak baryon number violation involves astonishingly beautiful physics, tying the topology of the classical configuration space of the gauge and Higgs fields to fermion level crossing, so that 'winding up' the gauge fields has the effect of dragging three baryons and three leptons out from the Dirac sea. And all this is in the boring old standard model! It is built in for very deep reasons too - first because the gauge group has nontrivial topology ($SU(2) \approx S^3$, $\pi_3(S^3) = Z$), and second because fermions couple chirally (i.e. only left handed fermions couple to the $SU(2)$ gauge fields). It is hard to resist the feeling that a phenomenon this fundamental must have 'been made use of' by nature, and what better use than to create the observed matter-antimatter asymmetry.

Most importantly, the B violating processes are potentially accessible to experiment. The energy required to pass over the 'sphaleron' barrier is well within the range of proposed accelerators (assuming they are not cancelled!), although the requirement of producing a coherent classical field configuration of $\sim 10^2$ weak bosons is certainly daunting. One possibility would be energetic collisions of heavy nuclei, where one could hope to see significant excitation of long wavelength pion modes which might in turn excite weak bosons.

The major uncertainties in electroweak baryogenesis are:

• What is the B violating rate in the unbroken phase? Existing calculations of Ambjørn et. al. are consistent with the naive expectation $\Gamma = \kappa(\alpha_W T)^4$, with $\kappa = 0.1 - 1.0$, but they are calculations in a *classical* field theory in which the cutoff (lattice spacing) cannot be removed. As Ambjørn stressed in his talk, it is far from clear that the correct effective action describing classical evolution of the long wavelength modes in a high temperature phase is the microscopic action of the fundamental theory.

• Detailed investigations of the effective potential governing the electroweak phase transition have only recently begun, since there was little motivation before electroweak baryogenesis was taken seriously. Lattice investigations so far seem to reveal one loop perturbation theory to be a poor guide to most important aspects of the phase transition.

• The nonequilibrium calculations needed in any scenario are still at a primitive state of development. Discussion of bubble formation and propagation, CP violating dynamics, the Boltzmann and diffusion equations has been very limited.

None of these obstacles seem insurmountable, although all require hard work.

The attractions of making electroweak baryogenesis work in the standard model are legion - for all its warts the single Higgs theory is light years ahead of multi-Higgs versions in terms of simplicity and economy. And the prospect of relating the magnitude and sign of the baryon asymmetry to the CP violation in the kaon system is compelling. Nevertheless, generating the observed asymmetry in the minimal standard model has been widely regarded as a preposterous proposition for some time, for two well known reasons. The CP violation due to the Kobayashi-Maskawa matrix is very small, suppressed by many powers of light quark masses and mixing angles. A naive estimate assuming analyticity in fermion masses indicates that CP violation should enter in the combination

$$d_{CP} = J\Pi_{i \neq j}(m_i^2 - m_j^2)\Lambda^{-12} \tag{1}$$

where the product is over light charged quarks, and $J \sim 10^{-5}$ is a product of KM angles. Taking the scale Λ to be the temperature $T \sim 100$ GeV, one finds $d_{CP} \sim 10^{-19}$ leading to a baryon symmetry well below the required value $\sim 10^{-10}$. Second, the requirement that the anomalous processes be suppressed after the phase transition imposes an upper bound on the mass of the Higgs boson [6] - estimated from the one loop potential and the bubble nucleation rate [7], to be $M_H < 35$ GeV, in conflict with the LEP bound $M_H > 60$ GeV. But recent lattice studies [8] indicate that the electroweak transition is far more strongly first order, relaxing the upper bound on M_H, perhaps up to 100 GeV. However, if Fermilab's recent value for the top quark mass $m_t \sim 174$ GeV is correct, the effective potential in the minimal standard model is unbounded below for large values of the Higgs field, for $M_H < 134$ GeV [9]. This isn't too disturbing, because no-one regards the standard model as an ultimate theory, and this problem is readily cured by introducing some extra field content e.g. supersymmetry. Likewise, a more subtle symmetry breaking mechanism than the standard Higgs mechanism may be at work.

Last year, Farrar and Shaposhnikov [10] made the intriguing suggestion that (1) might be incorrect. They considered the scattering of fermions off a bubble wall, carefully including one loop corrections to the real part of the fermion self energy. In an expansion in powers of momentum keeping only the first power, a modified Dirac equation results. According to this equation, there is a narrow band of incident (very low) momenta in which d quarks are transmitted but $\bar{d}$ quarks are reflected. The antiquarks are then destroyed by B violating processes in the unbroken phase, and a baryon asymmetry results. According to them, an acceptable asymmetry could be produced, with several powers of the scale Λ in (1) being effectively replaced with light quark masses. However (as they acknowledged) there are many questionable elements in their final estimate of the asymmetry, including:

• They ignored the (large) imaginary part of the quark self energies due to strong scattering. The conventional interpretation of these is that they yield the timescale for an excitation being returned to thermal equilibrium, in which quantum mechanical phase coherence is lost. Gavela et. al. [11], and Huet and Sather [12] included the imaginary parts in the calculation and find a huge suppression factor, of order (1). In effect the scale Λ is provided by $g_s^2 T$. One aspect of these latter calculations that is not totally clear to me is how the various global charges and flavour quantum numbers of a quasiparticle are lost as it 'decays' - presumably they somehow disappear into the heat bath, but one would like to see this spelled out. It may be for example that some additional diffusion of global charge is needed for the quasiparticle to really disappear.

• They calculate in one dimension. In three dimensions there is much less phase space at very low momentum, reducing the result by $(m_s/T)^2 \sim 10^{-4}$ at least.

• They assume the bubble wall is very thin, the diffusion length very large, and the wall velocity very slow so that all reflected antiparticles escape far in front of the wall and are destroyed by B violating processes. There are no calculations supporting these assumptions in the standard model. In making them, they effectively ignore suppression of the result by the B violation rate $\Gamma_S \propto \alpha_W^4 \sim 10^{-6}$. While no complete calculation has been performed, I think it's fair to say that if the Gavela et. al. and Huet/Sather calculations are correct, the Farrar-Shaposhnikov mechanism seems unlikely to yield a final value for η greater than 10^{-30}.

At the end of this talk I am going to discuss a quite different mechanism recently suggested by Nasser and myself [13] as to how a reasonable asymmetry might after all be generated in the standard model. But first let me discuss two simpler mechanisms which do seem to give an asymmetry 'in the right ballpark', which operate in extensions of the standard model involving two (or more) Higgs fields. In such theories there is CP violation in the classical field equations of the gauge/Higgs fields, through a C and CP-odd field which is the phase θ of $\varphi_1^\dagger\varphi_2$. As the system passes from the unbroken to the broken symmetry phase, this phase changes in a definite direction depending on certain quadratic and quartic terms in the Higgs potential. The coupling of this phase to the fermions via Yukawa terms induces a baryon asymmetry through both 'local' and 'nonlocal' mechanisms.

'Local' baryogenesis occurs via a one loop diagram in which one integrates out the fermions, introducing a parity-violating term in the gauge/Higgs equations of motion. The anomaly in the baryon number current is odd under C and P, so one needs to violate both in order to produce a net baryon asymmetry. The Higgs potential violates C, and the parity violating term violates P, so that in effect the Higgs potential drives the anomaly positive (or negative), producing a baryon asymmetry. This mechanism was originally suggested in [14] and developed in [15] and [3] - for a recent discussion including its application in the minimal standard model (where it doesn't work) see [16]. One persistent criticism (see e.g. [16]) of the 'local' mechanism is that since it only works on the wall, and the *equilibrium* rate for B violating processes turn off once ϕ gets larger than $\sim gT$, there should be some additional suppression. According to this argument, as one lowers the temperature at which the transition occurs, the baryon asymmetry produced should become exponentially suppressed. Simulations in 1+1 dimensions were conducted to investigate this, and [17] they showed precisely the opposite behaviour (i.e. the asymmetry increases as the starting temperature is lowered. While entirely unanticipated, the effect has a very simple explanation (of course!) which should also apply in 3+1 dimensions. Which is that the 'vacuum energy' stored in one gauge field correlation volume of the unbroken symmetry phase is more than enough to drive the gauge fields over the 'sphaleron' barrier for the parameter range of interest in the standard model (for a detailed discussion see [3]). The picture of 'nonlocal' baryogenesis that emerged from the simulations was not one of 'biased thermal fluctuations', but rather that of classical field dynamics with a driving force, *counteracted by* thermal fluctuations. It would be very interesting to perform classical simulations in the 3+1 case, although there are clearly problems simulating the Higgs fields as Ambjørn mentioned. I would suggest starting with a transition simulated at a very low starting temperature (as explained in [17]), so that the finite temperature does not cause huge statistical fluctuations. Our experience in 1+1 dimensions taught us the limitations of guesswork in regard to nonlinear field dynamics, and I suspect that as in many other fields of physics there is much to be learned from well thought out simulations.

'Nonlocal' baryogenesis occurs instead as a two stage process. First the CP violating phase $\theta(z)$ leads to a chiral flux of fermions (a deficit of left handed fermions, excess of right handed antifermions) being reflected off a bubble wall into the unbroken phase. There, the anomalous processes act to restore the system to equilibrium, creating left handed fermions and destroying right handed antifermions. (Only these couple to the $SU(2)$ gauge fields, which have the topology required for permanent winding number changes). As left handed fermions are created, so is a net baryon asymmetry, according to the equation

$$\dot{n_B} = -\frac{\Gamma_S}{T}(3\mu_{q_L} + \mu_{l_L}) \propto (n_{\overline{L}} - n_L) \tag{2}$$

where n_B is the baryon density, μ denotes the chemical potential and n_L is the density of left handed fermions, $n_{\overline{L}}$ that of right handed antifermions (the CP conjugate particles).

The 'nonlocal' scheme was originally suggested for top quarks in [18]. It's advantage over the 'local' mechanism is that the CP violation is transported away from the bubble wall deep into the unbroken phase, which gives unsuppressed B violation a longer time in which to operate. This picture is somewhat questionable in the case of top quarks, for which the diffusion length is not much greater than a bubble wall thickness, but clearly does apply in the case of leptons and particularly right handed leptons, as discussed recently in [19]. Prokopec discusses the τ lepton scenario in detail in his contribution to these Proceedings.

The Lagrangian of interest is a Dirac fermion coupled to a z dependent complex mass term.

$$\begin{aligned} \mathcal{L} &= \frac{1}{2}\overline{\psi} i\gamma^\mu \partial_\mu \psi - m(z)\overline{\psi}_L \psi_R + \text{h.c} \\ m(z) &= y|\phi(z)|e^{i\theta(z)} \end{aligned} \tag{3}$$

We are mainly interested in the flux of particles back into the unbroken phase where B violating processes can operate.

Let me briefly discuss the simplest aspect of the problem, namely how symmetry constrains fermion-wall scattering. Relations between the reflection probabilities r of left and right handed particles and antiparticles are given by applying the CPT operations:

$$\begin{aligned} \mathbf{C} &: r(\mathbf{L} \to \mathbf{R}, \theta) = r(\overline{\mathbf{R}} \to \overline{\mathbf{L}}, -\theta) \\ \mathbf{CP} &: r(\mathbf{L} \to \mathbf{R}, \theta) = r(\overline{\mathbf{L}} \to \overline{\mathbf{R}}, -\theta) \\ \mathbf{CPT} &: r(\mathbf{L} \to \mathbf{R}, \theta) = r(\overline{\mathbf{R}} \to \overline{\mathbf{L}}, \theta) \end{aligned} \tag{4}$$

(note $\overline{L}$ here denotes a *right* handed antiparticle). Conservation of angular momentum dictates that a left handed particle reflects into a right handed particle and vice versa. C changes particle to antiparticle, and reverses θ. The action of P is simplest to see by considering reflection from a spherical bubble, in which the symmetry operation reverses the momentum and coordinate of the incoming particle. Finally, T exchanges initial and final states, and involves complex conjugation so $\theta \to -\theta$.

Thus CPT symmetry dictates that equal numbers of $\overline{L}$'s and R's are reflected, and equal numbers of $\overline{R}$'s and L's. So the reflected flux violates CP but not C. The baryon number violating processes couple in a C violating, but CP conserving manner, and luckily not in an equal and opposite manner for L's and R's because in that case their effect would cancel. Instead, they involve only the L's and the $\overline{L}$'s. If $\theta(z)$ is not constant, so it cannot be removed by a global rotation, there is a difference in the reflected flux of L's and $\overline{L}$'s and B violating processes produce a net baryon asymmetry according to (2).

To calculate the asymmetry in reflected fluxes, we need to solve the Dirac equation following from (3). In a chiral basis, with $\psi = e^{-iEt}(\chi_1, \chi_2, \chi_3, \chi_4)$ the positive energy solutions obey

$$\begin{aligned} -i\partial_z \chi_1 &= E\chi_1 + m^* \chi_3 \\ -i\partial_z \chi_3 &= -E\chi_3 - m\chi_1 \end{aligned} \tag{5}$$

for the components with $S_z = +\frac{1}{2}$, and the same equations with $m \to m^*$, for (χ_4, χ_2) with $S_z = -\frac{1}{2}$. (As written the equation only applies if the momentum $p_\perp$ perpendicular to z vanishes. In the more general case one boosts to a frame in which $p_\perp$ vanishes, so that in (5), $E \to \sqrt{E^2 - p_\perp^2}$, which is just $|p_z|$ at infinity in the unbroken phase.)

It is straightforward to iterate (5) in powers of $m(z)$, and compute the reflection probabilities. One finds that [19]

$$r_{R\to L} - rL \to R = \approx 2\int_{-\infty}^{\infty} dz \text{Im}[m(z)m^*(\infty)]\frac{1}{|p_z|} \tag{6}$$

for unbroken phase momenta in the range $|m_l| < |p_z| < m_H$ (the latter condition arising from the thickness $L \sim m_H^{-1}$ of the wall). This expression is highly nonlocal as a result of quantum mechanical interference. The latter is substantial only at low momentum, $|p_z|L < 1$. At higher momentum the WKB approximation becomes good, and phase interference effects become exponentially suppressed. This has led to statements in the literature that quantum phase interference effects are necessary for a CP violating chiral flux to be produced. This is not true - even in the WKB regime gradients in the phase $\partial_z\theta$ produce a force which acts chirally, and when this is combined with the fact that left and right handed particles diffuse differently, a chiral flux results.

It is suggestive to rewrite equation (5) by rotating the phase out of the mass term $m(z) \equiv |m|e^{i\theta(z)}$. Setting $(\chi_1, \chi_3) = e^{-i\theta\sigma_3/2}(\hat{\chi}_1, \hat{\chi}_3)$, one finds (5) is replaced by

$$
\begin{aligned}
(-i\partial_z + \frac{1}{2}\partial_z\theta)\chi_1 &= E\chi_1 + |m|\hat{\chi_3} \\
(-i\partial_z - \frac{1}{2}\partial_z\theta)\chi_3 &= -E\chi_3 - |m|\hat{\chi_1}
\end{aligned}
\tag{7}
$$

in which $\partial_z\theta$ plays the role of a *gauge field*, which couples chirally, with $g_R = \frac{1}{2}$ and $g_L = -\frac{1}{2}$. The gauge field is purely longitudinal, and it can clearly be gauged away as long as $|m| = 0$, but not otherwise. This rewriting is useful in order to see that the gauge field induces a classical chiral force, calculated from the dispersion relation, which I shall derive below. And as mentioned above, even in the WKB limit the chiral force induces a net chiral flux into the unbroken phase (see the discussion in Joyce's contribution to these proceedings).

The analogy between $\partial_z\theta$ and a gauge field suggests how the standard model might itself provide CP violation on the bubble wall, independently of the KM matrix. Could it be that the electroweak gauge fields play the role of providing the CP violation on bubble walls? It is straightforward to check that the longitudinal Z boson plays exactly the same role as $\partial_z\theta$ in the Dirac equation (7). Of course if we ignore the KM matrix, CP is a good symmetry, there must be both signs of CP violation possible. It will be the role of CP violating phase in the KM matrix to distinguish the two.

I discussed above the strongest argument against standard model baryogenesis, that the CP violation is much too small. How might the tiny CP violation (1) be amplified into a reasonable asymmetry? The point is that (1) is a microscopic parameter - it governs the CP violation *per particle*. However if one has *spontaneous* CP violation in the theory, then there should appear macroscopic regions in which both signs of CP violation appear. When these regions compete, it is not the free energy per particle which matters, but differences between macroscopic parameters governing the behaviour of very large numbers of particles. This is the central idea in a recent suggestion by Nasser and myself of how baryogenesis might work in the standard model.

Calculations we have done show that dynamical CP violation does indeed occur with certain simplifying assumptions. Whether more detailed calculations enhance or destroy the effect remains to be seen. According to our calculations, top quark reflection from bubble walls leads to a term in the equation of motion for the Z boson with the correct sign to *destabilise* it, with a longitudinal Z boson condensate forming on bubble walls. This has the same effect as the phase $\theta(z)$ which appeared in two-Higgs baryogenesis scenarios, and produces a baryon asymmetry at least 'in the right ballpark'.

But the Z field can point either 'out' or 'in', so we produce both baryons and antibaryons. Patches of each phase initially cover each bubble, separated by one dimensional phase boundaries, which are in fact Z-magnetic flux tubes. As the bubble expands, small patches shrink away under the tension of the phase boundary, while larger patches grow with the bubble. The long time available before bubbles collide allows each bubble to be covered by *macroscopic* regions of the 'in' and the 'out' phases. Even if each bubble becomes completely ordered, macroscopic regions of both phases are placed in competition when large bubbles collide. Our suggestion is to use the competition between these macroscopic regions to amplify the tiny CP violation from the KM matrix.

As a bubble wall moves through the plasma, its terminal velocity is given by $v_t = \Delta P/\Gamma$ where ΔP is the pressure difference between the false and true vacua and where the frictional drag/unit area is given by $F_{\rm drag} = \Gamma v$.

As a result of the CP violation in the KM matrix, we expect the 'in' and 'out' phases to feel slightly different Γ's, $\delta\Gamma/\Gamma \sim d_{CP}$. It follows that the two phases move through the plasma with slightly different velocities $\delta v \sim d_{CP}$. Now consider the neighbourhood of a phase boundary. When the bubbles are large enough, we may take them to be planar as far as the local dynamics of the phase boundary go. As the bubbles grow, the 'out' phase begins to bulge over the 'in' phase, the height of the bulge being given by $\sim d_{CP}t$.

Even though d_{CP} is tiny, we have a very long time available. Calculations based on the perturbative potential [20] show that the phase transition completes after $\sim 10^{-4}$–10^{-5} expansion times. If the phase transition is more strongly first order, this could increase to a value closer to $t \sim 10^{16}T^{-1}$. The microphysical scale which enters the equations determining the shape of the propagating bubble wall is $l = \sigma/\Delta P$ with σ the surface tension. Ignoring all dimensional couplings, $l \sim T^{-1}$, so the bulge becomes significant when $d_{CP}t \sim T^{-1}$. This condition is fulfilled if we use (1) and take $\Lambda \sim 50$ GeV, which does not seem unreasonable.

The equation governing the motion of the bubble wall is (ignoring relativistic effects)

$$\Gamma\dot{\vec{x}}_N = \Delta P\vec{n} + \sigma\vec{x}'' \tag{8}$$

where $\vec{x}(t,s)$ is the worldsheet of the wall, s a spacelike parameter along it and we suppress one spatial coordinate. Dot denotes time derivative, prime s derivative. Further, we have picked the gauge $|\vec{x}'| = 1$, so s is the length along the wall. Finally, $\vec{n}$ denotes a unit vector normal to $\vec{x}'$, and subscript N denotes the component along the normal vector. One can easily find analytic steady state solutions to (8) representing sections of wall with some imposed boundary conditions. The simplest solution is a straight line propagating at $v_t = \Delta P/\Gamma$. But if the ends of a section of wall are constrained to move through the medium at a velocity different than this, then the solution is curved, being convex if $v < v_t$, concave if $v > v_t$. By matching such solutions at boundaries representing the phase boundaries between 'out' and 'in' regions we have found the following behaviour.

When the bulge becomes significant, i.e. when $\delta vt \sim l$, it gives rise to a net tangential force on the phase boundary, causing it to sweep across the bubble at a velocity of the order of the bubble wall velocity v_t. The faster moving phase then, in a time of the order of an expansion time, completely takes over the slower moving phase. As the electroweak phase transition completes, bubble surfaces of one CP violating variety predominate. Notice that the only place the small CP-violating parameter comes in is in ensuring that the 'out' and 'in' phases expand at slightly different rates.

Now let me repeat the argument of [13] that a longitudinal Z condensate can form on propagating bubble walls. We now consider the classical field equation for Z_μ, in background of a bubble wall propagating through the plasma. In the wall rest frame we take the Higgs field to have a fixed profile $\phi(z)$, its width L_w being of order the inverse of the Higgs mass m_H in the broken phase. The wall is treated as planar, which is a good approximation once a bubble gets large. We consider a Z field $Z^\mu = (0,0,0,Z(z))$.

We drop spatial derivative terms, which are suppressed by powers of m_H/m_Z, and m_H/m_{top} (which are small at temperatures near the phase transition) relative to the Z mass term and the top quark current which dominate. The Z equation of motion then becomes

$$\ddot{Z} = J_Z(Z) + J_F(Z) \tag{9}$$

where $J_Z(Z) = -4g_A^2\phi^2 Z$ is the current carried by the Z condensate and $J_F(Z) = g_A\overline{\psi}\gamma^3\gamma^5\psi + g_V\overline{\psi}\gamma^3\psi$ that carried by the fermions. Here $g_A = \frac{1}{4}(g_1^2+g_2^2)^{\frac{1}{2}}$, with g_1, g_2 the $U(1)_Y$ and $SU(2)_W$ gauge couplings.

Now the stability of the solution $Z = 0$. The condition for an instability is simply that $m_{eff}^2 = 4g_A^2\phi^2 - (\partial J_F/\partial Z)|_{Z=0} < 0$.

The fermion current J_F is calculated by solving the time independent Dirac equation in the presence of ϕ and Z condensates. Although the WKB approximation isn't particularly good mode by mode in the regime of interest (we are going to look at behaviour near a turning point), it *is* surprisingly accurate after integrating over the energies of the incident particles. We have checked it against results obtained with an exact linear response function. The WKB approximation has the additional advantage that J_F is a *local* function of Z, and using it one can easily compute the full 'effective potential' for Z.

The relevant Dirac equation is

$$(i\gamma^\mu(\partial_\mu + iZ_\mu(g_V + g_A\gamma^5)) - m(z))\psi = 0 \tag{10}$$

where $m(z) = y\phi(z)$ with y the Yukawa coupling. We ignore for the moment the one loop thermal contributions to the fermion self energies. ¿From (10) the following dispersion relation is found

$$\omega_\pm = \sqrt{p_\perp^2 + (\sqrt{(p_z - g_V Z)^2 + m^2(z)} \mp g_A Z(z))^2} \tag{11}$$

for the two eigenstates, with spin $S^z = \pm\frac{1}{2}$ in a frame where $p_\perp = 0$ (this dispersion relation may be derived from (7)). Note that when $m = 0$ one can 'gauge away' Z completely by shifting the momentum p_z.

In the rest frame of the wall, both the energy ω and the transverse momentum $p_\perp$ are conserved, and the local value of the momentum p_z varies according to (11). It is related to $p_{z,-\infty}$, the value of the momentum at $z = -\infty$, by the formula $E = \sqrt{p_\perp^2 + p_{z,-\infty}^2}$. Regions for which $p_{z,-\infty} < m(z) \mp g_A Z$ are classically disallowed: the $S^z = \pm\frac{1}{2}$ excitations see a 'barrier' $m(z) \mp g_A Z(z)$ respectively.

The fact that left and right handed fermions carry different charges now comes into play - particles of given S^z carry the same current when travelling in either direction! One can anticipate the destabilising effect - for positive Z there is a region where only $S^z > 0$ particles penetrate from the left, creating a positive chiral current. This acts through the equation of motion (9) to further destabilise the Z. Including currents from antiparticles, which are given by the substitutions $J \to -J$; $g_{A,V} \to -g_{A,V}$, multiplies the axial current by two, while cancelling the vector current. The parity violation in the Z-fermion coupling is crucial - the charged vector bosons $W^\pm$ do not destabilise the Z because the $Z/W^\pm$ equations of motion are parity invariant so that left and right moving $W^\pm$ modes carry opposite currents.

A more detailed analysis reveals a competing stabilising effect which exactly cancels the leading destabilising effect at zero wall velocity. The current carried per mode is given by the classical formula $J_z = -(\partial\omega/\partial Z)$. Integrating over momenta this is (for $S^z > 0$)

$$
\begin{aligned}
J_F &= -\int \frac{d^3p}{(2\pi)^3} f(p) \frac{\partial\omega_+}{\partial Z} \\
&= g_A \int \frac{d^3p_{-\infty}}{(2\pi)^3} \left(\frac{dp_z}{dp_{z,-\infty}}\right) f(p_{-\infty}) \frac{p_{z,-\infty}}{\sqrt{p_\perp^2 + p_{z,-\infty}^2}}
\end{aligned}
\tag{12}
$$

(the phase space density f is constant along particle trajectories). The only Z dependence is now in the Jacobian, which is just the ratio of the group velocity at infinity to that locally,

$$
v_z = \frac{\partial\omega}{\partial p_z} = \frac{\sqrt{(p_{z,-\infty} + g_A Z)^2 - m^2}}{p_{z,-\infty} + g_A Z} \left(\frac{p_{z,-\infty}}{w}\right)
\tag{13}
$$

The Jacobian represents the enhancement of the local particle density due to a 'slowing down' effect. For positive Z, $S^z > 0$ particles see a 'well' and speed up, decreasing the chiral current, and $S^z < 0$ particles see a 'barrier', slowing down and adding a negative chiral current. Thus particles passing 'over' the barrier (in either direction) act to stabilise the Z condensate, whereas particles 'bouncing' off the barrier tend to destabilise it.

Now we compute the net chiral current in the WKB approximation. We assume the barrier is monotonic ($|(dZ/dm)| < 1$), and sum over particles and antiparticles of both spins, incident from both sides of the wall with thermal distributions at $z = \pm\infty$. As already mentioned, the leading effect, a square root divergence, cancels at zero v_w - there is a factor of two in the 'bouncing' contribution, cancelled by a minus one from particles going over the barrier in each direction. But at finite v_w more particles are encountered from the left, enhancing the number of 'bouncing' particles. The dominant term occurs as $m(z)$ approaches m_∞, and to lowest order in v_w is given by

$$
J_F = \frac{2g_A N_C}{(2\pi)^2} \frac{m_\infty^2 v_w}{e^{\beta m_\infty} + 1} \sqrt{(m_\infty + g_A Z)^2 - m(z)^2} - (Z \to -Z)
\tag{14}
$$

where we sum over colours N_C. Note that

$$
\frac{\partial J_F}{\partial Z}|_{Z=0} \propto g_A^2 m_\infty^3 (m_\infty^2 - m(z)^2)^{-\frac{1}{2}} \propto g_A^2 m_\infty^2 e^{z/2L}
\tag{15}
$$

diverging as $z \to \infty$, so that (within the WKB, free particle approximation we have made) the fermionic contribution to the effective Z squared mass diverges, due to the 'bouncing' effect, and an instability always develops sufficiently far behind the wall. In [13] we report on an exact linear response calculation that confirms the WKB picture, and shows that for the measured value of the top mass, and for a reasonable wall thickness $\sim 20T^{-1}$ an instability does indeed occur.

One interesting effect we have ignored is the decay process $t \to W + b$, which occurs once the t gets a large enough mass on the wall. The tree level rate is $\frac{1}{16}\alpha_2 m_t^3/m_W^2(1 - m_W^2/m_t^2)^2(1 + 2m_W^2/m_t^2) \sim m_t/110$, a slow process compared to the time the top quarks spend on the wall, in the regime of interest. Nevertheless the decay process may contribute in an interesting way to the chiral current [21].

As I emphasised above, the most questionable aspect of the calculation is in treating the fermions bouncing off the wall as free particles, ignoring their interactions with the plasma. In doing this we are ignoring (at least) two opposing effects - the approach to local thermal equilibrium engendered by interactions diminishes the instability, but the 'pile-up' of top quarks [7] in front of the wall enhances it. More detailed calculations are needed to reveal whether the net effect is to strengthen or suppress the destabilising term. Note that a big difference from the work of Farrar and Shaposhnikov, which I mentioned has been criticised for failing to include dissipative strong interactions, is that the spontaneous CP violation here is essentially a *classical* effect, and does not rely on the quantum mechanical coherence. We expect it to be less susceptible to destruction by strong scattering. Also, our effect comes mainly from quarks with *large* perpendicular momenta $p_\perp \sim 2T$, and for these one would expect soft gluon interactions to be less important.

Assuming that strong scattering effects do not wash out the instability, a host of interesting questions remain. How does the CP violation from the KM matrix enter, and distinguish the 'in' and 'out' phases of the Z condensate? In a one dimensional treatment (which likely overestimates the effect), one can see from the calculations of [12] for example that a Z independent term $\sim d_{CP}$ enters in the equation of motion (9), and we expect to find similar CP violating terms at other even powers of Z. We are investigating the analogous three dimensional calculation. Finally, before the baryon asymmetry may be accurately calculated, we need to find the stable solutions for the Z condensate.

Let me close with two critical (and self-critical!) remarks. First, the attraction of claiming to predict (actually, 'postdict') η has tended to distract from attention to detail and the uncertainties in the arguments. The literature on two Higgs theories abounds with comparative statements on the merits and demerits of various mechanisms, but calculations believable to better than a couple of orders of magnitude are few and far between.

At this stage I think its important both to understand which qualitative physical effects could have played a role in determining the final asymmetry, without necessarily being able to calculate their effect with high precision, and at the same time to develop accurate calculational tools. For instance, the 'nonlocal' τ lepton transport scenario offers reasonable prospects for a relatively clean and complete calculation. But we are a long way from a believable top quark calculation which properly includes the effects of strong scattering.

In the long run, we should not delude ourselves - very complex physics is involved, the tools for calculating nonequilibrium processes in field theory are limited (though very well worth developing). Even if we do manage to bring calculations to a level of precision of a factor of a few, a single number is a fairly limited test of any theory. The connection to experiment seems tenuous at present, but I do think that is a crucial element of the subject and ultimately, its future.

ACKNOWLEDGEMENTS

I thank M. Joyce, T. Prokopec, and S. Nasser for collaboration, and the organisers for making this such an exciting and stimulating meeting. This work was partially supported by NSF contract PHY90-21984, and the David and Lucile Packard Foundation.

REFERENCES

[1] G. 't Hooft, Phys. Rev. Lett. **37**, 8 (1976); Phys. Rev. **D14**, 3432 (1976); A. Linde, Phys. Lett. **70B**, 306 (1977); S. Dimopoulos and L. Susskind, Phys. Rev. **D18**, 4500 (1978); F. Klinkhamer and N. Manton, Phys. Rev. **D30**, 2212 (1984).

[2] V. Kuzmin, V. Rubakov and M. Shaposhnikov, Phys. Lett. **155B** 36 (1985); P. Arnold and L. Mclerran, Phys. Rev. **D37**, 1020 (1987); J. Ambjørn, T. Askgaard, H. Porter and M. Shaposhnikov, Phys. Lett. **244B**, 497 (1990); Nucl. Phys. **353**, 346 (1991).

[3] N. Turok, in *Perspectives in Higgs Physics*, ed. G. Kane, pub. World Scientific, p. 300 (1992).

[4] A. Cohen, D. Kaplan and A. Nelson, Ann. Rev. Nucl. Part. Sci. **43**, 27 (1993).

[5] A. Sakharov, Pisma ZhETF **13**, 335 (1970).

[6] M. E. Shaposhnikov, JETP Lett, **44**, 465 (1986); Nucl. Phys. **B287**, 757 (1987); Nucl. Phys. **B299**, 797 (1988).

[7] M. Dine, R. Leigh, P. Huet, A. Linde, and D. Linde, Phys. Rev. **D46**, 550 (1992); B-H. Liu, L. Mclerran and N. Turok, Phys. Rev **D46**, 2668 (1992).

[8] K. Farakos, K. Kajantie, K. Rummukainen and M. Shaposhnikov, Nucl. Phys. **B407**, 356 (1993); preprint CERN-TH.7244/94.

[9] M. Sher, Willams and Mary preprint (1994).

[10] G. Farrar and M. Shaposhnikov, Phys. Rev. Lett. **70**, 2833 (1993); **71**, 210(E) (1993); preprint CERN-TH.6732/93.

[11] M.B. Gavela, P. Hernandez, J. Orloff and O. Pene, preprint CERN-TH.7081/93.

[12] P. Huet and E. Sather, preprint SLAC-PUB-6179 (1994).

[13] S. Nasser and N. Turok, Princeton preprint PUP-TH-1456 (1994).

[14] N. Turok and J. Zadrozny, Phys. Rev. Lett. **65**, 2331 (1990).

[15] L. Mclerran, M. Shaposhnikov, N. Turok and M. Voloshin, Phys. Lett. **256B**, 451 (1991).

[16] M. Dine and S. Thomas, Santa Cruz preprint (1994).

[17] D. Grigoriev, N. Turok and M. Shaposhnikov, Phys. Lett **B** (1992).

[18] A. Cohen, D. Kaplan and A. Nelson, Nucl. Phys. **B373**, 453 (1992).

[19] M. Joyce, T. Prokopec and N. Turok, Princeton preprint PUP-TH-1437 (1994).

[20] N. Turok, Phys. Rev. Lett. **68**, 1803 (1992).

[21] I. Kogan, private communication.

A NOTE ON SPONTANEOUS BARYOGENESIS

Michael Joyce

Princeton University
Joseph Henry Laboratories
Physics Department, PO Box 708
Princeton, NJ 08544, USA

Abstract: The original calculation of 'spontaneous' baryogenesis overlooked the role played by transport of particles onto the bubble walls. For typical 'adiabatic' wall thicknesses and velocities one can model the problem in a fluid approximation and the mechanism is best understood as another limit of 'charge transport' baryogenesis, and can produce a similar non-local effect in front of the wall.

In this brief talk I will make a few comments on 'spontaneous baryogenesis', a mechanism for the production of baryons at the electroweak phase transition in the 'adiabatic' limit of thick slowly moving bubble walls proposed by Cohen, Kaplan and Nelson (CKN) in [1]. My comments summarize results of a collaboration with Tomislav Prokopec and Neil Turok. The central point is that the constraints imposed on particle numbers in the thermodynamic equilibrium calculations which this mechanism involves are inappropriate for the relevant wall thicknesses. A correct calculation of the baryon production involves taking into account the effect on the surrounding plasma of what is effectively a background axial gauge field on the wall. This calculation should really be thought of as another limit of 'non-local' baryogenesis, advocated by CKN in the case of very thin bubble walls and discussed by my collaborators in other talks at this conference [2], rather than as a distinct mechanism.

Consider the familiar two Higgs doublet model in which the CP violating phase θ changes on the bubble wall during the phase transition. If, for simplicity, we couple only one of the two doublets to the fermions through Yukawa terms, the mass terms which result when the Higgs doublet acquires a vev contain the (gauge invariant) phase θ and may be written (in the appropriate gauge)

$$Yve^{i\theta}\bar{q}_L q_R + Y'v e^{-i\theta}\bar{q}'_L q'_R + h.c. \tag{1}$$

where q stands for the charge $\frac{2}{3}$ quarks, q' for the charge $-\frac{1}{3}$ quarks and the charged leptons and Y and v for the Yukawa coupling and vev.

The observation made by CKN is that if one does a hypercharge rotation (chosen because it is anomaly free) on the fermions to remove the phase θ and make the mass terms real, the kinetic terms generate

$$2\partial_\mu \theta \Sigma_i y_i \bar{\psi}_i \gamma^\mu \psi_i \tag{2}$$

where the sum is over all the fermions which have hypercharge y_i. If one now considers the case in which the spatial gradients $\partial_i\theta$ are negligible, (2) looks like a chemical potential term for fermionic hypercharge. Because it is not strictly a chemical potential - it does not arise from a constraint- CKN refer to it as a 'charge potential'. However if one does thermodynamic calculations with this Hamiltonian the effect of the term is just like that of a chemical potential e.g. the number density n_i of a massless fermion i in the presence of a chemical potential μ_i is

$$n_i = \frac{T^2}{6}(\mu_i + 2y_i\dot{\theta}). \tag{3}$$

So how does this bias electroweak anomalous processes? As the bubble wall passes through a region this charge potential is turned on and the system will approach the local thermal equilibrium attained in this background. CKN calculate this local equilibrium in [1] by imposing constraints on the quantum numbers in the system which are conserved by the perturbative processes which are slow relative to the timescale for the wall to pass by. They do this by introducing chemical potentials μ_A to constrain each of these charges to zero (their global value). Amongst these charges is the baryon number B (because the sphaleron processes are too slow to reach equilibrium as the wall passes) and a corresponding chemical potential μ_B. For example if we take the conserved charges, for simplicity, to be Y, the total hypercharge, B and $B-L$, where L is the lepton number, we find

$$\begin{aligned} Y &\propto 20\dot{\theta} + (10+n)\mu_Y + 8\mu_{B-L} + 2\mu_B = 0 \\ B-L &\propto 16\dot{\theta} + 8\mu_Y + 13\mu_{B-L} + 4\mu_B = 0 \\ B &\propto 2\dot{\theta} + \mu_Y + 2\mu_{B-L} + 2\mu_B = 0. \end{aligned} \tag{4}$$

using the relation $\mu_i = \Sigma_A q_i^A \mu_A$ and $n_i \propto k_i \mu_i$ for the particles which we assume massless. k_i is a counting factor which is one for fermions and two for bosons, q_i^A is the Q_A charge of species i, and n is the number of Higgs doublets in the theory. Solving, we find $\mu_B \propto \dot{\theta} n$.

To find the resulting B violation we use the fact that the rate of change of a 'conserved' charge X due to a 'slow' process is

$$\dot{X} = -\Gamma \frac{\delta F}{T} \delta X \tag{5}$$

where δF and δX are the changes in the free energy and in X per process respectively and Γ is its equilibrium rate. Applying this to the electroweak sphaleron processes we have

$$\dot{B} = -\Gamma_{sph} N_F^2 \frac{\mu_B}{T} \tag{6}$$

where Γ_{sph} is the equilibrium electroweak sphaleron rate and N_F the number of families. We conclude that B violation on the wall is directly driven by $\dot{\theta}$ and the final generated asymmetry is simply calculable from (6).

There is a simple explanation of the proportionality of the answer to the number of Higgs doublets. In the local thermal equilibrium in which each particle species S_i

has chemical potential μ_i the rate at which any particle density n_i changes due to some out of equilibrium process $\nu_j S_j \rightarrow 0$ (where, for example, $A \rightarrow 2B + C$ has $\nu_A = 1, \nu_B = -2, \nu_C = -1$) is, by (5) above,

$$\dot{n}_i = -\frac{\Gamma}{T}(\Sigma_j \nu_j \mu_j)\nu_i. \tag{7}$$

It follows from this that the particle densities are only affected directly when $\dot{\theta}$ is turned on by processes which have Higgs particles in the ingoing or outgoing state. Any other process obeys $\Sigma_i \mu_i \nu_i = 0$ simply by conservation of fermionic hypercharge, and therefore by (7) is not out of equilibrium when $\dot{\theta}$ is turned on. In CKN's calculation it is the top/Higgs coupling which is taken to be 'fast'. Forcing this process to be in equilibrium alters the abundances in such a way as to produce a non-zero rate for the out of equilibrium sphaleron processes.

It is now instructive to compare this with the case of a potential $\phi_Y(x)$ for *total* hypercharge which is turned on in some region of space. When such a potential is turned on no process is out of equilibrium and hence no density directly affected, simply because total hypercharge is conserved in all processes. However this does not mean that the densities of particles do not change if the potential is turned on in some small region. In fact the potential will be screened by drawing in charge from the plasma. The local equilibrium, which is a solution of the Boltzmann equation, is given by the particle phase space distributions f_i

$$f_i(p,x) = \frac{1}{e^{\beta E_i(x)} \pm 1} \tag{8}$$

where $E_i = |\vec{p}| + y_i \phi_Y(x)$ and, therefore, $n_i \propto y_i \phi_Y(x)$.

In the case of the charge potential $\dot{\theta}$ there is also precisely such an unconstrained equilibrium solution, except that E_i is now the energy level of species S_i in the presence of $\dot{\theta}$. If this equilibrium is attained neither the Higgs processes nor the sphaleron processes are out of equilibrium. In the same way as in the presence of a total hypercharge potential there can be transport processes which bring the system towards the static equilibrium (8).

So the central question is whether there can be significant transport of charge into the small region (the bubble wall during the first order phase transition) over the relevant timescale. A simple estimate is as follows. The distance a particle with diffusion constant D typically diffuses in a given direction in time t is $\sqrt{2Dt}$. In order that the transport processes be negligible the distance a particle diffuses in the time the wall of length L moving with velocity v_w takes to pass should be much less than the wall thickness i.e.

$$L >> \frac{2D}{v_w}. \tag{9}$$

For quarks $D \sim \frac{10}{T}$ and in order to allow the Higgs/top processes, for which $\Gamma \sim \frac{T}{50}$, time to equilibrate we need a small v_w for a typical 'thick' wall with $L \sim \frac{20-40}{T}$. Therefore for typical 'adiabatic' wall thicknesses and velocities the constrained equilibrium calculated by CKN is not attained.

A more quantitative analysis requires a full treatment which we will present elsewhere [3], but bears out this rough qualitative argument. The problem becomes that of the determination of the deviations caused by the motion of the wall from the equilibrium (8) when $\frac{v_w L}{D} < 1$. Given that the wall is thicker than the mean free path of the quarks it is valid to treat the problem in a fluid approximation. The effect of the charge potential $\dot{\theta}$ - or more generally of $\partial_\mu \theta$ - is to both alter the rates of processes coupling the fluids locally *and* to induce a force pulling or pushing the fluids around on the wall.

To conclude we return to the lagrangian after the hypercharge rotation which can be written

$$\bar{\psi}(i\partial\!\!\!/ + (y_R + y_L)\partial\!\!\!/\theta + (y_R - y_L)\gamma^5 \partial\!\!\!/\theta)\psi + m\bar{\psi}\psi. \tag{10}$$

The vector part of the induced term can be rotated away by making use of the surviving electromagnetic vector $U(1)$ [1]. One can see explicitly that the effect of the changing phase is to turn on an axial $U(1)$ gauge field, which despite the fact that it is pure gauge cannot be removed because the axial symmetry is explicitly broken by the mass term.

The physical effects of this field can be understood in various different limits. In fact the calculation of the solutions for a Dirac particle in this background is what is done to find the reflection coefficients for the different chirality quarks or leptons in the 'charge transport' mechanism. In the present case where the wall is thick and the particles cannot be treated as free we need to understand the dynamics of the particles while they are on the wall. Working in the WKB limit (which should be good for most of the particles on the wall if we take $\partial_z \theta \sim L^{-1} << T$) one can calculate an effective chiral 'classical' force, which one then uses in the fluid equations. This can lead to a chiral disturbance *in front of* the bubble wall, just as in the 'charge transport' mechanism. In contrast to that case the chiral asymmetry in front of the wall is not driven by a reflected flux of particles with $p_z \sim L^{-1}$ but by the effect on the fluid of a build-up of particle densities on the wall due to a classical force which acts on particles of *all momenta* as they cross the wall. Despite the fact that the reflection is very suppressed for thick walls the same $\partial_\mu \theta$ field can in this case produce a chiral asymmetry and 'non-local' baryogenesis in front of the wall through its effect on a different part of the phase space. We will explore this limit of 'classical non-local' baryogenesis in a forthcoming publication.

Acknowledgements

In addition to my collaborators T. Prokopec and N. Turok, I am grateful to S. Nasser and M. Shaposhnikov for useful discussions. This work was done with the support of a Charlotte Elizabeth Procter Fellowship at Princeton University.

References

[1] A. Cohen, D. Kaplan and A. Nelson, Phys. Lett. **B263**, 86 (1991).

[2] See the talks by N. Turok and T. Prokopec in this volume.

[3] A sketch of the fluid calculation is given in M. Joyce, T. Prokopec and N. Turok, Princeton Preprint PUP-TH-1436(1993) and will be presented in detail in a forthcoming publication.

[1] The corresponding ambiguity about the original rotation was ignored in our discussion above.

TAU LEPTONS AND TOP QUARKS IN A 'NON-LOCAL' BARYOGENESIS AT THE ELECTROWEAK PHASE TRANSITION

Tomislav Prokopec

Princeton University
Joseph Henry Laboratories
Physics Department, PO Box 708
Princeton, NJ 08544, USA

ABSTRACT [1]

Baryon asymmetry can be generated at the first order electroweak phase transition, provided there is a CP violation on the bubble wall. In this report we discuss the role of leptons and quarks in the 'non-local' baryogenesis mechanism.

It is quite a remarkable fact that baryogenesis may have occurred at the electroweak phase transition [1] with a minimal extension of the Higgs sector, in order to provide for the necessary CP violation [2].

CP violation in a two complex Higgs doublet model is secured by a CP-odd scalar field which changes in a definite manner on the bubble wall, as specified by the Higgs potential. This field couples through the anomaly, producing a bias in the baryon violating processes, and therefore the possibility of baryon production [3].

Cohen, Kaplan and Nelson [4] have proposed a non-local 'charge transport' mechanism: the *top* quark asymmetry generated by a CP-violating reflection off the wall propagates far into the unbroken phase, and biases the baryon production.

In comparison to a 'local' baryogenesis occurring on the wall, there are two main advantages to the 'nonlocal' baryogenesis:

(a) the anomalous baryon number violating rate is unsupressed in the unbroken phase, while it becomes exponentially suppressed on the wall,

[1]This lecture is based on a collaboration with Michael Joyce and Neil Turok. The lecture was delivered at the NATO Advanced Research Workshop on 'ELECTROWEAK PHYSICS AND THE EARLY UNIVERSE' held at Sintra, Portugal from March 22 to 25, 1994.

(b) the extent over which baryogenesis takes place is much larger, it is specified by the properties of the diffusion tail in front of the wall.

In this work we will focus on discussing the role of various particle species in a 'non-local' baryogenesis, in particular we will compare the roles of the *top* quark and τ-lepton. Before we proceed, we review the main steps involved in the calculation of the asymmetry.

In an analytic treatment, the problem of calculating the baryon number can be broken up into: (a) modeling the reflection off the wall, which sources (b) the diffusion equations, thus specifying the profiles of particle species in front of the wall; (c) the effects of the hypercharge screening on propagation of the asymmetry, and finally, (d) integration of the baryon rate equation.

(a) *Reflection.* Treatment of the reflection problem using the perturbative Dirac equation is valid only in the thin wall limit: $Lm \leq 1$, which for the *top* quark gives a bound: $L \leq 1/T$. Perturbative calculations suggest $L \approx 20 - 40/T$. For τ-leptons this bound is marginally satisfied. When the wall is much thicker than the mean free path of a particle, the reflected asymmetry is exponentially suppressed due to the loss of coherence [5]. In the thin wall limit the perturbative Dirac equation gives the reflected chiral particle asymmetry proportional to the mass squared, wall velocity (v_w), CP-violation in the wall ($\Delta\theta_{CP}$), and inversely proportional to the thickness of the wall (m_H):

$$J_L(0) = -J_R(0) \approx \frac{v_w m^2 m_H \Delta\theta_{CP}}{4\pi^2} \tag{1}$$

Note that this is a purely chiral current, which, after it thermalizes, diffuses in front of the wall. To calculate the distribution of the diffusing particles, we model the source to the diffusion equations by the chiral current 1.

(c) *Diffusion.* Propagation of the asymmetry in front of the wall is specified by the diffusion equations or particle densities n_i in the rest frame of the wall ($x = z - v_w t$)

$$D_i n_i'' + v_W n_i' \Sigma_A \Gamma_A \Sigma_j \mu_j = J_i'(x) \tag{2}$$

where v_w denotes the velocity of the wall, D_i diffusion constant of the particle species i, Γ_A is the rate for a process A, and $\Sigma_j \mu_j$ is the sum of the chemical potentials on the external legs of the process A. The extension of the source ξ_i is defined as $\xi_i J_i^0 = \int J_i$.

(d) *Baryon rate equation.* The rate of baryon production in the unbroken phase is determined by the following rate equation:

$$\dot{B} = -\frac{N_f \Gamma_s}{T} \Sigma_i \mu_i, \tag{3}$$

where $\Gamma_s = \kappa \alpha_w^4 T$ is the weak sphaleron rate ($\kappa \in [0.1, 1]$), $N_f = 6$ is the number of the fermionic flavors, and $\Sigma_i \mu_i$ is the sum over the chemical potentials for the left-handed particles.

Now we are ready to discuss the role of leptons and quarks in the 'non-local' baryogenesis. Because of their much larger tree-level mass, *top* quarks and τ-leptons are reflected much more abundantly, and therefore require no further consideration.

Let us first discuss the propagation of leptons. The relevant diffusion equations for the third family chiral leptons (L_L and L_R) in the wall rest frame are static:

$$D_L L_L'' + v_w L_L' - \Gamma_{LR}(aL_L - bL_R) = J_L' \tag{4}$$

$$D_R L_R'' + v_w L_R' + \Gamma_{LR}(aL_L - bL_R) = J_R' \tag{5}$$

In the above we also include the dominant decay channel, which converts the left-handed leptons into the right-handed ones *via* emission of a Higgs. For leptons: $\Gamma_{LR} \approx 0.3\alpha_w y_\tau^2 T$, where y_τ is the τ-lepton Yukawa coupling constant, α_w the $SU(2)$ coupling constant, $a \approx 1/2$, $b \approx 1$. (Here we assume that the Higgs particles do not significantly affect the lepton distribution function, as it was argued in[6].)

¿From eq. (3), we infer that the relevant quantity to calculate, which is proportional to the baryon number, is the integral of the left-handed lepton number in front of the wall, which can be obtained in quite a simple manner. After summing and integrating twice eqs. (5) we get

$$\int_0^{+\infty} L_L = \frac{b}{a+b} \frac{\int_{-\infty}^{+\infty}(J_L + J_R)}{v_w} \tag{6}$$

To obtain this expression we used the fact that the dominant contribution to the lepton integral comes from the diffusion tail in front of the wall, in which $aL_L = bL_R$, and also that any lepton number vanish at the spatial infinity. The above integral gives the correct result for the integrated lepton number for slow walls, for which the diffusion tail extends far into the unbroken phase: $v_w^2 << \Gamma_{LR} D_R < 1$.

Next we discuss the *top* quark case. We can write the analogous equations for the left- and right-handed baryons (B_L and B_R), with the decay term changed to $\Gamma_{ss}(B_L - B_R)$, where Γ_{ss} is the strong sphaleron rate. Nevertheless, the simple trick applied in solving the lepton equations does not work, because the diffusion constants for left and right handed quarks are roughly the same. This has as a consequence: $J_{B_L} \approx -J_{B_R}$, so that $B = B_L + B_R$ is not directly sourced by any current, and there is no diffusion tail for baryons in front of the wall. The main source for the baryon number comes from the strong sphaleron tail $(D_q/2\Gamma_{ss})^{1/2}$ for $B_L - B_R$. The result is

$$\int_0^{\infty}(B_L - B_R) = \frac{1}{4}\frac{\int(J_{B_L} - J_{B_R})}{\sqrt{2\Gamma_{ss}D_q}} \tag{7}$$

The ratio of the baryon numbers for the *top* quark and the τ-lepton cases then reads:

$$\frac{B_{top}}{B_\tau} = \frac{1}{4}\frac{a+b}{b}\frac{v_w}{\sqrt{2\Gamma_{ss}D_q}}\frac{\int(J_{B_L} - J_{B_R})}{\int(J_{L_L} - J_{L_R})} \approx \frac{3}{8}\frac{v_w}{\sqrt{2\Gamma_{ss}D_q}}\frac{D_q}{D_{\tau_R}}\left(\frac{m_t(T)}{m_\tau(T)}\right)^2 \tag{8}$$

There are two reasons for the *top* quark suppression. *Firstly*, there is no diffusion tail for *tops*; the strong sphalerons cut it off in front of the wall, giving a suppression factor of approximately 10. *Secondly,* the extent of the source current (before it thermalizes) is larger for the right handed τ-leptons at least by the ratio of the diffusion constants $D_{\tau_R}/D_q \sim 50$. In the case of thicker walls, this suppression tends to be stronger than linear, because the *tops* have the tendency to be thermalized before they escape the region of the wall, and thus be lost as a source for the diffusion equations.

Naively, these two sources of suppression might be outweighed by the large ratio of *top*-to-τ mass: in the standard model $m_t^2/m_\tau^2 \sim 10^4$, such that the *top* still contributes more to the asymmetry than the τ. However, in the two Higgs doublet theory, quarks and leptons may couple to the different Higgs, which may have different zero temperature *vevs*, giving the possibility that around the phase transition the *top*-to-τ mass ratio be much larger than at the zero temperature. In this case the τ-lepton produces more baryons.

And as a concluding remark, we announce the result of the recent work [6] on thicker walls (comparable to the perturbative result: $20 - 40/T$). We consider the case when the wall thickness is much larger than the mean free path of particles, which justifies the usage of the fluid equations approximation, but still thinner than the relevant diffusion length D/v_w. We find that, besides the reflection, there is a *classical force* due to the CP-violation on the bubble wall acting differentially on particle species, causing a diffusion tail in an axial particle number in front of the wall. We find that the effect is proportional to the mass squared of the particles in consideration and *independent* on the diffusion properties of the species, singling out the *top* quark as the main candidate for baryogenesis!

Acknowledgements. I thank Michael Joyce, Neil Turok and Larry McLerran for useful discussions. The work of T.P. is partially supported by NSF contract PHY90-21984, and the David and Lucile Packard Foundation.

References

[1] V. Kuzmin, V. Rubakov and M. Shaposhnikov, Phys. Lett. **155B**, 36 (1985); F. Klinkhamer and N. Manton, Phys. Rev. **D30**, 2212 (1984).

[2] N. Turok and J. Zadrozny, Phys. Rev. Lett. **65**, 2331 (1990); N. Turok and J. Zadrozny, Nuc. Phys. **B 358**, 471 (1991); L. McLerran, M. Shaposhnikov, N. Turok and M. Voloshin, Phys. Lett. **256B**, 451 (1991).

[3] N. Turok, in *Perspectives in Higgs Physics, ed.* by G. Kane, pub. World Scientific, p. 300 (1992); A. Cohen, D. Kaplan and A. Nelson, Ann. Rev. Nucl. Part. Phys. **43** 27-70 (1993).

[4] A. Cohen, D. Kaplan and A. Nelson, Nuc. Phys. **B373**, 453 (1992); A. Cohen, D. Kaplan and A. Nelson, Phys. Lett. **B294** (1992) 57.

[5] M. B. Gavela, P. Hernández, J. Orloff, and O. Pène, *Standard Model CP-violation and Baryon Asymmetry*, preprint CERN 93/7081, LPTHE Orsay-93/48, HUTP-93/A036, HD-THEP-93-45 (1993); Patrick Huet and Eric Sather, *Electroweak Baryogenesis and Standard Model CP-violation*, SLAC-PUB-6479, April, 1994.

[6] Michael Joyce, Tomislav Prokopec, and Neil Turok, *Classical Baryogenesis at the Electroweak Phase Transition*, PUPT preprint (1994).

PREVENTING BARYON WASHOUT WITH A Z'

M.C. Bento[1] and O. Bertolami[1]

CERN, Theory Division
CH-1211 Geneva 23
Switzerland

Recently much attention has been devoted to the possibility that the baryon asymmetry of the universe (BAU) is generated at the electroweak phase transition by sphaleron-induced transitions between inequivalent non-abelian vacua[1]. However, it is necessary to ensure that the BAU thus created will not be subsequently erased via the same processes; the requirement that B-violating transitions drop out of equilibrium after the phase transition sets an upper bound on the Higgs mass, $m_H < 64$ GeV[2], barely consistent with recent LEP searches, which establish that $m_H > 58.4$ GeV[3]. It seems, therefore, that a way out of this difficulty necessarily involves some extension of the Standard Model (SM) and, in fact, several models have already been analysed in the literature[4].

In this article, we assume that the necessary baryon asymmetry has been generated when the electroweak phase transition is completed, regardless of the scale at and the mechanism by which it has been produced, and we analyse the question of the preservation of the asymmetry from spharelon-induced baryon number violating processes after the electroweak phase transition in the simplest extension of the SM symmetry, namely the models with $SU(2)\times U(1)\times \tilde{U}(1)$ gauge symmetry at the TeV scale[5]. Indeed, many extensions of the SM predict the existence of (at least) one additional gauge boson, Z', and some models allow Z' boson masses detectable at present or future colliders. Typical models are the ones based either on a left-right symmetry (LR) or E_6 (superstring-motivated or not) or SO(10), e.g. the so-called η, ψ and χ models defined by the symmetry breaking schemes: $E_6 \to SU(3) \times SU(2) \times U(1) \times U(1)_\eta$, $E_6 \to SO(10) \times U(1)_\psi$ and $SO(10) \to SU(5) \times U(1)_\chi \times U(1)$, respectively[6]. The extra gauge boson would manifest itself because the Z mass and couplings are modified by mixing and Z' exchange effects. The latter would also be responsible for the deformation of the Z resonance shape; study of the shape of the resonance sets the limit $m_{Z'} \gtrsim 100$ GeV[7]. On the other hand, direct searches lead to $m_{Z'} \gtrsim 400$ GeV[8], assuming SM coupling strengths; for other models, e.g. the LR, η, ψ and χ models

---

[1] On leave from Departamento de Física, Instituto Superior Técnico, Av. Rovisco Pais, 1096 Lisboa, Portugal

Electroweak Physics and the Early Universe, Edited by
Romao and Freire, Plenum Press, New York, 1994

mentioned above, bounds become less stringent varying typically between 170 and 350 GeV[8]. In what follows, we shall not consider any model in particular but rather the class of models that admit a relatively "light" extra boson Z', i.e. with masses in the range 100 — 500 GeV, and assume that the remaining new scalar and fermionic states are heavy and decouple from the low energy physics. Under these assumptions, and, for the mass range we are considering, i.e 100 GeV $\leq m_{Z'} \leq$ 500 GeV, the finite temperature 1-loop effective potential, relevant to the discussion of the electroweak phase transition, is the following[9,10,11]:

$$V(\phi, T) = D(T^2 - T_o^2)\phi^2 - \frac{2}{3}ET\phi^3 + \frac{\lambda_T}{4}\phi^4, \quad (1)$$

where

$$D = \frac{1}{8\sigma^2}(2m_W^2 + m_Z^2 + m_{Z'}^2 + 2m_t^2), \quad (2)$$

$$E = \frac{1}{4\pi\sigma^3}(2m_W^3 + m_Z^3 + m_{Z'}^3), \quad (3)$$

$$T_o^2 = \frac{1}{4D}(m_H^2 - 8B\sigma^2), \quad (4)$$

$$B = \frac{3}{64\pi^2\sigma^4}(2m_W^4 + m_Z^4 + m_{Z'}^4 - 4m_t^4), \quad (5)$$

$$\lambda_T = \lambda - \frac{3}{16\pi^2\sigma^4}\left[2m_W^4 \ln\left(\frac{m_W^2}{a_B T^2}\right) + m_Z^4 \ln\left(\frac{m_Z^2}{a_B T^2}\right) + m_{Z'}^4 \ln\left(\frac{m_{Z'}^2}{a_B T^2}\right) - 4m_t^4 \ln\left(\frac{m_t^2}{a_F T^2}\right)\right] \quad (6)$$

with $m_H^2 = 2\lambda\sigma^2$, $\sigma = 246$ GeV, $\ln a_B \simeq 3.91$ and $\ln a_F \simeq 1.14$.

Notice that we have already added the ring diagram contributions, whose net effect is essentially to reduce the coefficient of the cubic term by a factor 2/3[9,10].

The Higgs mass bounds, obtained from the requirement that spharelon-induced baryon number violating processes drop out of thermal equilibrium, which translates into the condition $\left(\frac{\phi}{T}\right)_c \gtrsim 1$[12], are shown in Figure 1 as functions of the top quark mass, in the range 100 GeV $< m_t <$ 200 GeV, for $m_{Z'} = 100, 200, 300, 500$ GeV.

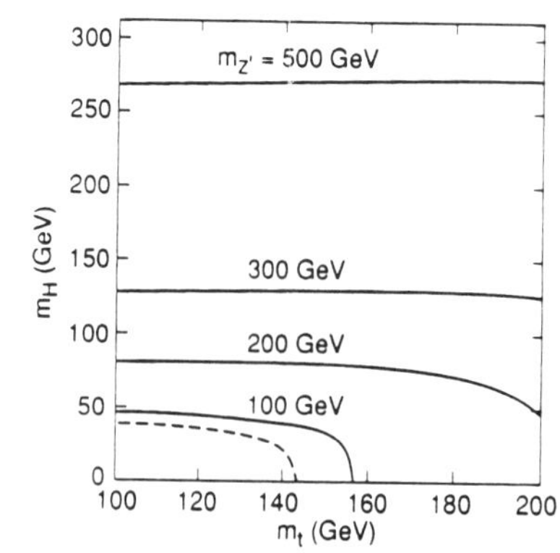

Figure 1. Comparison of Higgs mass bounds versus the top mass, for the Standard Model (dashed curve) and models with a Z', for $m_{Z'} = 100,\ 200,\ 300,\ 500$ GeV (full curves).

We find that one needs $m_{Z'} > 100$ GeV to obtain compatibility with $m_H \gtrsim 60$ GeV[3]. For $m_{Z'} = 200$ GeV, we get $m_H < 100$ GeV for all values of m_t in the above-mentioned range and, for $m_{Z'} \gtrsim 300$ GeV, bounds on m_H become fairly insensitive to the top quark mass. For $m_{Z'} = 500$ GeV, the Higgs mass bound is weakened up to about 270 GeV[5]. Furthermore, since the validity of perturbation theory used to obtain the effective potential (1) is restricted by the condition $m_H \lesssim m_W$, it follows from Figure 1 that this implies $m_{Z'} \lesssim 230$ GeV.

Notice that, allowing $m_{Z'} \gtrsim 300$ GeV leads to $T_0^2 \leq 0$ and $\lambda_T \leq 0$ for certain combinations of m_H and m_t. Negative values of T_o^2 and λ_T require that relevant expressions be rewritten in terms of $|T_o^2|$ and $|\lambda_T|$.

Let us now turn to the question of the validity of the thin wall approximation in the class of models we are considering. The relevant parameter is

$$\epsilon = \frac{T_c^2 - T_N^2}{T_c^2 - T_o^2}, \tag{7}$$

where T_N is the nucleation temperature. Thin walls satisfy the condition $\epsilon \ll 1$. As discussed in ref. 10, for typical values of SM parameters, the thin wall approximation is not valid; we find that this is also the case for the models we are considering. Hence, when discussing the rate of tunnelling from the false to the true vacuum per unit of four-volume[13]

$$\frac{\Gamma}{V_4} = T^4 \left(\frac{S_3}{2\pi T}\right)^{3/2} \exp\left(-\frac{S_3}{T}\right), \tag{8}$$

one cannot estimate the O(3)-symmetric bounce action, $\frac{S_3}{T}$, using the thin wall approximation and therefore should use the alternative formula of ref. 10:

$$\frac{S_3}{T} = \frac{87.3 D^{3/2}}{E^2} \theta^{3/2} f\left(\frac{4.5 \lambda_{T_o} D}{E^2} \theta\right), \tag{9}$$

where

$$f(y) = 1 + \frac{y}{4}\left[1 + \frac{2.4}{1-y} + \frac{0.26}{(1-y)^2}\right], \tag{10}$$

and $\theta(T) \equiv \frac{\Delta T}{T} = 1 - \frac{T_o}{T}$. One is interested in the nucleation temperature, T_N, at which $\Gamma(T_N) = 1$. In eq. (8) $V_4 \sim t^4 \sim H^{-4}$ and, from $H^2 = 8\pi\rho/3M_P^2$, with $\rho \approx \rho_o = (25 \text{ GeV})^4$, we get

$$\Gamma(T) = \frac{9 M_P^4}{64 \pi^2 \rho_o^2} T^4 \left(\frac{S_3}{2\pi T}\right)^{3/2} \exp\left(-\frac{S_3}{T}\right). \tag{11}$$

The fate of the generated baryon asymmetry will depend on the entropy release, Δ, ensued by the first order phase transition. It is necessary that $\Delta \lesssim 1$, i.e. that $T_R \lesssim T_N$ since

$$\Delta \equiv \frac{\eta_R T_R^3}{\eta_N T_N^3}, \tag{12}$$

and $\eta_R \gtrsim \eta_N$, T_R being the reheating temperature obtained from the condition: $\rho(\phi = 0, T_N) = \rho(\phi = \sigma, T_R)$, where $\rho(\phi, T) = V(\phi, T) - T\frac{\partial V(\phi,T)}{\partial T}$.

In the SM, excessive entropy release would represent a serious threat to the baryon asymmetry if the Higgs boson mass were close to the Coleman-Weinberg mass[14]. However, given the present lower bound on m_H, this problem no longer exists for the SM. In order to check whether this is the case for the models we are considering, we have computed the ratio between the reheating and nucleation temperatures

$$\frac{T_R}{T_N} = \frac{2D(1-\theta)}{\pi}\left[\frac{15}{\eta_R}\left(1+\sqrt{\frac{1+\gamma}{\delta}}\right)\right]^{1/2}, \tag{13}$$

where

$$\gamma = \frac{\pi^2}{30D^2}\eta_R\left[\frac{\delta}{(1-\theta)^2} + 8B - \lambda_{T_N} + \frac{\eta_N \pi^2 \delta^2}{120(1-\theta)^4 D^2}\right], \tag{14}$$

with $\delta = (m_H/\sigma)^2 - 8B$, $\eta_R = 92.5$ and $\eta_N = 73$, for various combinations of $m_{Z'}$, m_H and m_t and we conclude that realistic models with an extra U(1) gauge symmetry are indeed safe in what concerns the possibility of excessive supercooling[5].

REFERENCES

1. For a review see A.G. Cohen, D.B. Kaplan and A.E. Nelson, Ann. Rev. Nucl. Part. Phys. 43(1993)27.
2. M.E. Shaposhnikov, Phys.Lett B277(1992)324; B282(1992)483(E).
3. Aleph Collaboration, Preprint CERN-PPE/93-110.
4. M. Fukugita and T. Yanagida, Phys. Lett. B174(1986)45;
   M.E. Shaposhnikov, Nucl. Phys. B287(1987)757; B299 (1988)797;
   L. McLerran, Phys. Rev. Lett. 62(1989)1075;
   N. Turok and J. Zadrozny, Phys. Rev. Lett. 65(1990)2331; Nucl. Phys. B358(1991)471;
   M. Dine, P. Huet, R. Singleton and L. Susskind, Phys. Lett. B257(1991)351;
   G.R. Farrar and M.E. Shaposhnikov, Phys. Rev. Lett. 70(1993)2833; Preprint CERN-TH. 6734/93; RU-93-11,1993.
5. M.C. Bento and O. Bertolami, Phys. Lett. B323(1994)373.
6. L.S. Durkin and P. Langacker, Phys. Lett. B166(1986)436.
7. L3 Collaboration, Preprint CERN PPE/93-44.
8. F. Abe et al. (CDF Collaboration), Phys. Rev. Lett. 68(1992)1463.
9. M.E. Carrington, Phys. Rev. D45(1992)2933.
10. M. Dine, R.G. Leigh, P. Huet, A.D. Linde and D. Linde, Phys. Rev. D46(1992) 550; Phys. Lett. B283(1992)319.
11. G.W. Anderson and L.J. Hall, Phys. Rev. D45(1992)2685.
12. V.A. Kuzmin, V.A. Rubakov and M.E. Shaposhnikov, Phys. Lett. B155(1985)36.
13. A.D. Linde, Phys. Lett. B100(1981)132; Nucl. Phys. B216(1983)421.
14. E. Witten, Nucl. Phys. B177(1981)477.

BARYOGENESIS FROM A PRIMORDIAL LEPTON ASYMMETRY

S. A. Abel

Rutherford Appleton Laboratory
Chilton, Didcot
Oxon OX11 0QX
England

Any primordial asymmetry for baryogenesis is in danger of being washed out by sphaleron transitions unless it is protected. The possibility which will concern us here, is that the asymmetry was encoded in a conserved global quantum number such as $B - L$ number [1]. This was generated (or set by hand) primordially, and after possible processing at the electroweak phase transition, was converted into the presently observed baryon number. Recent work [2,3] has suggested that the three

$$L_i - B/3$$

numbers may play this role (i is a generation index). These could have been produced by generational (and CP violating) differences in the out of equilibrium decays of lepto-quarks [2,4]. The baryogenesis could have occurred *after* the electro-weak phase transition as a results of mass effects. For this to be possible it is necessary that sphalerons remained in equilibrium for some time implying a second-order transition. Campbell, Davidson, Ellis, and Olive [5] have taken this idea further by demonstrating how to avoid the m^2/T^2 suppression through lepton-violating interactions. Provided that some, but not all, lepton flavours are violated by $\Delta L \neq 0$ interactions in equilibrium, B may be regenerated without the m^2/T^2 lepton mass effects. Instead B depends only on the initial asymmetry of the non-equilibrating modes. For example, if lepton-violating interactions equilibrate $L_1 - B/3$ and $L_2 - B/3$ to zero, then the non-zero $L_3 - B/3$ biases baryon production via sphalerons:

$$B \propto B - L = B/3 - L_3 \neq 0 \ .$$

Here I shall discuss an alternative way in which these quantum numbers can furnish a net baryon number. This mechanism operates if the phase transition is robustly first order, as argued for in the case of the Standard Model in refs.[6,7], and as demonstrated in extensions of it in ref.[8].

Again we rely on some unspecified mechanism to generate a non-zero $L_3 - B/3$ number. The values of the other two numbers are unimportant, and it could be that $B - L = 0$. Once the temperature drops below the critical temperature, bubbles of true vacuum begin to nucleate and expand into the symmetric phase. As the

Electroweak Physics and the Early Universe, Edited by
Romao and Freire, Plenum Press, New York, 1994

walls advance through the plasma, a fraction (proportional to their physical mass) of the τ-leptons are reflected back into the symmetric phase. Since the $L_3 - B/3$ number is non-zero, more taus than anti-taus are reflected, and a small hypercharge builds up on the outside of the bubbles with an opposing hypercharge on the inside. On the outside sphaleron transitions convert some of the lepton number into baryons, whilst on the inside sphaleron transitions are suppressed by a large exponential factor, resulting in a net number of baryons being produced. The resulting baryon number is proportional to the initial τ-lepton asymmetry ρ_τ. We stress that we do not make any assumption about enhanced (or maximal) CP violation in the dynamics of the phase transition (which plays no role in our mechanism).

The starting point of our analysis is the rate equation describing the approach of baryon number to equilibrium [9];

$$\dot{\rho}_B = -\frac{\Gamma_B}{T}\frac{\partial F}{\partial B} . \tag{1}$$

The partial derivative of the free energy is taken with all conserved quantum numbers held fixed and is simply proportional to the hypercharge;

$$\frac{\partial F}{\partial B} = \xi\frac{\rho_Y}{T^2} , \tag{2}$$

where the parameter ξ depends on the model under consideration and will be determined explicitly later. Integrating the rate equation ahead of the advancing bubble wall,

$$\begin{aligned} \rho_B &= -\frac{\Gamma_B}{T}\int dt\,\frac{\partial F}{\partial B} = -\frac{\xi\Gamma_B}{T^3}\int_{-\infty}^{z/u} dt\ \rho_Y(z-ut) \\ &= -\frac{\xi\Gamma_B f_Y \tau_T}{uT^3} , \end{aligned} \tag{3}$$

where f_Y is the reflected hypercharge flux and τ_T the thermal transport time ($\sim$ thermal scattering length). Scaling by the entropy density and recalling the rate for anomalous baryon violation in the symmetric phase (for three generations) [10,11],

$$\Gamma_B = 3\kappa\alpha_{\rm W}^4 T^4 , \tag{4}$$

we arrive at our final expression for the observed baryon number,

$$\frac{\rho_B}{s} = -\frac{3\alpha_{\rm W}^4\kappa\xi}{u}\left(\frac{f_Y\tau_T T}{s}\right) . \tag{5}$$

We consider the ranges $0.1 < \kappa < 1$ [11] and $0.1 < u < 1$ [12]. In what follows I shall describe the computation ξ, τ_T and f_Y, which has been done in detail in ref.[13]. For the remainder of this discussion, I shall assume for concreteness that $B - L = 0$, although this is not necessarily the case as stated previously.

At high temperature or density, the fundamental excitations do not coincide with the elementary particles. As a result, perturbation theory in terms of the bare fields does not properly describe effects essentially due to the ambient plasma. Therefore, in order to proceed in a valid perturbative calculation, we first determined the thermal masses of the quasiparticle modes that exist in a relativistic plasma at high temperature.

In order to compute f_Y we have adopted the techniques used in ref.[7] in which the quasi-particles were assumed to undergo coherent scattering at the wall. As discussed in ref.[14], the quasi-particle damping rate is probably fast enough to invalidate this assumption for strongly interacting particles, due to interactions with gluons. For leptons however, we believe that these techniques are still valid, since the scattering rate of the quasi-particles in the plasma is significantly reduced, and as we shall find later, their thermal length is of the order of $100/T$ (c.f. the quark scattering length of order few/T). This should be compared with the typical wall thickness of $20 - 40/T$. In addition the long thermal scattering implies that the quasi-particle modes decay slowly (inversely as the scattering length), and indeed, substituting the relevant quantities into the reflection coefficients of ref.[14], we find that there is no damping. Specifically, the the reflected wave is "depleted" by a factor $300M_\tau/T$; i.e. not at all. Therefore the assumption of coherent scattering is still a good one for leptons. Given this, the flux should also be proportional to the physical mass of the τ-lepton, since total reflection occurs in an energy interval of this width.

Because the reflected flux $\propto M_l/T$, where M_l is the lepton flavour mass and $T \sim 100$ GeV is the critical temperature of the phase transition, the τ-lepton flux will predominate. We find that f_Y is simply proportional to the primordial τ-lepton asymmetry. Our final result for the reflected hypercharge flux is

$$f_Y \approx -\frac{27}{16\pi} \cdot \frac{(\Omega_L - \Omega_R)^2 M_\tau}{T^3} \cdot \frac{\rho_\tau}{1 + \cosh\left(\frac{\omega_0}{T}(1 - u/3)\right)} . \tag{6}$$

where $\omega_0 \sim (\Omega_L + \Omega_R)/2$ and $\Omega_L - \Omega_R \approx 0.1T$ are the thermal contribution to the lepton masses. We believe this expression to be a conservative lower bound, since we have truncated the flux integral in the region of total reflection where the leptons are moving perendicular to the wall.

The thermal scattering length of the leptonic quasiparticles gives an indication of how long the reflected leptons remain in the unbroken phase before being absorbed by the advancing front of broken phase. It is defined as the average time that a reflected τ-lepton spends in the plasma prior to absorption by the bubble of broken phase. Capture occurs when the wall intercepts the randomly walking particle. The thermal transport time is found to be given by [13]

$$\tau_T \equiv \frac{x}{u^2 T} \sim \frac{100}{u^2 T} . \tag{7}$$

This estimate for the thermal time is in agreement with the calculations of leptonic diffusion lengths in ref.[15]. We note that this time is much longer than that for strongly interacting particles as stated earlier.

The quantity $\partial F/\partial B$ depends on the interactions and species in equilibrium. The important timescale for the problem is the average time that the lepton remain in the unbroken phase after reflection, τ_T. All quark mixings were taken to be in equilibrium which is true in the limit as $\tau_T \to \infty$ (or wall velocity $u \to 0$). We therefore take all $SU(3)_C \otimes SU(2)_L \otimes U(1)_Y$ interactions to be in equilibrium (including quark mixing and light-quark Yukawa interactions, which are assumed to be out of equilibrium in the analysis of ref.[16]), and only the anomalous $B + L$-violation to be not in equilibrium. Thus we take L_j to be conserved and $B_j = B/3$, and the constraints we included were $B = 0 = L$, so that $B_j = 0$ although $L_j \neq 0$ is allowed. Imposing these conditions, we found that

$$\frac{\partial F}{\partial B} = \mu_B = \frac{\rho_Y}{(5+n)T^2} \equiv \frac{\xi \rho_Y}{T^2} \,. \tag{8}$$

Finally, inserting these estimates into eq.(9) we obtain

$$\frac{\rho_B}{s} \approx 2.1 \cdot 10^{-6} \cdot \kappa \left(\frac{6}{5+n}\right)\left(\frac{0.1}{u}\right)^3 \left(\frac{100\,\mathrm{GeV}}{T}\right)\left(\frac{2}{1+\cosh \frac{\omega_0}{T}(1-u/3)}\right)\left(\frac{x}{100}\right)\cdot\left(\frac{\rho_\tau}{s}\right) \,, \tag{9}$$

where the various parameters have been scaled by their typical values. We believe this to be a conservative lower bound on the effect of this mechanism since we have underestimated the flux integrals, and have reduced the integral over phase space to those parts where the quasi-particle modes are travelling perpendicular or nearly perpendicular to the wall. In doing so we have neglected the huge volume of phase space where particles graze the wall. This contribution we expect to be suppressed because of the loss of coherence in these modes.

By requiring the observed baryon number of $\rho_B \sim 10^{-10}$ to be generated in this manner, we obtain an estimate for the primordial lepton asymmetry,

$$\frac{\rho_\tau}{s} \sim 5 \cdot 10^{-5} \,, \tag{10}$$

for the typical values of parameters ($\kappa = 1$, $u = 0.1$, $x = 100$). This corresponds to

$$\frac{n_l - n_{\bar{l}}}{n_l + n_{\bar{l}}} \sim 0.005 \,. \tag{11}$$

It is of interest to compare this constraint with that obtained from equilibrium scenarios for lepton-to-baryon conversion. As a generic example, we consider the analysis of Kuzmin, Rubakov, and Shaposhnikov [2]. Taking the requisite large Higgs mass to be $M_H \sim 100$ GeV, and consequently the sphaleron freeze-out temperature (approximately the critical temperature) to be $T_* \sim 150$ GeV, we estimate the generated baryon asymmetry as

$$\frac{\rho_B}{s} \approx -\frac{4}{13\pi^2}\frac{m_\tau^2}{T_*^2} \cdot \left(\frac{\rho_\tau}{s}\right) \approx -4.4 \cdot 10^{-6} \cdot \left(\frac{\rho_\tau}{s}\right) \,. \tag{12}$$

Hence the observed baryon asymmetry may be accounted for in this scheme for a primordial lepton asymmetry of $\rho_\tau/s \sim 2 \cdot 10^{-5}$, which is remarkably similar to the value found above despite the m_τ^2/T^2 suppression. This is not surprising since, as we have emphasised, the assumption that baryon violation is in equilibrium (on the timescale of interest) is an overly optimistic one.

A remark is in order about the behaviour of eq.(9) as $u \to 0$. In contrast to ref.[16], our result diverges unashamedly at low wall velocity as $\sim 1/u^3$. To see why this is acceptable, consider the extreme case of a stationary wall. In this case the τ-leptons simply diffuse away from the wall, finally establishing equilibrium once the hypercharge profile is constant (and different) on either side. In this sense the wall is behaving like a semi-permeable membrane since it forces different partial pressures of leptons on the outside and inside of the bubble. Intuitively, one expects the partial pressure of leptons in the symmetric phase to create baryons via sphaleron transitions until it is fully depleted. However in this calculation we have made an approximation by setting the chemical potentials to be constant. This means an unlimited supply of lepton number – hence the divergence. Our approximation breaks down once the wall velocity is slow enough so that the lepton numbers are depleted significantly. This condition can be gauged by the ratio of generated density of baryons to initial density of leptons. This is a small number for $u \gtrsim 0.001$ which gives a sufficient range of values for first order phase transitions (where $u \gtrsim 0.01$ is more reasonable)[12]. Clearly, in this scenario, the slower the wall velocity the more efficient the mechanism.

References

[1] M. Fukugita and T. Yanagida, *Phys. Lett.* **B174** (1986) 45; J. A. Harvey and M. S. Turner, *Phys. Rev.* **D42** (1990) 3344

[2] V. A. Kuzmin, V. A. Rubakov, and M. E. Shaposhnikov, *Phys. Lett.* **B191** (1987) 171

[3] H. Dreiner and G. G. Ross, *Nucl. Phys.* **B410** (1993) 188

[4] M. Luty, *Phys. Rev.* **D45** (1991) 455

[5] B. A. Campbell, S. Davidson, J. Ellis, and K. A. Olive, *Phys. Lett.* **B297** (1992) 118

[6] K. Kajantie, K. Rammukainen and M. E. Shaposhnikov, Report No. CERN-TH.6901/93; M. E. Shaposhnikov, Report No. CERN-TH.6981/93

[7] G. R. Farrar and M. E. Shaposhnikov, *Phys. Rev.Lett.* **70** (1993) 2833; Report No. CERN-TH.6732/93

[8] M. Pietroni, *Nucl. Phys.* **B402** (1993) 27

[9] M. Dine, O. Lechtenfeld, B. Sakita, W. Fischler, and J. Polchinski, *Nucl. Phys.* **B342** (1990) 381

[10] P. Arnold and L. McLerran, *Phys. Rev.* **D36** (1987) 581; *Phys. Rev.* **D37** (1988) 1020; S. Y. Khlebnikov and M. E. Shaposhnikov, *Nucl. Phys.* **B308** (1988) 885; L. Carson, X. Li, L. McLerran, and R. T. Wang, *Phys. Rev.* **D42** (1990) 2127

[11] J. Ambjorn, T. Askaard, H. Porter, and M. E. Shaposhnikov, *Phys. Lett.* **B244** (1990) 479; *Nucl. Phys.* **B353** (1991) 346; J. Ambjorn and K. Farakos, *Phys. Lett.* **294B** (1992) 248

[12] B. H. Liu, L. McLerran and N. Turok, *Phys. Rev.* **D46** (1992) 2668

[13] K. E. C. Benson, *D.Phil. Thesis* (1993); S. A. Abel and K. E. C. Benson, Report No. RAL-93-103

[14] M. B. Gavela, P. Hernández, J. Orloff and O. Pène, CERN-TH.7081/93

[15] M. Joyce, T. Prokopec and N. Turok, Report No. PUP-TH-1436 (1993); Report No. PUP-TH-1437 (1993)

[16] A. E. Nelson, D. B. Kaplan and A. G. Cohen, *Nucl. Phys.* **B373** (1992) 453

ELECTROWEAK PHASE CHANGES IN TWO HIGGS MODELS

A.T. Davies*, C.D. Froggatt, G. Jenkins, and R.G. Moorhouse

Department of Physics and Astronomy
University of Glasgow
Scotland

INTRODUCTION

As reviewed by Espinosa[1] in his contribution, a strongly first order electroweak phase transition is required in order that the baryon asymmetry, however generated, be not washed out by sphaleron transitions. In the standard one Higgs doublet model, perturbatively treated, this implies an unacceptably low Higgs mass of < 35 GeV. There are indications that non-perturbative calculations can push the bound up to around the current experimental lower limit of 60 GeV, but in view of the possibility that this may be only a temporary remission we investigate the extent to which it may be evaded in a general two doublet model. This has the advantage compared with other extensions of the standard model of allowing CP violation in the Higgs sector, a desideratum of several baryogenesis scenarios.

Our work differs from earlier studies of this model [2,3] by including the CP violating term, by working in the Landau gauge rather than a suspect unitary gauge, by incorporating Debye suppression of longitudinal gauge boson modes (which weakens the first order transition), and incorporating a heavy top.

We find that a two doublet model can coexist with the preservation of baryon asymmetry, perhaps not surprisingly in view of the many experimentally ill-determined parameters in the potential. The baryogenesis constraint rules out about 95% of otherwise acceptable sets of coupling constants. The lightest physical scalar corresponding to potentials fulfilling all our criteria is in the range 50-200 GeV. The presence of a CP violating term in the potential makes it easier to satisfy the required conditions.

Finally, as baryogenesis places non-trivial constraints on the potential, we inquire whether the acceptable sets include the special case of the Minimal Supersymmetric Standard Model.

---

* Presented by A.T. Davies

TWO HIGGS DOUBLET EFFECTIVE POTENTIAL

To suppress flavour changing neutral currents we adopt the usual tree-level scalar potential with a softly broken discrete symmetry $\Phi_2 \to -\Phi_2$ with Φ_2 having a Yukawa coupling g_t only to the t quark, and Φ_1 to the b quark and τ lepton. Couplings to lighter particles are neglected. In the notation of Sher [4] this is

$$
\begin{aligned}
V_0 = -\mu_1^2\Phi_1^*\Phi_1 - \mu_2^2\Phi_2^*\Phi_2 - \mu_3^2\Phi_1^*\Phi_2 - \mu_3^{2*}\Phi_2^*\Phi_1 + \lambda_1(\Phi_1^*\Phi_1)^2 + \lambda_2^2(\Phi_2^*\Phi_2)^2 \\
+ \lambda_3(\Phi_1^*\Phi_1)(\Phi_2^*\Phi_2) + \lambda_4\left|\Phi_1^*\Phi_2\right|^2 + \frac{1}{2}\left[\lambda_5(\Phi_1^*\Phi_2)^2 + \lambda_5^*(\Phi_2^*\Phi_1)^2\right]
\end{aligned} \tag{1}
$$

where μ_1^2, μ_2^2 and the dimensionless λ_{1-4} are real. The relative phase of the Φ can be adjusted to make λ_5 real, and then a complex μ_3^2 implies CP violation. (In the MSSM case, $\lambda_5 = 0$ and then μ_3^2 can always be made real.) Φ_1 and Φ_2 contain a total of 8 real fields ϕ_i.

The one loop effective potential in the high temperature approximation is

$$
V(\phi,T) = V_0(\phi) - \frac{N_{eff}\pi^2T^4}{90} + \frac{T^2}{24}Tr\,M_0^2 - \frac{T}{12\pi}Tr\,M_T^3 \tag{2}
$$

where

$$
\begin{aligned}
Tr\,M_0^2 = -4(\mu_1^2 + \mu_1^2) + (6\lambda_1 + 2\lambda_3 + \lambda_4 + 3g_b^2 + g_\tau^2 + \frac{9}{4}g^2 + \frac{3}{4}g'^2)\sum_1^4\phi_i^2 \\
+(6\lambda_2 + 2\lambda_3 + \lambda_4 + 3g_t^2 + \frac{9}{4}g^2 + \frac{3}{4}g'^2)\sum_5^8\phi_i^2
\end{aligned} \tag{3}
$$

and

$$
Tr\,M_T^3 = \sum_1^8 m_i^3(\phi,T) + 4\left[\frac{g^2}{4}\sum_1^8\phi_i^2\right]^{\frac{3}{2}} + 2\left[\frac{g^2+g'^2}{4}\sum_1^8\phi_i^2\right]^{\frac{3}{2}}. \tag{4}
$$

The scalar masses used in Tr M_0^2 are the eigenvalues of the mass matrix derived from V_0. This T^2 term provides thermal mass corrections, which are included in the Tr M_T^3 term where they are needed to keep V real near its minimum. This term has to be evaluated numerically, as the mass matrix consists of 4x4, not 2x2, blocks, due to the complex μ_3^2 terms.

RESULTS FOR THE GENERAL TWO DOUBLET CASE

We survey the parameter space of the potential.

In the first step we restrict to a basis space of parameters which are acceptable if baryogenesis is not considered. The potential involves 9 real parameters: 4 μ_i and 5 λ_i. We replace μ_1 and μ_2 by one unknown β, where $\tan\beta = v_2/v_1$ at T=0 and $v_1^2 + v_2^2 = (246\text{ GeV})^2$. At each point on an 11x11 grid of $\mu_3^2 = \mu_R^2 + i\,\mu_I^2$, with $0 \le \mu_R^2, \mu_I^2 \le 1.5 \times 10^4\text{ GeV}^2$, we take random values of the λ_i, restricted to $|\lambda_i| < 0.5$ so that perturbation theory may apply, and of $\tan\beta > 0.9$ so that the top Yukawa coupling is not much greater than unity.

We assume a T=0 vacuum structure

$$
\Phi_1 = \frac{1}{\sqrt{2}}\begin{pmatrix} 0 \\ v_1 e^{i\theta} \end{pmatrix}, \qquad \Phi_2 = \frac{1}{\sqrt{2}}\begin{pmatrix} 0 \\ v_2 \end{pmatrix} \tag{5}
$$

which puts further constraints on λ_i, as does the requirement that the potential is bounded below. Many λ_i have also to be discarded to ensure that the masses of the 5 physical scalars are real and compatible with experimental bounds, which, somewhat approximately, we take as > 50 GeV. Acceptable parameters also need the high T minimum at $\phi_i = 0$ to become unstable as T falls below a critical value T_C corresponding to the onset of the electroweak transition. For each set of parameters which passes these tests, we minimize $V(\phi,T_C)$ and determine the vacuum expectation values $v_1(T_C)$, $v_2(T_C)$ and $\theta(T_C)$. At this stage, some points in the basis space have to be discarded because they lead to some $m_i(\phi,T_C)/T_C$ too large for the high T expansion of V to be safe.

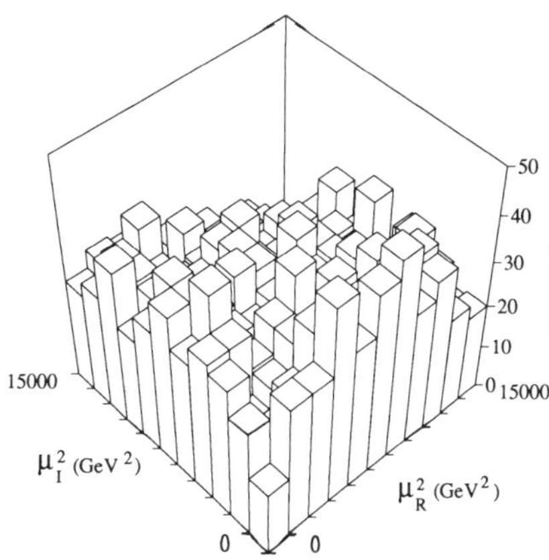

Figure 1a. Frequency distribution of baryon number preserving cases in the complex $\mu_3{}^2$ plane. Each block represents 500 cases. Note that $\mu_3{}^2 = 0$ is the least favourable case, with a success rate of only 3%

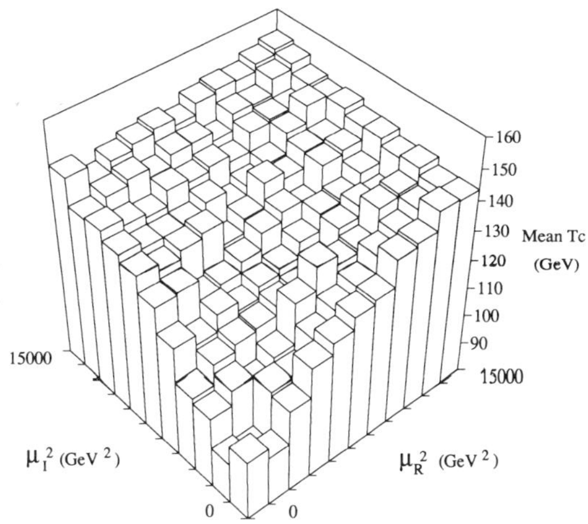

Figure 1b. The average critical temperature $\overline{T}_c$ for the same cases. $\overline{T}_c$ ranges from 90 -160 GeV, being lowest for $\mu_3{}^2 = 0$.

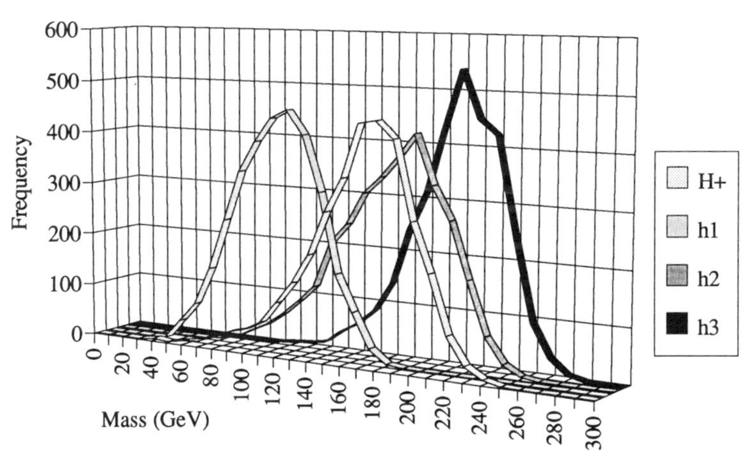

Figure 2. The scalar masses for the baryon preserving cases, showing that the 35 GeV perturbative bound of the Standard Model can be easily evaded. The high mass cutoff is primarily due to the restriction to small λ_i. The neutral scalars are denoted h_1, h_2 and h_3 in order of increasing mass.

Now we can assess what fraction of the basis space satisfies the baryon preservation criterion[1] $v(T_c)/T_c > \xi$, as well as having scalar masses > 50 GeV. The successful cases are presented in Figures 1 and 2 for $\xi = 1$, and for M_{top} = 170 GeV. Of 60,000 points in the basis space, overall only about 5% preserve baryon asymmetry.

MINIMAL SUPERSYMMETRIC STANDARD MODEL

The requirement of baryon number preservation thus places significant but not improbable restrictions on the potential. Is the very special case of the MSSM acceptable?

Unbroken SUSY determines all λ_i in terms of g and g' the SU(2) x U(1) coupling constants,

$$\lambda_1 = \lambda_2 = \frac{g^2 + g'^2}{8}, \quad \lambda_3 = \frac{g^2 - g'^2}{4}, \quad \lambda_4 = \frac{-g^2}{2}, \quad \lambda_5 = 0$$

but SUSY is of course broken at the energy of the EW phase transition.

We link the SUSY and EW scales by making the drastic approximation that SUSY is exact above a certain temperature TS, but that below TS there are only the experimentally established particles (plus top), i.e. we take M_{stop} = TS. We use the two doublet renormalization group equations (RGE) to run g and g' up from their values at the EW scale to those at TS, and hence find the λ_i. at TS. The RGE are then used to run the λ_i down to the EW scale. For a particular TS, for which we try several choices in the range 500-2500 GeV, there remain only 2 adjustable parameters, β and the real μ_3^2.

If we adhere to our earlier restriction tan $\beta > 0.9$ the condition for baryon preservation implies unacceptably low scalar masses. If however we allow smaller β and a high SUSY breaking scale TS, acceptable solutions can be found for a range of μ_3^2; to quote one example, TS = 2.5 TeV , tan β = 0.67, M_{top} = 150 GeV and μ_3 = 58 GeV gives a baryon preserving solution with neutral scalars of mass 70, 87 and 156 GeV and charged scalars of mass 113 GeV.

This conclusion is more optimistic for coexistence of the MSSM and baryogenesis than that of Brignole et al.[5] Supersymmetry breaking has been treated very differently from these authors, but whether this accounts for our different conclusion remains to be investigated. Our solutions do however require tan $\beta < 1$, which implies a large Yukawa coupling to give a heavy top, thus endangering perturbation theory and also MSSM unification predictions such as b-τ mass unification at high energy.

REFERENCES

1. J.R. Espinosa, contribution to this meeting, and references therein
2. A.I. Bochkarev, S.V. Kuzmin, M.E. Shaposhnikov, Phys. Lett. B244 (1990) 275
3. N. Turok and J. Zadrozny, Nucl. Phys. B358 (1991) 471
4. M.Sher, Phys. Rep. 179 (1989) 273
5. A. Brignole, J.R. Espinosa, M. Quiros,and F. Zwirner, Phys. Lett. B307 (1994) 106

ON PRIMORDIAL MAGNETIC FIELDS

KARI ENQVIST
Nordita
Blegdamsvej 17, DK-2100 Copenhagen, Denmark

Abstract

A primordial magnetic field could be responsible for the observed magnetic fields of the galaxies. One possibility is that such a primordial field is generated at the electroweak phase transition because of the fluctuations in the Higgs field gradients. I describe a statistical averaging procedure which gives rise to a field of a correct magnitude. Another possibility, where the Yang-Mills vacuum itself is ferrromagnetic, is also discussed.

1 Introduction

The very early universe is in notoriously short supply of observables that would have survided until today. It is likely that the baryon number of the universe and the density perturbations are one. Another, but a much more speculative possibility, might be the magnetic fields of spiral galaxies, the origin of which still largely remains a puzzle. The nearby galaxies have magnetic fields of the order of $B \simeq 10^{-6}$ G [1], which can be deduced from observations of the syncrotron radiation put out by electrons travelling through the fields, assuming equipartition of magnetic and particle energies[1]. Recently, a field of a similar magnitude has been observed also in a object with z=0.395 [2].

The model for galactic magnetic fields most studied is the galactic dynamo [4], where differential rotation and turbulence of the ionized gas amplifies a weak seed field by several orders of magnitude. Not much is known about the seed field. As the dynamo growth time of the magnetic field cannot be smaller than the galactic rotation period $\tau \simeq 3 \times 10^8$ yrs, this gives a lower limit of $B_{\text{seed}} \gtrsim 10^{-19}$ G on a comoving scale of the protogalaxy, about 100 kpc. In the Milky Way and the Andromeda Nebula the dynamos appear to be rather weak and the growth time as long as $\tau \simeq 10^9$ yrs which would imply that $B_{\text{seed}} \gtrsim B \exp(-t/\tau) \simeq 10^{-10}$ G. Moreover, in the Milky Way the magnetic field changes its direction by about 180° between the Sagittarius and the Orion spiral arms [5], and it has been argued [6] that such a reversal implies a

[1]Equipartition may not to be valid for certain irregular galaxies [3].

Electroweak Physics and the Early Universe, Edited by
Romao and Freire, Plenum Press, New York, 1994

stringent lower bound of $B_{\rm seed} \gtrsim 10^{-7}$ G on the seed field. As such a reversal has only been observed in the Milky Way, it might not be a generic feature.

One interesting possibility is that the seed field is truly primordial, with an origin that predates nucleosynthesis. In that case the protogalaxy collapsed with a frozen-in magnetic field, which enhanced the primordial cosmological field by a factor of 10^4 [7]. Thus at the scale of 100 kpc the dynamo mechanism requires a primordial field somewhere in the ballpark of of 10^{-18} G, with an uncertainty of a few orders of magnitude.

From a theoretical point of view, however, the generation of a sufficiently large persistent magnetic field in the early universe is rather difficult. There are various attempts, relying on phase transitions such as the cosmic inflation or the QCD phase transition [8], but the field often comes out to be too small to be of cosmological interest. It has been suggested that a large field might actually be generated at the electroweak phase transition because of random fluctuations in the Higgs field [9]. If one assumes a stochastic, uncorrelated distribution of the Higgs field gradients, or of the magnetic field itself, one finds [10] today at 100 kpc a root–mean–square field of the order of 10^{-19} G, which could well serve as the origin of the seed field. This positive result is based on calculating the statistical averages along an arbitrary curve. This is not the only possibility, but averaging over areas or volumes would produce a field far too small to be of relevance for the dynamo effect.

In Yang-Mills theories there is also the possibility [11] that the vacuum is an analog of the ferromagnet with a non-zero background magnetic field. This is a non-perturbative effect, and the resulting field is typically very small. If one is willing, however, to go up all the way to the GUT scale one finds that a typical GUT phase transition could have given rise to a background field large enough to serve as the seed field [12].

2 EW magnetic fields

Electromagnetism first occurs when the standard electroweak $SU(2)\otimes U(1)_Y$ theory is broken down to $U(1)_{\rm em}$. It is therefore particularly attractive that Vachaspati [9] has explained the origin of a primordial field in terms of the cosmological boundary condition that all physical quantities should be uncorrelated over distances greater than the horizon distance. Since we start with the group $SU(2)\otimes U(1)_Y$ before the electroweak phase transition, the resulting electromagnetic field can be constructed in a way which is different from the usual $F_{\mu\nu} = \partial_\mu A_\nu - \partial_\nu A_\mu$. The result is [9]

$$\begin{aligned} F_{ij} &= -i(V_i^\dagger V_j - V_i V_j^\dagger) \ , \\ V_i &= \frac{2}{|\phi|}\sqrt{\frac{\sin\theta}{g}}\, \partial_i \phi \ , \end{aligned} \tag{1}$$

where ϕ is the Higgs field. At the electroweak phase transition the correlation length in the broken phase is $\sim 1/m_W$ (assuming that the Higgs mass is comparable to m_W). The field F_{ij} is thus constant over a distance $\sim 1/m_W$, but it varies in a random way over larger distances. Its variation is due to the fact that the field ϕ makes a random walk on the vacuum manifold of ϕ. The problem then is to estimate the field F_{ij} over a length scale $\sim N/m_W$. If $N = 1$, it then follows that on dimensional grounds $F_{ij} \sim m_W^2 \sim 10^{24}$ G, with probably an uncertainty of ± 1 in the exponent. For N large, one should use an approriate statistical argument. The issue is, which is the appropriate way to average over the random fields.

Let us consider random fields walking around in space in a certain number of steps. Thus we replace the continuum by a lattice, where the points are denoted by greek letters $\alpha, \ldots$. I wish to estimate the magnetic field over a *linear* scale (which at most is equal to the horizon scale). Thus, let us consider a curve consisting of N steps in the lattice and define the mean value

$$\overline{B} = \frac{1}{N} \sum_{i=1}^{N} B^{\alpha_i} \quad , \tag{2}$$

where B is a component of the magnetic field, and where the lattice points α_i are on the curve.

Now this curve is arbitrary, and we could take any other curve. Let us therefore define the average $\langle \ldots \rangle$, which averages over curves spanning an N^3 lattice, i.e. over all space. Then, for example,

$$\langle \overline{B} \rangle = \frac{1}{N} \langle \sum_{i=1}^{N} B^{\alpha_i} \rangle \quad , \tag{3}$$

which means that for each curve with N steps the mean value $\overline{B}$ is computed, and this is done for a set of curves which span an N^3-lattice, and the average is then computed. Therefore $\langle \overline{B} \rangle$ depends in general on N, but for simplicity of notation we shall leave out the explicit reference to this dependence. It should be emphasized that the ensemble average (3) takes into account the field value at each lattice point, so that *the average is really over the whole lattice volume* [10].

Similarly, one can define higher moments such as

$$\langle \overline{B}^2 \rangle = \frac{1}{N^2} \sum_{i,j=1}^{N} \langle B^{\alpha_i} B^{\alpha_j} \rangle \quad , \tag{4}$$

together with quantities like $\langle (\overline{B} - \langle \overline{B} \rangle)^2 \rangle$. Note that in (4) the sum is over curves of length N steps of the non-local quantity $\langle B^{\alpha_i} B^{\alpha_j} \rangle$.

In [9] the stochastic variable was taken to be the Higgs field itself which varies over the vacuum manifold. Vachaspati argued that the gradients are of order $1/\sqrt{N}$, since ϕ makes a random walk on the vacuum manifold with $\Delta\phi \sim \sqrt{N}$, and since $\Delta x \sim N$. Thus V_i is, in a root mean square sense, of the order $1/\sqrt{N}$, and hence F_{ij} is of order $1/N$. Taking further into account that the flux in a co-moving circular contour is constant, the field must decrease like $1/a(t)^2$, where $a(t)$ is the scale factor. Using the fact that in the early universe a goes like the inverse temperature, the field was then estimated to behave like $\langle F_{ij}\rangle_T \sim T^2/N$. For a scale of 100 kpc this leads to $\langle F_{ij}\rangle_{now} \sim 10^{-30}$ G, which is far too small to explain the galactic fields (unless there exists some large scale amplification mechanism). One should also point out that this assumption presumes that the total magnetic flux through a given surface is a stochastic sum of the fluxes through the unit cells of that surface, or in other words, that the fluxes through two adjoining unit cells are uncorrelated. Whether this is true or not is an open question.

It is however natural to assume that also the gradient vectors V_i are stochastic, as was done in [10]. This is because they directly specify whether there is a magnetic field or not, whereas this is only true indirectly for the Higgs field itself. Also, the vectors V_i are relevant for questions of alignment between neighbouring domains.

3 Random Higgs gradients

Consider the expression (1) for the magnetic field in terms of the Higgs gradients V_i. It is convenient to split these fields into real and imaginary parts,

$$V_i(\mathbf{x}) = R_i(\mathbf{x}) + iI_i(\mathbf{x}) \ , \tag{5}$$

where R_i and I_i are real vectors. Let us consider the system at a fixed time. The cosmological boundary condition is then that R_i and I_i are random fields. Let us make the following assumptions:

(i) The random fields have a Gaussian distribution. Thus, the mean value of some quantity Q is given by

$$\langle Q\rangle = \prod_{\alpha,i} \int \frac{d^3R_i^\alpha}{D}\frac{d^3I_i^\alpha}{D} Q\ e^{-\lambda(R_i^\alpha-\langle\overline{R}_i\rangle)^2-\lambda(I_i^\alpha-\langle\overline{I}_i\rangle)^2} \ , \tag{6}$$

where D is a normalization factor defined such that $\langle 1\rangle = 1$, and λ is a measure of the inverse width. The quantities $\overline{R}_i$ and $\overline{I}_i$ are the mean values of R_i and I_i defined along a curve of length N steps.[2] Thus, eq. (6) is relevant for a 3-dimensional world which is an N^3 lattice.

(ii) I assume that the mean values are isotropic, i.e. $\langle\overline{R}_1\rangle = \langle\overline{R}_2\rangle = \langle\overline{R}_3\rangle$ and $\langle\overline{I}_1\rangle = \langle\overline{I}_2\rangle = \langle\overline{I}_3\rangle$.

Assumption (i) is certainly the simplest way of implementing lack of correlation of the gradient vectors over distances compatible with the horizon scale, whereas assumption (ii) is natural as there is no reason to expect any preferred direction.

[2]Note that this implies that the mean value in a point is assumed to be equal to the mean value computed along all curves of length N. Thus the mean values can depend on N.

It should be noted that the distribution (6) factorizes into an I-part and an R-part. Thus, for any expectation value consisting of I's and R's one has factorization,

$$\langle \overline{R}_{i_1}^{\alpha_1} \ldots \overline{R}_{i_n}^{\alpha_n}\ \overline{I}_{j_1}^{\beta_1} \ldots \overline{I}_{j_m}^{\beta_m} \rangle = \langle \overline{R}_{i_1}^{\alpha_1} \ldots \overline{R}_{i_n}^{\alpha_n} \rangle \langle \overline{I}_{j_1}^{\beta_1} \ldots \overline{I}_{j_m}^{\beta_m} \rangle \quad . \tag{7}$$

This property turns out to be very useful in computing the higher moments.

The expectation value of a component B_i of the magnetic field can now easily be found. From the expression (1) one finds that

$$B_i = \frac{1}{2} \varepsilon_{ijk} F_{jk} = -i\ \varepsilon_{ijk} V_j^\dagger V_k = 2\varepsilon_{ijk} R_j I_k \quad . \tag{8}$$

Thus

$$\overline{B}_j = \frac{1}{N} \sum_{i=1}^{N} B_j^{\alpha_i} = 2\varepsilon_{jlk} \frac{1}{N} \sum_{i=1}^{N} R_l^{\alpha_i} I_k^{\alpha_i} \quad . \tag{9}$$

Hence

$$\begin{aligned} \langle \overline{B}_j \rangle &= \frac{2}{N} \varepsilon_{jlk} \langle \sum_{i=1}^{N} R_l^{\alpha_i} I_k^{\alpha_k} \rangle \\ &= \frac{2}{N} \varepsilon_{jlk} \langle \sum_{i=1}^{N} (R_l^{\alpha_i} - \langle \overline{R}_l \rangle)(I_k^{\alpha_i} - \langle \overline{I}_k \rangle) + N \langle \overline{R}_l \rangle \langle \overline{I}_k \rangle \rangle \quad . \end{aligned} \tag{10}$$

Now, due to the factorization (7), the first term on the right-hand side of the last Eq. (10) vanishes[3], and hence

$$\langle \overline{B}_j \rangle = 2\varepsilon_{jlk} \langle \overline{R}_l \rangle \langle \overline{I}_k \rangle = \varepsilon_{jlk} \left(\langle \overline{R}_l \rangle \langle \overline{I}_k \rangle - \langle \overline{R}_k \rangle \langle \overline{I}_l \rangle \right) = 0 \tag{11}$$

because of the isotropy assumption (ii). Consequently *the mean value of the magnetic field vanishes.*

The second order moment is given by

$$\begin{aligned} \langle \overline{B}_i^2 \rangle &= \frac{4}{N^2} \sum_{\alpha\beta} \langle R^\alpha R^\beta \cdot I^\alpha I^\beta - R^\alpha I^\beta \cdot I^\alpha R^\beta \rangle \\ &= \frac{4}{N^2} \sum_{\alpha\beta} \left\{ \langle R^\alpha R^\beta \rangle \langle I^\alpha I^\beta \rangle - \langle R_i^\alpha R_j^\beta \rangle \langle I_i^\beta I_j^\alpha \rangle \right\} \quad , \end{aligned} \tag{12}$$

using the factorization (7). Now

$$\begin{aligned} \langle R_i^\alpha R_j^\beta \rangle &= \prod_\gamma \int \frac{d^3 R^\gamma}{D}\ R_i^\alpha R_j^\beta e^{-\lambda (R^\gamma - \langle \overline{R} \rangle)^2} \\ &= \frac{1}{2\lambda} \delta_{ij} \delta^{\alpha\beta} + \prod_\gamma \int \frac{d^3 R^\gamma}{D} \left[\langle \overline{R}_i \rangle R_j^\beta + \langle \overline{R}_j \rangle R_i^\alpha - \langle \overline{R}_i \rangle \langle \overline{R}_j \rangle \right] e^{-\lambda (R^\gamma - \langle \overline{R} \rangle)^2} \end{aligned} \tag{13}$$

and similarly for $\langle I_i^\alpha I_j^\beta \rangle$. Further we have e.g.

$$\begin{aligned} \prod_\gamma \int \frac{d^3 R^\gamma}{D}\ R_j^\beta e^{-\lambda (R^\gamma - \langle \overline{R} \rangle)^2} &= \prod_\gamma \int \frac{d^3 R^\gamma}{D}\ (R_j^\beta - \langle \overline{R}_j \rangle) e^{-\lambda (R^\gamma - \langle \overline{R} \rangle)^2} + \langle \overline{R}_j \rangle \\ &= \langle \overline{R}_j \rangle \ , \end{aligned} \tag{14}$$

[3]Because $\langle R_l^{\alpha_i} - \langle \overline{R}_l \rangle \rangle = \langle I_k^{\alpha_i} - \langle \overline{I}_k \rangle \rangle = 0$ for symmetry reasons.

i.e., the mean value in a given arbitrary point β on the lattice is equal to the mean value computed over all curves. Using Eqs. (13) and (14) in Eq. (12) we get

$$\langle \overline{B}_i^2 \rangle = \frac{4}{N^2} \sum_\alpha \left(\frac{3}{2\lambda^2} + \frac{1}{\lambda}(\langle \overline{I} \rangle^2 + \langle \overline{R} \rangle^2) \right) + \frac{4}{N^2} \sum_{\alpha\beta} \left(\langle \overline{R} \rangle^2 \langle \overline{I} \rangle^2 - (\langle \overline{R} \rangle \langle \overline{I} \rangle)^2 \right) \quad . \tag{15}$$

The first term is $\mathcal{O}(N/N^2) = \mathcal{O}(1/N)$. The last term, being the square of the mean value, actually vanishes because of isotropy. Thus we conclude that the *rms* value of the magnetic field scales like

$$\sqrt{\langle \overline{B}_i^2 \rangle} = \frac{2}{N} \sqrt{\sum_\alpha \left(\frac{3}{2\lambda^2} + \frac{1}{\lambda}(\langle \overline{I} \rangle^2 + \langle \overline{R} \rangle^2) \right)} \sim \mathcal{O}(\frac{1}{\sqrt{N}}) \quad . \tag{16}$$

The reason for this slow decrease is the fact that isotropy prevents the mean value from entering in $\langle \overline{B}_i \rangle$ and $\langle \overline{B}^2 \rangle$, and that the correlations of the gradient vectors are of short range.

4 Consequences of the electroweak magnetic field

Let us now assume that at the time of the electroweak phase transition, a magnetic field with a coherence length ξ_0 is generated, with a scaling as given by Eq. (16). Such a field evolves according to usual magnetohydrodynamics

$$\frac{\partial \mathbf{B}}{\partial t} = \nabla \times (\mathbf{v} \times \mathbf{B}) - \sigma^{-1} \nabla \times \nabla \times \mathbf{B}, \tag{17}$$

where the conductivity $\sigma \sim \infty$ in the early universe. Accordingly, the field is then imprinted on the charged plasma[4].

At later times the original coherence length is redshifted by the expansion according to

$$\xi(t) = \frac{a(t)}{a_0} \xi_0 \quad . \tag{18}$$

The frozen–in magnetic field is also redshifted by the expansion of the universe. Thus at later times at the physical distance scale $L = N\xi$ one finds,

$$B_{\rm rms}(t, L) = B_0 \left(\frac{a_0}{a(t)} \right)^2 \frac{1}{\sqrt{N}} = B_0 \left(\frac{t_0}{t_*} \right)^{\frac{3}{4}} \left(\frac{t_*}{t} \right) \left(\frac{\xi_0}{L} \right)^{\frac{1}{2}}, \tag{19}$$

where $T_0^2 t_0 = 0.301\ M_P/\sqrt{g_*(T_0)}$ with g_* the effective number of degrees of freedom, and $t_* \simeq 1.4 \times 10^3 (\Omega_0 h^2)^{-2}$ yrs is the time when the universe becomes matter dominated; for definiteness, we shall adopt the the value $\Omega_0 h^2 = 0.4$, which is the upper limit allowed by the age of the universe.

It is not obvious what the coherence length ξ actually is. It is likely that it is macroscopic and much larger than the interparticle separation. Let me however assume

[4]There is a possible caveat here: in very large magnetic fields such as considered here the velocity of the plasma might depend on the background magnetic field. In the following I will neglect this.

for brevity that $B_0 \simeq 10^{24}$ G and $\xi \simeq 1/T$. It is then easy to find from Eq. (19) the size of the cosmological field today. Taking $t = 1.5 \times 10^{10}$ yrs and $L = 100$ kpc (corresponding to $N = 1.0 \times 10^{24}$), we find that today the electroweak magnetic field at the scale of intergalactic distances is

$$B_{\rm rms} = 4 \times 10^{-19} \text{ G} \ . \qquad (20)$$

This seems to be exactly what is required for the numerical dynamo simulations to produce the observed galactic magnetic fields of the order 10^{-6} G. The inherent uncertainties in the estimate (20) are: the value of $\Omega_0 h^2$ used for computing t_*; the time at which the magnetic field froze, or T_0; the actual value of the field B_0. Therefore one should view (20) as an order–of–magnitude estimate only.

We should also check what other possible cosmological consequences the existence of the random magnetic field, Eq. (19), may have. Let us first note that the energy density ρ_B in the *rms* field is very small. In the radiation dominated era we find that the energy density within a horizon volume V is

$$\rho_B = \frac{1}{2V}\int_0^{r_H} d^3\mathbf{r} B_{\rm rms}^2 = \frac{3}{4}B_0^2\left(\frac{T}{T_0}\right)^4 \frac{1}{r_H T} \ . \qquad (21)$$

The horizon distance is $r_H = 2t$ so that $\rho_B \sim T^5/M_P \ll \rho_\gamma$, and the magnetic field contribution to the total energy density is negligible.

In principle, magnetic fields could modify primordial nucleosynthesis, as discussed in [13]. However, the electroweak magnetic field at the time of nucleosynthesis and at the horizon scale is only $B_{\rm rms} \simeq 1500$ G which is far too small to give rise to any modifications of the Big Bang nucleosynthesis.

5 Ferromagnetic universe

There is also another, more exotic possibility for producing magnetic fields, which is based on the observation that, due to quantum fluctuations, the Yang–Mills vacuum is unstable in a large enough background magnetic field [11]. There are indications from lattice calculations that this is a non-perturbative result [14]. Such magnetic field fluctuations in the early universe could be sufficient to trigger the phase transition to a new, ferromagnet–like ground state with a magnetic field made permanent by the charged plasma. In this scenario the primordial field is thus generated as a non-perturbative quantum effect.

The new vacuum results provided the β–function has a Landau singularity:

$$\left|\int_g^\infty \frac{dx}{\beta(x)}\right| < \infty. \qquad (22)$$

Then the effective Lagrangian has a minimum away from the perturbative ground state Tr $F^2 = 0$, given by

$$\frac{1}{2}g^2 \text{Tr } F_{\mu\nu}^2|_{\rm min} = \Lambda^4, \qquad (23)$$

where Λ is the renormalization group invariant scale

$$\Lambda = \mu \exp\left(-\int_{\infty}^{g} \frac{dx}{\beta(x)}\right), \tag{24}$$

where μ is a subtraction point associated with the definition of g.

The condition for the minimum can be realized in many ways. One of them is a constant non-abelian magnetic field $B_i^a = \epsilon_{ijk} F_{jk}^a$ with a non-zero component only in one direction in the group space, and with a length given by

$$g\sqrt{B^a B^a} = \Lambda^2. \tag{25}$$

Consider now SU(N) at the one-loop level. We then have the one-loop, zero temperature effective energy for a constant background non-abelian magnetic field which in pure SU(N) theory reads [11]

$$V(B) = \frac{1}{2}B^2 + \frac{11N}{96\pi^2} g^2 B^2 \left(\ln \frac{gB}{\mu^2} - \frac{1}{2}\right) \tag{26}$$

with a minimum at

$$gB_{\text{min}} = \mu^2 \exp\left(-\frac{48\pi^2}{11Ng^2}\right) \tag{27}$$

and $V_{\text{min}} \equiv V(B_{\text{min}}) = -0.029(gB_{\text{min}})^2$. Thus the ground state (the Savvidy vacuum) has a non-zero non-abelian magnetic field, the magnitude of which is exponentially suppressed relative to the renormalization scale, or the typical momentum scale of the system. Thus, for example, for $SU(2)_L$ at the electroweak scale the vacuum magnetic field would be very small. In the early universe, however, where possibly a grand unified symmetry is valid, the exponential suppression is less severe. It is also attenuated by the running of the coupling constant. For a set of representative numbers, one might consider a (susy) SU(5) model with $\alpha_{\text{GUT}} \simeq 1/25$ and $T_{\text{GUT}} \simeq 10^{15}$ GeV, as in the supersymmetric Standard Model. This yields $B \simeq 5 \times 10^{-8}\mu^2$, which turns out to be a magnitude which is relevant for the dynamo mechanism.

In the early universe the effective energy picks up thermal corrections from fermionic, gauge boson, and Higgs boson loops. In SU(2) these are obtained by summing the Boltzmann factors $\exp(-\beta E_n)$ for the oscillator modes

$$E_n^2 = p^2 + 2gB(n + \frac{1}{2}) + 2gBS_3 + m^2(T), \tag{28}$$

where $S_3 = \pm 1/2$ (± 1) for fermions (vectors bosons). In Eq. (28) I have included the thermally induced mass $m(T) \sim gT$, corresponding to a ring summation of the relevant diagrams. Numerically, the effect of the thermal mass turns out to be very important.

The detailed form of the thermal correction depends on the actual model, but we may take our cue from the SU(2) one-loop calculation, which for the fermionic and scalar cases can be extracted from the real–time QED calculation in [15]. The result is

$$\delta V_T^f = \frac{(gB)^2}{4\pi^2}\sum_{l=1}^{\infty}(-1)^{l+1}\int_0^\infty \frac{dx}{x^3}e^{-K_l^a(x)}\left[\, x\coth(x) - 1\right],$$

$$\delta V_T^s = \frac{(gB)^2}{8\pi^2}\sum_{l=1}^{\infty}\int_0^\infty \frac{dx}{x^3}e^{-K_l^a(x)}\left[x\frac{1}{\sinh(x)} - 1\right], \qquad (29)$$

where the normalization is such that the correction vanishes for zero field, and

$$K_l^a(x) = \frac{gBl^2}{4xT^2} + \frac{m_a^2 x}{gB} \qquad (30)$$

where $a = f,\ b$ stands for fermions or bosons.

For vector bosons there is the added complication that there exists a negative, unstable mode, which gives rise to an imaginary part[5]. At high temperatures the instability is absent for fields such that $gB < m^2(T)$, which is the case we are interested in here, so that no regulation of the unstable $n = 0,\ S_3 = -1$ mode is needed. Thus we find [12]

$$\delta V_T^v = \frac{(gB)^2}{8\pi^2}\sum_{l=1}^{\infty}\int_0^\infty \frac{dx}{x^3}e^{-K_l^b(x)}\left[x\frac{\cosh(2x)}{\sinh(x)} - 1\right]. \qquad (31)$$

At high temperature, the bosonic contributions are more important than the fermionic ones. When $B \ll T^2 \simeq m^2(T)$, we find numerically that $\delta V_T^v \simeq 0.016 \times (gB)^2$. This gives rise to a small correction to the magnitude of the field at the minimum as obtained from Eq. (27). We may thus conclude that the Savvidy vacuum exists for all T.

The transition to this new ferromagnet-like vacuum is triggered by local fluctuations. Charged particles in the primeval plasma generate current $\mathbf{j} = \nabla \times \mathbf{B}$. The typical interparticle distance is $L \sim 1/T$ and a typical curl goes like $1/L$ so that $B \sim jL$ where j is like charge density with one charge in the volume L^3. Thus the Maxwell equations imply that $B \sim 1/L^2 = T^2$, indicating that the creation of the Savvidy vacuum can take place locally. A constant non–abelian magnetic field, given by Eq. (27), is then imprinted on the plasma of particles carrying the relevant charges. The Maxwell magnetic field B_{em} is a projection in the space of non–abelian magnetic fields, and we take it to be of the size comparable to B in Eq. (32).

The magnetic flux remains conserved (recall that the primodial plasma is an extremely good conductor), and we may write

$$B(T) = g_{\rm GUT}^{-1}\mu^2 \exp\left(-\frac{48\pi^2}{11Ng^2}\right)\left(\frac{T^2}{\mu^2}\right) \simeq 3\times 10^{42}{\rm G}\left(\frac{a(t_{\rm GUT})}{a(t)}\right)^2, \qquad (32)$$

where $\mu \simeq T$ and $a(t)$ is the scale factor of the universe, and the last figure is for susy SU(5). This expression is valid because the energy E of the vacuum is redshifted by $1/a(t)$. Now in the minimum E is proportional to VB^2, where V is the volume. Since

[5]Physically the imaginary part is an indicator that the vacuum also contains vector particles [16].

V is proportional to a^3, we get $B \sim 1/a^2$. Hence the magnetic energy per horizon is much less than the radiation energy.

As time passes, the universe undergoes a number of phase transitions. Each of these correspond to new types of ferromagnetic vacua, which in general have decreasing field strengths. However, the original GUT vacuum has existed for a time that is long enough for the plasma to interact with the vacuum field B given by Eq. (32). This interaction does not allow for the GUT flux to decrease once it has been created since it has become a feature of the plasma, which conserves the flux in the sense that the magnetic lines of force are frozen into the fluid. Even if the original field is suddenly removed by the creation of a new vacuum, the magnetic field will survive in a perfect conductor (see e.g. pp. 186-189 in ref. [17]).

From Eq. (32) we find that the Maxwell magnetic field at $t_{now} \simeq 10^{10}$ yr is given by $B_{now} \simeq 3 \times 10^{42}\mathrm{G}(t_{\mathrm{GUT}}/t_*)(t_*/t_{now})^{4/3} \simeq 10^{-14}\mathrm{G}$. Such a magnetic field appears comparable to what is needed for the seed field in galactic dynamo models. Note also that at nucleosynthesis one obtains $B \sim 10^4$ G, which is well below the nucleosynthesis bound on magnetic fields [13].

6 Discussion

The GUT causal domain l_0 has today the size of only about 1 m. Obviously, during the course of the evolution of the universe, domains with magnetic fields pointing to different directions have come into contact with each other. One might think that this results in domain walls. However, here it is important that the magnetic flux lines follow the plasma particles and cease to be homogenous. If inside each GUT horizon a magnetic field line at a certain time passes through two plasma particles, then this is true at any later time. The magnetic field thus "aligns" with the plasma. When two horizon bubbles collide, the two plasmas rearrange and become one plasma, and the same is true for the magnetic field lines, which are part of the new plasma [18]. The field "realigns" with the "new" plasma and the root mean square of the magnetic field remains of the same order as before. Because there are no domains, no random walk factor appears at large distances.

In this argument it is important that B_{rms} is much smaller than the square of the rms momentum, since otherwise the electrical conductivity would depend on B. This condition implies that the radius of curvature of a typical plasma particle is very large compared to the mean free path, or that the magnetic energy is much less than the kinetic energy of the plasma. This certainly is the case in the ferromagnetic universe model.

The size of the field is determined by the scale at which the ferromagnet vacuum is created, and the earlier this happens, the bigger the field. If there is a period of cosmic inflation, then the relevant field would be created after reheating. If the reheating temperature is comparable to GUT scales, the strength of the magnetic field would be given as in (32), with interesting consequences for the formation of galactic magnetic fields. In this scenario the primordial seed field is thus a relic from the GUT era.

If the origin of the magnetic field is the electroweak phase transition, the situation might be different because at scales $1/T$ the magnetic energy equals to the energy in radiation. In such fields the plasma might be trapped by the field, rather than the field being imprinted on the moving plasma. Very likely this would result in a domain structure, but in the absence of any true dynamical calculation, the details must remain unclear. It nevertheless seems clear that a primordial magnetic field would have many intriguing consequences, some of which might actually be observable. It would of course be of great interest to detect this relic field directly in the intergalactic space.

Acknowledgements

I wish to thank Poul Olesen for many enjoyable discussions on primordial magnetic fields.

References

[1] Ya.B. Zeldovich, A.A. Ruzmaikin and D.D. Sokoloff, *Magnetic Fields in Astrophysics* (McGraw-Hill, New York, 1980); E.N. Parker, *Cosmological Magnetic Fields* (Oxford Univ. Press, Oxford, 1979); A.A. Ruzmaikin, A.A. Shukurov and D.D. Sokoloff, *Magnetic Fields of Galaxies* (Kluwer, Dordrecht, 1988).

[2] P.P. Kronberg, J.J. Perry and E.L.H. Zukovski, Astrophys. J. **387** (1992) 393.

[3] X. Chi and A. W. Wolfendale, Nature **362** (1993) 610.

[4] See e.g. A. Brandenburg *et al.*, Astron. Astrophys. **271** (1993) 36, and references therein.

[5] M. Simard-Normandin and P.P. Kronberg, Nature **229** (1979) 115; Astrophys. J. **242** (1980) 74.

[6] A. Poedz, A. Shukurov and D. Sokoloff, Mon. Not. R. Astron. Soc. **264** (1993) 285.

[7] K. Enqvist, V. Semikoz, A. Shukurov and D. Sokoloff, Phys. Rev. **D48** (1993) 4557.

[8] C.J. Hogan, Phys. Rev. Lett. **51** (1983) 1488; M.S. Turner and L.W. Widrow, Phys. Rev. **D37** (1988) 2743; B. Ratra, Astrophys. J. Lett. **391** (1992) L1; W. Garretson, G.B. Field and S.M. Carroll, Phys. Rev. **D46** (1992) 5346;

[9] T. Vachaspati, Phys. Lett. **B265** (1991) 258.

[10] K. Enqvist and P. Olesen, Phys. Lett. **B319** (1993) 178.

[11] G.K. Savvidy, Phys. Lett. **B71** (1977) 133.

[12] K. Enqvist and P. Olesen, preprint NORDITA-94/6 P, to appear in Phys. Lett. **B**.

[13] B. Cheng, D.N. Schramm and J.W. Truran, preprint FERMILAB-PUB-93-259-A.

[14] H.D. Trottier and R.M. Woloshyn, Phys. Rev. Lett. **70** (1993) 2053; A.R. Levy and J. Polonyi, MIT-CTP-2161 (1993).

[15] W. Dittrich, Phys. Rev. **D19** (1979) 2385; P. Elmfors, D. Persson and B.-S. Skagerstam, preprint NORDITA-93/78 P.

[16] N.K. Nielsen and P. Olesen, Nucl. Phys. **B144** (1978) 376.

[17] L.D.Landau and E.M.Lifshitz, *Electrodynamics of Continuous Media* (Pergamon, Oxford, 1960)

[18] K. Enqvist and P. Olesen, preprint NORDITA-94/23 P.

ANOTHER LOOK AT THE ELECTROWEAK VORTEX SOLUTION

F. R. Klinkhamer

CHEAF / NIKHEF–H
Postbus 41882
NL–1009 DB Amsterdam

Many, if not most, of the topics discussed at this workshop are non-perturbative in origin. For this reason it may be of interest to report here on one minor, but rigorous, non-perturbative result obtained in collaboration with P. Olesen [1]. Similar ideas have been presented at this meeting by M. James. The goal of this talk then is to suggest a new point of view for an old classical solution, namely the well-known vortex solution [2] as embedded in the electroweak standard model [3]. Specifically, we would like to know where in configuration space this solution fits in.

Let us start by reviewing what limited knowledge we have of configuration space, i. e. the abstract space of static, finite energy field configurations, with the gauge freedom eliminated. For simplicity, we consider only the bosonic fields of the electroweak standard model (the response of the fermionic fields is certainly important for the physics applications, but the crucial dynamics is beleived to be carried by the bosonic fields). The classical vacuum, with vanishing gauge fields and a constant Higgs field (energy $E_{\mathrm{V}} = 0$), corresponds to a single point V in configuration space. The energy surface at this point V is stationary, in other words the vacuum configuration solves the classical field equations. It turns out that the topology of configuration space is highly non-trivial and this leads to the existence of other stationary points, i. e. new classical solutions. We mention two of them.

First, there exists a non-contractible loop (NCL) in configuration space, parametrized by $\mu \in [-\pi, +\pi]$, which for $\mu = \pm\pi$ passes through the vacuum V and for $\mu = 0$ through a new classical solution, the sphaleron S [4]. This NCL captures some of the topology of configuration space, because it is based on a non-trivial map $S_1 \times S_2 \rightarrow SU(2) \sim S_3$, where S_1 refers to the loop parameter μ, S_2 to the angles on the sphere at spatial infinity and SU(2) to the non-abelian gauge group of the electroweak standard model. In short, the NCL wraps around a "hole" in configuration space. The energy of the corresponding sphaleron S is a function of the mass ratios M_H/M_W and M_Z/M_W, its order of magnitude being $E_{\mathrm{S}} = \mathrm{O}(M_W/\alpha) \sim 10$ TeV. Second, there exists a non-contractible sphere (NCS) in configura-

Electroweak Physics and the Early Universe, Edited by
Romao and Freire, Plenum Press, New York, 1994

tion space, parametrized by $\mu, \nu \in [-\pi, +\pi]$, which for $\mu = \pm\pi$ or $\nu = \pm\pi$ passes through V and for $\mu = \nu = 0$ through another sphaleron S* [5]. The essential non-trivial map is now $S_2 \times S_2 \rightarrow SU(2)$, where the first S_2 refers to the sphere with parameters μ and ν and the second to sphere at spatial infinity. The energy of the sphaleron S* is a little less than twice that of the sphaleron S. As to the physics applications, the crucial observation is that both the NCL and the NCS are related to anomalies, respectively the chiral U(1) anomaly and the global SU(2) anomaly. More concretely, the sphaleron S seems to play a role in B+L violating processes at high temperatures [4, 6, 7], whereas the sphaleron S*, or rather its related constrained instanton I*, may have to do with the asymptotics of perturbation theory [5, 8, 9].

These considerations were for 3-dimensional configurations of finite energy E, but we can also restrict ourselves to field configurations which are constant in, say, the x_3 direction. [Cylindrical coordinates may be defined in terms of the cartesian coordinates by $(\rho\cos\phi, \rho\sin\phi, z) \equiv (x_1, x_2, x_3)$.] The dimensionless "energy" for these 2-dimensional configurations is then given by

$$\begin{aligned}\epsilon &\equiv \frac{1}{\pi v^2}\int dx_1 dx_2\, e \\ &= \frac{1}{\pi v^2}\int dx_1 dx_2 \left[\frac{1}{4\,g^2}\left(W^a_{mn}\right)^2 + \frac{1}{4\,g'^2}\left(B_{mn}\right)^2 + |D_m\Phi|^2 + \lambda\left(|\Phi|^2 - v^2/2\right)^2\right], \quad (1)\end{aligned}$$

with field strengths and covariant derivatives

$$\begin{aligned} W_{mn} &\equiv W^a_{mn}\,\tau^a \equiv \partial_m W_n - \partial_n W_m + [W_m, W_n] \\ B_{mn} &\equiv \partial_m B_n - \partial_n B_m \\ D_m\Phi &\equiv \left(\partial_m + B_m/(2\,i) + W_m\right)\Phi \\ W_m &\equiv W^a_m\,\tau^a \\ \tau^a &\equiv \sigma^a/(2\,i)\,. \end{aligned}$$

Here W and B are the SU(2) and U(1) gauge fields and Φ is the complex doublet Higgs field. The semiclassical masses of the $W^\pm$ and Z^0 vector bosons are $M_W = \frac{1}{2}\,g\,v$ and $M_Z = M_W/\cos\theta_w$, with the weak mixing angle θ_w defined as $\tan\theta_w \equiv g'/g$, and the mass of the physical Higgs scalar is $M_H = \sqrt{8\,\lambda/g^2}\,M_W$. Our goal now is to investigate the topology of this different configuration space, i. e. the space of finite ϵ field configurations. Again, we will do this by constructing a non-contractible sphere.

Introducing for the parameters $\mu, \nu \in [-\pi, +\pi]$ the notation $[\mu\nu] \equiv \max(|\mu|, |\nu|)$, our NCS of 2-dimensional configurations is given by

$$\begin{aligned} \pi/2 \le [\mu\nu] \le \pi \;:\;\; & W = 0 \\ & B = 0 \\ & \Phi = \left(1 - (1-h)\sin[\mu\nu]\right)\frac{v}{\sqrt{2}}\begin{pmatrix}0\\1\end{pmatrix} \\[1em] 0 \le [\mu\nu] < \pi/2 \;:\;\; & W = -f\,G^a\,\tau^a \\ & B = f\,\sin^2\theta_w\,F^3 \\ & \Phi = h\,\frac{v}{\sqrt{2}}\,U\begin{pmatrix}0\\1\end{pmatrix}, \end{aligned} \qquad (2)$$

with the following Lie algebra valued 1-forms

$$\begin{aligned} F^a\,\tau^a &\equiv U^{-1}\,\mathrm{d}U \\ G^a\,\tau^a &\equiv U\left[F^1\,\tau^1 + F^2\,\tau^2 + \cos^2\theta_w\,F^3\,\tau^3\right]U^{-1}, \end{aligned}$$

$SU(2)$ matrices

$$
\begin{aligned}
M(\mu,\nu,\phi) &= \begin{pmatrix} \sin\mu \\ \cos\mu\ \sin\nu \\ \cos\mu\ \cos\nu\ \sin\phi \\ \cos\mu\ \cos\nu\ \cos\phi \end{pmatrix} \cdot \begin{pmatrix} -i\sigma_1 \\ -i\sigma_2 \\ -i\sigma_3 \\ 1 \end{pmatrix} \\
U(\mu,\nu,\phi) &= M(\mu,\nu,0)^{-1}\ M(\mu,\nu,\phi)
\end{aligned}
$$

and boundary conditions for the axial functions $f(\rho)$ and $h(\rho)$

$$
\begin{aligned}
f(0) = h(0) &= 0 \\
\lim_{\rho\to\infty} f, h &= 1\,.
\end{aligned}
$$

The topology of configuration space is captured by the non-trivial map $M : \mathrm{S}_2 \times \mathrm{S}_1 \to \mathrm{SU}(2)$, where S_2 refers to the sphere with parameters μ, ν and S_1 to the circle at spatial infinity. This map with winding number $m = 1$ can be generalized by replacing $\sin\phi$ and $\cos\phi$ by $\sin m\phi$ and $\cos m\phi$. Also note that there is no longer the possibility of having a non-contractible loop, since the map $\mathrm{S}_1 \times \mathrm{S}_1 \to \mathrm{SU}(2)$ is always trivial.

With these configurations the resulting dimensionless energy density is for $[\mu\nu] < \pi/2$

$$
\begin{aligned}
e &= \cos^2\mu\ \cos^2\nu\ \left(1 - \cos^2\mu\ \cos^2\nu\ \sin^2\theta_w\right)\ \cos^{-2}\theta_w\ \left[\rho^{-2}\,(\partial_\rho f)^2\right] + \\
&\quad \cos^2\mu\ \cos^2\nu\ \left[\rho^{-2}\,h^2\,(1-f)^2\right] + (\partial_\rho h)^2 + \frac{1}{4}\left(\frac{M_H}{M_Z}\right)^2 \left(h^2 - 1\right)^2
\end{aligned} \tag{3}
$$

and for $\pi/2 \leq [\mu\nu] \leq \pi$

$$
e = (\partial_\rho k)^2 + \frac{1}{4}\left(\frac{M_H}{M_Z}\right)^2 \left(k^2 - 1\right)^2, \tag{4}
$$

with $k \equiv 1 - (1-h)\sin[\mu\nu]$. Perhaps the main subtlety in the ansatz (2) is the way the SU(2) and U(1) gauge fields are distributed, in order to keep their total kinetic energy density (3) down by partially cancelling the $\cos^{-2}\theta_w$ factor.

We now observe that the NCS configuration at $\mu = \nu = 0$ gives precisely the electroweak vortex solution (Z-string). Just recall that the fields of the Z-string can be written in the following form [2, 3]

$$
\begin{aligned}
W &= -\cos^2\theta_w\, f_Z\, \mathrm{d}V\ V^{-1} \\
B &= -\tan^2\theta_w\, W^3 \\
\Phi &= h_Z\,\frac{v}{\sqrt{2}}\,V\begin{pmatrix} 0 \\ 1 \end{pmatrix},
\end{aligned} \tag{5}
$$

where the $SU(2)$ matrix V is simply

$$
V = \cos\phi\, 1 - \sin\phi\, i\sigma_3 \equiv U(0,0,\phi)
$$

and the functions f_Z and h_Z are the solutions of the field equations with the standard boundary conditions. The corresponding energy density of the Z-string

$$
e_Z = \rho^{-2}\,(\partial_\rho f_Z)^2 + \rho^{-2}\,h_Z^2\,(1-f_Z)^2 + (\partial_\rho h_Z)^2 + \frac{1}{4}\left(\frac{M_H}{M_Z}\right)^2 \left(h_Z^2 - 1\right)^2 \tag{6}
$$

Table 1. Numerical results for the Z-string solution: the dimensionless energy ϵ_Z, the integrals a–d and the critical mixing angle $\theta_w^{\rm NCS}$.

	M_H/M_Z						
	1/8	1/4	1/2	1	2	4	8
ϵ_Z	0.47	0.59	0.76	1.00	1.34	1.79	2.35
a	0.16	0.18	0.20	0.21	0.21	0.20	0.18
b	0.19	0.22	0.25	0.29	0.35	0.43	0.52
c	0.50	0.43	0.36	0.29	0.22	0.16	0.12
d	0.15	0.17	0.19	0.21	0.22	0.21	0.18
$\sin^2\theta_w^{\rm NCS}$	0.69	0.69	0.69	0.70	0.73	0.76	0.79

is independent of the value of the weak mixing angle θ_w.

With the axial functions of the Z-string $f = f_Z$ and $h = h_Z$, the energy profile over the NCS (2) is for $0 \leq [\mu\nu] < \pi/2$

$$\epsilon(\mu,\nu) = \epsilon_Z \left[\cos^2\mu \, \cos^2\nu \left(\frac{1-\cos^2\mu\,\cos^2\nu\,\sin^2\theta_w}{\cos^2\theta_w}\, a + b \right) + c + d \right] \tag{7}$$

$$= \epsilon_Z \left[1 + (\mu^2+\nu^2)\left(\tan^2\theta_w - 1 - b/a\right) a + {\rm O}(\mu^4,\,\nu^4,\,\mu^2\nu^2) \right] , \tag{8}$$

in terms of the following integrals

$$\begin{aligned} a &\equiv \epsilon_Z^{-1} \int_0^\infty d\rho\, \rho^{-1}\, (\partial_\rho f_Z)^2 \\ b &\equiv \epsilon_Z^{-1} \int_0^\infty d\rho\, \rho^{-1}\, h_Z^2\, (1-f_Z)^2 \\ c &\equiv \epsilon_Z^{-1} \int_0^\infty d\rho\, \rho\, (\partial_\rho h_Z)^2 \\ d &\equiv \epsilon_Z^{-1} \int_0^\infty d\rho\, \rho\; 1/4\; (M_H/M_Z)^2 \left(h_Z^2 - 1\right)^2 , \end{aligned}$$

which add up to 1. From (8) there is manifest instability of the Z-string for $\theta_w \leq \pi/4$ and arbitrary Higgs mass. In short, the Z-string of the electroweak interactions (with the experimental value $\theta_w \sim \pi/6$) is the 2-dimensional sphaleron of a non-contractible sphere.

Numerical results allow us to extend the range of instability beyond $\sin^2\theta_w = 1/2$. In Table 1 we give the maximum allowed value θ_w for instability according to (8). The Z-string solution has for $\theta_w < \theta_w^{\rm NCS}$ two negative modes, made explicit by the fields (2) of the NCS. This range of instability is consistent with the results of [10].

These numerical results give also the complete energy profile (7) of the non-contractible sphere through the Z-string, as a function of both θ_w and M_H/M_Z. For the case of perturbative stability [10] of the Z-string, we find an energy barrier for decay towards the vacuum. Of course, this is not yet the minimal energy barrier, but it may be indicative. Taking the parameter values $M_H/M_Z = 1/8$ and $\sin^2\theta_w = 63/64$, for example, we easily find the $\mu = 0$ section of the energy profile $\epsilon(\mu,\nu)$ over the NCS : starting at $\nu = 0$, with $\epsilon = \epsilon_Z = 0.47$, the energy rises to a maximum value $\epsilon = 1.57$ at $\nu \sim \pi/4$ and then drops to zero as $\nu \to \pi$.

To conclude, we have obtained a topological understanding for the existence and generic instability of the electroweak vortex solution.

Acknowledgements

It is a pleasure to thank F. Freire, J. C. Romão and the other organizers for bringing the Early Universe (and us) to Sintra.

References

[1] F. Klinkhamer and P. Olesen, Nucl. Phys. **B** in press

[2] H. Nielsen and P. Olesen, Nucl. Phys. **B61** (1973), 45

[3] Y. Nambu, Nucl. Phys. **B130** (1977), 505

[4] F. Klinkhamer and N. Manton, Phys. Rev. **D30** (1984), 2212

[5] F. Klinkhamer, Nucl. Phys. **B410** (1993), 343

[6] P. Arnold and L. McLerran, Phys. Rev. **D36** (1987), 581

[7] M. Dine, O. Lechtenfeld, B. Sakita, W. Fisschler and J. Polchinski, Nucl. Phys. **B342** (1990), 381

[8] F. Klinkhamer, Nucl. Phys. **B407** (1993), 88

[9] V. Rubakov and O. Shvedov, *Sphalerons and large order behaviour of perturbation theory in lower dimension*, preprint hep-ph/9404328

[10] M. James, L. Perivolaropoulos and T. Vachaspati, Nucl. Phys. **B395** (1993), 534

ELECTROWEAK BARYOGENESIS WITH TOPOLOGICAL DEFECTS

A.C. Davis[1], R.H. Brandenberger[2] and M. Trodden[2]

[1]DAMTP
Cambridge University
Silver Street
Cambridge CB3 9EW
UK
[2]Physics Department
Brown University
Providence
RI 02912
USA

1. INTRODUCTION

Electroweak baryogenesis has become a topic of much recent activity [1]. Here we discuss a new scenario which has the advantage of being insensitive to the order of the electroweak phase transition. We briefly review a mechanism [2] using unstable electroweak strings [3] and discuss in detail a mechanism [4] using topological defects (in particular cosmic strings) left behind after a previous phase transition.

In the standard electroweak theory all three necessary conditions to generate a net baryon number are satisfied. Sphaleron transitions violate baryon number [5]. The electroweak theory explicitly violates C invariance and in extensions of the standard model with non-minimal Higgs structure there is explicit CP violation [6]. Finally, out of equilibrium field configurations may result as remnants of the phase transition.

The key issue is how the 'out of equilibrium' condition is realised. In most previous work [7,8] use was made of bubble walls which form if the electroweak phase transition is first order. However, at present it is unclear [9] whether the electroweak phase transition is sufficiently strongly first order for baryogenesis mechanisms involving bubble walls to be effective. In [2] it was pointed out that topological or non-topological defects may play a role similar to bubble walls in triggering electroweak baryogenesis. Such a mechanism is independent of the order of the phase transition.

In the next section we show how electroweak baryogenesis can be implemented in models with a second order phase transition and compare it with the first order case. A mechanism with metastable electroweak strings is then briefly reviewed. A more robust mechanism using topological defects produced in a previous phase transition is discussed and the resulting baryon asymmetry estimated. Finally, we discuss future extensions to our work and conclusions.

Electroweak Physics and the Early Universe, Edited by
Romao and Freire, Plenum Press, New York, 1994

2. ELECTROWEAK BARYOGENESIS WITH A SECOND ORDER PHASE TRANSITION

Let us review how Sakharov's conditions are realised in electroweak baryogenesis scenarios and compare the implementations in first and second order phase transitions.

Baryon number violation occurs via sphaleron transitions. The transition rate is exponentially suppressed in the broken phase. However, in the symmetric phase transitions are copious. Their rate per unit volume is [10]

$$\Gamma_B \sim \alpha_W^4 T^4 \tag{2.1}$$

where $\alpha_W = g^2/4\pi$, g being the $SU(2)$ gauge coupling constant. In extensions of the standard electroweak theory with non-minimal Higgs structure containing explicit CP violation, there exists a CP violating phase which changes by an amount $\Delta\theta$ during the phase transition.

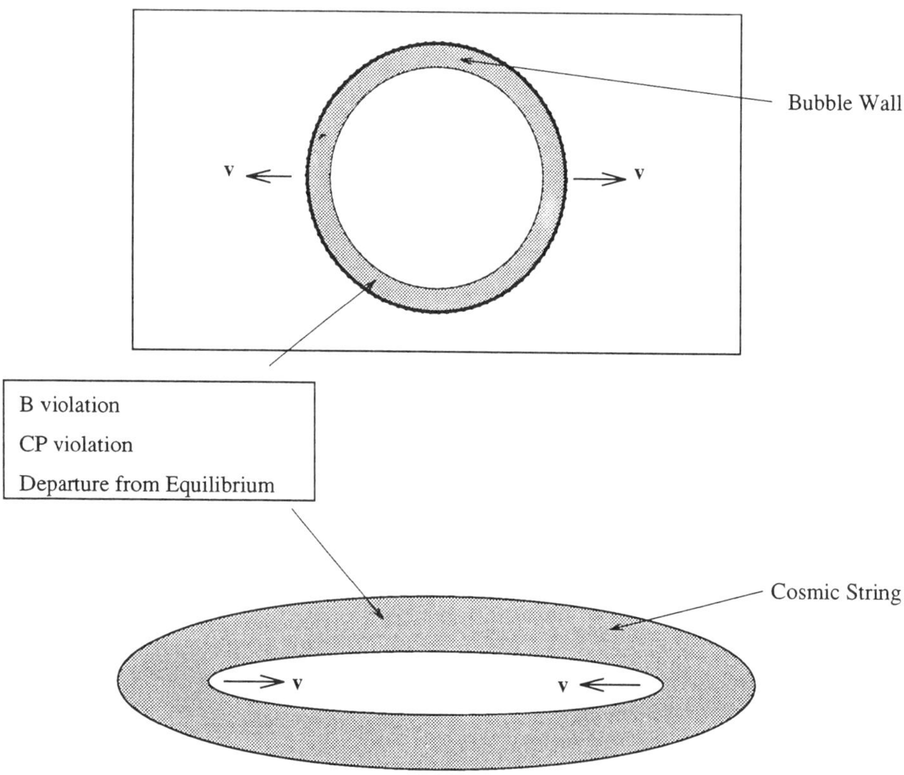

Figure 1. A comparison between the electroweak baryogenesis mechanisms using first order phase transitions (top) and our mechanism (bottom). The contracting topological defect (bottom) plays a role similar to that of the expanding bubble wall (top) in that it is the location of extra CP violation and of baryon number violating processes taking place out of equilibrium.

In Fig.1 we compare the ways in which the out of equilibrium condition is realized in models with first and second order phase transitions. The key role is played by expanding bubble walls and contracting topological defects respectively. In the scenarios of Refs.1,2, baryogenesis takes place in the outer edge of the bubble wall, ie. where

$$|\phi| < g\,\eta_{EW} \tag{2.2}$$

$|\phi|$ being the order parameter of the transition. The amplitude $|\phi|$ is increasing at any point in space which the bubble wall crosses. This may be related to the change of the CP-odd phase and hence CP violation has a preferred sign. Finally, as long as the bubble wall moves at relativistic speed, there will be no time to establish thermal equilibrium inside the walls.

With a second order phase transition, the role of the bubble wall is played by topological defects. Baryogenesis takes place inside the core of the defect and the surrounding region where (2.2) is satisfied. If the defects contract (and eventually evaporate), then there will be an overall increase in $|\phi|$ and hence net baryon number generation. The field configurations within contracting topological defects are out of thermal equilibrium.

In the next section we consider a specific implementation of this using electroweak strings.

3. BARYOGENESIS WITH ELECTROWEAK STRINGS

In [2] it was suggested that electroweak strings could provide the out of equilibrium condition necessary to generate a baryon asymmetry. The electroweak theory doesn't admit topologically stable strings. However, it has been shown that it is possible to embed the Nielsen-Olesen vortex [11] in the electroweak theory [12]. Such strings could be metastable, in which case their formation would be similar to that of topological strings. The core of the string would play a similar role to the bubble wall in first order phase transitions. In the core of the string the electroweak symmetry is restored and baryon violating processes are unsuppressed with rate as in (2.1). Outside the string the electroweak symmetry is broken and baryon violating transitions are exponentially suppressed. Since the strings are at best metastable and of finite length, terminating on a monopole anti-monopole pair, they will collapse along their axis. This collapse of the string provides the out of thermal equilibrium condition. If this is the case in the 2-doublet model then in the tip of the string the CP violating phase is rapidly changing. Hence, we have all the necessary conditions to generate a baryon asymmetry.

An optimistic estimate of the baryon asymmetry can be obtained by assuming that the mean length and average separation of electroweak strings at t_G, the time corresponding to the Ginsburg temperature of the phase transition, is the correlation length, $\xi(t_G) \sim \lambda^{-1}\eta^{-1}$. The strings will collapse along their axes and decay in time interval

$$\Delta t_s \sim v^{-1}(\lambda\eta)^{-1} \tag{3.1}$$

where v is the velocity of collapse, taken to be ~ 1. By considering the rate of change of the volume in which CP violation is effective we can estimate the rate of baryon number generation per string to be

$$\frac{dN_B}{dt} \sim w^2 v \Gamma_B \Delta\theta \Delta t_c \tag{3.2}$$

where $w \sim \lambda^{-\frac{1}{2}}\eta^{-1}$ is the string width and Δt_c is the length of time a fixed point in space is in the transition region. Taking one string per correlation volume $\xi(t_G)^3$ and integrating from the Ginsburg time t_G to $t_G + \Delta t_s$ gives

$$\frac{n_B}{s} \sim \frac{9}{2\pi^2 g^*} \frac{\lambda}{\gamma(v)v} \Delta\theta g^3 \alpha_W^4 \tag{3.3}$$

where g^* is the effective number of degrees of freedom. In our estimate we have included a suppression for the fact that baryon violating processes are only unsuppressed for $|\phi| < g\eta$.

For our mechanism to work we require the core radius to be large enough to support sphaleron processes. In addition, sphaleron transitions must be suppressed in the broken phase for $T = T_G$. Finally, we require the electroweak string to be metastable in the 2-doublet model. Unfortunately, it has been shown [13] that this is not the case for physical Weinberg angle [14]. Attempts to stabilise it with quark and lepton condensates have been unsuccessful [15]. Hence, a more robust method of electroweak baryogenesis could involve topological defects formed at a phase transition at a scale above the electroweak scale.

4. BARYOGENESIS WITH TOPOLOGICAL DEFECTS

In order to obtain topological defects – cosmic strings to be specific – we assume that at an energy scale η, larger than η_{EW}, there is another symmetry breaking which produces strings. One possibility is to embed $SU(2) \times U(1)$ in some larger simply connected group G such that at a scale η

$$G \longrightarrow SU(2) \times U(1)$$

and

$$\Pi_1 \left(G/(SU(2) \times U(1)\right) \neq 1$$

A second possibility is to assume that electroweak symmetry breaking is induced dynamically by having a technifermion condensate form at the scale η_{EW}:

$$\langle \bar{\psi}_{TC} \psi_{TC} \rangle \neq 0 \,, \quad T \leq \eta_{EW} \,.$$

Here, ψ_{TC} denotes the technifermion. In this case we can assume that fermion masses are induced by a second phase transition in the technifermion sector at a scale η which in general is only slightly higher than η_{EW}. It is possible that strings form in this transition. In the core of these strings the fermion condensates vanish, the electroweak symmetry is unbroken, and hence baryon number violating processes are unsuppressed.

The region of electroweak symmetry restoration is actually larger than the core of the string. In [16] it was shown that cosmic strings formed at a previous phase transition and coupled to the Weinberg-Salam model restore the electroweak symmetry out to a region of order η_{EW}^{-1}. This is a result of the coupling between the string fields and the electroweak gauge fields. If the string is superconducting then the region of symmetry restoration is much larger, being given by [17,16].

$$R_s \sim I/\eta_{EW}^2 \tag{4.1}$$

where I is the current carried by the string. The maximal current is of order η. (For grand unified strings this region is macroscopic, being of order 10^{-5} m!)

Let us now give a rough estimate of the baryon to entropy ratio which can be generated using the proposed mechanism and compare the result with that obtained by the mechanisms of Refs.7,8 which rely on bubble wall expansion. To simplify the calculations we assume that the phase transition is rapid and that the strings move relativistically (in order that the out-of-equilibrium condition is satisfied).

The important parameters in our calculation are the total volume V, the volume V_{BG} in which net baryon number violating processes are taking place, the rate Γ_B of these processes (see (2.1)), and the net change

$$\Delta\theta = \int dt\dot{\theta}$$

in the CP violating phase θ. We are making the plausible assumption that the electroweak symmetry is restored inside the core of the string. In this case, the mean value of the CP violating phase vanishes in the core. In the broken phase, the distinguished value of θ will be nonvanishing. Hence, for points in space initially inside the string core, the net change in θ will have a preferred direction. In this respect there is no difference between our mechanism and the ones of Refs.7,8.

The net baryon number density Δn_B is then given by

$$\Delta n_B = \frac{1}{V}\frac{\Gamma_B}{T} V_{BG}\Delta\theta \,. \tag{4.2}$$

The volume V_{BG} is determined by the mean separation ξ of the strings and the radius R_s of the string core (strictly speaking the part of the core where (2.2) is satisfied and hence n_B violating processes are unsuppressed).

The key to the calculation is a good estimate of V_{BG}. Note that the translational motion of a topological defect does not lead to any net baryogenesis since $\Delta\theta = 0$ integrated over time. At the leading edge of the moving defect, a baryon number with one sign will be produced, but at the trailing edge baryogenesis will have the opposite sign. We will return to this point later. Contraction, on the other hand, does produce a net Δn_B. Integrated over time, there is a net $\Delta\theta \neq 0$ in the entire volume corresponding to the initial defect configuration. Hence, a lower bound on the volume V_{BG} is obtained by taking the volume occupied initially by the collapsing defects at the time when baryogenesis commences (see below). We focus on string loops. Their mean separation is $\xi(t)$. Hence, in one horizon volume

$$V = \frac{4\pi}{3} t^3 \tag{4.3}$$

the corresponding volume where net baryon number generation takes place is

$$V_{BG} \sim R_s^2 \xi(t) \left(\frac{t}{\xi(t)}\right)^3 . \tag{4.4}$$

The last factor on the right hand side is the number of string loops per horizon volume, the second factor is the length of a loop.

Most of the contribution to the baryon to entropy ratio is generated at a time t_U soon after t_{EW} when sphaleron processes cease to be thermally excited in the true vacuum. To simplify the equations, we will set the two times equal in the following. Thus, to obtain an order of magnitude estimate of the strength of our baryogenesis mechanism we will evaluate all quantities at t_{EW}. Combining (4.2)-(4.4) and (2.1) yields

$$\Delta n_B(t_{EW}) \sim \alpha_W^4 \Delta\theta \left(\frac{R_s}{\xi(t_{EW})}\right)^2 T_{EW}^3 \tag{4.5}$$

or

$$\frac{\Delta n_B}{s} \sim g^{*-1} \alpha_W^4 \Delta\theta \left(\frac{R_s}{\xi(t_{EW})}\right)^2 \equiv g^{*-1} \alpha_W^4 \Delta\theta\,(SF)\,, \tag{4.6}$$

with g^* being the number of spin degrees of freedom in radiation, and with

$$(SF) = \left(\frac{R_s}{\xi(t_{EW})}\right)^2 . \tag{4.7}$$

Apart from the factor (SF), this is the same order of magnitude as obtained in the mechanisms using a first order phase transition [7,8]. Hence, we call (SF) the "suppression factor".

Above, we implicitly assumed that all strings have the same radius. This is a good approximation for strings in the friction dominated epoch [18], but not for a string network in the scaling regime. In the latter case we need to integrate over all loop sizes to obtain (SF).

The above analysis does not depend on the topology of the defect in a key way. For collapsing domain walls, our mechanism is stronger since the suppression factor (SF) would be

$$(SF) \sim \frac{R_c}{\xi(t_{EW})} ,$$

R_c being proportional to the domain wall core radius. For collapsing monopoles, however, the mechanism is weaker since

$$(SF) \sim \left(\frac{R_c}{\xi(t_{EW})}\right)^3 .$$

There are two ways to increase (SF): either we decrease $\xi(t_{EW})$ or we increase R_s. The obvious way to decrease $\xi(t_{EW})$ is to decrease the scale η of the string producing phase transition. The earlier we are in the friction dominated epoch, the closer the strings are relative to the horizon since [18]

$$\xi(t) \sim \xi(t_f) \left(\frac{t}{t_f}\right)^{5/4} , \tag{4.8}$$

t_f being the time of string formation (given by η). According to the Kibble mechanism

$$\xi(t_f) \simeq \lambda^{-1}\eta^{-1}, \tag{4.9}$$

where λ is the string scalar field self coupling constant. As mentioned previously, a cosmic string coupled to the Weinberg-Salam model restores the electroweak symmetry in a region

$$R_s \sim \eta_{EW}^{-1} \tag{4.10}$$

and therefore

$$(SF) \sim \left(\frac{\eta_{EW}}{\eta}\right)^3 . \tag{4.11}$$

The second way to increase (SF) is to make the strings superconducting. For maximal string current this region is

$$R_s \sim \frac{\eta}{\eta_{EW}^2} \tag{4.12}$$

In general, however, the initial current on a superconducting string will be much less than η and thus R_s smaller than (4.12). To obtain a more realistic estimate consider the superconductivity to be bosonic and the winding of the boson condensate at formation to be $\sim$1. The winding at later times is $\sim N^{\frac{1}{2}}$, where N is the ratio of comoving volumes corresponding to loop sizes at t and t_f respectively. This winding in the field giving rise to superconductivity induces a current of order

$$I \sim I_{\max}(T/\eta)^{\frac{1}{4}} \tag{4.13}$$

giving

$$R_s \sim (T/\eta)^{\frac{1}{4}}\,\frac{\eta}{\eta_{EW}^2} \tag{4.14}$$

and resulting suppression factor of

$$SF \sim \lambda^2 \left(\frac{\eta_{EW}}{\eta}\right)^{\frac{3}{2}} \tag{4.15}$$

For η just above the electroweak transition and λ close to unity the strength of this mechanism for baryogenesis is comparable to the first order one.

5. FUTURE DEVELOPMENTS

In the previous section we ignored the effect of the string's transverse motion and just concentrated on the initial collapse of the loop. For superconducting strings this is a gross underestimate. In this case the region of symmetry restoration, R_s, is significantly larger than the mean free path of baryons within the string. As the string moves, the CP violating phase, $\Delta\theta$, is equal and opposite on the leading and trailing edges of the string. At the leading face anti-baryons are produced. However, before the trailing face arrives the anti-baryons have already decayed to anti-leptons and thus do not annihilate with the baryons produced at the trailing edge. The result of this is that the volume of baryogenesis is

$$V_{BG} \sim \xi(t)^2 R_s \left(\frac{t}{\xi(t)}\right)^3 \tag{5.1}$$

rather than (4.4). The resulting suppression factor becomes

$$SF \sim \frac{R_s}{\xi(t_{EW})} \sim \lambda \left(\frac{\eta_{EW}}{\eta}\right)^{\frac{3}{4}} \tag{5.2}$$

This makes our mechanism even more attractive.

Finally, we note that the baryogenesis mechanism discussed by Barriola [19] in the context of electroweak strings can also be applied to topological defects. Barriola noted that integrating the anomaly in the baryon current gives the change in baryon number in a given volume and time interval

$$\begin{aligned}\Delta B = \frac{N_f}{32\pi^2}\int d^4x[g^2 E_W.B_W + G^2\cos(2\theta_W)E_Z.B_Z \\ + \frac{G^2}{2}\sin(2\theta_W)(E_Z.B_A + E_A.B_Z)]\end{aligned} \tag{5.3}$$

where E and B refer to the electric and magnetic fields of the physical fields after symmetry breaking, N_f refers to the number of families and $G^2 = g^2 + g'^2$, where g and g' are the $SU(2)$ and $U(1)$ coupling constants respectively. In (5.3) the first term is small in the broken phase, the second term is only non-zero in defects with a non-zero helicity [20], but the third term could be large. In the region R_s around a cosmic string there is a non-trivial Z-flux [16]. As discussed at this meeting [20,21] the electroweak phase transition produces a background magnetic field of order $B_A \sim gT^2$. In the two-doublet model this gives a contribution to the action of

$$\Delta S = \frac{N_f G^2}{64\pi^2}\sin(2\theta_W)\int d^4x\theta(E_A.B_Z + B_A.E_Z) \tag{5.4}$$

Integrating by parts this gives a contribution to the free energy density

$$F_B = -\dot{\theta}B \tag{5.5}$$

In the core of collapsing topological defects $\dot{\theta} > 0$ and so B is driven positive to minimise the free energy.

As a first estimate the rate per unit volume of baryon violation for electroweak strings is

$$\Gamma_B \sim \gamma_v v_t g^2 T^4 \tag{5.6}$$

where v_t is the speed at which the loop is collapsing and γ_v is the associated Lorentz factor. This rate is considerably larger than that resulting from sphaleron processes in (2.1).

The change in baryon number is again given by (4.2), but with rate Γ_B as in (5.6) and volume V_{BG} of (5.1). Unlike the mechanism discussed in section 4, this increased rate of baryon violation allows the exciting possibility of electroweak baryogenesis with GUT strings. By the time of the electroweak phase transition the GUT string network has reached a scaling solution. Considering the scaling solution for the number of loops per unit volume and integrating over loop size, we have estimated the suppression factor to be $\sim 10^{-1}$ for a scale of symmetry breaking of 10^{15} GeV. This allows the exciting possibility that the same cosmic strings that may participate in the formation of large scale structure could also mediate electroweak baryogenesis [22].

To conclude, we have discussed a new mechanism for electroweak baryogenesis which operates even if the phase transition is second order. Topological defects produced in a previous phase transition can play a role analogous to bubble walls. We have discussed two mechanisms for generating an asymmetry of order $n_B/s \sim 10^{-10}$. The first involves a transition shortly before the electroweak one and sphaleron processes. The second, and more speculative, involves a new, stronger mechanism of baryogenesis and allows the exciting possibility of GUT strings mediating electroweak baryogenesis.

ACKNOWLEDGEMENTS

We wish to thank Tomislav Prokopec for discussions on section 5. ACD thanks the organizers of the Sintra meeting, and particularly Filipe Freire, for inviting her to participate in such a stimulating meeting.

REFERENCES

(1.) N Turok - these proceedings.

(2.) R Brandenberger and A C Davis - Phys Lett B308 (1993).

(3.) T Vachaspati - Phys Rev Lett 68 (1992) 1977.

(4.) R Brandenberger, A C Davis and M Trodden - Phys Lett B(in press).

(5.) N Manton - Phys Rev D28 (1983) 2019, F Klinkhammer and N Manton - Phys Rev D30 (1984) 2212.

(6.) J F Gunion, H E Haber, G L Kane and S Dawson, 'The Higgs Hunters' Guide' (Addison-Wesley, 1989).

(7.) N Turok in 'Perspectives on Higgs Physics' , ed G Kane (World Scientific, 1992) and references therein.

(8.) A Cohen, D Kaplan and A Nelson - Ann Rev Nucl and Part Sci 43 (1993) 27 and references therein.

(9.) M Dine, R Leigh, P Huet, A Linde and D Linde - Phys Rev D46 (1992) 550

(10.) J Ambjorn, M Laursen and M Shaposhnikov, Phys Lett B197 (1989) 49.

(11.) H B Nielsen and P Olesen - Nucl Phys B61 (1973) 61.

(12.) Y Nambu - Nucl Phys B130 (1977) 505.
T Vachaspati - Phys Rev Lett 68 (1992) 1977.

(13.) M A Earnshaw and M James - Phys Rev D48 (1993) 5818.

(14.) M James, L Perivolaropoulos and T Vachaspati - Phys Rev D46 (1992) 5232.
W B Perkins - Phys Rev D47 (1993) 5224.

(15.) M A Earnshaw and W B Perkins - Phys Let B (in press).

(16.) W B Perkins and A C Davis - Nucl Phys B406 (1993) 377.

(17.) J Ambjorn and P Olesen - Int J Mod Phys A5 (1990) 4525 and references therein.

(18.) T W B Kibble - Nucl Phys B252 (1985) 227, J Phys A9 (1976) 1387.

(19.) M Barriola - these proceedings.

(20.) T Vachaspati - these proceedings.

(21.) K Enqvist - these proceedings.

(22.) R Brandenberger, A C Davis, T Prokopec and M Trodden - in preparation.

ELECTROWEAK STRINGS, SPHALERONS AND MAGNETIC FIELDS

Tanmay Vachaspati

Tufts Institute of Cosmology
Department of Physics and Astronomy
Tufts University, Medford, MA 02155

1. INTRODUCTION

In this talk I would like to discuss three main topics. First, I will connect electroweak strings and the sphaleron in a more elaborate manner than that described in my paper with George Field [1] and discuss the important issue of the stability of electroweak strings. Secondly, I would like to initiate a discussion of the formation of electroweak strings and their evolution by painting a heuristic picture of the phase transition if the phase transition is weakly first order or second order. Inextricably tied to this picture is a possible scenario for baryogenesis and it is hoped that the present discussion will inspire more rigorous work to either confirm or reject the scenario. Finally, I will make a few comments on the generation of magnetic fields at the electroweak phase transition. Kari Enqvist has already described the basic idea of how this scenario works [2] and the purpose of these comments is to clarify some of the issues that have not been addressed in the literature. I will also describe the scenario in an alternate way that makes an interesting connection between electroweak strings and primordial magnetic fields.

2. THE Z-STRING AND THE SPHALERON

The bosonic sector of the standard model of the electroweak interactions is described by the Lagrangian:

$$L = L_W + L_B + L_\Phi - V(\Phi) \tag{2.1}$$

where,

$$L_W = -\frac{1}{4} W_{\mu\nu a} W^{\mu\nu a} \tag{2.2}$$

$$L_Y = -\frac{1}{4} Y_{\mu\nu} Y^{\mu\nu} \tag{2.3}$$

where $W^a_{\mu\nu}$ and $Y_{\mu\nu}$ are the field strengths for the $SU(2)$ and $U(1)$ gauge fields W^a_μ and Y_μ respectively. Also,

$$L_\Phi = |D_\lambda \Phi|^2 \equiv \left| \left(\partial_\lambda - \frac{1}{2} i g \tau^a W^a_\lambda - \frac{1}{2} i g' Y_\lambda \right) \Phi \right|^2 \tag{2.4}$$

$$V(\Phi) = \lambda (\Phi^\dagger \Phi - \eta^2/2)^2 , \tag{2.5}$$

where, Φ is a complex doublet. In addition, we define

$$\vec{Z} \equiv cos\theta_W n^a \vec{W}^a - sin\theta_W \vec{Y} \ , \quad \vec{A} \equiv sin\theta_W n^a \vec{W}^a + cos\theta_W \vec{Y} \ , \tag{2.6}$$

Electroweak Physics and the Early Universe, Edited by
Romao and Freire, Plenum Press, New York, 1994

where, n^a is the unit vector

$$n^a = -\frac{\Phi^\dagger \tau^a \Phi}{\Phi^\dagger \Phi} \qquad (2.7)$$

$$tan\theta_W \equiv \frac{g'}{g} , \quad \alpha \equiv (g^2 + g'^2)^{1/2} . \qquad (2.8)$$

The electroweak model has two different string solutions [3,4], namely, the $W-$string and the $Z-$string [5,6,7]. The latter is the lighter of the two and is better studied and hence we will not discuss the $W-$string but only comment that some of the ensuing discussion directly applies to the $W-$string as well.

The $Z-$string is a solution to the classical equations of motion following from the Lagrangian in (2.1). In cylindrical coordinates (r, θ, z) it is:

$$\Phi = \frac{\eta}{\sqrt{2}} f(r) e^{im\theta} \begin{pmatrix} 0 \\ 1 \end{pmatrix} , \quad Z_\mu = -\frac{v(r)}{\alpha r} \delta_{\mu\theta} \qquad (2.9)$$

with all the other gauge fields equal to zero. The functions $f(r)$ and $v(r)$ can be evaluated numerically (for example, see Ref. 8). The string solution in (2.9) is precisely a Nielsen-Olesen vortex [9] that has been embedded in the electroweak model [3,4].

The field configuration in (2.9) describes a tube of energy density localized around the z-axis. At the center of this tube the Higgs field vanishes. The tube contains a flux of Z-magnetic field whose value is:

$$F_Z = \frac{4\pi}{\alpha} = \frac{4\pi}{e} sin\theta_W \ cos\theta_W . \qquad (2.10)$$

where, $e = g sin\theta_W$ is the electric charge on an electron. The energy per unit length of the string is:

$$\mu = \pi \eta^2 M(\beta) \qquad (2.11)$$

where $M(1) = 1$ and the tendency of M is to increase with increasing β. But the dependence of M on β is not too strong in the neighbourhood of $\beta = 1$ and so we shall set $M = 1$ whenever we need numerical estimates. (For a plot of μ as a function of β, see Ref. 8.)

The solution in (2.9) describes an infinite $Z-$string along the z-axis but one can think of finite string configurations in the form of closed loops [10] or open segments. Finite string configurations in vacuum would not be static but would oscillate. And if the configuration is large compared to the string thickness, the dynamics is described quite accurately by the Nambu-Goto action. With time such oscillating loops and segments will radiate away their energy and convert into particles. If the loops and segments are not in vacuum but in a plasma that strongly interacts with the strings, the dynamics will be very different and one expects that the oscillations will be damped and the radiation suppressed.

A finite segment of $Z-$string terminates on magnetic monopoles [5]. The electromagnetic flux emanating from a monopole is:

$$F_A = \frac{4\pi}{\alpha} tan\theta_W = \frac{4\pi}{e} sin^2\theta_W . \qquad (2.12)$$

One way to understand the presence of monopoles at the end of $Z-$strings is to note that the Z gauge field is a linear superposition of the W^3 and Y fields as given in eqn. (2.6). Then, when the string terminates, the Y flux cannot terminate because it is a $U(1)$ gauge field and the Y magnetic field is divergenceless. Therefore some field must continue even beyond the end of the string. This has to be the massless field of the theory, that is, the electromagnetic field.

Following Nambu, the asymptotic Higgs field configurations of a monopole and an antimonopole are[5]:

$$\Phi_m = \begin{pmatrix} cos(\theta_m/2) \\ sin(\theta_m/2)\ e^{i\phi} \end{pmatrix} , \quad \Phi_{\bar{m}} = \begin{pmatrix} sin(\theta_{\bar{m}}/2) \\ cos(\theta_{\bar{m}}/2)\ e^{i\phi} \end{pmatrix} \tag{2.13}$$

where θ_m and $\theta_{\bar{m}}$ are spherical angles defined with the monopole and antimonopole at the origin respectively and we have rescaled the Higgs field so that the vacuum manifold is given by $\Phi^\dagger\Phi = 1$. The gauge fields are taken to be so that the covariant derivative of the Higgs field vanishes.

$$gW_\mu^a = -\epsilon^{abc}n^b\partial_\mu n^c + icos^2\theta_W n^a(\Phi^\dagger\ \partial_\mu\Phi - \partial_\mu\Phi^\dagger\ \Phi) \tag{2.14a}$$

$$g'Y_\mu = -isin^2\theta_W(\Phi^\dagger\ \partial_\mu\Phi - \partial_\mu\Phi^\dagger\ \Phi)\ . \tag{2.14b}$$

where, n^a is defined in (2.7). For the special case $\theta_W = 0$, this is equivalent to the usual

$$W_\mu = -i\frac{2}{g}(\partial_\mu U)U^{-1} \tag{2.15}$$

where, U is a 2×2 unitary matrix defined by

$$\Phi = U\,(\,1,0\,)^T \tag{2.16}$$

Note that the monopole configuration has a singularity along the negative z-axis since the Higgs field becomes multi-valued when we set $\theta_m = \pi$. Similarly the antimonopole has a singularity along the positive z-axis ($\theta_{\bar{m}} = 0$). These singularities tell us the location of the $Z-$string that is attached to the monopole and the antimonopole.

Once we have the monopole and antimonopole configurations, we can patch them together to get the field configuration for a finite segment of $Z-$string:

$$\Phi_{m\bar{m}} = \begin{pmatrix} cos(\Theta/2) \\ sin(\Theta/2)\ e^{i\phi} \end{pmatrix} \tag{2.17}$$

where,

$$cos\Theta \equiv cos\theta_m - cos\theta_{\bar{m}} + 1\ . \tag{2.18}$$

It is straightforward to check that (2.17) yields the monopole field configuration close to the monopole ($\theta_{\bar{m}} \to 0$) and the antimonopole configuration close to the antimonopole ($\theta_m \to \pi$). It also yields a string singularity along the straight line joining the monopole and antimonopole ($\theta_m = \pi$, $\theta_{\bar{m}} = 0$).

What is important for us is that there are other Higgs field configurations that also describe monopoles and antimonopoles. These are given by global $U(1)$ transformations of (2.13). Therefore we will write

$$\Phi_m = e^{i\gamma}\begin{pmatrix} cos(\theta_m/2) \\ sin(\theta_m/2)\ e^{i\phi} \end{pmatrix} , \quad \Phi_{\bar{m}} = e^{i\gamma}\begin{pmatrix} sin(\theta_{\bar{m}}/2) \\ cos(\theta_{\bar{m}}/2)\ e^{i\phi} \end{pmatrix} \tag{2.19}$$

This seemingly trivial observation is very useful because it allows us to construct $Z-$string segments which are twisted. Consider the Higgs field configuration:

$$\Phi_{m\bar{m}}(\gamma) = \begin{pmatrix} sin(\theta_m/2)sin(\theta_{\bar{m}}/2)e^{i\gamma} + cos(\theta_m/2)cos(\theta_{\bar{m}}/2) \\ sin(\theta_m/2)cos(\theta_{\bar{m}}/2)e^{i\phi} - cos(\theta_m/2)sin(\theta_{\bar{m}}/2)e^{i(\phi-\gamma)} \end{pmatrix} \tag{2.20}$$

together with the gauge fields given by eqn. (2.14). When we take the limit $\theta_{\bar{m}} \to 0$ we find the monopole configuration of (2.13) and when we take $\theta_m \to \pi$ the configuration is that of the antimonopole of eqn. (2.19) provided we perform the rotation $\phi \to \phi+\gamma$. The monopole and antimonopole in (2.20) also have the usual string singularity joining them. This means that the configuration in (2.20) describes a monopole and antimonopole pair that are joined by a Z−string segment that is twisted by an angle γ.

Now we will calculate the Chern-Simons number (which is, loosely speaking, the baryon number) of the twisted segment of string described in (2.20) and with gauge fields in (2.14).

Let me assume that γ is a rational fraction of 2π so that we can write $\gamma = 2\pi p/q$ where p and q are integers. Then we can take q twisted segments, each of which is described by eqns. (2.20) and (2.14), and join them up - the antimonopole of one segment can be brought to annihilate the monopole of another segment - to form a closed loop. In this way we will get a loop of Z−string that is twisted by an angle $2\pi p$. Now we need to calculate the Chern-Simons number of this loop. The Chern-Simons number is defined as

$$CS = \frac{N_F}{32\pi^2}\int d^3x\epsilon_{ijk}\left[g^2\left(W^{aij}W^{ak} - \frac{g}{3}\epsilon_{abc}W^{ai}W^{bj}W^{ck}\right) - g'^2Y^{ij}Y^k\right] . \quad (2.21)$$

where, we have included the number of families N_F in the definition. For a closed loop of Z−string, this expression simplifies considerably since the only non-vanishing gauge field is the Z gauge field. In this circumstance, (2.21) reduces to

$$CS = N_F\frac{\alpha^2}{32\pi^2}cos(2\theta_W)\int d^3x\ \vec{Z}\cdot(\vec{\nabla}\times\vec{Z}) \quad (2.22)$$

We can now easily calculate the Chern-Simons number of the twisted loop by using the result that if we have a twisted flux loop of a gauge field $\vec{A}$ with flux F and twist $2\pi p$, then,

$$\int d^3x\ \vec{A}\cdot(\vec{\nabla}\times\vec{A}) \ = \ 2\ F^2\ p\ . \quad (2.23)$$

This result is well-known to people working in hydrodynamics and astrophysics [11] but less known in the particle physics community. So we give a quick sketch of the derivation.

We first note that the internal twisting of a flux tube is equivalent to a linking of two different flux tubes [12]. Hence we can restrict ourselves to evaluating the left-hand side of (2.23) for two untwisted flux tubes that are linked p times.

We work in the gauge $\vec{\nabla}\cdot\vec{A} = 0$ and the magnetic field is given by $\vec{B} = \vec{\nabla}\times\vec{A}$. Therefore, if we are given a magnetic field configuration, the gauge field can be found from

$$\vec{A}(\vec{x}) = -\frac{1}{4\pi}\int d^3x'\vec{B}(\vec{x}')\times\frac{\vec{x}-\vec{x}'}{|\vec{x}-\vec{x}'|}\ . \quad (2.24)$$

Denoting the integral on the left-hand side of (2.23) by I, we then have:

$$I = -\frac{1}{4\pi}\int d^3x\ \vec{B}(\vec{x})\cdot\int d^3x'\vec{B}(\vec{x}')\times\frac{\vec{x}-\vec{x}'}{|\vec{x}-\vec{x}'|} \quad (2.25)$$

and then perform integrations over the cross-section of the flux tubes. This has the effect

$$I = (2F^2)\left[-\frac{1}{4\pi}\oint d\vec{f}\cdot\oint d\vec{g}\times\frac{\vec{f}-\vec{g}}{|\vec{f}-\vec{g}|}\right] \quad (2.26)$$

where, we have assumed that both flux tubes carry the same flux F and that their locations are given by $\vec{f}$ and $\vec{g}$. The expression in square brackets is the Gauss linkage formula [13] for the curves $\vec{f}$ and $\vec{g}$ and this proves (2.23).

Now, using (2.23) (with $\vec{A}$ replaced by $\vec{Z}$) in (2.22) and inserting the value of the Z flux in the string (eqn. (2.10)), we get the Chern-Simons number of the $Z-$string loop that is twisted by $2\pi p$:

$$CS = N_F cos(2\theta_W)p \tag{2.27}$$

Since the loop was built out of q segments and $\gamma = 2\pi p/q$, the Chern-Simons number of one segment is

$$CS = N_F cos2\theta_W \frac{\gamma}{2\pi} . \tag{2.28}$$

So far we have been working with string segments having arbitrary twist. But now consider the case, $\gamma = \pi/cos(2\theta_W)$. With this twist, the Chern-Simons number is $N_F/2$ - precisely that of the sphaleron [14]!

Given that the segment with twist $\pi/cos(2\theta_W)$ has Chern-Simons number equal to that of the sphaleron, it is natural to ask if some deformation of it will yield the sphaleron. This deformation is not hard to guess for the $\theta_W = 0$ case. In this case, if we let the segment size shrink to zero, we have $\theta_m = \theta_{\bar{m}} = \theta$ and the Higgs field configuration of (2.20) gives:

$$\Phi_{m\bar{m}}(\gamma = \pi) = \begin{pmatrix} cos\theta \\ sin\theta \ e^{i\phi} \end{pmatrix} . \tag{2.29}$$

And this is exactly the field configuration of Nick Manton's $SU(2)$ sphaleron [6]. (Note that the gauge fields continue to be given by (2.14) or, equivalently in this case, by (2.15).)

Encouraged by this successful connection between the twisted string segment and the sphaleron in the $\theta_W = 0$ case, we can conjecture that the twisted segment of $Z-$string with Chern-Simons number $N_F/2$ will collapse into the sphaleron *for any* θ_W. Therefore the asymptotic Higgs field configuration for the sphaleron can be obtained by letting $\theta_m = \theta_{\bar{m}} = \theta$ in (2.20) and taking $\gamma = \pi/cos(2\theta_W) = \gamma_S$. Denoting the sphaleron asymptotic Higgs field configuration by Φ_S, we conjecture

$$\Phi_S = \begin{pmatrix} sin^2(\theta/2) - cos^2(\theta/2) \ e^{i\gamma_S} \\ sin(\theta/2)cos(\theta/2) \ e^{i\phi}(1 - e^{-i\gamma_S}) \end{pmatrix} \tag{2.30}$$

and the asymptotic gauge fields are given by (2.14). If this conjecture is true, we should be able to find a solution to the classical equations of motion with this ansatz for the asymptotic fields. Of course, we would first need to insert a suitable radial dependence in the ansatz and then solve the field equations.

On physical grounds it seems reasonable that there should be a critical value of twist at which one can get a static solution for a $Z-$string segment. This is because the segment likes to shrink under its own tension but the twist prevents the shrinkage and is equivalent to a repulsive force between the monopole and antimonopole. Then, if the string is sufficiently twisted, the attractive force due to the tension and the repulsive force due to the twist will balance and a static solution can exist. So far we have been assuming that the only dynamics of the segment is towards collapsing or expanding. However, since we are dealing with twisted segments, we should also include the rotational dynamics associated with twisting and untwisting. So, while any twist greater than a certain critical twist will successfully prevent the segment from collapsing, only a special value of the twist will give a static solution to the rotational dynamics. Furthermore, we expect that this solution will be unstable towards rotations that twist and untwist the string segment. This would be the unstable mode of the sphaleron.

A question that I have frequently been asked is that how can an object have fractional Chern-Simons number? This question is easily answered: only the vacuua have integer Chern-Simons number; outside the vacuua, the Chern-Simons number can be anything. But then, what would happen when this object having fractional Chern-Simons number decays into particles? We know that the change in the Chern-Simons number of the object equals the change in the baryon number, then how can fractional baryon number be produced? The answer to this question is currently under investigation by Eddie Farhi and collaborators [15] and requires a quantum treatment of the decay of the fractionally charged object. The belief is that the equality of baryon number change to the change in Chern-Simons number should be viewed as an equality of expectation values. So the change in baryon number will always be integral and only upon averaging will one get a fractional change in baryon number.

This completes what I wanted to say about the connection of the $Z-$string and the sphaleron. (Other discussions of this connection may be found in Refs. 4, 16.) Just to summarize, twisted segments of $Z-$string carry Chern-Simons number in proportion to their twist and when the twist is $\pi/cos(2\theta_W)$, the Chern-Simons number is $N_F/2$. We conjecture that the string segment with this special value of twist is precisely an extended sphaleron.

3. $Z-$STRING STABILITY

A non-trivial and very important aspect of studying electroweak strings is to understand their (meta-)stability. In an exhaustive analysis done in collaboration with Margaret James and Leandros Perivolaropoulos, we plotted the region of parameter space in which the $Z-$string is stable towards small perturbations [17]. The conclusion is that one can only have stable strings for $sin^2\theta_W \gtrsim 0.9$, that is, for $g' \gtrsim 3g$. The experimentally determined value of $sin^2\theta_W$ is 0.23 and is deep inside the instability region.

Recently there have been a few papers that have shed more light on the issue of stability for θ_W not too large. First there was a paper by Warren Perkins that showed that there is an instability towards developing W fields [18] for $sin^2\theta_W \lesssim 0.8$. Then Manuel Barriola, Martin Bucher and I used a simple argument to show that the string is unstable in the case $\theta_W = \pi/4$. In this instability, the upper component of the Higgs field grows, the lower component diminishes in magnitude and the gauge field spreads [4]. To see the explicit form of the instability, denote the unperturbed electroweak string solution by Φ_0 and $Z_j^{(0)}$ and consider the sequence of field configurations labeled by a parameter ξ:

$$\begin{aligned} \Phi(\vec{x};\xi) &= \cos\xi\ \Phi_0(\cos\xi\ \vec{x}) + \sin\xi\ \Phi_\perp \\ Z_j(\vec{x};\xi) &= \cos\xi\ Z_j^{(0)}(\cos\xi\ \vec{x}) \end{aligned} \tag{3.1}$$

where, Z_j is given in terms of the $SU(2)_L \times U(1)_Y$ gauge fields in eqn. (2.6) together with $n^a = (0,0,1)^T$ (note!), and,

$$\Phi_\perp = \frac{\eta}{\sqrt{2}}\begin{pmatrix}1\\0\end{pmatrix} . \tag{3.2}$$

For $\xi = 0$, the configuration is the embedded defect solution and for $\xi = \pi/2$, the configuration describes the vacuum. We then have,

$$\begin{aligned} \Phi^\dagger(\vec{x};\xi)\Phi(\vec{x};\xi) &= \cos^2\xi\ \Phi_0{}^\dagger\Phi_0 + \sin^2\xi\ \eta^2 \\ W_{ij}^a(\vec{x};\xi)W_{ij}^a(\vec{x};\xi) &= \cos^4\xi\ W_{ij}^a(\vec{x})W_{ij}^a(\vec{x}), \\ Y_{ij}^a(\vec{x};\xi)Y_{ij}^a(\vec{x};\xi) &= \cos^4\xi\ Y_{ij}^a(\vec{x})Y_{ij}^a(\vec{x}), \\ D_i\Phi(\vec{x};\xi) &= \cos^2\xi\left[\partial_i + i\frac{\alpha}{2}Z_i(\vec{x})T^Z\right]\Phi_0(\vec{x}) \\ V[\Phi(\vec{x},\xi)] &= \cos^4\xi\ V[\Phi_0(\vec{x})] \end{aligned} \tag{3.3}$$

where,

$$T^Z = \begin{pmatrix} 0 & 0 \\ 0 & 1 \end{pmatrix} \tag{3.4}$$

¿From (3.3) we get

$$[D_i\Phi(\vec{x};\xi)]^\dagger \; [D_i\Phi(\vec{x};\xi)] = \cos^4\xi \; (D_i\Phi_0(\vec{x}))^\dagger (D_i\Phi_0(\vec{x})) \; , \tag{3.5}$$

and hence, the total energy of the configuration is

$$E(\xi) = \cos^2\xi \; E(\xi = 0) \; . \tag{3.6}$$

This explicitly shows a sequence of configurations that start at the string ($\xi = 0$) and end up in the vacuum ($\xi = \pi/2$) and with monotonically decreasing energy.

Although in the sequence of configurations considered above, only the Z field is written as being non-zero, this is really misleading because Φ is no longer proportional to $(0,1)^T$. So the gauge fields in the above configurations actually include $W^\pm$ fields and the instability demonstrated above is towards developing W fields. If the parameter ξ is replaced by a parameter along the length of the string (z), the above sequence of configurations are slices of a string along the z axis that has terminated in a monopole. So the instability described above is the instability of a string to break-up by forming monopoles and antimonopoles.

Frans Klinkhamer and Poul Olesen[19], and, independently, Margaret James[20], have constructed a two parameter family of configurations - a two sphere in configuration space - with the string on top of the sphere for $sin^2\theta_W \lesssim 0.7$. This picture implies that the string has at least two unstable modes for small θ_w. However, it seems likely that there is only one unstable mode of the string solution under *small* perturbations since the two modes that Klinkhamer and Olesen, and, James, have found are gauge equivalent when the perturbations are taken to be small at the solution[21]. If this is true, we might reasonably guess that the only unstable perturbative mode is towards breaking up of strings by forming monopole-antimonopole pairs*.

There is a simple way to understand why the W fields develop[22] and destabilize the Z string[23,18]. For this consider the energy of a spin s particle in a uniform magnetic field $\vec{B}$ along the z-direction:

$$E^2 = k_z^2 + m^2 + (2n+1)eB - 2e\vec{B}\cdot\vec{s} \tag{3.7}$$

where, the charge on the particle is e and the g-factor of the particle (in the last term) is taken to be 2. The right-hand side contains the z-momentum and mass contribution to the energy. Then there is the contribution of the various Landau levels labeled by the integer n and finally there is the coupling of the magnetic moment (spin) to the magnetic field. Clearly if $s = 1$, it is possible for the right-hand side to be negative if

$$B > \frac{m^2}{e} \tag{3.8}$$

and hence there is an instability of the magnetic field to condensing spin one particles or, to "W condensation".

In the case of the $Z-$string, the magnetic field $\vec{B}$ in (3.7) is the $Z-$magnetic field within the string and the spin-1 particles are the W bosons. It can be checked that the g-factor is indeed 2 as has been assumed in writing (3.7) and the Z charge on the W boson is $e_Z = gcos\theta_W$. Therefore the conditions are right for the W condensation

*The physics of the two unstable modes is not completely clear to me. Perhaps the constructions of Ref. 19,20 are counting the instabilities towards forming monopoles that are twisted in the way described in Sec. 2.

argument to apply to the Z string. The only condition that is not immediately satisfied is the assumption in (3.7) that the magnetic field is uniform. But if the string is not too thin relative to the W boson Compton wavelength, one is justified in thinking of the Z magnetic field of the string as being uniform. This is precisely the case for low values of $sin^2\theta_W$. For small values of θ_W, we also see that $e_Z = gcos\theta_W$ is not small and so the critical Z magnetic field required for W condensation (eqn. (3.8)) can be relatively small. These facts show that Z strings are unstable towards W condensation for small θ_W.

A point to remember is that we have only been considering infinite strings so far. The instability towards break-up will not apply when the string length is less than a certain critical length since the formation of monopoles requires certain gradients along the length of the string. The critical length of string below which break-up is not possible is likely to be a few times the size of the monopole ($\sim m_W^{-1}$). Such short segments are not string-like in appearence but can still be important as they can carry Chern-Simons number just as in the case of longer segments.

Are there any new instabilities of a finite segment that are absent in the case of the infinite string? As the string segment is not topological, there is a possibility that the string segment could destabilize as a whole and disappear into the vacuum. But this does not appear to be possible since the evolution of the string segment has to conserve magnetic charge and so the only way a monopole can disappear is to annihilate with an antimonopole. Therefore the only new instability of a finite segment of string is towards dynamical collapse. If the segment has angular momentum or is twisted, this dynamical instability may not be too severe. (Nambu has given rough estimates for the life-time of a long string segment with angular momentum towards radiative decay [5].)

There are some circumstances under which the stability of the $Z-$string improves. Thermal effects can be shown to improve stability [24] but the improvement is more so that strings with higher values of β become stable and strings with small values of θ_W remain unstable. Rick Watkins and I showed that if there are particles bound to the string, the string can be stable down to lower values of $sin^2\theta_W$. In our analysis [25], we found stable strings down to about $sin^2\theta_W = 0.5$ but did not find stable $Z-$strings with $sin^2\theta_W = 0.23$.

The picture then is the following: if we start with a long (infinite) $Z-$string in the case when θ_w is small, the string will rapidly break up into small segments of a critical length - of the order of a few times the string thickness - which will then survive until they radiate away their energy and angular momentum.

If we are willing to consider extensions of the standard model, stable strings are possible. The popular two Higgs model does not yield stable strings [26] but more complicated models, such as, left-right models can yield stable strings [24]. Other extensions can also give topologically stable strings with $Z-$flux in them [27].

4. ELECTROWEAK STRINGS AT THE PHASE TRANSITION

I will assume that the electroweak phase transition is second order or weakly first order - after hearing the various talks at this conference, this seems likely. This means that the correlation length at the phase transition is of the order T_c^{-1} where T_c is the critical temperature at which the phase transition occurs. Under this assumption the temperature (T) dependent vacuum expectation value of Φ is given by:

$$< |\Phi| > = \frac{\eta}{\sqrt{2}} \left(1 - \frac{T^2}{T_c^2} \right)^{1/2} . \tag{4.1}$$

Therefore the mass per unit length of the $Z-$string follows from (2.11):

$$\mu(T) \approx \pi M(\beta) < |\Phi| >^2 = \frac{\pi\eta^2}{2} \left(1 - \frac{T^2}{T_c^2} \right) M(\beta) . \tag{4.2}$$

Let us now define the "Hagedorn temperature" T_H as the temperature at which

$$T_H = \sqrt{\frac{3\mu(T_H)}{2\pi}} . \tag{4.3}$$

This gives,

$$T_H = T_c \sqrt{\frac{x}{1+x}} , \quad \text{with} \quad x \equiv \frac{3M(\beta)\eta^2}{4T_c^2} . \tag{4.4}$$

For $\beta = 1$ (that is, $m_H = m_Z$), $\eta = 250GeV$ and $T_c = 100GeV$, this gives $T_H \approx 0.9T_c = 90GeV$. Note that, while T_H is quite close to T_c, the time required for the universe to cool from T_c to T_H is about $10^{16}T_H^{-1}$.

The Hagedorn temperature is significant in the study of string statistical mechanics because it is the temperature that marks a phase transition in a string network. Below the Hagedorn temperature, a string network prefers to break up into the smallest possible loops or segments. Above the Hagedorn temperature, the network is dominated by long strings. These results have been obtained by studying the statistical mechanics of a box of strings and it is believed that the results apply to fundamental strings as well as strings that arise as defects in a phase transition.

How can we understand these results? The basic point is that the number of states of a string grows exponentially with length while the Boltzmann distribution decays exponentially with length. It is found that the partition function depends on an integrand proportional to

$$exp\left[\left(\sqrt{\frac{2\pi}{3\mu(T)}} - \frac{1}{T}\right) E\right] \tag{4.5}$$

where E is the energy of the string. Above the Hagedorn temperature, the number of states associated with the long strings wins over the Boltzmann suppression and the network settles into long (infinite) strings.

There are various subtleties in studying string statistical mechanics and the above argument is only meant to suggest that something strange should happen above the Hagedorn temperature which is still below the critical temperature. For those wishing to pursue this line of argument further, a paper summarizing earlier work on the statistical analysis of strings as well as applying string statistical mechanics to cosmic strings, is the paper by Dave Mitchell and Neil Turok [28]. In this paper, the authors also treat the statistical mechanics of open strings where the monopoles at the end of strings and the strings themselves form during the same cosmological phase transiton - exactly the case relevant to electroweak strings! However, Mitchell and Turok ignore string-string interactions in their anlaysis and this may be a bad approximation to make. In particular, important processes such as string break-up and reconnections have not been taken into account in the analysis. For this reason, we cannot import the results of Ref. 28 wholesale but can only take them as an indication that the density of strings may be significant at temperatures near the Hagedorn temperature but still below the critical temperature.

There is another way of seeing that something bizzare may happen above the Hagedorn temperature. For $T_c > T > T_H$, the tension in the string is smaller than the thermal excitations due to interaction with the plasma. Therefore the strings are "hot" and the tension in the string can be ignored compared to thermal excitations. This would mean that points on a string segment would diffuse and the distance between any two points on a string with grow with time. So the length of the segment would grow and a given segment would not be able to shrink and disappear. All that a hot segment of electroweak string can do is to grow in length and break up. The smaller pieces that are formed due to the fragmentation would also not be able to shrink and would keep growing bigger. Then the density of strings would become high and the likelihood of strings merging to form bigger strings would increase. In this way one is led to conclude that, at temperatures lower than T_c but higher than T_H, the strings will be relatively long. This is in contrast to the picture where long strings are Boltzmann suppressed.

These arguments lead us to the picture that, for $T_c > T > T_H$, the plasma contains a significant density of long strings. As the temperature falls, the separation between strings increases. If the phase transition is first order, the strings get separated by the process of bubble formation whereas if the phase transition is second order, the inter-string separation grows larger continuously. At $T = T_H$ there is a phase transition in the string network itself and long strings begin to break-up at this temperature. If the strings are metastable, the strings break-up by nucleating monopoles and antimonopoles and this would be a first order phase transition in the string network itself. If the strings are unstable, the break-up is more like a second order phase transition in the string network. In this picture, the long electroweak strings at T_H are genuine relics of the unbroken phase and are not due to thermal fluctuations (since these ceased at T_c).

Now we follow the evolution to temperatures below the Hagedorn temperature. As the universe cools below the Hagedorn temperature, the infinite strings will break up and yield an exponential distribution of string segments and the number of string segments of length l will be proportional to $\exp[-al]$ where $a > 0$ is a constant. With further cooling, the tension in the strings starts becoming important and the segments start shrinking and decaying into radiation. All this will happen relatively quickly and so we can say that the string network disappears at a temperature $T \sim T_H$. As the network disappears, it will produce baryons and anti-baryons since, from our results in Sec. 2, twisted and linked strings carry baryon number. If there is sufficient CP violation in the dynamics of the string network, the decay of strings would lead to a production of net baryon number.

The above scenario is a scenario of baryogenesis at the Hagedorn temperature T_H and it is a concern that subsequent sphaleron transitions might erase any baryon number produced at T_H. To check if this happens, we find the temperature T_S at which the sphaleron transition rate first falls below the Hubble expansion rate[29]:

$$T_S \exp\left[\frac{-M_S(T_S)}{T_S}\right] = \frac{T_S^2}{m_{Pl}} . \tag{4.6}$$

where, $M_S(T)$ is the temperature dependent mass of the sphaleron[30,31] and m_{Pl} is the Planck mass. Now we know that, for $\theta_W = 0$,

$$M_S(T) = 2B(\lambda/\alpha_W)\frac{M_W(T)}{\alpha_W} \tag{4.7}$$

where, B is a weakly dependent function of the coupling constant ratio λ/α_W, $\alpha_W = g^2/4\pi$ and $M_W(T)$ is the temperature dependent mass of the $W-$boson. Writing $m_W = M_W(0)$ and approximating $\ln(T_S)$ by $\ln(T_c)$ we get,

$$T_S = T_c\left[1 + \left\{\frac{\alpha_W T_c}{2Bm_W}\ln\left(\frac{m_{Pl}}{T_c}\right)\right\}^2\right]^{-1/2} . \tag{4.8}$$

Now comparing T_S to T_H (eqn. (4.4)), we find that T_S is larger than T_H provided

$$\left[\frac{\alpha_W \eta}{4m_W}\ln\left(\frac{m_{Pl}}{T_c}\right)\right]^2 3M(\beta) < [B(\lambda/\alpha_W)]^2 \tag{4.9}$$

With $\alpha_W = 1/30$, $\eta = 250GeV$, $m_W = 80GeV$ and $\ln(m_{Pl}/T_c) = 17\ln(10)$, this condition gives

$$3M(\beta) < [B(\lambda/\alpha_W)]^2 . \tag{4.10}$$

The value of B ranges from 1.5 to 2.7 - at least for the $SU(2)$ sphaleron (that is, for the $\theta_W = 0$ case) - and is roughly 1.9 when $\beta = 1$ ($m_H = m_Z$) at which point $M(\beta) = 1$. Therefore the condition (4.9) is satisfied in the parameter range of interest and we have $T_H < T_S$. For $\theta_W \neq 0$ too, we expect that this condition will be satisfied for an interesting range of parameters.

The above argument assures us that sphaleron transitions cannot completely erase the baryon number that would be generated by the string network at the Hagedorn temperature. The exact fraction of baryon number that survives can be found by studying the equations of detailed balance in an expanding universe.

While the above scenario seems plausible to me, it has several weaknesses. The first weakness is that we do not understand the formation of electroweak strings at the electroweak phase transition. The arguments given above suggest that electroweak strings should be present in significant numbers above the Hagedorn temperature but we still do not have a method to get a rigorous estimate of quantities such as the number density of strings, or, the length distribution. The second weakness is that the production of baryons over antibaryons in the process of string decay requires CP violation which is thought to be very weak in the standard model. This, however, is a problem with any scenario of electroweak baryogenesis and it is quite common to consider more strongly CP violating extensions of the standard model. One could also take this approach with electroweak strings. On the other hand, it would be more satisfying if there were some unusual source of CP violation in the monopole-antimonopole system within the standard model that may not be present in the usual particle interactions. These issues are presently being investigated.

5. MAGNETIC FIELDS

In this last section, I would like to discuss the generation of cosmological magnetic fields arising at the electroweak phase transition[32]. Kari Enqvist has discussed the scenario in some detail and my purpose is to make a few clarifying remarks about some of the assumptions that go into the scenario. Then I would like to describe a connection between electroweak strings and primordial magnetic fields.

The basic idea is that, as the Higgs field acquires a vacuum expectation value, currents are produced that lead to a magnetic field. Specifically, one defines the electromagnetic gauge field as in eqn. (2.6) but the electromagnetic field strength as:

$$F^{em}_{\mu\nu} = sin\theta_W n^a W^a_{\mu\nu} + cos\theta_W Y_{\mu\nu} \\ \qquad - i4g^{-1}\eta^{-2} sin\theta_W [(D_\mu\Phi)^\dagger D_\nu\Phi - (D_\nu\Phi)^\dagger D_\mu\Phi] \ . \tag{5.1}$$

So far, all that we have done is to define the electromagnetic field strength and have not included any dynamics. The dynamics comes in when we estimate the various terms in (5.1) in a plasma at high temperature. Suppose we are interested in a macroscopic volume of size L which is much larger than the thermal wavelength T^{-1}. We expect that the field strengths $W^a_{\mu\nu}$ and $Y_{\mu\nu}$, when averaged over the volume of size L, will be exponentially decreasing with increasing L since the plasma is neutral on such scales and has no net currents either. However, the covariant derivative is expected to fall off as a power law in L as it is simply the covariant gradient of a scalar field. For example, the covariant gradient may be estimated as the change in the value of Φ occuring over the scale L - therefore, $|D_i\Phi| \sim \eta/L$. This is the hinge on which the production of magnetic fields at the electroweak phase transition rests and one can give several arguments why this seems reasonable. (Though none of the arguments rigorously proves the assumption.)

We now give the arguments that indicate that the covariant derivative falls off as a power law and not exponentially with L. The motivation for expecting this behaviour comes from the Kibble argument. If the covariant derivative vanishes, there is complete correlation between distant parts of the universe which have never been in causal contact and this is inadmissible. This argument is strictly valid for global symmetries where the phases in Φ have physical meaning and so the argument can legitimately be questioned for gauge symmetries. On the other hand, note that we are dealing with a system at high temperatures when the energy is distributed among the various degrees of freedom. These include the covariant derivative of Φ and it is natural to assume that the covariant derivative takes on a value given by energy equipartition. Whereas $W^a_{\mu\nu}$ and $Y_{\mu\nu}$ fall off exponentially with L due to vanishing currents in the plasma, there is no current density in the particles (fermions etc.) in the plasma that is responsible for the last term in (5.1) and so a power law fall off is possible.

Here we will give another argument for why magnetic fields should be produced at the electroweak phase transition. This argument is consistent with the theme of the talk as it suggests that there is a source for the magnetic field and the source is none other than the electroweak monopoles present at the ends of electroweak strings. As in Sec. 4, we expect electroweak monopoles to be produced at the electroweak phase transition and then to go away at a somewhat lower temperature. Therefore, imagine a distribution of monopoles and antimonopoles in a plasma with the magnetic lines of force running from monopoles to antimonopoles. In addition, the monopoles are connected to antimonopoles via electroweak strings. With time, the strings shrink and the monopoles and antimonopoles at the ends of the string annihilate. However, the magnetic lines of force are glued to the plasma because the plasma is a very good electrical conductor. So the magnetic lines of force survive even when the monopoles themselves annihilate and disappear. If the lines were present only on very small scales, they too would eventually disappear (as the conductivity of the plasma decreases with time). But this is not likely; instead we expect that the magnetic lines of force will percolate much like the percolation of a cosmic string network [33]. With time, the small scale curvature on the magnetic lines of force will disappear and the lines will straighten out but there will always be lines of force present on large scales where they are frozen in the plasma. Hence, at any epoch a relic magnetic field will be present.

The magnetic field strength is roughly given by $B \sim T^2$ where T is the temperature of the universe provided we assume that the field is frozen-in on the smallest scales (of order T^{-1}). However, this is an incorrect assumption as the frozen-in scale is much larger than the thermal wavelength. If the frozen-in scale is denoted by l_f, we have, $l_f \sim 10^{12}T^{-1}$. Below this scale, the plasma is unimportant for the evolution of the magnetic field and so the field will smooth itself out. Using a flux average to estimate the field strength, we find $B \sim 10^{-12}T^2$ and this field is coherent on a scale l_f. At the electroweak scale, this gives $B \sim 10^{12}G$ with a coherence scale of $10^{-5}cms$.

The presence of strong magnetic fields in the very early universe leads to very interesting physics and is currently under investigation by several groups. Here, I would only like to remark that, if the above arguments connecting electroweak strings with magnetic fields is correct, the direct or indirect observation of such a magnetic field would immediately yield information about the cosmological electroweak phase transition and about electroweak strings!

Acknowledgements

I thank the organizers of this meeting and especially Filipe Freire for hosting this conference and making it so enjoyable. This work was supported by the National Science Foundation.

References

1. T. Vachaspati and G. B. Field, TUTP-94-1.
2. K. Enqvist, at this conference.
3. T. Vachaspati and M. Barriola, Phys. Rev. Lett. 69, 1867 (1992).
4. M. Barriola, T. Vachaspati and M. Bucher, "Embedded Defects", Phys. Rev. D, to be published; TUTP-93-7.

5. Y. Nambu, Nucl. Phys. B130, 505 (1977).
6. N. S. Manton, Phys. Rev. D28, 2019 (1983).
7. T. Vachaspati, Phys. Rev. Lett. 68, 1977 (1992); 69, 216(E) (1992); Nucl. Phys. B397, 648 (1993).
8. P. Laguna-Castillo and R. A. Matzner, Phys. Rev. D 36, 3663 (1987).
9. H. B. Nielsen and P. Olesen, Nucl. Phys. B 61, 45 (1973).
10. M. B. Einhorn and R. Savit, Phys. Lett. B 77, 295 (1978).
11. J. J. Moreau, C. R. Acad. Sci. Paris 252, 2810 (1961); H. K. Moffat, J. Fluid Mech. 35, 117 (1969); M. Berger and G. Field, J. Fluid Mech. 147, 133 (1984).
12. For example, see L. H. Kauffman, "On Knots", Princeton University Press (1987).
13. For example see H. Flanders, "Differential Forms" (Academic Press, 1963).
14. M. Axenides and A. Johansen, NBI-HE-93-74.
15. E. Farhi, V. V. Khoze and R. Singleton Jr., Phys. Rev. D 47, 5551 (1993).
16. M. Hindmarsh and M. James, hep-ph/9307205 (1993).
17. M. James, L. Perivolaropoulos and T. Vachaspati, Phys. Rev. D46 (1992) R5232; Nucl. Phys. B395, 534 (1993).
18. W. Perkins, Phys. Rev. D 47, R5224 (1993).
19. F. R. Klinkhamer and P. Olesen, NIKHEF-H/94-02.
20. M. James, DAMTP-HEP-94-13.
21. M. James (private communication).
22. For a review see J. Ambjorn and P. Olesen, Int. J. Mod. Phys. A 5, 4525 (1990).
23. P. Olesen (private communication).
24. R. Holman, S. Hsu, T. Vachaspati and R. Watkins, *Phys. Rev.* **D46**, 5352 (1992).
25. T. Vachaspati and R. Watkins, Phys. Lett. B 318, 163 (1993).
26. M. A. Earnshaw and M. James, hep-ph/9308223 (1993).
27. G. Dvali and G. Senjanovic, Phys. Rev. Lett. 71, 2376 (1993).
28. D. Mitchell and N. Turok, Nucl. Phys. B 294, 1138 (1987).
29. M. E. Shaposhnikov, JETP Lett. 44, 465 (1986).
30. F. R. Klinkhamer and N. S. Manton, Phys. Rev. D30, (1984) 2212.
31. J. Kunz, B. Kleihaus and Y. Brihaye, Phys. Rev. D 46, 3587 (1992).
32. T. Vachaspati, Phys. Lett. B 265, 258 (1991).
33. T. Vachaspati and A. Vilenkin, Phys. Rev. D 30, 2036 (1984).

ELECTROWEAK STRINGS PRODUCE BARYONS

Manuel Barriola

Harvard-Smithsonian CFA
60 Garden st.
Cambridge, MA 02138
USA

INTRODUCTION

One implication of the standard hot big bang cosmology is that from the relics of the early universe we should in principle expect the universe to contain the same abundance of baryons and antibaryons. However there exist compelling empirical evidence that the universe is made out of matter with relative few antimatter. From the present asymmetry we can extrapolate that at the energy scale of the electroweak phase transition there was one part in 10^8 more matter than antimatter in the universe. The goal of baryogenesis is to explain why there is this asymmetry between the amount of matter and antimatter i.e. there is a nonbanishing ratio between the net baryon number density and entropy density (n_B/s) in the universe.

In this article we propose a new mechanism to generate baryon asymmetry. at the time of the electroweak phase trasition[1]. First we would like to briefly review electroweak strings (in particular Z-strings[10]) and different mechanisms that imply magnetic fields at the time of the electroweak phase transition. We shall show that electroweak strings produced at the electroweak phase transition will introduce anomalous currents through interactions of the strings with the background electromagnetic field and through changes in the helicity of the string network. These anomalous currents will produce fluctuations in the baryon and lepton number. Finally we shall study their implications as a possible mechanism for baryogenesis.

STRINGS AND MAGNETIC FIELDS

It has been shown[10] that in the standard electroweak model there are solutions that look like Nielsen-Olesen strings with magnetic Z-flux along their cores. Nambu[10] has also shown that these strings can connect monopole-antimonopole pairs. The strings are metastable for some range in parameter space even though the standard electroweak theory is topologically trivial. However they are not stable for the physical value of the Weinberg angle θ_w. We should expect that Z-string solutions should survive for extensions of the standard electroweak model. For example Refs.11and 12have shown that this is the case in the two-Higgs model extension. Ref.11has also studied the stability of this solutions to small perturbations. They find that for realistic values of the Higgs mass and the Weinberg angle the string is unstable.

Dvali and Senjanovic[14]have also shown that if we consider the electroweak model with two Higgs doublets and we add an extra global $U(1)_{gl}$ symmetry to the theory, then there are topologically stable string solutions that also carry magnetic Z-flux along their core. This result also suggest that this Z-flux tubes may be metastable when the extra $U(1)_{gl}$ is an approximate symmetry. In Ref.15these authors have also shown that the minimal supergravity extension of the standard model has these topological solution if the hidden sector has an exact R-symmetry. These strings have different characteristics than 'ordinary' embedded Z-strings produced at the time of the phase transition. We shall not consider them in this paper.

In what follows we will consider the embedded Z-strings solutions of Ref.10. We shall assume extensions of the minimal standard model in which these strings are stable or meta-stable. We should also point out that these strings will be stabilized by the nature of this baryogenesis mechanism. As we shall see, at the core of the strings massless baryons will be produced while outside they would be massive. Ref.13has shown that bound states stabilize non-topological solitons.

Since the electroweak model has monopole solutions we expect that when our string network is produced we will have $(M - \bar{M})$ pairs connected by strings and small string loops. The $M - \bar{M}$ and strings will interact with the background plasma and the monopoles will contract with some speed v_c. If l is the typical length of the strings in the network then after the time interval $\delta t_d = l/v_c$, the strings will decay. The strings will also have some transverse velocity with respect to the background that we will denote by v_t.

There are several mechanism that will generate primodial magnetic fields in the early universe. Some of them relay on physical effects during an inflationary phase transition[16]. It has also been suggested[17]that in the electroweak phase transition the fields have to be uncorrelated at distances larger than the initial correlation length(ξ). Therefore the gradients of the Higgs fields can not be compensated by the gauge fields for scales bigger than ξ. These gradients imply that we will have electromagnetic fields. Recently Ref.18have used the Savvidy vacuum[19]to generate magnetic fields at the scale of grand unification. They propose that magnetic field fluctuations at the GUT scales produce a phase transition to a new ground state with a non zero magnetic field. All these mechanisms imply strong magnetic fields at the time of the electroweak phase transition.

ANOMALOUS CURRENTS AND BARYOGENESIS

In the electroweak model all the currents that couple with the gauge fields are free of anomalies so that the theory is renormalizable. However the baryon and lepton currents have anomalies of the form

$$\partial_\mu J^\mu_B = \partial_\mu J^\mu_L = N_f \left(\frac{g^2}{32\pi^2} W^{\mu\nu}_a \tilde{W}^a_{\mu\nu} - \frac{g'^2}{32\pi^2} Y_{\mu\nu} \tilde{Y}^{\mu\nu} \right). \tag{1}$$

Where N_f is the number of families, $W^{\mu\nu}_a$ $a = 1, 2, 3$ and $Y_{\mu\nu}$ are the $SU(2)$ and $U_Y(1)$ field strengths respectively, with the tilde referring to the usual definition of the dual of the field strength. g and g' are the associated gauge couplings. Note that there is not an anomaly for the difference of the baryon to the lepton currents.

If we integrate both sides of Eq.(1)over a volume V and assume that the currents vanish at the surface of V we obtain that the baryon number $B = \int d^3x J_B^0$ changes in some time interval by,

$$\Delta B = \frac{N_f}{32\pi^2} \int dt \int d^3x \left(g^2 W_a^{\mu\nu} \tilde{W}_{\mu\nu}^a - g'^2 Y_{\mu\nu} \tilde{Y}^{\mu\nu} \right). \tag{2}$$

We see that if the r.h.s. of Eq.(2)is non zero we shall have a change in the baryon number.

After the phase transition the W_3^μ and Y^μ mix to form the massive Z^μ and the massless electromagnetic A^μ vector bosons.

In terms of these new vector fields Eq.(2)takes the form,

$$\begin{aligned} \Delta B = \frac{N_f}{32\pi^2} \int d^4x [\quad & g^2 \vec{E}_W^{\bar{a}} \cdot \vec{B}_W^{\bar{a}} + \alpha^2 \cos 2\theta_w \vec{E}_Z \cdot \vec{B}_Z + \\ & \frac{\alpha^2}{2} \sin 2\theta_w (\vec{E}_A \cdot \vec{B}_Z + \vec{E}_Z \cdot \vec{B}_A) + i.t.] \end{aligned} \tag{3}$$

where we have used the standard definitions for non-abelian electric and magnetic fields and $\bar{a} = 1, 2$. The last term of Eq.(3)represent interaction terms of the electric and magnetic components of $\vec{Z}$ and $\vec{A}$ with $\vec{W}_{\bar{a}}$ and self-interactions of $\vec{W}_{\bar{a}}$. From Eq.(3)we see that there is no coupling between the electric and magnetic components of $\vec{A}$.

Once the W^1, W^2 and Z bosons become massive, unless there is some topological obstructions in their configurations, we should expect that they are exponentially suppressed. Equation(3)can be reduced to

$$\Delta B = \frac{N_f}{32\pi^2} \int d^4x \left[\alpha^2 \cos 2\theta_w \vec{E}_Z \cdot \vec{B}_Z + \frac{\alpha^2}{2} \sin 2\theta_w (\vec{E}_A \cdot \vec{B}_Z + \vec{E}_Z \cdot \vec{B}_A) \right]. \tag{4}$$

$\vec{B}_Z$ and $\vec{E}_Z$ survive because we assume that after the phase transition we have Z-strings. At the core of the strings we have false vacuum and the Z bosons will be massless. From Eq.(4)we can infer a neat physical interpretation for the change in the baryon number. The first term in the r.h.s. of Eq.(4)is the change in the helicity of the string network[20]. The second term can be easily understood if we realize that the background magnetic fields transform to electric fields in the frame of the moving strings. Therefore this term reflects the interactions of the magnetic Z flux along the core of the Z-strings with these electric fields. The last term will have the inverse interpretation of the previous one. The transformation of $\vec{B}_Z$ from the string to the background frame will give us an $\vec{E}_Z$ field that will interact with the background magnetic field.

To quantify the effect of the first term of Eq.(4)would require first to know, using a computer simulation, the percentage of loops that are linked in the string network. We shall not do this calculation[21]. In what follows we shall assume that the contribution from this term is not greater than the contributions coming from the other two terms.

Now that we have established a new mechanism to introduce anomalous currents that will violate baryon number, we need a process that biases these currents towards the production of a net baryon number. We are going to consider the two-Higgs model as an example. We shall show that the classical equations of motion for the strings are modified by the quantum corrections coming from the effective potential. For collapsing strings, this corrections will favor the production of baryons over antibaryons.

The relevant term in the effective action that biases the baryon number is given by the coupling of the relative phase θ of the two Higgs doublets with the W^a gauge bosons through the triangle diagram[22]. For slowly varying θ this term can be expanded in a power series in the momenta[23]. We also do the same diagram but replacing the W^a by the Y gauge fields. When we include both diagrams we find that the effective action contains the term,

$$\Delta S = \int d^4x \kappa\theta \frac{m_t^2}{T^2}\left(g^2 W_a^{\mu\nu}\tilde{W}^a_{\mu\nu} - g'^2 Y_{\mu\nu}Y^{\mu\nu}\right) \tag{5}$$

where $\kappa = \frac{14}{3\pi^2\zeta(3)}$ and the expansion is up to quadratic order in m_t/T. Using the standard definitions of Z^μ, A^μ vector bosons and assuming that the W^1 and W^2 are exponentially suppressed we obtain

$$\Delta S \sim \frac{N_f\alpha^2}{64\pi^2}\kappa \sin 2\theta_w \frac{m_t^2}{T^2}\int d^4x\theta\left(\vec{E}_A\cdot\vec{B}_Z + \vec{E}_Z\cdot\vec{B}_A\right). \tag{6}$$

This term in the action will modify the classical equations of the string network interacting with the background electromagnetic fields introducing a bias towards the production of a net baryon number. This term violates C and P but conserves CP. However the evolution of θ will not conserve CP because of explicit breaking of CP in the two-Higgs potential.

As the string network is formed Eq.(6)will bias the evolution of the interaction of the magnetic flux of the strings with the electromagnetic background. But in order for the mechanism to work we need that the angle θ changes in time in a definite direction so we favor the production of a net baryon number. This directionality is given by the fact that the string network is collapsing. In particular we have seen that we will have strings bounded by monopole-antimonopole pairs and small loops that will be collapsing. At the moving ends of the strings, θ and the Higgs fields that describe the strings will evolve in a definite way in time. The $M - \bar{M}$ pairs and string loops will collapse at relativistic speeds, this will give production of baryons that are out of thermal equilibrium.

REFERENCES

1. For a review see Ref.2,3and 4
2. A.G. Cohen, D.B. Kaplan and A.E. Nelson, Annu. Rev. Nucl. Part. Sci **43** 27 (1993).

3. N. Turok, in 'Perspectives on Higgs Physics', ed. G. Kane (World Scientific, Singapore, 1992).

4. A.D. Dolgov UM-AC.93-91 preprint (1993).

5. N. S. Manton Phys. Rev. D **28**, 2019 (1983): F. R. Klinkhammer and N. S. Manton, Phys Rev. D **30**, 2212 (1984).

6. R. H. Brandenberger and A.C. Davis, Phys. Lett. **B308**, 79 (1993).

7. See 2and references therein.

8. R. Brandenberger and A. C. Davis, Phys. Lett. **B308**, 79 (1993).

9. R. Brandenberger, A. C. Davis and M. Trodden, BROWN-HET-935 preprint (1994).

10. T. Vachaspati, Phys. Rev. Lett. **68**, 1977 (1992). For earlier work see Y. Nambu, Nucl. Phys. **B130**, 505 (1977); M.B. Einhorn and A. Savit, Phys. Lett. **77B**, (1978) 295; N.S. Manton, Phys. Rev. **D28**, (1983) 2018.

11. M. A. Earnshaw and M. James Phys. Rev. **d48**, 5818 (1993).

12. L. Perivolaropoulos Phys. Lett. **B316**, 528 (1993).

13. T. Vachaspati and R. Watkins, Phys. Lett. **B318**, 163 (1993); T.D. Lee, Phys. Rep. **23C**, 254 (1976).

14. G. Dvali and G. Senjanovic, Phys. Rev. Lett. **71**, 2376 (1993).

15. G. Dvali and G. Senjanovic IC/94 preprint (1994).

16. C.J. Hogan, Phys. Rev. Lett. **51** (1983) 1488; M.S. Turner and L.W. Widrow, Phys. Rev. **D37** (1988); W. Garretson, G.B. Field and S.M. Carroll, Phys. Rev. **D46** (1992) 5346.

17. T. Vachaspati, Phys. Lett. **B265**, 258 (1991).

18. K. Enqvist and P. Olesen NORDITA-94/6 preprint (1994).

19. G.K. Savvidy, Phys. Lett. **B71** (1977) 133.

20. I want to thank George Field for pointing this out when I showed him the calculation.

21. T. Vachaspati and G. Field, TU/94 preprint (1994).

22. N. Turok and J. Zadrozny, Phys. Rev. Lett. **65**, 2331, (1990).

23. M. Dine and S. Thomas SCIPP 94/01 preprint (1994).

ELECTROWEAK STRINGS AND TOPOLOGY

Margaret James

Department of Applied Mathematics and Theoretical Physics
University of Cambridge
Silver Street
Cambridge CB3 9EW

INTRODUCTION

It is well known that the sphaleron [1,2] is the configuration of maximal energy on a non-contractible loop (NCL) in the configuration space of electroweak thory in three-dimensions. Forgács and Horváth [3] suggested that the existence of non-contractible spheres of higher dimension may also be connected with solutions of the field equations. We show that the electroweak Z-string [4] and W-strings [5,6] are solutions of this type: each lies on a non-contractible two-sphere (NCS) in configuration space.

THE NCS PASSING THROUGH THE STRINGS

We consider static bosonic configurations in two-dimensional electroweak theory: there is no time dependence and we work in the temporal gauge where the time components of the gauge fields are set to zero. The energy functional is given by

$$E = \int \left[\tfrac{1}{4}F^a_{ij}F^a_{ij} + \tfrac{1}{4}f_{ij}f_{ij} + (D_i\Phi)^\dagger(D_i\Phi) + \lambda(\Phi^\dagger\Phi - \tfrac{1}{2}\eta^2)^2\right] d^2x \; . \tag{1}$$

We work in the polar gauge so that $W_r = 0$ and $a_r = 0$. For finite energy configurations, we may define

$$\Phi^\infty(\varphi) = \lim_{r\to\infty} \Phi(r,\varphi) \quad \text{with} \quad |\Phi^\infty| = \frac{\eta}{\sqrt{2}} \; . \tag{2}$$

We can use the global gauge freedom to fix

$$\Phi^\infty(\varphi = 0) = \frac{\eta}{\sqrt{2}}\begin{pmatrix}0\\1\end{pmatrix} . \tag{3}$$

There remains a global $U(1)$ freedom corresponding to the unbroken gauge group; however, this is irrelevant. The vacuum configuration is now completely fixed. Equation (2) implies that we can regard Φ^∞ as a map $\Phi^\infty : S^1 \to S^3$. We parameterise a two-sphere in configuration space by (μ, ν) with $\mu \in [0, \pi]$ and $\nu \in [0, \pi]$. The boundaries of the square, $\mu = 0$ or π and $\nu = 0$ or π, give a single point in configuration space. The asymptotic Higgs field of the configurations on the two-sphere then define a family of maps, $\Phi^\infty(\mu, \nu) : S^1 \to S^3$, which vary continuously with μ and ν. We now wish to find a single map, $\Psi : S^3 \to S^3$ which is topologically equivalent to this family of maps.

We start with the parameterisation of S^3 (the domain of the map Ψ) given by

$$p(\mu, \bar{\theta}, \bar{\varphi}) = (\sin\mu \sin\bar{\theta} \cos\bar{\varphi}, \sin\mu \sin\bar{\theta} \sin\bar{\varphi}, \sin^2\mu \cos\bar{\theta} + \cos^2\mu, \sin\mu \cos\mu(\cos\bar{\theta} - 1)) \,. \tag{4}$$

p is a continuous function of $(\mu, \bar{\theta}, \bar{\varphi})$. The dependence of p on $\bar{\theta}$ and $\bar{\varphi}$ is consistent with them being spherical polar coordinates. Each point p on S^3 occurs for a unique triple $(\mu, \bar{\theta}, \bar{\varphi})$ except that $p(\mu = 0, \bar{\theta}, \bar{\varphi}) = p(\mu = \pi, \bar{\theta}, \bar{\varphi}) = (0, 0, 1, 0)$ for all $\bar{\theta}, \bar{\varphi}$, and $p(\mu, \bar{\theta} = 0, \bar{\varphi}) = (0, 0, 1, 0)$ for all μ.

This parameterisation of S^3 is a particular case of the identity

$$S^1 \wedge S^n \simeq S^{n+1} \tag{5}$$

where $\wedge$ denotes the smash product. The smash product of two topological spaces X and Y is defined by

$$X \wedge Y = \frac{X \times Y}{X \vee Y} \tag{6}$$

where

$$X \vee Y = X \times y_0 \cup x_0 \times Y \subset X \times Y \tag{7}$$

for fixed points $x_0 \in X$ and $y_0 \in Y$. The particular case we have used is

$$S^1(\mu) \wedge S^2(\bar{\theta}, \bar{\varphi}) \simeq S^3(\mu, \bar{\theta}, \bar{\varphi}). \tag{8}$$

The fixed points of the product are $\bar{\theta} = 0$ on S^2 and $\mu \in \{0, \pi\}$ on S^1.

We now make a change of co-ordinates: for fixed μ we reparameterise $S^2(\bar{\theta}, \bar{\varphi})$ as a smash product $S^1(\nu) \wedge_1 S^1(\varphi)$, where the fixed points are $\nu \in \{0, \pi\}$ and $\varphi = 0$. The $\nu = \pi/2$ slice makes an angle δ with the positive x-axis. δ is fixed and takes its value in the range $[0, 2\pi]$. Explicitly

$$\begin{pmatrix} \sin\bar{\theta}\cos\bar{\varphi} \\ \sin\bar{\theta}\sin\bar{\varphi} \\ \cos\bar{\theta} \end{pmatrix} = \begin{pmatrix} \cos\delta \sin\varphi \sin\nu - \sin\delta \cos\nu \sin\nu(1 - \cos\varphi) \\ \sin\delta \sin\varphi \sin\nu + \cos\delta \cos\nu \sin\nu(1 - \cos\varphi) \\ \sin^2\nu \cos\varphi + \cos^2\nu \end{pmatrix} . \tag{9}$$

so that

$$\begin{aligned} p(\mu, \nu, \varphi, \delta) = (&\sin\mu(\cos\delta \sin\varphi \sin\nu - \sin\delta \cos\nu \sin\nu(1 - \cos\varphi)), \\ &\sin\mu(\sin\delta \sin\varphi \sin\nu + \cos\delta \cos\nu \sin\nu(1 - \cos\varphi)), \\ &\sin^2\mu \sin^2\nu(\cos\varphi - 1) + 1, \sin\mu \cos\mu \sin^2\nu(\cos\varphi - 1)) \end{aligned} \tag{10}$$

for fixed δ. Note that each point p on S^3 occurs for a unique triple (μ, ν, φ) except for $p = (0, 0, 1, 0)$ where $p(\mu = 0, \nu, \varphi) = p(\mu, \nu = 0, \varphi) = p(\mu, \nu, \varphi = 0) = (0, 0, 1, 0)$. We now have

$$S^3(\mu, \nu, \varphi) = S^1(\mu) \wedge_2 (S^1(\nu) \wedge_1 S^1(\varphi)) \tag{11}$$

where the fixed points of $\wedge_2$ are $\mu = 0$ on $S^1(\mu)$ and $\varphi = 0$ on $S^2(\nu, \varphi)$. Hence

$$S^3(\mu, \nu, \varphi) = S^2(\mu, \nu) \wedge_3 S^1(\varphi) \tag{12}$$

where the fixed points of $\wedge_3$ are $\mu = 0$ or $\nu = 0$ on $S^2(\mu, \nu)$ and $\varphi = 0$ on $S^1(\varphi)$.
If we write the Higgs field

$$\Phi = (\Phi_1, \Phi_2)^T \ , \tag{13}$$

then we can define the related four component real vector

$$\Phi_{Re} = (Re\Phi_1, Im\Phi_1, Re\Phi_2, Im\Phi_2)^T. \tag{14}$$

Asymptotically $\frac{\sqrt{2}}{\eta}\Phi_{Re}$ must lie on the unit 3-sphere (the target of the map Ψ). Then a two-sphere of configurations whose asymptotic Higgs fields are given by

$$\Phi^{\infty}_{Re}(\mu, \nu, \varphi, \delta) = \frac{\eta}{\sqrt{2}}\Psi\left(p(\mu, \nu, \varphi, \delta)\right) \tag{15}$$

will be non-contractible if Ψ has non-zero degree.
We set $\delta = 0$ and choose

$$\Psi = \begin{pmatrix} 0 & 0 & 0 & 1 \\ 0 & 1 & 0 & 0 \\ 0 & 0 & 1 & 0 \\ 1 & 0 & 0 & 0 \end{pmatrix} \tag{16}$$

which has degree one. Note that this is compatible with our gauge choice since $\Phi^{\infty}_{Re} = \frac{\eta}{\sqrt{2}}(0, 0, 1, 0)$ for all (μ, ν, φ) such that $p(\mu, \nu, \varphi) = (0, 0, 1, 0)$. We now define

$$F_\varphi = F^a_\varphi \frac{\tau^a}{2i} = (U^\infty)^{-1}\partial_\varphi U^\infty \tag{17}$$

where

$$U^\infty = \frac{\sqrt{2}}{\eta}\begin{pmatrix} \Phi_2^{\infty *} & \Phi_1^\infty \\ -\Phi_1^{\infty *} & \Phi_2^\infty \end{pmatrix} . \tag{18}$$

Then the following field configurations

$$\Phi = h(r)U^\infty \begin{pmatrix} 0 \\ 1 \end{pmatrix} ,$$

$$gW_\varphi = -U^\infty\left(\left(F^1_\varphi\frac{\tau^1}{2i} + F^2_\varphi\frac{\tau^2}{2i}\right)f_1(r) + F^3_\varphi\frac{\tau^3}{2i}f_3(r)\right)(U^\infty)^{-1} , \qquad g'a_\varphi = f_0(r)F^3_\varphi ,$$

$$gW_r = g'a_r = 0 \tag{19}$$

are defined over all space, have finite energy and satisfy the correct asymptotic behaviour provided that the fields obey the boundary conditions that as $r \to 0$ $h(r) = f_1(r) = f_3(r) = f_0(r) = 0$, $r \to \infty$ $h(r) = \frac{\eta}{\sqrt{2}}$, $f_1(r) = 1$, $f_0(r) = 1 - c$, $f_3(r) = c$ where c is a constant. If we now set

$$f_3(r) = g^2\frac{Z(r)}{2\alpha} \qquad f_0(r) = g'^2\frac{Z(r)}{2\alpha}, \tag{20}$$

and h and Z satisfy the equations for the Nielsen-Olesen profile functions [7] then at $\mu = \pi/2$, $\nu = \pi/2$ the configuration (19) is the electroweak Z-string.

We now take Ψ to be the identity map (again of degree one). Then the following configurations

$$\Phi_\delta = h(r) U_\delta^\infty \begin{pmatrix} 0 \\ 1 \end{pmatrix}, \quad W_{\delta\varphi} = -\frac{Z(r)}{2} \partial_\varphi U_\delta^\infty (U_\delta^\infty)^{-1} \tag{21}$$

are defined over all space, have finite energy and satisfy the correct asymptotic boundary conditions provided that as $r \to 0$, $h \to 0$, $Z \to 0$ and as $r \to \infty$, $h \to \frac{\eta}{\sqrt{2}}$ and $Z \to 2/g$. The configuration (21) at $\mu = \nu = \pi/2$ is the W-string configuration .

DISCUSSION

We have shown that the Z-string and W-strings are connected to the existence of a non-contractible two-spheres in configuration space. The sphaleron lies on a NCL and has one unstable mode (for $M_H \leq 12 M_W$). The Z-string and W-strings, solutions on NCSs, have (at least) two unstable modes for the physical value of the parameters $\beta = (M_H/M_Z)^2$ and Θ_w [8]. However, unlike the case of the sphaleron, where the existence of the non-contractible loop in configuration space is connected to the process of baryon-number violation, the non-trivial topology that we have discussed is not linked to an anomaly and there does not seem to be any related new physics. However, in this theory, we have realised Forgács and Horváths' conjecture, that the existence of non-contractible spheres of dimension higher than one suggests the possible existence of solutions, albeit in reverse. This may lend encouragement to attempts to find new solutions by this process.

REFERENCES

1 N. Manton *Phys. Rev.* **D28** (1983), 2019
2 F.R.Klinkhamer, N.S.Manton *Phys. Rev.* **D30** (1984), 2212
3 P. Forgács, Z. Horváth *Phys. Lett.* **B138** (1984), 397
4 T. Vachaspati *Phys. Rev. Lett.* **68** (1992), 1977; *Nucl. Phys.* **B397** (1993), 648
5 T. Vachaspati, M. Barriola *Phys. Rev. Lett.* **69** (1992), 1867
6 M. Barriola, T. Vachaspati, M. Bucher preprint hep-th/9306120
7 H. Nielsen, P. Olesen *Nucl. Phys.* **B61** (1973), 45
8 M. James, T. Vachaspati, L. Perivolaropoulos *Phys. Rev.* **D46** (1992), 5232; *Nucl. Phys.* **B395** (1993), 534

SPHALERONS AND STRINGS

Mark Hindmarsh

School of Mathematical and Physical Sciences
University of Sussex
Brighton BN1 9QH

INTRODUCTION

This work is based on a paper with Margaret James.[1] In it we showed that the dipole moment of the sphaleron has its origin in two components: a ring of electric current circulating around the edge of the sphaleron; and also two regions of opposite magnetic charge above and below the ring. This magnetic charge has a partly topological explanation, arising from the fact that the sphaleron is axisymmetric and parity invariant. Here, I shall make a few remarks that are complementary to our paper. Firstly, I discuss the definition of the electromagnetic field and its sources in the Standard Model, comparing ours with the better-known one of 't Hooft.[2] Both allow magnetic charges and currents, but 't Hooft's magnetic charge is designed purely to count zeroes of the Higgs field and is not a very satisfactory dynamical quantity. Secondly, I summarize the results of our paper,[1] which uses our definition of the electromagnetic field to look inside the sphaleron and find the distribution of charges and currents which give rise to its dipole moment. Finally, I go into a bit more detail about the resemblance between the sphaleron and Nambu's "dumb-bell"[3] – a segment of Z-string which connects a monopole to an antimonopole. It has been conjectured that the string segment is a kind of "stretched" sphaleron:[4,5] however, it is possible to estimate the Chern-Simons number $N_{\rm CS}$ using the topology of the Higgs field alone, providing the gauge is chosen correctly. In this gauge it can be shown that the string segment lacks a crucial twist in its field configuration that in the sphaleron results in $N_{\rm CS} = \frac{1}{2}$.

ELECTROMAGNETISM IN THE STANDARD MODEL

In the Standard Model there is no unambiguous definition of the electromagnetic field. The first step, on which all agree, is to use a unit isovector

$$\hat{\phi}^a = \Phi^\dagger \sigma^a \Phi / |\Phi|^2, \tag{1}$$

to project out the massless component of the gauge potential:

$$A_\mu = -W^a_\mu \hat{\phi}^a \sin\theta_{\rm W} + Y_\mu \cos\theta_{\rm W}, \tag{2}$$

where W^a_μ and Y_μ are the SU(2) and U(1) gauge fields respectively. This reduces to the usual expression when the Higgs takes its conventional vacuum expectation value $\Phi = (0,1)^{\rm T} v/\sqrt{2}$, for then $\hat{\phi}^a = -\delta^a_3$. We can also project out the massless component of the gauge field strength tensor:

$$F^{\rm em}_{\mu\nu} = -F^a_{\mu\nu}\hat{\phi}^a \sin\theta_{\rm W} + F^0_{\mu\nu}\cos\theta_{\rm W}. \tag{3}$$

However, there is a remnant of the non-Abelian theory in which it is embedded, for the relationship between $F^{\rm em}_{\mu\nu}$ and A_μ does not take the simple form of ordinary electromagnetism. In fact,

$$F^{\rm em}_{\mu\nu} = \partial_{[\mu}A_{\nu]} + \frac{\sin\theta_{\rm W}}{g}\left[\hat{\phi}^a\partial_\mu\hat{\phi}^b\partial_\nu\hat{\phi}^c - \hat{\phi}^a D_\mu\hat{\phi}^b D_\nu\hat{\phi}^c\right]\epsilon^{abc}, \tag{4}$$

where $D_\mu\hat{\phi}^a = \partial_\mu + g\epsilon^{abc}W^b_\mu\hat{\phi}^c$. (A sign error in Ref. 1 has been corrected here.) This means that the Bianchi identity is not satisfied, and there is a magnetic current given by the right hand side of

$$\tfrac{1}{2}\epsilon^{\mu\nu\rho\sigma}\partial_\nu F^{\rm em}_{\rho\sigma} = -\tfrac{1}{2}\sin\theta_{\rm W}\epsilon^{\mu\nu\rho\sigma}F^a_{\nu\rho}D_\sigma\hat{\phi}^a. \tag{5}$$

Note that $D_\sigma\hat{\phi}^a$ is orthogonal to $\hat{\phi}^a$, so that the magnetic current is comprised of gauge fields orthogonal both to the electromagnetic field and to the Z field, $F^{\rm Z}_{\mu\nu} = -F^a_{\mu\nu}\cos\theta_{\rm W} - F^0{\mu\nu}\sin\theta_{\rm W}$. In other words, magnetic currents in the Standard Model are made of W fields. W fields also make electric currents, for one can show that

$$\partial^\nu F^{\rm em}_{\mu\nu} = -\sin\theta_{\rm W}F^a_{\mu\nu}D^\nu\hat{\phi}^a. \tag{6}$$

The Higgs field does not contribute directly to this expression. This is to be expected, for the physical Higgs field is neutral.

A definition that many authors use is that of 't Hooft, who gives the electromagnetic field as

$$\mathcal{F}^{\rm em}_{\mu\nu} = F^{\rm em}_{\mu\nu} + \frac{\sin\theta_{\rm W}}{g}\hat{\phi}^a D_\mu\hat{\phi}^b D_\nu\hat{\phi}^c\epsilon^{abc}. \tag{7}$$

The purpose of adding the extra term is to force the electromagnetic field to satisfy the Bianchi identity almost everywhere, for

$$\tfrac{1}{2}\epsilon^{\mu\nu\rho\sigma}\partial_\nu\mathcal{F}^{\rm em}_{\rho\sigma} = \frac{\sin\theta_{\rm W}}{2g}\epsilon^{\mu\nu\rho\sigma}\partial_\nu\hat{\phi}^a\partial_\rho\hat{\phi}^b\partial_\sigma\hat{\phi}^c\epsilon^{abc}. \tag{8}$$

The right hand side of this expression vanishes everywhere except along world lines around which ϕ^a takes "hedgehog" configurations.[2] Thus the 't Hooft magnetic flux out of a closed surface serves merely to count zeroes of the Higgs field.

The two expressions (3) and (7) agree in the Higgs vacuum, where $D_\mu\hat{\phi}^a = 0$. Away from the vacuum they disagree, and there is no absolute standard by which to judge them. However, the first definition has the advantage that the energy density in the magnetic field is always $F^{\rm em}_{ij}F^{{\rm em}\,ij}$, whereas the 't Hooft magnetic field energy density must be corrected with derivatives of the isovector $\hat{\phi}^a$. Furthermore, there seems no reason to try and satisfy the Bianchi identities almost everywhere – we know that there is magnetic charge in non-Abelian theories, so why not have it spread out and created out of physical fields of the theory?

Accordingly, we shall use definition (3) in what follows. With it, we can look inside the sphaleron and find that at non-zero Weinberg angle the W fields do provide both electric currents and magnetic charges.

INSIDE THE SPHALERON

The sphaleron at zero Weinberg angle is a spherically symmetric solution of the classical field equations of the Standard Model, which can be written in the following form, in the radial gauge $W_r = 0$:

$$\Phi = U(\hat{\boldsymbol{x}})\begin{pmatrix}0\\1\end{pmatrix}h(r)\frac{v}{\sqrt{2}}, \qquad W = -\frac{2i}{g}dUU^{-1}f(r), \tag{9}$$

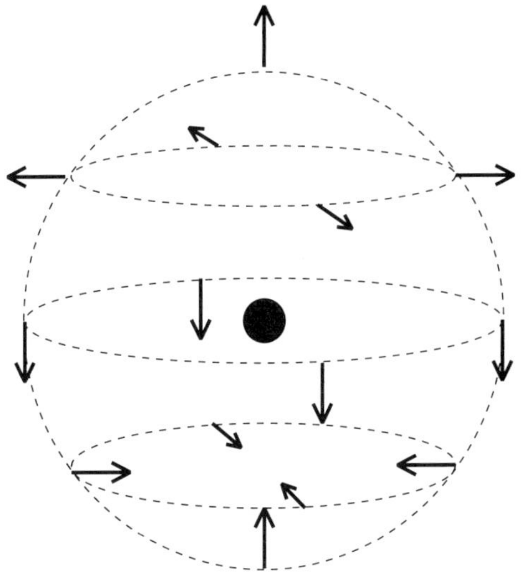

Figure 1. Isovector field configuration $\hat{\phi}^a$ around the sphaleron at $\theta_W = 0$.

where $U = i\hat{\boldsymbol{x}} \cdot \boldsymbol{\sigma}\sigma_2$. The isovector field around this configuration is

$$\hat{\phi}^a = 2\hat{x}^a\hat{x}^3 - \delta^{a3}. \tag{10}$$

This field is illustrated in Figure 1. In the core of the sphaleron, the fields leave the vacuum and there is non-zero $D_i\Phi$ and F^a_{ij}. The spherical symmetry of the energy density is in fact guaranteed by the custodial SU(2) symmetry of the Standard Model, so it is understandable that when $\theta_W \neq 0$ the equal-energy contours of the sphaleron solution become prolate.[6] The sphaleron also develops a magnetic dipole moment, which has been calculated perturbatively in the small θ_W limit,[7,8] and also numerically for general values of θ_W.[6]

With the definition (3) of the electromagnetic field we can understand this dipole moment in terms of magnetic charges and electric current defined in a gauge invariant way. Let us evaluate them to first order in θ_W, where the perturbation to the background configuration does not appear. Using equations (5) and (6) we find that the magnetic charge density $\tilde{\rho}$ and the electric current J_i are given by

$$\tilde{\rho} = \frac{g'}{g} 8\cos\theta \frac{f'(1-f)}{gr^2}, \qquad \boldsymbol{J} = \frac{g'}{g} 8\sin\theta \frac{f(1-f)^2}{gr^4} \hat{\boldsymbol{e}}_\varphi, \tag{11}$$

where $\hat{\boldsymbol{e}}_\varphi$ is the azimuthal unit vector in spherical polar coordinates. The magnetic dipole moment of this charge and current distribution is

$$\frac{g'}{g}\frac{32\pi}{3g} \int_0^\infty \left[rf'(1-f) + f(1-f)^2 \right] dr. \tag{12}$$

When the function f is found numerically,[8] we find that the charge and the current contribute 70% and 30% respectively to the dipole moment. Moreover, this expression can be shown to be *identical* to that derived by Klinkhamer and Manton.[7] The total magnetic charge in each hemisphere is $4\pi\theta_W/g$, which is θ_W multiplied by the charge of a 't Hooft-Polyakov SU(2) monopole[2,9]. Indeed, the isovector field configuration looks as if it could be the Higgs field around a monopole-antimonopole pair in the Georgi-Glashow model. This might make one wonder if the presence of the magnetic charge was in some way topological.

The topology lies in the subspace of field configurations which are restricted to be axisymmetric and parity invariant. On the equatorial plane these configurations have $\hat{\phi}^a$ constant, which can be chosen to be $-\delta^a_3$, and the SU(2) gauge field takes the form[6]

$$gW_i dx^i = if_1(r)(-\sin\varphi\sigma_1 + \cos\varphi\sigma_2)d\theta + if_2(r)\sigma_3 d\varphi, \tag{13}$$

where f_1 and f_2 are functions which tend to 1 at infinity and vanish at the origin. The magnetic charge $\tilde{Q}$ in the region $z \geq 0$ can be expressed as an integral of the magnetic field over a surface consisting of a hemisphere at infinity N plus the equatorial plane E:

$$\tilde{Q}_{z\geq 0} = \int_{N+E} dx^i B_i^{\mathrm{em}}. \tag{14}$$

Using the expression (4) for the electromagnetic field tensor, we see that the only contribution to this surface integral comes from the "non-Abelian" parts involving derivatives of the isovector field $\hat{\phi}^a$. On the equatorial plane, both $\partial_i\hat{\phi}^a$ and $\hat{\phi}^a D_i\hat{\phi}^b D_j\hat{\phi}^c\epsilon^{abc}$ vanish. Thus the magnetic charge $\tilde{Q}$ is given in terms of the integral

$$\tilde{Q}_{z\geq 0} = \frac{\sin\theta_{\mathrm{W}}}{g}\int_N \hat{\phi}^a\partial_i\hat{\phi}^b\partial_j\hat{\phi}^c\epsilon^{abc}. \tag{15}$$

Recall that $\hat{\phi}^a$ is constant on the boundary of the hemisphere: thus the hemisphere is effectively a 2-sphere, and integral on the right hand side of this equation measures the winding number of the isovector field around the unit sphere $\hat{\phi}^2 = 1$. Therefore, *for axisymmetric, parity-invariant configurations*, the magnetic charge in the region $z \geq 0$ is quantized in units of $4\pi\sin\theta_{\mathrm{W}}/g$. The same considerations apply to the region $z \leq 0$. The parity operation reverses magnetic fields, and so switches the sign of magnetic charge. Thus the charge in the other hemisphere must be equal and opposite.

SPHALERONS AND ELECTROWEAK STRINGS

The Standard Model contains another non-trivial classical solution, which is a vortex carrying Z-flux.[3,10] It is not topologically stable, and is dynamically stable only in the unphysical region $\sin\theta_{\mathrm{W}} \simeq 1$.[11] Because of their topological triviality, they can end, and Nambu[3] showed that they terminate on monopoles, which have magnetic charge $\pm 4\pi\sin\theta_{\mathrm{W}}/g$. Nambu proposed that there was a massive long-lived excitation in the Standard Model: a spinning segment of electroweak string, with oppositely charged monopoles at each end. This "dumb-bell" bears some resemblance to the sphaleron,[1,5] which we saw in the last section also has a quantized magnetic dipole within it. This section explores the connection between the two field configurations, amplifying some remarks made in Ref. 1.

The Higgs field in Nambu's dumb-bell configuration vanishes along a line segment, taking the form away from this line

$$\Phi = \begin{pmatrix} \cos\Theta \\ \sin\Theta e^{i\varphi} \end{pmatrix}\frac{v}{\sqrt{2}}, \qquad \cos 2\Theta = \cos\theta_+ - \cos\theta_- + 1, \tag{16}$$

where $\theta_\pm$ are polar angles measured from the ends of the line segment (see Figure 3). The gauge fields can be written in terms of the Higgs field, by solving the equation $D_i\Phi = 0$:

$$gW_i^a = -\epsilon^{abc}\hat{\phi}^b\partial_i\hat{\phi}^c - i\cos^2\theta_{\mathrm{W}}\hat{\phi}^a(\hat{\Phi}^\dagger\overleftrightarrow{\partial}_i\hat{\Phi}), \tag{17}$$

$$g'Y_i = -i\sin^2\theta_{\mathrm{W}}(\hat{\Phi}^\dagger\overleftrightarrow{\partial}_i\hat{\Phi}). \tag{18}$$

This solution actually requires two extra pieces of information: that vortices carry only Z flux (and no electromagnetic flux); and a gauge choice, which sets a possible arbitrary abelian part $(\alpha_i\hat{\phi}^a, \alpha_i)$ to zero. The second point is a crucial one, for the sphaleron is usually exhibited in the radial gauge, so direct comparisons can be misleading. In Nambu's paper the fact that a gauge choice has been made is rather implicit: a clearer discussion (for the Georgi-Glashow model) was given by Manton,[12] who showed that in this gauge $\hat{\phi}^a$ is constant along lines of magnetic flux, or

$$B_i^{\mathrm{em}}\partial_i\hat{\phi}^a = 0. \tag{19}$$

This is illustrated in Figure 2, which depicts the flux lines originating from the ends of the line segment. An important point about this configuration is that away from the singular line

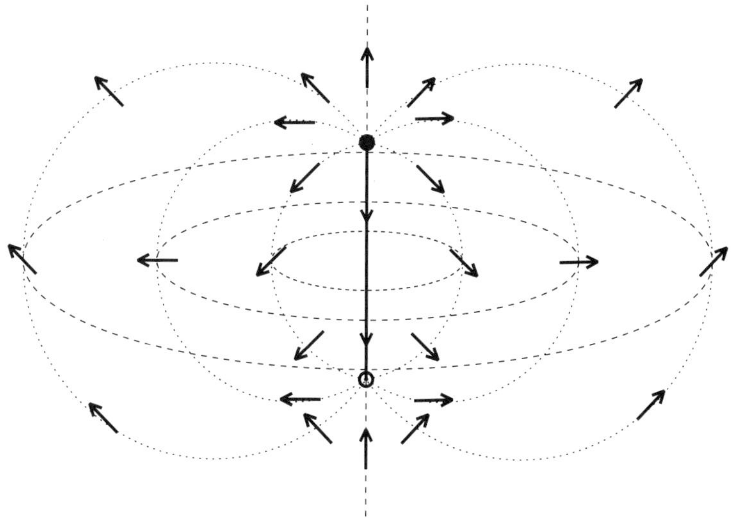

Figure 2. Isovector field configuration around Nambu' dumb-bell configuration. The thick solid line is the line of zeroes of the Higgs field, dotted lines are lines of magnetic flux. Dashed lines guide the eye on the equatorial plane

the fields obey the equations of motion: that is, not only does the covariant derivative of the Higgs vanish, but also $\epsilon_{ijk}\partial_j B_k^{\rm em} = 0$.

Now, for the purposes of comparison, let us try and construct a sphaleron-type configuration in this gauge, which also vanishes on a line segment. The most convenient way to do this is to use prolate spheroidal coordinates (μ, β, φ).[13] The surfaces of constant μ are increasingly prolate as $\mu \to 0$, collapsing to a line of length $2d$ at $\mu = 0$, while becoming more spherical as $\mu \to \infty$. (β, φ) are polar coordinates on the spheroids. If $r_\pm$ are the distances from the ends of the line segment then $d\cosh\mu = (r_+ + r_-)/2$, and $\cos\beta = (r_+ - r_-)/2d$. Then we may adapt (9) by redefining U:

$$U(\hat{\boldsymbol{\xi}}) = [\cos\chi(\mu) + i\sin\chi(\mu)\boldsymbol{\sigma}\cdot\hat{\boldsymbol{\xi}}]\sigma_2, \tag{20}$$

where $\hat{\boldsymbol{\xi}} = (\sin\beta\cos\varphi, \sin\beta\sin\varphi, \cos\beta)$; and $\chi(0) = \pi/2$, $\chi(\infty) = 0$. In the limit that $d \to 0$, $\hat{\boldsymbol{\xi}} \to \hat{\boldsymbol{x}}$, and away from the origin the configuration is just a gauge transformation $\Omega = \exp(i(\pi/2 - \chi)\boldsymbol{\sigma}\cdot\hat{\boldsymbol{x}})$ of the sphaleron. The isovector field can be used to trace the magnetic flux lines:

$$\hat{\phi}^a = \cos 2\chi(\delta^{a3} - \hat{\xi}^a\hat{\xi}^3) + \sin 2\chi\epsilon^{a13}\hat{\xi}^i + \hat{\xi}^a\hat{\xi}^3. \tag{21}$$

Figure 3 shows the isovector field configuration around this "stretched" sphaleron. It is clear that the magnetic field has an azimuthal component: the lines of flux twist through π as they travel from one pole to another. This twist is the source of the difference between the sphaleron and the dumb-bell, for it turns out to be crucial in supplying the Chern-Simons number.[14] Twisting the field is an unnatural thing to do, for the magnetic field cannot sustain such a twist without a current. The current can only exist in the core of the string segment, where the Higgs field is off the vacuum and $F^a_{ij}D_j\hat{\phi}^a$ does not vanish. Thus, without an external electromagnetic current, the twist can exist only within the string.

The true significance of these field configurations is rather obscure, for they are not solutions of the field equations. However, they may be solutions to the equations subject to a constraint: that the Higgs field is forced to vanish on a line. This is certainly a well-defined variational problem, for the constraint is gauge-invariant. However, to find which (if either) of the dumb-bell or the stretched sphaleron is a solution seems exceedingly hard.

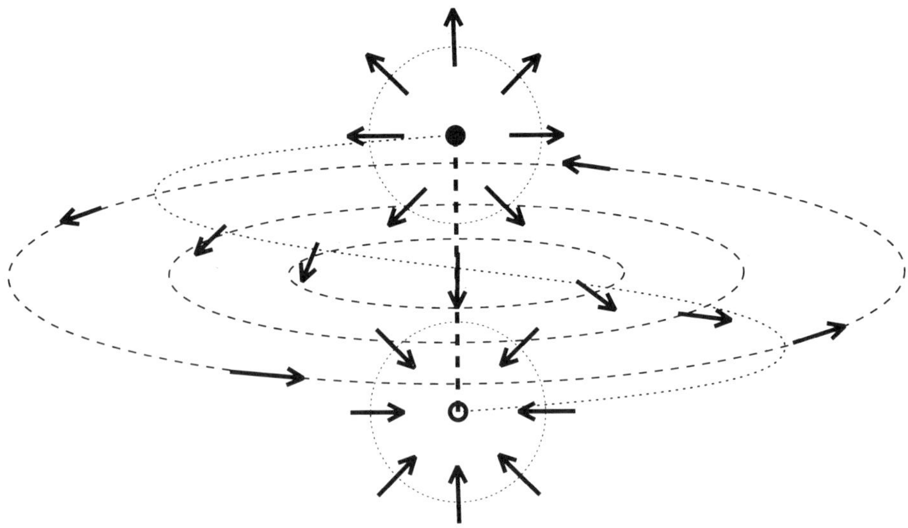

Figure 3. Isovector field configuration around the stretched sphaleron. Only one line of magnetic flux is shown. Note the twist as it travels between the ends of the line of Higgs field zeroes.

REFERENCES

1. M. Hindmarsh and M. James, *Phys. Rev. D*, (to appear, 1994).
2. G. 't Hooft, *Nucl. Phys.* **B79**, 276 (1974).
3. Y. Nambu, *Nucl. Phys.* **B130**, 505 (1977).
4. T.Vachaspati, "Electroweak Strings: a Progress Report", in: proceedings of Texas/-PASCOS 92: Relativistic Astrophysics and Particle Cosmology, *Ann. N. Y. Acad. Sci.* Vol. 688, (1993).
5. M. Barriola, T. Vachaspati and M. Bucher, Embedded Defects, *Phys. Rev. D* (to appear, 1994).
6. B. Kleihaus, J. Kunz and Y. Brihaye, *Phys. Rev.* **D46**, 3587 (1992).
7. F. Klinkhamer and N. Manton, *Phys. Rev.* **D30**, 2212 (1984).
8. M. James, *Zeit. Phys.* **C55**, 515 (1992).
9. A.M. Polyakov, *JETP Lett.* **20**, 194 (1974).
10. T. Vachaspati, *Phys. Rev. Lett.* **68**, 1977 (1992).
11. M. James, L. Perivolaropoulos and T. Vachaspati, *Nucl. Phys.* **B395**, 534 (1993).
12. N. Manton, *Nucl. Phys.* **B126**, 525 (1977).
13. P. Morse and H. Feschbach, "Methods of Mathmatical Physics," McGraw-Hill, New York (1953).
14. T. Vachaspati and G.B. Field, *Tufts U. preprint* (1994).

EXISTENCE AND STABILITY OF SPINNING EMBEDDED VORTICES

Leandros Perivolaropoulos[1,2]

Division of Theoretical Astrophysics
Harvard-Smithsonian Center for Astrophysics
60 Garden St., Cambridge, Mass. 02138, USA

This work focuses on a new class of embedded vortices which are non-static but stationary and carry non-zero angular momentum. They may therefore be called *spinning embedded vortices*. It will be shown that these vortices have improved stability properties compared to their non-spinning counterparts (electroweak vortices with in a hypothetical model with Weinberg angle $\theta_w = \pi/2$[1, 2, 3]) whose properties are discussed in the contributions of Klinkhamer and Vachaspati in this volume.

Let's first clarify the motivation for generalizing the concept of the non-spinning electroweak (hereafter EW) vortex. As is well known the vacuum manifold of the standard EW model is a three sphere S^3. This implies that the model does not support *any* stable topological defects. However, almost a couple of years ago, Vachaspati[2] pointed out that there is a vortex-like coherent state in the EW model which appears as an exact solution to the field equations. In a later paper[3] we showed that this coherent state, called the embedded electroweak vortex[4], is in fact stable for a finite sector in parameter space. We also showed that the stability sector does not include the physically realized values of parameters. This was unfortunate but raised an exciting new question: *Are there different types of embedded vortices with improved stability properties?*.

Until recently there had been two main classes of attempts to address this question. The first class originated from a work of Vachaspati and Watkins[5] who showed that bound states on the embedded vortex may indeed improve its stability. Due to the complexity of the problem however, it was not clear if this improvement of stability was enough to stabilize the electroweak vortex for the physical values of parameters. The second class focused on embeddings in *extensions* of the standard EW model. There have been several interesting works on this subject[6, 7, 8, 9] but of particular interest is the work of Dvali and Senjanovic[8] who discovered *topologically stable* vortices in the two Higgs doublet EW model.

[1]E-mail address: leandros@cfata3.harvard.edu
[2]Also, Visiting Scientist, Department of Physics, Brown University, Providence, R.I. 02912.

Here I focus on a new stabilization mechanism which is applicable to vortices of the standard, minimal EW model and introduces *spin* to improve the stability of the embedded vortices. An efficient way to introduce angular momentum to the vortex configuration is to embed it in a background of charge density which may be provided for example, by a charged external field with coherence length much larger than the width of the vortex.

Consider a toy model with symmetry breaking $SU(2)_{global} \times U(1)_{local} \longrightarrow U(1)_{global}$. For our purpose, this model is identical to the bosonic sector of the standard EW model with $\theta_w = \pi/2$. The vacuum manifold in this model is S^3. It has been shown[1, 12] that the Nielsen-Olesen vortex[10] is supported as an exact solution and that it is stable for a limited parameter range. Consider now an embedding ansatz[11] which is a stationary generalization[13] of the Nielsen-Olesen ansatz

$$\Phi = \begin{pmatrix} 0 \\ f(r)e^{im\theta}e^{i\omega_0 t} \end{pmatrix} \tag{1}$$

$$A_\theta = \frac{v(r)}{r}, \qquad A_0 = \alpha(r) \tag{2}$$

The differences between the ansatz (1), (2) and the embedded Nielsen-Olesen vortex[2] (EW string with $\theta_w = \pi/2$) is the linear time dependence of the Goldstone field and the allowed possibility of having $A_0 \neq 0$. The Lagrangian density which supports this ansatz as a solution is a generalized Abelian Higgs Lagrangian where the scalar field Φ has been promoted to an $SU(2)$ doublet and also a background charge density has been introduced and coupled to the gauge field A_μ. It is of the form

$$\mathcal{L} = -\frac{1}{4}F_{\mu\nu}F^{\mu\nu} + \frac{1}{2}|D_\mu\Phi|^2 - \frac{\lambda}{4}(|\Phi|^2 - \eta^2)^2 - A_\mu J^\mu \tag{3}$$

where $J^\mu = (\rho(r), 0, 0, 0)$ is a background charge density. Let $\rho(r) \rightarrow \rho_0$ asymptotically, where ρ_0 is a constant. The two crucial parameters that determine the stability of the embedded vortex (1), (2) will be shown to be β and ρ_0. The angular velocity ω_0 is fixed in terms of those parameters. The total charge of the vortex configuration plus the background is

$$Q = \int_V d^2x \, [f^2\,(\omega_0 - \alpha) + \rho(r)] \tag{4}$$

where the first term in the integral is due to the vortex field and the second is due to the background. This charge is in general non-zero, finite and conserved. The angular momentum $\vec{M}_v$ of the vortex configuration may be shown to be proportional to the charge Q_v and the winding number m of the vortex field

$$\vec{M}_v = \int_V d^2x \, \vec{r} \times (\vec{E}_v \times \vec{B}) = -2 \int_V d^2x \, f(r)^2 \, (\omega_0 - \alpha)\, v(r)\, \hat{e}_z \simeq -2\, m\, Q_v\, \hat{e}_z \tag{5}$$

The crucial question that needs to be addressed is, *what are the stability properties of the vortex and how do they compare with the corresponding properties of its non-spinning counterpart (embedded EW vortex with $\theta_w = \pi/2$)?*

Consider a perturbation to the ansatz (1), (2) of the form

$$\delta\Phi = \begin{pmatrix} g(r)e^{in\theta} \\ \delta f(r,\theta) \end{pmatrix}, \qquad \delta A_\mu(r,\theta) \tag{6}$$

The energy of the perturbed configuration decouples as follows

$$E = E_0(f, v, \alpha) + \delta E(\delta f, \delta A_\mu) + E_1(g) \tag{7}$$

where the energy E_0 of the unperturbed configuration is identical in form to the energy of the topologically stable configuration obtained when the complex doublet in the ansatz is replaced by a complex singlet. The term δE is identical in form to the perturbation of the topologically stable spinning vortex and must be positive definite since the topological vortex is stable. Thus the term whose sign determines the stability is the term $E_1(g)$. For those parameter values (β, ρ_0) for which $E_1(g)$ can be negative, the embedded vortex (1), (2) is unstable. But E_1 may be shown[11] to be positive definite *iff* the eigenvalue equation

$$-g'' - \frac{g'}{r} + \frac{(v-n)^2}{r^2} g + \beta(f^2 - 1)g + \alpha^2 g = \omega^2 g \tag{8}$$

has no negative eigenvalues. For the range of parameters (β, ρ_0) for which this happens, the embedded spinning vortex is stable.

The unperturbed fields, f, v, α and therefore the sign of E_1, also depend on the detailed form of $\rho(r)$ close to the core of the vortex. I will therefore consider two opposite extreme forms of $\rho(r)$ and show that stability is dramatically improved in both cases. For the first type of background, the *soft* background, $\rho(r)$ adjusts in a way to completely neutralize the charge induced by the spin of the vortex *i.e.*

$$\rho(r) = -\omega_0 f^2(r), \qquad \alpha = 0 \tag{9}$$

For such a background, the field equations allow $\alpha = 0$ everywhere. This implies that the Schroedinger-type potential of the eigenvalue equation (8), looses a positive definite contribution: the term α^2. This makes the equation more receptive to negative eigenvalues. We are therefore led to the conjecture that the soft background is the background of minimum stability.

The other opposite extreme form is that of a *hard* background where $\rho(r)$ does not feel the presence of the vortex and is constant everywhere

$$\rho(r) = \rho_0, \qquad \alpha \neq 0 \tag{10}$$

In that case, A_0 (or α) can not vanish[11] and we expect better stability properties for this background type.

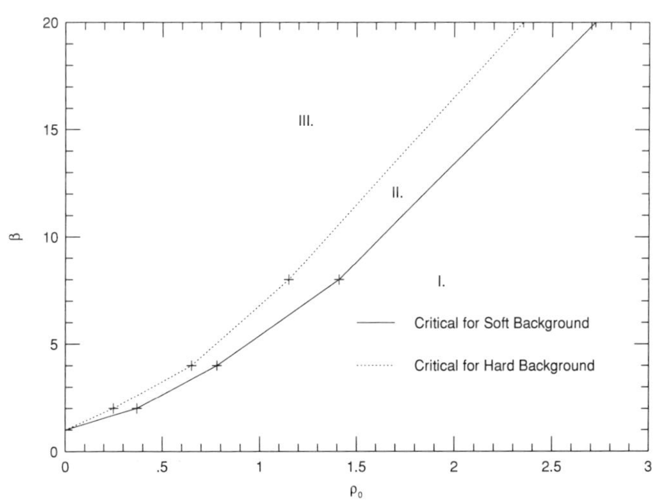

Figure 1. The stability map is parameter space for the embedded spinning vortex. Sector III. (II. and III.) is the stability sector for the case of a soft (hard) background.

It is straightforward to solve the eigenvalue problem numerically using relaxation methods[14], for the two types of backgrounds and map the stability sectors in parameter space (β, ρ_0) for each case. The stability sector for the soft (hard) background is on the right of the continous (dotted) line in Fig. 1.

The case of the non-spinning EW vortex is obtained for $\rho_0 = 0$ and has been studied previously [12, 3] (stability only for $\beta < 1$). Clearly the range of β corresponding to stability increases dramatically when spin is introduced ($\rho_0 \neq 0$). In fact it may be shown analytically[11] that for $\rho_0 \to \infty$ the embedded spinning vortex is stable for *any* β.

In conclusion, I have shown that angular momentum induced by a charged background can indeed stabilize the embedded EW vortex. This result raises some interesting new questions. First, *can spin stabilize the EW vortex for physical values of parameters?*. Second, *can global embedded vortices (which are normally always unstable[15]) be stabilized by the introduction of spin?* These issues are currently under investigation.

References

[1] T. Vachaspati and A. Achucarro, *Phys. Rev.* **D44**, 3067 (1991).

[2] T. Vachaspati, *Phys. Rev. Lett.* **68**, 1977 (1992).
T. Vachaspati, *Nucl. Phys.* **B397**, 648 (1993).

[3] M. James, L. Perivolaropoulos and T. Vachaspati, *Phys. Rev.***D46**, 5232 (1992)
M. James, L. Perivolaropoulos and T. Vachaspati, *Nucl. Phys.* **B395**, 534 (1993).

[4] T. Vachaspati and M. Barriola, *Phys. Rev. Lett.* **69**, 1867 (1992).

[5] T. Vachaspati and R. Watkins, *Phys. Lett.* **B318**, 163 (1993).

[6] H. S. La, hep-ph/9302220 (1993).

[7] L. Perivolaropoulos, *Phys. Lett.* **B316**, 328 (1993).

[8] G. Dvali and G. Senjanovic, *Phys. Rev. Lett.* **71**, 2376 (1993).

[9] M. Earnshaw and M. James, *Phys. Rev.* **D48**, 5818 (1993).

[10] H. B. Nielsen and P. Olesen, *Nucl. Phys.* **B61**, 45 (1973).
see also, L. Perivolaropoulos, *Phys. Rev.* **D48**, 5961 (1993) for a correction in the vortex asymptotic behavior presented by Nielsen-Olesen and in most subsequent papers.

[11] L. Perivolaropoulos, *Phys. Rev.* **D** in press.

[12] M. Hindmarsh, Phys. Rev. Lett. **68**, 1263 (1992)
A. Achúcarro, K. Kuijken, L. Perivolaropoulos T. Vachaspati, *Nucl. Phys.* **B388**, 435 (1992).

[13] R. Davis, *Mod. Phys. Lett.* **A5**, 955 (1990).

[14] W. Press, B. Flannery, S. Teukolsky, W. Vetterling, *Numerical Recipes, 2nd Edition*, Cambridge Univ. Press (1993).

[15] M. Barriola, L. Perivolaropoulos and T. Vachaspati, unpublished.

EFFECTIVE POTENTIAL FOR THE WILSON LINE IN THE STANDARD MODEL

Chris Korthals-Altes[1], Kimyeong Lee[2], and Robert D. Pisarski[3]

[1]Centre Physique Theorique au CNRS
Section 2, B.P. 907, Luminy
F 13288 Marseille, France

[2]Physics Department
Columbia University
New York, NY 10027

[3]Department of Physics
Brookhaven National Laboratory
P.O. Box 5000
Upton, New York 11973-5000

ABSTRACT

We compute the effective potential for the Wilson line at nonzero temperature to one loop order in the standard model. Besides the usual perturbative vacua, there are metastable states in the direction of $U(1)$ hypercharge. Assuming that the universe starts out in such a metastable state at the Planck scale, it easily persists to the scale of electroweak symmetry breaking, where it serves as a source of CP violation - as required for the generation of baryon number asymmetry - that is naturally of order one.

1. THE WILSON LINE IN THE STANDARD MODEL

The following is a description of work in progress by C. P. Korthals Altes, K. Lee, and myself [1]. I am taking advantage of the proceedings of this workshop to give a summary of our results, since the essential features are clear already.

At a nonzero temperature T in the imaginary time formalism, consider the Wilson line which runs in the direction of imaginary time, $\Omega(x) = exp(ig \int_0^{1/T} A_0(x,\tau)d\tau)$ [2 - 11]. In an abelian theory Ω is just a phase factor and so automatically gauge invariant under strictly periodic gauge transformations. In a nonabelian theory Ω is a matrix, where only the eigenvalues are gauge invariant [4,5,8,9]. These eigenvalues lie in the space of the mutually commuting generators, which is the Cartan subalgebra of the group. To parametrize a theory where $\langle\Omega\rangle \neq 1$, at one loop order it suffices to calculate the effective potential in a background field of constant A_0. Notice that we deal not just with the sum of the eigenvalues, as arises for the Polyakov line $= tr(\Omega)$, but all eigenvalues.

The nature of the effective potential for the Wilson line of $SU(3)$ color has been studied previously. In the absence of dynamical fermions this potential allows one to compute the interface tension between the $Z(3)$ degenerate vacua, $\Omega = 1$ and $\Omega = exp(\pm 2\pi i/3)$, by semiclassical means [5]. Adding dynamical fermions lifts the $Z(3)$ degeneracy, so the trivial vacuum at $\Omega = 1$ is an absolute minimum. Dixit and Ogilvie [6] noticed, however, that while no longer degenerate with the trivial vacuum, the states at $\Omega = exp(\pm 2\pi i/3)$ are often metastable. Assuming that the universe starts out with $\Omega = exp(\pm 2\pi i/3)$ at the Planck scale, Ignatius, Kajantie, and Rummukainen [7] find that the universe decays into $\Omega = 1$ by temperatures of order 10 TeV, well before the electroweak phase transition.

Our basic point is that the quantity of significance in the standard model is not just the effective potential for any *single* gauge group, such as $SU(3)$ color, but that for the *full* gauge group, $U(1) \times SU(2) \times SU(3)$. This is a four dimensional space: one coordinate for the $U(1)$ of hypercharge, one for the Cartan subalgebra of weak $SU(2)$, and two for the Cartan subalgebra of $SU(3)$ color. We give the timelike component of each vector potential a constant expectation value,

$$\mathcal{A}_0 = \frac{2\pi T}{g_{st}} \begin{pmatrix} q/3 + r/2 & 0 & 0 \\ 0 & q/3 - r/2 & 0 \\ 0 & 0 & -2q/3 \end{pmatrix},$$

$$A_0 = \frac{2\pi T}{g} \begin{pmatrix} s/2 & 0 \\ 0 & -s/2 \end{pmatrix}, \quad B_0 = \frac{2\pi T}{g'} t, \tag{1}$$

$\mathcal{A}_0$ is the field for $SU(3)$ color, with coupling constant g_{st}, A_0 is the field for weak $SU(2)$, with coupling g, and B_0 is the field for hypercharge $U(1)$, with coupling g'.

The Wilson line, and so the potential thereof, is a function of four variables, q, r, s, and t. The variables q and r refer to the color field $\mathcal{A}_0$, with the term in q proportional to the $SU(3)$ matrix λ_8, and the term in r part proportional to λ_3. The weak $SU(2)$ field is proportional to the matrix σ_3, with coefficient s, while the $U(1)$ hypercharge field is proportional to t. A Wilson line formed from these gauge fields is

$$\Omega(q,r,s,t) =$$

$$exp\left\{ \pi i \left[\frac{t}{3} + \begin{pmatrix} s & 0 \\ 0 & -s \end{pmatrix} + \begin{pmatrix} 2q/3 + r & 0 & 0 \\ 0 & 2q/3 - r & 0 \\ 0 & 0 & -4q/3 \end{pmatrix} \right] \right\}. \tag{2}$$

The contribution to the Wilson line for the $SU(2)$ and $SU(3)$ gauge fields are the standard expressions for the fundamental representation of each group, where of course each matrix acts in a different space. For the $U(1)$ of hypercharge, however, we choose the Wilson line which corresponds to the *smallest* hypercharge. In the standard model the hypercharges are $+1/6$, $-1/3$, $-1/2$, $+2/3$, and -1 times g', so we take the $U(1)$ part of the Wilson line as $exp(ig'/6 \int_0^{1/T} B_0 d\tau) = exp(\pi i t/3)$. The smallest charge enters because that fixes the Wilson line to be periodic in t with period 6; the larger charges all have smaller periods which are integral divisors of 6. Besides t, the Wilson line is periodic in q with period 3, and periodic in r and s with period 2. Without loss of generality, then, we can consider just the region $0 \leq q \leq 3$, $0 \leq r \leq 2$, $0 \leq s \leq 2$, and $0 \leq t \leq 6$.

At one loop order the potential involves the function

$$v(x) = [x]^2(1-[x])^2 \ , \quad [x] = |x|_{mod\,1} \ . \tag{3}$$

$v(x)$ depends on x through $[x]$, the absolute value of x modulo 1. The absolute value of x enters because the overall system is charge neutral, and so the sign of x doesn't matter. Modulo 1 enters because if the Wilson line is $exp(2\pi i x)$, then $x \to x+1$ is a symmetry. In the standard model the total potential is

$$\int d^3x \ \pi^2 T^2 \left(-\frac{427}{360} + \mathcal{V}(q,r,s,t) \right) \ . \tag{4}$$

The first term, $-427/360$, is proportional to the free energy of the standard model to lowest order in the coupling constants, as a sum of ideal gases. To include the effects of colored fields in the fundamental representation we introduce the function

$$v_c(q,r,x) = v\left(\frac{q}{3}+\frac{r}{2}+x+\frac{1}{2}\right) + v\left(\frac{q}{3}-\frac{r}{2}+x+\frac{1}{2}\right) + v\left(-\frac{2q}{3}+x+\frac{1}{2}\right) \tag{5}$$

The total potential for the Wilson line is

$$\begin{aligned} \mathcal{V}(q,r,s,t) &= \frac{4}{3}v(s) + \frac{4}{3}\left(v(r)+v\left(q\pm\frac{r}{2}\right)\right) \\ &+ \frac{2}{3}v\left(\pm\frac{s}{2}+\frac{t}{2}\right) - 2v\left(-t+\frac{1}{2}\right) - 2v\left(\pm\frac{s}{2}-\frac{t}{2}+\frac{1}{2}\right) \\ &- 2v_c\left(q,r,-\frac{t}{3}\right) - 2v_c\left(q,r,+\frac{2t}{3}\right) - 2v_c\left(q,r,\pm\frac{s}{2}+\frac{t}{6}\right) + 30\,v\left(\frac{1}{2}\right) \ . \end{aligned} \tag{6}$$

In this expression $v(q \pm r/2)$ denotes $v(q+r/2)+v(q-r/2)$, *etc.* The terms in (6) can be read off from the charge assignments of the standard model. We assume that there is a single isodoublet Higgs field and three generations of quarks and leptons. We compute in the limit of very high temperature, where the mass of the Higgs field, and so spontaneous symmetry breaking, can be ignored. The contributions in (6) are due to, respectively: the three $SU(2)$ gauge bosons, the eight $SU(3)$ gluons, the Higgs field, right handed leptons, left handed leptons, right handed down, strange, and bottom quarks, right handed up, charm and top quarks, and left handed quarks. Lastly there is an overall constant to fix $\mathcal{V}(0,0,0,0)=0$.

Because of left handed quarks, which transform nontrivially under each part of $U(1)\times SU(2)\times SU(3)$, the full potential $\mathcal{V}(q,r,s,t)$ is not separable, and the determination of all metastable points is not elementary. Here we discuss two examples.

The simplest thing to do is to consider subspaces where most of the coordinates vanish. For example, the case of pure $SU(3)$ color is given by setting $s=t=0$. Then there are the points of Dixit and Ogilvie [6],

$$q_1 = (1,0,0,0) \ , \quad q_2 = (2,0,0,0) \ , \tag{7}$$

for which $\Omega(q_1) = exp(2\pi i/3)$ and $\Omega(q_2) = exp(-2\pi i/3)$. These points are metastable in both the q and r directions; they are also metastable in the s direction. We find, however, that they are *un*stable in the t direction. Thus they are not possible candidates for an alternative cosmology.

Next we consider the direction of $U(1)$ hypercharge, $(0,0,0,t)$. While it is not obvious from (6), graphically one can see that there is quite a bit of structure in this direction. This is easy to understand physically. Structure in the effective potential for the Wilson line arises when there are fields with different charges. In the $SU(2)$ and $SU(3)$ directions, there are only fields in the fundamental and adjoint representations of the gauge group. For example, for $SU(3)$ the fundamental fields have charges $+1/3$ and $-2/3$ times the coupling constant, versus ± 1 for the adjoint fields. Thus the only place where something interesting can happen is for $q = 1$ or 2.

For $U(1)$ hypercharge, however, in the standard model there are fields with charge $+1/6$, $-1/3$, $-1/2$, $+2/3$, and -1 times the coupling constant. Thus it is very natural to develop a nontrivial potential in the t-direction. Numerical analysis shows that there are two metastable points in this direction,

$$t_1 = (0,0,0,1.28751...) \quad , \quad t_2 = (0,0,0,1.74843...) \, . \tag{8}$$

These points are metastable in *all* directions: q, r, s, and t.

There are several other metastable points where both q and t are nonzero. These points represent alternate cosmologies, like (8), and can be treated analogously to t_1 and t_2.

Assuming that the universe starts out in t_1 or t_2 in (8), the probability for the universe to decay is computed by standard techniques [13]. Adding the potential of (6) to the classical action [5], we obtain the effective action

$$\mathcal{S}_{eff} = \frac{32\pi^3}{(g')^3} \, \widetilde{\mathcal{S}}_{eff} \, ,$$

$$\widetilde{\mathcal{S}}_{eff} = \int_0^\infty d\tilde{r} \, \tilde{r}^2 \left(\frac{1}{2} \left(\frac{dt}{d\tilde{r}} \right)^2 + \mathcal{V}(0,0,0,t) \right) \, . \tag{9}$$

To obtain this expression we have rescaled the radial coordinate r to the dimensionless variable $\tilde{r} = r(g'T)/2$. The stationary point of (9) represents the tunneling path for an $O(3)$ symmetric bounce, as appropriate at high temperature. Of course by letting only t vary we are implicitly assuming that is the minimal path.

Since there are two metastable points, there are three possible tunneling paths; these were found numerically, without recourse to the thin wall approximation. For example, $\widetilde{\mathcal{S}}_{eff}(t_2 \to 0)$ represents a path which starts near $t = 0$ at $r = 0$ ($t_{initial} \sim .03729...$), and ends at $t = 1.74843...$ at $r = \infty$ (in practice $r = 10$). We find

$$\widetilde{\mathcal{S}}_{eff}(t_2 \to 0) \sim 1.83 \quad , \quad \widetilde{\mathcal{S}}_{eff}(t_1 \to 0) \sim .278 \quad , \quad \widetilde{\mathcal{S}}_{eff}(t_1 \to t_2) \sim .083 \, . \tag{10}$$

To estimate the temperature at which tunneling occurs we use a naive analysis, much simpler than that of Ignatius, Kajantie, and Rummukainen [7]. We assume that the decay rate per unit space time volume is $\Gamma \sim T^4 \, exp(-\mathcal{S}_{eff})$, and that tunneling occurs when $\Gamma t^4 \sim 1$, where t is the cosmological time. In a radiation dominated era $t \sim m_{Planck}/T^2$, so tunneling occurs for

$$T_{tunneling} \sim m_{Planck} \, exp \left(-\frac{\mathcal{S}_{eff}}{4} \right) \, . \tag{11}$$

Choosing the hypercharge coupling $g' = .535$, as appropriate at the GUT scale [14], we find that the most likely tunneling, $t_1 \to t_2$, does not occur until $T_{tunneling} \sim 10^{-39} GeV$, well below the temperature for the electroweak phase transition. The temperatures for the other tunnelings are even lower.

It is not difficult to understand why a metastable state for $U(1)$ hypercharge, (8), does not readily decay, while (ignoring its instability in the t direction) the metastable state for $SU(3)$ color, (7), does. The action for the two tunnelings is about equal at the GUT scale: for $t_1 \to t_2$, $\mathcal{S}_{eff} \sim 82/(g')^3$, while for $q_1 \to (0,0,0,0)$, $\mathcal{S}_{eff} \sim 92/(g_{st})^3$. In the analysis of ref. [7], the tunneling for $SU(3)$ color occurs because the the strong coupling, g_{st}, increases as the temperature decreases, and is large enough by $T \sim 10\,TeV$ to trigger tunneling. In constrast, the hypercharge coupling g' only gets smaller with decreasing temperature; its running is clearly negligible, since that only acts to make tunneling even more unlikely.

2. THE ELECTROWEAK PHASE TRANSITION

Well before the tunneling computed above occurs, the electroweak phase transition will happen, and dramatically alter the physics of the metastable states. To understand how the phase transition occurs, consider the toy model of a $U(1)$ gauge field coupled to a scalar field,

$$\mathcal{L} = |(\partial_\mu - ieA_\mu)\phi|^2 - m^2|\phi|^2 + (|\phi|^2)^2 , \tag{12}$$

Here the scalar field has a negative mass term, which at zero temperature drives spontaneous symmetry breaking.

Let the gauge field have a condensate appropriate to that for a nontrivial Wilson line, $A_0 = 2\pi Tu/e$. Then the mass term for the scalar field is $-m^2 + (2\pi Tu)^2$, and the phase transition for spontaneous symmetry breaking occurs for

$$T_c = \frac{m}{2\pi Tu} . \tag{13}$$

If nothing else, this is surely the easiest calculation of a critical temperature possible. In the usual theories, with $A_0 \sim u = 0$, it is necessary to compute the temperature dependent mass term for the scalar, which is typically of order $m_\phi^2 \sim e^2T^2$ and λT^2. Balancing this against the negative mass term at tree level gives a critical temperature that is of order $\sim m/e$ or $\sim m/\sqrt{\lambda}$. In contrast, in (13) one is balancing two terms at *tree* level; the resulting T_c is of order m times numbers of order one, and so is significantly smaller, by powers of e or $\sqrt{\lambda}$, than that for $A_0 = 0$.

What happens after spontaneous symmetry breaking? We argue that when the scalar field acquires a nonzero expectation value, the condensate A_0 dynamically relaxes to zero. Consider the scalar potential: solving for the vacuum expectation of ϕ, $\phi_0 = \sqrt{(m^2 - (2\pi Tu)^2)/(2\lambda)}$,

$$V(u) = -\frac{(m^2 - (2\pi Tu)^2)^2}{4\lambda} . \tag{14}$$

The potential is minimized for $u = 0$, when the Wilson line is equal to one. Thus once spontaneous symmetry breaking occurs, the theory will realize that it can lower its energy by relaxing $u \to 0$.

In the standard model, the phase transition is more complicated. Assume that we start out in one of the metastable states, t_1 or t_2. After spontaneous symmetry breaking this produces nonzero Wilson lines for the timelike components of the Z_0 field, Z_0^0, and for the A_0 field of electromagnetism. The condensate for the Z_0^0 decays, that for A_0 does not.

3. THE WILSON LINE IN ELECTROMAGNETISM

We are lead to compute the potential for the Wilson line in a theory at temperatures above the QCD scale, yet below that for the electroweak phase transition. At such temperatures we can work with quarks and leptons; for ease of discussion we assume that all of the quarks are massless. This obviously is wrong for the top quark, and quickly becomes inadequate for the other quarks. We ignore the contributions of the W and Higgs bosons completely, assuming that they are too massive to matter.

A condensate for B_0 produces an electromagnetic A_0,

$$A_0 = \frac{2\pi T}{e} u , \tag{15}$$

where if the Weinberg angle is θ_w, $A_0 = cos\theta_w B_0$, $e = cos\theta_w g'$, and so $u = cos^2\theta_w t$.

The computation of the effective potential for the Wilson line is straightforward: in this temperature regime it is a function of three variables, u for the $U(1)$ of electromagnetism, and q and r for $SU(3)$ color. We give the result for $q = r = 0$, and comment that u_1 in (17) is metastable in all directions. For $q = r = 0$, per massless family of quarks and leptons the potential for the $U(1)$ Wilson line is

$$\int d^3x \, \pi^2 T^3 \left(- \frac{7}{24} + \widetilde{\mathcal{V}}(u) \right) ,$$

where

$$\widetilde{\mathcal{V}}(u) = -4 \left(v\left(u + \frac{1}{2} \right) + 3v\left(\frac{2u}{3} + \frac{1}{2} \right) + 3v\left(\frac{u}{3} + \frac{1}{2} \right) - 7v\left(\frac{1}{2} \right) \right) \tag{16}$$

For the first generation, the terms on the right hand side are due to, respectively, electrons, up quarks, and down quarks.

Plotting this potential out one finds a metastable point at

$$u_1 = 1.23... . \tag{17}$$

Given the value of $\theta_w \sim .503$, if the theory starts out in the metastable points t_1 or t_2 of (8), then after the electroweak phase transition it will roll down into u_1. The action for tunneling is $\mathcal{S} = 32\pi^3(1.47...)/e^3$; given the small value of e, if the theory rolls into u_1, it will not decay before the QCD phase transition. This is why we could be so sloppy about computing the potential for the electromagnetic Wilson line: all we really need to know is that metastable states last to the time hadrons form.

As quarks and gluons coalesce into hadrons, only integrally charged states contribute to the potential for the electromagnetic Wilson line. Such a potential is trivial, with minima for $u = 0$, $u = 1$, etc. Thus if the theory finds itself in u_1 at the electroweak phase transition, by the time of the QCD transition is just rolls down into $u = 1$, which is then equivalent to the trivial vacuum, $\Omega = 1$.

4. CONCLUSIONS, SPECULATIONS

We have shown that if the universe falls into a metastable state of the Wilson line, such as t_1 or t_2, then it persists in such a state to the electroweak phase transition, disappearing eventually at some temperature about that for the QCD phase transition.

Are there any physical consequences? Two suggest themselves. For an abelian theory (or a field in the Cartan subalgebra of a nonabelian theory), a nonzero value of A_0 implies that this state is not symmetric under charge conjugation, $A_0 \to -A_0$. (For the Wilson line, under C symmetry $\Omega \to \Omega^\dagger$.) Consequently, the relaxation of the Z_0 condensate during the electroweak phase transition will *naturally* generate CP violation that is of order one, $\sim 2\pi t_{1,2}$. It is possible that such CP violation could help generate the baryon asymmetry of the universe at the electroweak scale. Of course much more analysis is needed to show this: for example, it is necessary for the electroweak phase transition is of first order. At nonzero A_0, the above mean field analysis indicated a second order phase transition. Perhaps the same type of loop effects that drive the transition first order for $A_0 = 0$ are operative for $A_0 \neq 0$; alternately, maybe the lack of infrared stable fixed points drives the transition first order.

Even more speculative is the following. About the QCD phase transition, the electromagnetic A_0 dynamically relaxes to zero. A time dependent A_0 generates a time dependent electric field, and thereby magnetic fields: could these "seed" primordial magnetic fluctuations, as discussed by Enqvist at this workshop?

In all of this I have overlooked one crucial point: are metastable states in the effective potential of the Wilson line thermodynamically accessible? This question was raised in ref. [10], who noted that the metastable states of the Wilson line can have negative "pressure"; for further discussions, see refs. [11] and [12]. I refer to "pressure" in quotes because the partition function which corresponds to nontrivial values of the Wilson line is one in which there is an *imaginary* chemical potential for global charge [3,8]. Since the chemical potential is imaginary, standard thermodynamic properties, such as positivity of the "pressure", do not apply.

The only consistent way out of this quandry is to assert that what is being computed is not a pressure *per se*, but something like a pressure, in a more general form of thermodynamics, special to gauge theories. Of course this is really begging the question; the onus is upon us to demonstrate this, which we haven't done yet.

A second question is what is the potential for the Wilson line at temperatures above the Planck scale, in the underlying string theory. Do the infinite tower of states in string theory smooth the potential out, or make it even bumpier? If the former, it is reasonable to imagine the theory, as it cools, falling into one of the metastable states discussed herein. If the latter, however, presumably the universe is always in a state in which the Wilson line is trivial.

In the end, our purpose here is incendiary: we have shown that the metastable states of the Wilson line *can* have dramatic consequences for cosmology.

Acknowledgements

R.D.P. would like to thank Filipe Freire and the other organizers of the workshop for the invitation to speak at such a stimulating meeting in such a stunning locale. The work of R.D.P. was supported by a DOE grant at Brookhaven National Laboratory (DE-AC02-76CH00016).

REFERENCES

[1] C. P. Korthals Altes, K. Lee, and R. D. Pisarski, manuscript in preparation.

[2] D. J. Gross, R. D. Pisarski, and L. G. Yaffe, *Rev. Mod. Phys.*, 53:43 (1981); N. Weiss, *Phys. Rev.* D24:475 (1981) and D25:2667 (1982).

[3] A. Roberge and N. Weiss, *Nucl. Phys.* B275[FS17]:734 (1986).

[4] V. M. Belyaev, *Phys. Lett.* B254:153 (1991).

[5] T. Bhattacharya, A. Gocksch, C. P. Korthals Altes, and R. D. Pisarski, *Phys. Rev. Lett.* 66:998 (1991); *Nucl. Phys.* B383:497 (1992).

[6] V. Dixit and M. C. Ogilvie, *Phys. Lett.* B269:353 (1991).

[7] J. Ignatius, K. Kajantie, and K. Rummukainen, *Phys. Rev. Lett.* 68:737 (1992).

[8] A. Gocksch and R. D. Pisarski, *Nucl. Phys.* B402:657 (1993).

[9] C. P. Korthals Altes, Marseille preprint CTP-P-2991 (Nov., 1993), to appear in *Nucl. Phys.*

[10] V. M. Belyaev, I. I. Kogan, G. W. Semenoff, and N. Weiss, *Phys. Lett.* B277:331 (1992); W. Chen, M. I. Dobroliubov, and G. W. Semenoff, *Phys. Rev.* D46:R1223 (1992).

[11] O.A. Borisenko, V.K. Petrov, G.M. Zinovjev, *Phys. Lett.* B264:166 (1991); V.V. Skalozub, *Mod. Phys. Lett.* A7:2895 (1992); L.Averchenkova, O.A.Borisenko, V.K.Petrov, and G.M.Zinovjev, *Yadernaya Fizika* 55:535 (1992); V.V.Skalozub, to appear in *Int. Jour. of Mod. Phys.*; V.V. Skalozub, Trieste preprint IC/92/405, to appear in *Phys. Rev.*; O.A.Borisenko, J.Bohàčik, V.V.Skalozub, hep-ph/9405208

[12] A. V. Smilga, Bern University preprint BUTP-93-03, (May, 1993) and Santa Barbara preprint NSF-ITP-93-120, (Dec., 1993); I. I. Kogan, Princeton University preprint PUPT-1415, (Nov. 1993); T. H. Hansson, H. B. Nielsen, and I. Zahed, USITP-94-09, hep-ph/9405234, (May, 1994).

[13] S. Coleman, *Phys. Rev.* D15:292 (1977); C. Callan and S. Coleman, *Phys. Rev.* D16: 1762 (1977); A. D. Linde, *Phys. Lett.* 100B:37 (1981); A. D. Linde, *Nucl. Phys.* B216:421 (1983).

[14] U. Amaldi, W. de Boer, P. H. Frampton, H. Furstenau, and J. T. Liu, *Phys. Lett.* B281:374 (1992).

FLUCTUATIONS AT PHASE TRANSITIONS

R.J. Rivers

Blackett Laboratory
Imperial College
Prince Consort Road
London SW7 2BZ U.K.

INTRODUCTION

It has become increasingly important to understand the nature of the field fluctuations that are a consequence of the phase transitions that occurred in the very early universe. On the one hand, fluctuations at the GUT era are assumed to have seeded the large-scale structure visible today. On the other hand, fluctuations at the slightly later electroweak transition are most likely the basis of baryogenesis.

In this talk I shall consider three aspects of fluctuations. The first is combinatorical, establishing a path-integral in terms of which fluctuations can be evaluated. The second is concerned with the initial-value problem, exemplified by both equilibrium and non-equilibrium processes for a real scalar field. The third provides a non-trivial application, the creation by fluctuations of (global) vortices in a symmetry-broken $U(1)$ theory of a complex scalar field.

This application is highly relevant, on two counts. Firstly, while there is no unambiguous mechanism for large-scale structure formation in the universe, arguably one of the least artificial (originally proposed by Kibble [1]) is to attribute it to the spontaneous creation of cosmic strings (vortices) by fluctuations at the GUT transitions. [Primordial magnetic fields are created by field fluctuations in a related way. See Enqvist, these proceedings and elsewhere [2]].

Secondly, very recent experiments [3] show that vortices are naturally generated in superfluid ^4He on quenching it through its critical density. In the Ginzburg-Landau theory of superfluidity the vortices of ^4He are the counterparts of global U(1) cosmic strings (albeit non-relativistic). These results have been argued [4] as providing direct support for the creation of cosmic strings in the early universe. We have a partial answer to this.

This highly enjoyable meeting arrived a little too early for some of the work presented here to be completed and my conclusions are rather more qualitative than I would have liked. A fuller discussion will be given elsewhere, as indicated in the text.

COMBINATORICS OF FLUCTUATIONS

To be concrete, we begin with the simplest possible theory, that of a real scalar field ϕ, with order parameter $\langle\phi\rangle$. A natural measure of the fluctuations of $\phi(t,\vec{x})$ is given by the field averages

$$\phi_v(t) = \frac{1}{v}\int_v d\vec{x}\;\phi(t,\vec{x}) \tag{1.1}$$

over fixed volumes $v = O(l^3)$. [The symbol v denotes both the position of the volume in question and its size]. The system as a whole is taken to have a large volume $V = O(L^3) \gg v$, in which V is ultimately taken infinite. Since, at any time, the field ϕ is correlated over some length ξ, it is sufficient to take $l \geq \xi$.

To simplify the notation, assume that the possible configurations $\Phi_j(\vec{x})$ of ϕ in the volume V are denumerable (e.g. we impose periodic boundary conditions). Particular values of ϕ_v then single out particular Φ_j.

In describing the quantum theory of ϕ, it is convenient to adopt the field representation. Suppose, at some initial time $t = t_{in}$, the system, described by the state functional $|\Phi_k, t_{in}\rangle$, is an eigenstate of $\hat{\phi}$, eigenvalue Φ_k. That is,

$$\hat{\phi}|\Phi_k, t_{in}\rangle = \Phi_k(\vec{x})|\Phi_k, t_{in}\rangle. \tag{1.2}$$

At a later time $t > t_{in}$, the state has evolved to

$$|\Phi_k, t\rangle = \sum_j c_{kj}|\Phi_j, t\rangle \tag{1.3}$$

a superposition of field eigenstates. The probability that the measurement of $\hat{\phi}$ in $|\Phi_k, t\rangle$ gives Φ_j is $|c_{kj}|^2$, most conveniently written in terms of path integrals as

$$|c_{kj}|^2 = \int_{\phi_1(t_{in},\vec{x})=\Phi_k(\vec{x})}^{\phi_1(t,\vec{x})=\Phi_j(\vec{x})} D\phi_1 \exp\{iS[\phi_1]\} \int_{\phi_2(t_{in},\vec{x})=\Phi_k(\vec{x})}^{\phi_2(t,\vec{x})=\Phi_j(\vec{x})} D\Phi_2 \exp\{-iS[\phi_2]\}. \tag{1.4}$$

The first term in (2.4) is c_{kj}, the second c_{kj}^*, where $S[\phi] = \int d^4x\mathcal{L}(\phi)$ denotes the classical action that determines the evolution of the system.

The probability $p(\phi_v(t) = \bar{\phi})$ that the coarse-grained field $\phi_v(t)$ takes some chosen value $\bar{\phi}$ is obtained by checking which $\Phi_j(\vec{x})$ have their average as $\bar{\phi}$, and multiplying by the probability for finding such a Φ_j. That is,

$$p(\phi_v(t) - \bar{\phi}) = \sum_j |c_{kj}|^2 \delta\left(\frac{1}{v}\int_v d^3\vec{x}'\Phi_j(\vec{x}') - \bar{\phi}\right) \tag{1.5}$$

$$= \sum_j \int_{\phi_1(t_{in})=\phi_2(t_{in})=\Phi_k}^{\phi_1(t)=\phi_2(t)=\Phi_j} D\phi_1 D\phi_2\; \delta\left(\frac{1}{v}\int_v d^3\vec{x}'\phi_1(t,\vec{x}') - \bar{\phi}\right) \exp\{iS[\phi_1] - iS[\phi_2]\}$$

$$= \int_{\phi_1(t_{in})=\phi_2(t_{in})=\Phi_k}^{\phi_1(t_{fin})=\phi_2(t_{fin})} D\phi_1 D\phi_2\; \delta\left(\frac{1}{v}\int_v d^3\vec{x}'\phi_1(t,\vec{x}') - \bar{\phi}\right) \exp\{iS[\phi_1] - iS[\phi_2]\}. \tag{1.6}$$

It might seem that the upper bound of the field integrals in (2.6) should be $\phi_1(t,\vec{x}) = \phi_2(t,\vec{x})$, but we can extend the time contour to any final time value $t_{fin} > t$ without changing the integral. We shall take t_{fin} infinite.

The expression (2.6) is written more simply as the closed-time path integral of Schwinger and Keldysh

$$p(\phi_v(t) - \bar{\phi}) = \int_C D\phi\, \delta\left(\frac{1}{v}\int_v d^3\vec{x}'\phi_1(t,\vec{x}') - \bar{\phi}\right)\exp\{iS_{ctp}\} \tag{1.7}$$

where $C = C_1 \oplus C_2$ is the closed-time path (see below) in which C_1 runs from t_{in} to t_{fin} and C_2 runs backward from t_{fin} to t_{in}, infinitesimally beneath C_1.

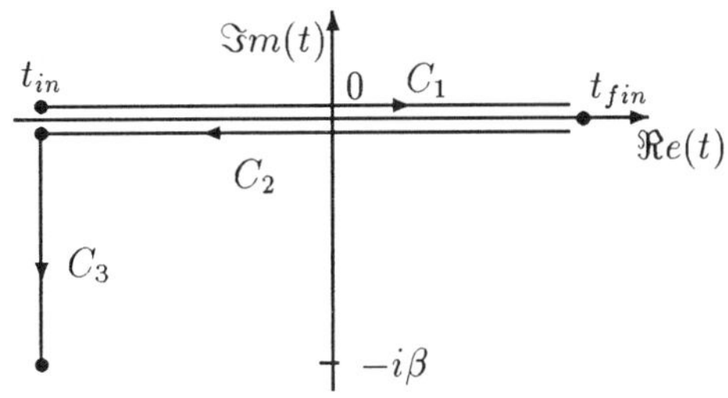

The relevant action $S_{ctp}[\phi]$, in the presence of a source $j(t,\vec{x})$, is described in terms of $\mathcal{L}(\phi)$ by

$$S_{ctp}[\phi, j] = \int_c dt d\vec{x}(\mathcal{L}(\phi) + j(t)\phi(t)) \tag{1.8}$$

$$= \int_{-\infty}^{\infty} dt d\vec{x}\left(\mathcal{L}(\phi_1) - \mathcal{L}(\phi_2) + j_1(t)\phi_1(t) + j_2(t)\phi_2(t)\right). \tag{1.9}$$

Spatial labels have been suppressed. The doublets are defined to be

$$\phi_a(t) = \begin{cases} \phi(t) & t \in C_1 \quad \text{if } a = 1. \\ \phi(t - i\epsilon) & t - i\epsilon \in C_2 \quad \text{if } a = 2. \end{cases} \tag{1.10}$$

and

$$j_a(t) = \begin{cases} j(t) & t \in C_1 \quad \text{if } a = 1. \\ -j(t - i\epsilon) & t - i\epsilon \in C_2 \quad \text{if } a = 2. \end{cases} \tag{1.11}$$

Note that, notionally, we have the second leg C_2 of our curve running back ϵ below the first.

This takes account of the quantum uncertainty. However, in general, the initial field $\Phi(t_{in}, \vec{x})$ is only specified statistically. Let $P_{in}[\Phi_k]$ be the probability that, at time $t = t_{in}$, the field takes configuration $\Phi_k(\vec{x})$. Then

$$p(\phi_v(t) - \bar{\phi}) = \sum_k P_{in}[\Phi_k]\int_{\phi_1=\phi_2=\Phi_k} D\phi_1 D\phi_2\, \delta\left(\frac{1}{v}\int_v d^3\vec{x}'\phi_1(t,\vec{x}') - \bar{\phi}\right)\exp\{iS_{ctp}\}$$

$$= \int D\Phi P_{in}[\Phi]\int_{\phi_1=\phi_2=\Phi} D\phi_1 D\phi_2\, \delta\left(\frac{1}{v}\int_v d^3\vec{x}'\phi_1(t,\vec{x}') - \bar{\phi}\right)\exp\{iS_{ctp}\}. \tag{1.12}$$

where we have now relaxed the (artificial) denumerability of the initial state. The generality of (2.12) is usually too much for us. To be tractable, simple initial conditions are required. The most convenient assumption is that $P_{in}[\Phi]$ is Boltzmann distributed as

$$P_{in}[\Phi] = N\exp\{-\beta_{in}H_{in}[\Phi]\} \tag{1.13}$$

at some initial temperature $T_{in} = \beta_{in}^{-1}$. This permits [5] us to introduce an initial Lagrangian density $\mathcal{L}_{in}(\phi)$ in terms of which P_{in} can be expressed as the Euclidean-time path integral (the diagonal density-matrix element)

$$P_{in}[\Phi] = \int_{\phi_3(t_{in},\vec{x})=\Phi(\vec{x})} D\phi_3\exp\{-S_{in}[\phi_3]\} \tag{1.14}$$

where $S_{in}[\phi] = \int dx \mathcal{L}_{in}(\phi(x))$ is independent of T_{in}. The time contour for the integral (2.14) is C_3 (see above), running from time t_{in} to $t_{in} - i\beta_{in}$, over which ϕ_3 is periodic. We stress that $\mathcal{L}_{in}(\phi)$ exists only to parametrise the initial conditions and, in principle, has nothing to do with the $\mathcal{L}(\phi)$ of later times. However, in practice we have in mind a situation in which $\mathcal{L}_{in}$ and $\mathcal{L}$ of the closed-time path have the same form, but in which the parameters have a time-dependence [6]. The other obvious initial condition, in which the field ϕ is localised at a particular value, in effect, can also be taken as a particular choice of Boltzmann distribution, as we shall see. [In principle we can invert (2.14) to deduce S_{in} from P_{in}, whether P_{in} describes a Boltzmann distribution or not. In practice, it is all but impossible.]

With this proviso, p of (2.12) can be written as

$$p(\phi_v(t) - \bar{\phi}) = \int_C D\phi_1 D\phi_2 D\phi_3 \delta\left(\frac{1}{v}\int_v d^3\vec{x}'\phi_1(t,\vec{x}') - \bar{\phi}\right)\exp\{iS_C\} \qquad (1.15)$$

where C is the contour $C_1 \oplus C_2 \oplus C_3$ and $S_C = S_{in}$ on C_3.

Let $I(\vec{x})$ be the window function (indicator function) for the volume v(i.e. $I(\vec{x}) = 1, \vec{x} \in v; I(\vec{x}) = 0, \vec{x} \notin v$). On using an exponential representation of the δ-function (2.15), rewritten as

$$p(\phi_v(t) = \bar{\phi}) = \int d\alpha \int_C D\phi_1 D\phi_2 D\phi_3 \exp\{iS_C[\phi] - \int_v d\vec{x}\alpha(\bar{\phi} - \phi_1(t,\vec{x}))\} \qquad (1.16)$$

can be further recast as

$$p(\phi_v(t) = \bar{\phi}) = \int d\alpha \exp\{-i\alpha v\bar{\phi}\} Z[\alpha I, 0, 0] \qquad (1.17)$$

$Z[j_1, j_2, j_3]$, defined by

$$Z[j_1, j_2, j_3] = \int D\phi_1 D\phi_2 D\phi_3 \exp\{iS_C[\phi_a, j_a]\}. \qquad (1.18)$$

is the generating functional for Green functions in the presence of sources on all three contours (the extension of $S_{ctp}[\phi_a, j_a]$ of (2.8) to include C_3). In (2.17) $j_1 = \alpha I(\vec{x})$ is a source coupled to the field on C_1. The boundary conditions are now $\phi_1(t_{in}) = \phi_3(t_{in} - i\beta_{in})$, $\phi_1(t_{fin}) = \phi_2(t_{fin})$

This splitting into triplets is merely a matter of formal definition. The reason for doing this is that it manages to encode the initial conditions into an expression (2.18) that looks familiar, and is hence amenable to our usual tricks. For the circumstances that we shall consider here, a Gaussian approximation to the fluctuations is a useful first step. Adopting the notation

$$\langle F[\phi] \rangle = \int_C D\phi_1 D\phi_2 D\phi_3 \; F[\phi_1] \exp\{iS_C[\phi, j = 0]\} \qquad (1.19)$$

for all $F[\phi]$, it follows from (2.17) that

$$p(\phi_v(t) = \bar{\phi}) = \int d\alpha \exp\{-i\alpha v\bar{\phi}\}\langle \exp\{i\alpha v\phi_v\}\rangle. \qquad (1.20)$$

The Gaussian approximation is obtained by curtailing the cumulant expansion

$$\langle \exp\{i\alpha v\phi_v\}\rangle = \exp\{i\alpha v\langle\phi_v\rangle - \tfrac{1}{2}\alpha^2 v^2[\langle\phi_v\phi_v\rangle - \langle\phi_v\rangle^2] + O(\alpha^3)\} \qquad (1.21)$$

at $O(\alpha^2)$. Performing the α integration gives

$$p(\phi_v(t) = \bar{\phi}) = N\exp\{-\bar{\phi}^2/2\langle\phi_v\phi_v\rangle\} \qquad (1.22)$$

where, for simplicity, we have assumed that $\langle\phi_v\rangle = 0$. Not surprisingly, the variance in $\bar{\phi}$ is given by the coarse-grained two-point correlation function.

More generally, we might wish to calculate the probability $p(f_v[\phi] = \bar{f})$. where $f_v[\phi]$ is some other coarse-grained functional of ϕ. In the same approximation

$$p(f_v[\phi] = \bar{f}) = N \exp\{-\bar{f}^2/2\langle f_v f_v\rangle\} \tag{1.23}$$

again assuming $\langle f_v\rangle = 0$ for simplicity.

THE INITIAL VALUE PROBLEM

As we have observed, the hard work has all been neatly tucked away in the boundary conditions. The familiarity of the path integral formalism makes it easy to lose the distinction between the contour C_3 (the initial condition) and $C_1(C_2)$ that carries the dynamics. However, for our first case, thermal equilibrium, the distinction has been intentionally removed.

Thermal Equilibrium

This is the most extreme case, in which there is no dynamical evolution of the system. Although an unlikely occurrence in the very early universe, thermal equilibrium has been examined in great detail, in particular through the thermal effective potential $V(\phi)$ [7]. It is the behaviour of $V(\phi)$ that determines the order of the phase transitions in which we are interested. Our approach, given in detail elsewhere [8], is close to that of Jona-Lasinio [9], who defined the effective potential in terms of probabilities for coarse-grained fluctuations of the type proposed in the previous section. With the probability $p(\phi_v(t) = \bar{\phi})$ now time-independent, the contours C_1 and C_2 can be shrunk to nothing, leaving the familiar Euclidean-time formalism on contour C_3. We stress that, since thermal equilibrium corresponds to making the choice $P_{in}[\Phi]$ of (2.14), where $S_{in}[\phi]$ is now the classical action $S[\phi]$ of the theory, we have no further freedom in our initial conditions, To cite one common misconception, we cannot now use the loop-expanded effective potential following from (2.14) as a 'classical' potential in which we subsequently consider the evolution of field configurations centred on metastable minima. Such an action corresponds to imposing two incompatible boundary conditions simultaneously. However, something different, but not wholly dissimilar, may make sense. [See Weinberg [10], these proceedings].

Let us return to $P[\Phi]$ of (2.14), corresponding to equilibrium at temperature $T = \beta^{-1}$. (With no evolution, all suffices have been dropped). Because of the reality of the integrand we can improve upon the Gaussian approximation. The integrated probability satisfies the Chebycheff inequality

$$p(\phi_v \geq \bar{\phi}) = \int D\Phi \exp\{-\beta H[\Phi]\}\, \theta\left(\frac{1}{v}\int_v d\vec{x}\; \Phi(\vec{x}) - \bar{\phi}\right) \tag{2.1}$$

$$\leq \int D\Phi \exp\{-\beta H[\Phi]\}\, \exp\{\beta\alpha\left(\frac{1}{v}\int_v d\vec{x}\; \Phi(\vec{x}) - \bar{\phi}\right)\} \tag{2.2}$$

for all $\alpha \geq 0$. Define $\omega_v(\alpha)$ by

$$\exp\{\beta v \omega_v(\alpha)\} = \int D\Phi \exp\{-\beta H[\Phi] + \beta \int j\Phi\} \tag{2.3}$$

where $j(\vec{x}) = \alpha I(\vec{x})$, as before. Then

$$p(\phi_v \leq \bar{\phi}) \leq \exp\{\beta v(\omega_v(\alpha) - \alpha\bar{\phi})\} \tag{2.4}$$

an inequality minimised for $\alpha = \bar{\alpha}$,

$$\frac{\partial \omega_v(\bar{\alpha})}{\partial \alpha} = \bar{\phi}. \tag{2.5}$$

That is

$$p(\phi_v \leq \bar{\phi}) \leq \exp\{-\beta v V_v(\bar{\phi})\} \tag{2.6}$$

where $V_v(\phi)$, the Legendre transform of $\omega_v(\alpha)$, is the coarse-grained effective potential.

More specifically, when $\beta \ll m$ we identify $H[\Phi]$ with $S_3[\Phi]$, the 3-dimensional action for the light Matsubara mode, obtained by integrating over all heavy modes (see Kajantie [11], these proceedings). Thus, but for light-mode self-interactions, $V_v(\bar{\phi})$ gives the usual thermal effective potential $V(\bar{\phi})$ (the convex hull of the loop-expanded potential) in the large-v limit, when the inequality (3.6) becomes an equality. We note that, if we were considering a gauge theory, rather than just this simple scalar theory, the interesting features of the effective potential (e.g.the order of the transition) are largely determined by the light gauge field modes, and hence are correctly incorporated in V_v.

We shall be much simpler, initially restricting ourselves to a free scalar field, mass m, temperature $T \gg m$. If $\tilde{I}(\vec{k})$ is the Fourier transform of $I(\vec{x})$ (normalised to $\tilde{I}(\vec{0}) = v = O(l^3)$) then

$$\langle \phi_v \phi_v \rangle = \frac{1}{v^2} \int d^3k |\tilde{I}(\vec{k})|^2 G(\vec{k}) \tag{2.7}$$

where

$$G(\vec{k}) \simeq \frac{T}{\vec{k}^2 + m^2} \tag{2.8}$$

is the free-field thermal propagator in this regime.

What we see from this is that, as we might have anticipated, the effect of coarse-graining is, through $|I(\vec{k})|^2$, to impose a cut-off at $|\vec{k}| < l^{-1}$. $\langle \phi_v, \phi_v \rangle$ decreases from $T/m^2 v$ at large v ($\tilde{I}(\vec{k}) \to \delta(\vec{k})$) to $T/Am^2 v = mT/A$ with $A \simeq 10$, when $l = \xi = O(m^{-1})$. That is, the variance in ϕ is

$$(\Delta\phi)^2 \simeq mT/A \tag{2.9}$$

for correlation-volume fluctuations. The value of A depends a little on the shape of V. What matters is that it reduces $(\Delta\phi)^2$ by an order of magnitude from our first guess [12]. This provides a very useful guide as to when fluctuations are important. [We note that $V_v(\phi)$ gets steeper as v diminishes].

We are interested in phase transitions, and for a real scalar theory the only possibility is in the breaking of reflection invariance $\phi \to -\phi$. Let us now consider the case when the scalar field theory has broken symmetry when cold. We approximate $H[\Phi]$ by the one-loop form

$$H[\Phi] = \int d\vec{x} \, [\tfrac{1}{2}(\nabla\Phi)^2 - \tfrac{1}{2}m^2(T)\Phi^2 + \frac{1}{4}\lambda\Phi^4]. \tag{2.10}$$

The correlation length $\xi = m(T)^{-1}$, where $m(T)$ is the effective mass

$$m^2(T) = m^2 \left(1 - \frac{T^2}{T_c^2}\right) \tag{2.11}$$

diverges at the critical temperature T_c, at which the theory undergoes a second-order transition.

It is not possible to calculate the coarse-grained $V_v(\phi)$ of (2.29) analytically. In the symmetric phase ($T > T_c$) perturbation theory is appropriate and we can expand about the Gaussian to include a quartic term $\lambda_v \phi^4$ in the coarse-grained potential $V_v(\phi)$. Its strength λ_v is, not surprisingly, given in terms of the coarse-grained four-point correlation function $\langle\phi_v\phi_v\phi_v\phi_v\rangle$. Very near to the phase transition the Gaussian approximation breaks down, but even then it remains a useful qualitative guide.

In the symmetry-broken phase perturbation theory breaks down, and even the Gaussian approximation is difficult. A Gaussian bound can be made by weakening the inequality (3.6) further by restricting the spatial integral to $\vec{x} \in v$ in (3.10). Without worrying about the details the end result is the highly sensible bound [8]

$$(\Delta\phi)^2 = \langle\phi_v\phi_v\rangle < \sigma^2 + \frac{1}{\beta v m^2 T} \tag{2.12}$$

The first term on the right hand side of (3.12) describes the long-range field correlations in the symmetry-broken phase, the second the fluctuations about the minima at $|\phi| = \sigma$. For correlation volumes $v = O(\xi^3) = O(m(T)^{-3})$ the free field result suggests that the inequality is improved by replacing $Tm(T) = (\beta v m^2(T))^{-1}$ by $Tm(T)/A$. When $Tm(T)/A = O(\sigma^2(T))$ there is a significant probability that correlation-volumes of false vacuum can be created. This is more transparent if written as the constraint on the effective thermal coupling strength λ_3, that

$$\lambda_3 = \lambda T/m(T) = O(1) \tag{2.13}$$

This condition is seen to be just the Ginzburg criterion [13], the condition that the effective thermal coupling λ_3 is becoming large, and signalling that the one-loop form (3.10) is beginning to break down. [The failure of the Gaussian approximation in the symmetric phase is now seen to correspond to $|\lambda_3| = O(1)$].

If $H[\Phi]$ describes a theory experiencing a first-order transition, as the EW transition is expected to do, the situation is more complicated. There has been an extensive discussion of the relative role of fluctuations versus quantum tunnelling by Gleiser, Kolb and others [14] (although our earlier caveat about incompatible boundary conditions is valid here). The consensus that thermal fluctuations are inadequate to populate the symmetry-broken vacuum state is largely confirmed by our analysis, which will be published elsewhere [15].

Non-Equilibrium Behaviour

The picture that we have given above is of correlation volumes of false vacuum produced with significant probability in the Ginzburg regime. As we cool below this thin temperature range (to $\lambda_3 < 1$) the vacuum is characterised by large domains in which $|\phi| = \sigma$ is constant (the 'true' vacuum). Thermal equilibrium is of little help in seeing how this happens. In fact, thermal equilibrium may be positively misleading. The expansion of the very early universe is so rapid that particles are separated from one another before they have undergone sufficient collisions to attain it. Even approximate thermal equilibrium is inappropriate. To exemplify the nature of out-of -equilibrium behaviour, and to offer the strongest contrast to thermal equilibruim, we consider the case of rapid quenching of the scalar theory from an initially symmetric state at time $t = t_{in}$ to a symmetry-broken state, described by the classical action

$$S[\phi] = \int d^4x \; [\tfrac{1}{2}(\partial\phi)^2 + \tfrac{1}{2}m^2\phi^2 + \frac{1}{4}\lambda\phi^4]. \tag{2.14}$$

More details are given by Lee (these proceedings, and work by Lee, Boyanovsky, de Vega [16]) in which a similar calculation is performed, and from which we borrow results. [Although motivated by early universe considerations, for simplicity we stay in flat spacetime. For a more realistic temporal evolution in the early universe, see de Vega [17] (these proceedings).]

At $t = t_{in}$ we would like the initial state to be symmetric, with P_{in} strongly peaked about $\phi = 0$. In the absence of more information our earlier observations suggest that this is most simply achieved by taking a Boltzmann-distributed probability distribution $P_{in}[\Phi] = N\exp\{-\beta_{in}H_{in}[\Phi]\}$. $H_{in}[\Phi]$ is derived, as in (2.14), from an action

$$S_{in}[\phi] = \int dx\ [\tfrac{1}{2}(\partial\phi)^2 - \tfrac{1}{2}m_{in}^2\phi^2] \tag{2.15}$$

for some m_{in}. While S_{in} is a phenomenological action that induces the required initial condition the effect is the same as if the system were in thermal equilibrium at temperature $T = T_{in} = \beta_{in}^{-1}$ for all time $t < t_{in}$. The dispersion in ϕ at this initial time is given by (31) as $(\Delta\phi)^2 = O(m_{in}T_{in})$ on coarse-graining, chosen as we wish. For $t > t_{in}$ the evolution of the initial state is taken to be determined by $S[\phi]$ of (3.14). We take $t_{in} = 0$ for convenience.

In effect, we are taking the theory to be determined for all time by the action

$$S_t[\phi] = \int d^4x\ [\tfrac{1}{2}(\partial\phi)^2 + \tfrac{1}{2}m^2(t)\phi^2 + \frac{1}{4}\lambda(t)\phi^4] \tag{2.16}$$

where

$$m^2(t) = \begin{cases} m_{in}^2 & t < 0 \\ -m^2 & t > 0 \end{cases} \tag{2.17}$$

and

$$\lambda(t) = \begin{cases} 0 & t < 0 \\ \lambda & t > 0 \end{cases} \tag{2.18}$$

subject to the condition that, for $t < 0$, it was in thermal equilibrium. In this extreme case it is apparent that the equilibrium thermal effective potential V and its coarse-grained descendents V_v have no role to play and it makes little sense to talk about the order of the transition.

We could have taken λ constant for all time, but have not done so in order to delineate between its role in establishing the initial condition (for which it is an artefact) and its role in the subsequent dynamics, in which it is ultimately a constant of the universe. Unsurprisingly, taking interactions into account properly is hard. In practice, because of the failure of perturbation theory, it is difficult to do better than a self-consistent (Hartree) linearisation of (3.16). Despite our preamble, for our purposes it is sufficient to set $\lambda(t) = 0$ in (3.18) (whence the classical potential is an upturned parabola) but only to consider the evolution of the system to times at which the field configuration would have spread from $\phi = 0$ to the point of inflexion $\phi_{inf} = O(m/\sqrt{\lambda})$. Since the characteristic growth rate is $\phi/m = e^{mt}$, this means times for which $t = O(m^{-1}ln(1/\lambda)$.

The spinodal dispersion in ϕ is now given by the 'free-field' equal time correlation function (explicitly dependent on t)

$$\langle\phi_v\phi_v\rangle = \frac{1}{v^2}\int_{\vec{x},\vec{x}'\in v} d\vec{x}\ d\vec{x}'\ \langle\phi(t,\vec{x})\phi(t,\vec{x}')\rangle \tag{2.19}$$

$$= \frac{1}{v^2}\int d\vec{x}\ d\vec{x}'\ I(\vec{x})I(\vec{x}')G(\vec{x}-\vec{x}';t) \tag{2.20}$$

$$= \frac{1}{v^2}\int d\vec{k}\ |\tilde{I}(\vec{k})|^2 G(\vec{k};t) \tag{2.21}$$

in terms of the Fourier transform

$$G(\vec{k};t) = \int d\vec{x} G(\vec{x};t)e^{-i\vec{k}.\vec{x}}. \tag{2.22}$$

At equal-time all correlation functions are the same, once the boundary condition $G(\vec{x};t_{in}) = G(\vec{x};t_{in} - i\beta_{in})$ has been implemented.

$G(\vec{k};t)$ possesses both oscillatory modes and exponentially growing (fading) modes, as solutions to

$$(\partial_t^2 + \vec{k}^2 + m^2(t))\mathcal{U}_k(t) = 0 \tag{2.23}$$

from which it is constructed. The dominant contributions to (3.21) are the exponential modes, and if we take them only then [16]

$$G(\vec{k};t) \simeq \frac{\theta(m^2 - \vec{k}^2)}{2\omega_{in}(\vec{k})}\left(1 + \tfrac{1}{2}\left(\frac{m^2 + m_{in}^2}{m^2 - \vec{k}^2}\right)(cosh(2t\sqrt{m^2 - \vec{k}^2}) - 1)\right) coth[\tfrac{1}{2}\beta_{in}\omega_{in}(\vec{k})] \tag{2.24}$$

where $\omega_{in}(\vec{k}) = \sqrt{\vec{k}^2 + m_{in}^2}$. If we consider fluctuations coarse-grained to volumes $v = O(l^3)$ the momentum is cut off at $|\vec{k}| = O(l^{-1})$, whereas the θ-function imposes a cut-off at $|\vec{k}| = m$. Since we are only interested in $l \geq m^{-1}$ the θ-function is irrelevant. On converting to spherical polars in (3.22) $k^2G(\vec{k}:t)$ has, for large values of mt, a sharp peak at $k^2 = O(m/t)$. For this to be inside the integral, so that the probability is significant, requires $l^2 < t/m$. Thus the field is correlated on a scale $\xi = O(\sqrt{t/m})$. At the relevant time $t = m^{-1}ln(1/\lambda)$ at which domain formation rapidly slows to a halt, the domain size is then

$$\xi = O(m^{-1}(ln(1/\lambda))^{\frac{1}{2}}) \tag{2.25}$$

This is in contrast to $\xi = O(m^{-1}/\sqrt{\lambda})$ in the equilibrium case at the Ginzberg temperature (3.13). In practice, unless λ is extremely small the sizes can be comparable when prefactors are taken into account properly. Further details are given by Lee. [There are differences in detail. Our coarse-graining is somewhat different, but the qualitative behaviour is the same].

VORTEX CREATION FROM FLUCTUATIONS

Consider a field theory invariant under a group G. Suppose the effect of spontaneous symmetry-breaking is to reduce G to its subgroup H. The vacuum manifold M can be identified with the coset space G/H. If M has a non-trivial first homotopy group $\Pi_1(M)$ (i.e. the embeddings of loops in M), then vortices (strings) can form [1]. [If the second homotopy group $\Pi_2(M)$ is non-trivial we can have monopoles, and if the third homotopy group $\Pi_3(M)$ is non-trivial we can have textures]. As we said earlier, the production of vortices in the early universe has the capacity to seed large-scale structure through their gravitational effects. In general, GUT vortices are assumed to arise from the breaking of a local gauge theory. However, because of their relative simplicity, a considerable effort has gone into understanding global vortices, and it is they that we shall mainly consider here. Given that the vortices of superfluid ^4He are global vortices, they are of more than theoretical interest.

The simplest theory permitting such vortices is the global $U(1)$ theory of a complex scalar field ϕ, with Minkowski space-time action

$$S[\phi] = \int d^4x\, [\tfrac{1}{2}(|\partial\phi|)^2 + \tfrac{1}{2}m^2|\phi|^2 - \frac{1}{4}\lambda|\phi|^4]. \tag{3.1}$$

With $m^2 > 0$ the $U(1)$ symmetry is spontaneously broken, and the classical vacuum determined by $|\phi|^2 = \sigma^2 = m^2/\lambda$. . The vacuum manifold $M = S^1$ has homotopy group $\Pi_1(S^1) = Z$, and vortices can exist with integer winding number n, Vortices with winding number $|n| > 1$ are unstable to decay into vortices with $|n| = 1$.

Consider an open surface S with oriented boundary ∂S. The line integral

$$N_S(t) = \frac{-i}{2\pi}\int_{\partial S} \underline{dl}.\frac{\phi^\dagger \overleftrightarrow{\partial} \phi}{|\phi|^2} \tag{3.2}$$

measures the winding number of the field configuration on S at time t. (i.e. $2\pi N_S$ is the change in phase of the field ϕ as it is taken once around ∂S).

If $N_S \neq 0$, continuity of ϕ requires that it vanish at some point or points of S. A vortex of winding number n is a tube of 'false' vacuum ($\phi \simeq 0$) containing a line of zeros of ϕ for which N_S of (4.2) has value n for any closed path enclosing this line. For a large loop ∂S, N_S measures the net winding number of the vortices that pass through it. At a distance r from the vortex centre $|\phi|$ approaches its vacuum value σ as $|\phi| = \sigma(1 - O(e^{-mr}))$. In practice it is more convenient to evaluate the related quantity

$$\bar{N}_S(t) = \frac{-i}{2\pi}\int_{\partial S} \underline{dl}.\frac{\phi^\dagger \overleftrightarrow{\partial} \phi}{\sigma|\phi|} \tag{3.3}$$

$$= \frac{-i}{2\pi\sigma^2}\int_S \underline{dS}'.(\underline{\partial}\phi^\dagger \wedge \underline{\partial}\phi)\frac{\sigma}{|\phi|} \tag{3.4}$$

$$= \frac{2}{2\pi\sigma}\int_S \underline{dS}'.(\underline{\partial}\rho \wedge \underline{\partial}\chi) \tag{3.5}$$

where, in (3.5), we have used the radial/angular decomposition

$$\phi = \rho e^{i\chi} \tag{3.6}$$

of the field. For a large loop ∂S the difference between N_S and $\bar{N}_S$ (not integer) is vanishingly small if no vortices pass close to ∂S, and $\bar{N}_S$ remains a good indicator of vortex production.

On further decomposing the radial field ρ as $\rho = \sigma + h$ for Higgs field h, S of (4.1) becomes

$$S[h,\chi] = \int dx\ [\tfrac{1}{2}(\partial h)^2 + \tfrac{1}{2}\sigma^2(\partial\chi)^2 - m^2h^2 - \lambda\sigma h^3 - \frac{1}{4}\lambda h^4] \tag{3.7}$$

In the Gaussian approximation the probability that $\bar{N}_S(t)$ takes the value n is, from (2.23)

$$p(\bar{N}_S(t) = n) = \exp\{-n^2/2\langle\bar{N}_S(t)\bar{N}_S(t)\rangle\} \tag{3.8}$$

On further defining the Goldstone mode g by $g = \sigma\chi$, from (4.5) it follows that

$$\langle\bar{N}_S(t)\bar{N}_S(t)\rangle = (\frac{2}{2\pi\sigma^2})^2\int\int_S dS'dS''\langle(\partial h' \wedge \partial g')(\partial h'' \wedge \partial g'')\rangle \tag{3.9}$$

All fields are at time t. The primes (doubleprimes) denote fields in the infinitesimal areas dS', dS'' of S respectively. For economy of notation we have not made ths scalar products explicit. Without loss of generality we take S in the 1-2 plane, whence

$$\langle\bar{N}_S(t)\bar{N}_S(t)\rangle = (\frac{2}{2\pi\sigma^2})^2\int\int_{i,j=1.2} dS'dS''\langle\partial_i h'\partial_i h''\partial_j g'\partial_j g'' - \partial_i h'\partial_j h''\partial_j g'\partial_i g''\rangle \tag{3.10}$$

It is convenient to refine our notation further, decomposing space-time as $x = (t, \vec{x}) = (t, \vec{x}_L, x_T)$ where $\vec{x}_L = (x_1, x_2)$ denotes the co-ordinates of S, and $x_T = x_3$ the transverse direction to S. Similarly, we separate 4-momentum p as $p = (E, \vec{p}_L, p_T)$.

Let $G_h(t,\vec{x}'-\vec{x}'') = \langle h(t,\vec{x}')h(t,\vec{x}'')\rangle$, $G_g(t,\vec{x}'-\vec{x}'') = \langle g(t,\vec{x}')g(t,\vec{x}'')\rangle$ be the Higgs field and Goldstone mode correlation functions respectively. As a first step we ignore correlations between Higgs and Goldstone fields. That is, we retain only the disconnected parts of $\langle \bar{N}_S \bar{N}_S\rangle$. Eqn. (4.10) then simplifies to

$$\langle \bar{N}_S(t)\bar{N}_S(t)\rangle = (\frac{2}{2\pi\sigma^2})^2 \int\int_{i,j=1.2} dS'dS''[\langle\partial_i h'\partial_i h''\rangle\langle\partial_j g'\partial_j g''\rangle - \langle\partial_i h'\partial_j h''\rangle\langle\partial_j g'\partial_i g''\rangle] \tag{3.11}$$

which can be written as

$$\langle \bar{N}_S(t)\bar{N}_S(t)\rangle = (\frac{2}{2\pi\sigma^2})^2 \int\int d^3p' d^3p''\ G_h(\vec{p}',t)G_g(\vec{p}'',t)|\tilde{I}(\vec{p}_L''-\vec{p}_L')|^2[(\vec{p}_L')^2(\vec{p}_L'')^2-(\vec{p}_L'.\vec{p}_L'')^2] \tag{3.12}$$

In (4.12) $\tilde{I}(\vec{p}_L)$ is the Fourier transform of the window function $I(\vec{x}_L)$ of the surface S (i.e. $I(\vec{x}_L) = 1$ if $\vec{x}_L \in S$, otherwise zero). The $G(\vec{p},t)$ are defined as in (3.22).

We coarse-grain in the transverse and longitudinal directions by imposing a cut-off in three-momenta at $|p_i| < \Lambda = l^{-1}$, for some l, as before. Thus $\bar{N}_S$ is now understood as the average value over a closed set of correlation-volume 'beads' through which ∂S runs like a necklace. However, we leave the notation unchanged.

For large loops ∂S, $\tilde{I}(\vec{q}_L) \simeq \delta(\vec{q}_L)$, enabling us to write

$$\langle \bar{N}_S(t)\bar{N}_S(t)\rangle = (\frac{2}{2\pi\sigma^2})^2 \int dq_T \int d^3p\ G_h(\vec{p}+\vec{q}_T,t)G_g(\vec{p},t) \int d^2q_L|\tilde{I}(\vec{q}_L)|^2[(\vec{p}_L)^2(\vec{q}_L)^2-(\vec{p}_L.\vec{q}_L)^2] \tag{3.13}$$

By $\vec{p}+\vec{q}_T$ we mean $(\vec{p}_L, p_T+q_T)$. The dependence on the contour ∂S is contained in the final integral

$$\mathcal{J} = \int d^2q_L\ |\tilde{I}(\vec{q}_L)|^2[(\vec{p}_L)^2(\vec{q}_L)^2 - (\vec{p}_L.\vec{q}_L)^2] \tag{3.14}$$

$$\simeq \pi p_L^2 \int^\Lambda dq_L\ q_L^3|\tilde{I}(q_L)|^2 \tag{3.15}$$

If this is evaluated for a circular loop of radius R, we find

$$\mathcal{J} = p_L^2\ O(2\pi R/l) \tag{3.16}$$

as we might have anticipated. The rms winding number behaves with path length as $\Delta n \propto \mathcal{J}^{\frac{1}{2}} = O(L^{\frac{1}{2}})$, where L is the number of steps of length l. This is consistent with the caorse-grained field phases of different volumes v being randomly distributed.

The final step is to relate the magnitude of the fluctuations in winding number N_S to the magnitude of the radial (Higgs) field fluctuations and angular (Goldstone) field fluctuations. To see this requires a specific choice of initial conditions, and we repeat those of the previous section, thermal equilibrium and instantaneous quenching.

Thermal Equilibrium

At temperature T the equilibrium propagators are time-independent, read off from (3.7) as

$$G_h(\vec{p}) = \frac{T}{\vec{p}^2 + m_H^2(T)} \tag{3.17}$$

$$G_g(\vec{p}) = \frac{T}{\vec{p}^2} \tag{3.18}$$

where $m_H(T) = \sqrt{2}m(T)$ is the effective Higgs mass at temperature T. If we take $\vec{p}_L^2 = 2\vec{p}^2/3$ in the integral then, up to numerical factors, we can approximate (3.10) as

$$\langle \bar{N}_S \bar{N}_S \rangle \simeq \mathcal{J}\left(\frac{2}{2\pi\sigma^2}\right) m(T)T \int_{|\vec{p}|<m(T)} d\!\!\!{}^-\vec{p}\, G_h(\vec{p}) \tag{3.19}$$

where we have coarse-grained to the Higgs correlation length $\xi = m_H^{-1}(T)$. [We have further assumed that the q_T integration can be approximated by setting q_T to zero in the integrand. Qualitatively this is a reasonable simplification]. The integral in (4.19) is essentially the integral (3.21). The end result is that, after substitution,

$$\langle \bar{N}_S \bar{N}_S \rangle = O\left(\left(\frac{m_H(T)T}{\sigma^2(T)}\right)^2\right) O(L) \tag{3.20}$$

where L is the length of the path in units of ξ. Equivalently, on using our previous results for equilibrium

$$\langle \bar{N}_S \bar{N}_S \rangle = O\left(\left(\frac{\langle h_v h_v \rangle}{\sigma^2(T)}\right)^2\right) O(L) \tag{3.21}$$

One power of $\langle h_v h_v \rangle$ comes from G_h, the other from residual factors. That is, the phase fluctuations are scaled by the Higgs fluctuations. In the Ginzburg regime, when correlation-volumes of the Higgs field can fluctuate to the false vacuum with significant probability, the fluctuations in field phase on the same distance scale are of order unity and, from the $O(L)$ term, are distributed randomly. [It is unclear whether the inclusion of connected correlation functions - linking the Higgs and Goldstone fields - would change the results qualitatively in this regime, and is under examination. The inclusion of four-point correlation functions in the real scalar field calculations had no dramatic effect, although the circumstances were somewhat different.] This ability not to just produce 'beads' of false vacuum, but to string them as vortices, is the Kibble mechanism referred to earlier [1], a common starting-point for numerical calculations [18].

A similar analysis can be performed for ^4He, treated as a quantum-mechanical system of non-relativistic bosonic point-particles with pair-wise potentials $V(\vec{x}_i - \vec{x}_j)$. In thermal equilibrium it is well-known [19] that the grand canonical partition function for such a system is identical to the partition function

$$Z = \int D\Phi\ \exp\{-S[\Phi]\} \tag{3.22}$$

of a complex non-relativistic quantum field Φ. The action in Euclidean space-time (with Euclidean time $0 \leq \tau \leq \beta$) for particles of mass m and (effective) chemical potential μ is

$$\begin{aligned} S[\Phi] &= \int d^4x \Phi^\dagger(x)\left(-\frac{1}{2m}\nabla^2 - \mu + \frac{\partial}{\partial\tau}\right)\Phi(x) \\ &+ \tfrac{1}{2}\int d^4x \int d^4x'\ |\Phi(x)|^2 V(x-x')|\Phi(x')|^2 \end{aligned} \tag{3.23}$$

$V(x) = \delta(t)V(\vec{x})$ is the time-instantaneous generalisation of the two-body potential introduced earlier.

Equation (4.23) is exact. If we adopt a local approximation for the two-body forces that enables us to replace the final term in (4.23) by $\lambda|\Phi|^4$, and integrate out the τ-dependent modes, the end result is a three-dimensional Landau-Ginzburg theory with effective Hamiltonian

$$H[\Phi] = \int d\vec{x}\ [\frac{1}{2m}|\nabla\Phi|^2 - \mu(T)|\Phi|^2 + \tfrac{1}{2}\lambda|\Phi|^4]. \tag{3.24}$$

like that of (3.10). At this level of approximation, in which Euclidean time has disappeared, the distinction between relativistic and non-relativistic theories has also disappeared (except for the details of $\mu(T)$). But for a slight change of definition, repeating our analysis will give a variance $(\Delta n)^2$ in winding number (along ∂S) proportional to the path length and to the radial field fluctuation. In thermal equilibrium there is, therefore, a true identity between the vortex fluctuations of a relativistic scalar field of the type that could occur in the early universe and those of superfluid ^4He. In the Ginzburg regime, when a random phase distribution is relevant, the vortex density will be large.

As a further example of scalar fields in thermal equilibrium we observed earlier that if $\Pi_2(M)$ is non-trivial we can have monopoles. In a cosmological context they are an embarrassment [1], since they would most likely dominate the energy of the universe if produced at the GUT phase transition and allowed to survive. As a final generalisation of our result (4.21) for winding number we note that it is straightforward to repeat our analysis for the production of global monoples. Consider the simplest case of a vector triplet of scalar fields in a broken global $O(3)$ theory. The variance in the monopole charge Q_v in a volume v is, in the same approximation as the above, a product of three two-point correlation functions. The end result is [20] that, in thermal equilibrium at temperature T, for coupling constant λ and effective Higgs mass $m(T)$,

$$\langle Q_v Q_v \rangle = O((\lambda T/m(T))^3) O((m(T)^3 v)^{2/3}) \tag{3.25}$$

where $m^3 v$ is the volume v measured in terms of correlation volumes, correlation length $\xi = m(T)^{-1}$. In the Ginzburg regime this is understood as the field taking random directions in field space in different correlation volumes.

Out-Of-Equilibrium Behaviour

Our discussion above has shown that, in thermal equilibrium close to the phase transition, field fluctuations are large enough to produce local topological charge (winding number, monopole charge) at high density, as anticipated by Kibble. However, this is of little use unless the future evolution of the system is able to freeze this charge in. Let us return to relativistic vortices. The random distribution of field phases shown above leads to macroscopic vortex fluctuations, crossing the system volume V. Numerical simulations may take an initial configuration of vortices based on random phases and then, switching to zero (or low) temperature solve for their classical evolution. This freezing in of defects goes beyond anything that we have said so far. As with the real scalar field of the previous section it is simplest to adopt a quenched approximation in which the closed time-path action for $t > 0$ is S of (4.1). In the light of our previous comments one of the most interesting choices would be to take the initial state as one of thermal equilibrium in the Ginzburg regime with its large fluctuations and watch the vortices freeze. This is too difficult. The simplest choice is to move instantaneously from a symmetric to a spontaneously-broken theory, rather as in (3.16), but with the differences outlined below.

Again we assume an initial symmetric state, centred at $\phi = 0$. Our first guess might be to take field configurations Boltzmann-distributed as $P_{in}[\Phi] = N\exp\{-\beta_{in}H_{in}[\Phi]\}$, where $H_{in}[\Phi]$ is derived from the free-field action

$$S_{in}[\phi] = \int d^4x\, [\tfrac{1}{2}|\partial\phi|^2 - \tfrac{1}{2}m_{in}^2|\phi|^2] \tag{3.26}$$

the $U(1)$ generalisation of (2.38). However, when (4.22) is decomposed into radial and angular fields as

$$S_{in}[\rho,\chi] = \int d^4x\,[\tfrac{1}{2}(\partial\rho)^2 + \tfrac{1}{2}\rho^2(\partial\chi)^2 - \tfrac{1}{2}m_{in}^2\rho^2] \tag{3.27}$$

the coupling between them (the second term) cannot be ignored since $\langle\rho\rangle = 0$. In the absence of any compelling reason (initial thermal equilibrium in the symmetric phase is hardly likely in the universe, although it would be appropriate for superfluid ^4He) we may as well make a choice of initial condition that is solvable. In this spirit, we take

$$S_{in}[\rho,\chi] = \int d^4x\,[\tfrac{1}{2}(\partial\rho)^2 + \tfrac{1}{2}\sigma^2(\partial\chi)^2 - \tfrac{1}{2}m_{in}^2\rho^2] \tag{3.28}$$

where $\sigma^2 = m^2/\lambda$ determines the vacuum in the subsequent symmetry-broken phase.

This S_{in}, which characterises the initial probabilities, is to be contrasted to the action which determines the subsequent spinodal evolution of the field. A similar problem arises for times $t > t_{in}$. Our first thought, the quadratic (in ϕ) action with downturned potential

$$S[\rho,\chi] = \int d^4x\,[\tfrac{1}{2}(\partial\rho)^2 + \tfrac{1}{2}\rho^2(\partial\chi)^2 + \tfrac{1}{2}m^2\rho^2] \tag{3.29}$$

with $m_H = \sqrt{2}m$, involves interactions between the Higgs and Goldstone fields after the radial/angular field decomposition. In the spirit of the Gaussian approximation we replace it by

$$S[\rho,\chi] = \int d^4x\,[\tfrac{1}{2}(\partial\rho)^2 + \tfrac{1}{2}\sigma^2(\partial\chi)^2 + \tfrac{1}{2}m^2\rho^2] \tag{3.30}$$

in which the decoupling between ρ and χ is enforced by putting $\rho^2 = \sigma^2$ in the second term of (3.29) The identity of the massless χ action in (3.28) and (3.30) guarantees that it remains in thermal equilibrium at all times. Although S of (3.30) is artificial, it is realistic in that we would not expect the χ modes to be dominantly exponential, but oscillatory. This is in contrast to the quenched behaviour of the radial Higgs field ρ as it falls off the hill, with its exponentially growing long-wavelength fluctuations. The Higgs two-point correlation function is, under these circumstances, identical to that of the scalar field in (2.21) and (2.24). The Goldstone field, always being massless, possesses the IR singularity for small $\vec{p}$,

$$G_g(\vec{p};t) \simeq \frac{T_{in}}{\vec{p}^2} \tag{3.31}$$

With this $1/\vec{p}^2$ effectively cancelling the p_L^2 in $\mathcal{J}$, $\langle\bar{N}_S(t)\bar{N}_S(t)\rangle$ behaves as

$$\begin{aligned}\langle\bar{N}_S(t)\bar{N}_S(t)\rangle &\simeq \mathcal{J}\left(\frac{2}{2\pi\sigma^2}\right)mT_{in}\int_{|\vec{p}|<m} d\!\!\!{}^-\vec{p}\,G_h(\vec{p};t) &(3.32)\\ &= O(L)\left(\frac{2}{2\pi\sigma^2}\right)mT_{in}\langle h_v(t)h_v(t)\rangle &(3.33)\end{aligned}$$

where L is the path length in units of l which, now, we take to be ξ of (2.25).

As before, the fluctuations in winding number vary as the square root of the step number, a consequence of random phase fluctuations, with scale set by the fluctuations of the Higgs field at the cessation of domain growth. Details will be given elsewhere [21]. How these spinodal domains then aggregate to produce the true vacuum is beyond our calculations, as is the subsequent evolution of the vortex network. What is interesting is the similarity between the equilibrium result (3.21) and (3.33). Whether the vortices are produced in equilibrium or strongly out of equilibrium the Kibble conjecture seems substantially correct.

Global monopoles can be treated in the same way. If the Goldstone modes are allowed to stay in equilibrium the fluctuations in monopole charge retain the equilibrium volume dependence of (4.25), scaled by the Higgs fluctuations. This is unfortunately not the case for ^4He. The form (4.23) for the ^4He Φ-field partition function relied on thermal equilibrium for the particle system to be identical to a field theory. We do not know to what extent a non-equilibrium system of point particles can be put into correspondence with non-equilibrium field theory. However, if it were possible to take a real-time continuation of (4.23) as a basis for non-equilibrium calculations there would be no difficulty in principle in adopting the same approximations, although the difference between non-relativistic and relativistic thermal field theories would now have to be addressed.

Local Vortices

Our experience of continuous symmetries in particle physics has been that all such symmetries have been made local by the presence of gauge fields. In that sense the global vortices that we have been discussing are unnatural.

The local gauge extension of the $U(1)$ theory (4.1) has action

$$S[\phi, A] = \int d^4x \, [-\frac{1}{4}F_{\mu\nu}F^{\mu\nu} + |\partial_\mu\phi - ieA_\mu|^2 + m^2|\phi|^2 - \frac{1}{4}\lambda|\phi|^4]. \tag{3.34}$$

(changing factors of $\frac{1}{2}$ for convenience). This still permits relativistic vortices, the simplest candidates for local cosmic strings. [Whereas the non-relativistic global strings are the vortices of superfluids, the non-relativistic counterparts of local strings are the vortices of superconductors]. For $e^2/\lambda \ll 1$ we have a Type-II theory and the strings are approximately global, and our previous results should apply.

The gauge-invariant expression (4.2) for winding number is still valid, but it is more convenient to define it through the gauge field A_μ as the line integral

$$N_S(t) = \frac{e}{2\pi}\int_{\partial S} \underline{dl}.\underline{A} \tag{3.35}$$

As before, field fluctuations will create local winding number. In the Gaussian approximation its variance is

$$\langle N_S(t)N_S(t)\rangle = \left(\frac{e}{2\pi}\right)^2 \int_{\partial S} dl'_i \int_{\partial S} dl''_j \, \langle A_i(t,\vec{x}')A_j(t,\vec{x}'')\rangle \tag{3.36}$$

where $\vec{x}', \vec{x}''$ denote the positions of the line increments on ∂S. In one sense this is significantly simpler than its scalar counterpart (4.10) since we do not have to worry about disconnected and connected parts. The difficulty lies in the more complicated correlation function. From our knowledge of gauge field thermal propagators there is no intrinsic problem in calculating $\langle N_S N_S\rangle$ in thermal equilibrium, at least in an approximation that takes thermal mass self-energies constant (in momentum space) and ignores the magnetic mass. However, as of this moment the work is not complete, but we hope to give the results later. This seems a good place to stop.

Acknowledgements

This talk is based, in large part, on collaborations with Mark Hindmarsh of Sussex (equilibrium fluctuations), Tim Evans of I.C. (non-equilibrium behaviour and gauge propagators) Luis Bettencourt of I.C. (first-order transitions and gauge fluctuations) and Alistair Gill (monopoles and ^4He). I thank them all.

References

1. T. W. B. Kibble, J. Phys. A9, 1387 (1976).
2. K. Enqvist, Magnetic fields of electroweak origin, these proceedings. K. Enqvist and P. Oleson, Phys. Lett. B319, 178 (1993)
3. P. C. Hendry, N. S. Lawson, R. A. M. Lee, P. V. E. McClintoch and C. D. H. Williams, Nature 368, 315 (1994)
4. W. H. Zurek, Nature 368, 292 (1994)
5. G. W. Semenoff and N. Weiss, Phys. Rev.D31, 689 (1985); Phys. Rev. D31, 699 (1985).
6. I. D. Lawrie, Phys. Rev. D40, 3330 (1989); J. Phys. A25, 2493 (1992).
7. D. A. Kirzhnits and A. D. Linde, Phys. Lett. 42B, 471 (1972). L. Dolan and R. Jackiw, Phys. Rev. D9, 3320 (1974); S. Weinberg, Phys. Rev. D9, 3357 (1974).
8. M. B. Hindmarsh and R. J. Rivers, Fluctuations at first order transitions, to be published in Nucl. Phys.B (1994).
9. G. Jona-Lasinio, in Scaling and Self-Similarity in Physics ed. J. Frolich, Progress in Physics, Vol.7, 11 (Birkhauser Press, 1983).
10. E. Weinberg, Bubble Nucleation rates, these proceedings.
11. K. Kajantie, Three dimensional physics and the EW phase transition, these proceedings.
12. M. Dine, R. G. Leigh, P. Huet, A. Linde and D. Linde, Phys. Rev. D46, 550 (1992).
13. V. I, Ginzburg, Fiz. Tverd. Tela 2, 2031 (1960); [Sov. Phys. Solid State 2, 1826 (1961)].
14. M. Gleiser, Phys. Rev. D42, 3350 (1990), M. Gleiser, E. W. Kolb and R. Watkins, Nucl. Phys. B364, 411 (1991).
15. L. Bettencourt and R. J. Rivers, Imperial College preprint, in preparation.
16. D. Boyanovsky and H. J. de Vega, Phys. Rev. D47, 2343 (1993); D. Boyanovsky, Da-Shin Lee and A. Singh, Phys. Rev. D48, 800 (1993).
17. H. J. de Vega, Quantum non-equilibrium evolution and the phase transition in the early universe, these proceedings.
18. T. Vachasparti and A. Vilenkin, Phys. Rev. D10, 2036 (1984).
19. F. Wiegel, Introduction to Path-Integral Methods in Physics and Polymer Science (World Scientific, Singapore, 1986).
20. A. Gill, Imperial College preprint, in preparation.
21. T. S. Evans and R. J. Rivers, Imperial College preprint, in preparation.

HIGH TEMPERATURE FIELD THEORY BEYOND PERTURBATION THEORY

Christof Wetterich

Institut für Theoretische Physik
Universität Heidelberg
Philosophenweg 16
D-6900 Heidelberg, FRG

ABSTRACT

For realistic values of the Higgs boson mass the high temperature electroweak phase transition cannot be described perturbatively. The symmetric phase is governed by a strongly interacting $SU(2)$ gauge theory. Typical masses of excitations and scales of condensates are set by the "high temperature confinement scale" $\approx 0.2\ T$. For a Higgs boson mass around 100 GeV or above all aspects of the phase transition are highly nonperturbative. Near the critical temperature strong electroweak interactions are a dominant feature also in the phase with spontaneous symmetry breaking. Depending on the value of the Higgs boson mass the transition may be a first order phase transition or an analytical crossover.

1. RUNNING COUPLINGS NEAR PHASE TRANSITIONS

Phase transitions in the very early universe may be tumultuous periods out of thermodynamic equilibrium. Some traces of such a transition may still be observable today, giving a hint on what happend shortly after the big bang. Remnants of the electroweak phase transition may help to understand the universe at the age of about $10^{-12}s$. One of the most prominent features of the standard model of electroweak interactions is spontaneous symmetry breaking. The masses of the gauge bosons and fermions are proportional to the vacuum expectation value of the Higgs doublet. According to the original argument of Kirzhnits and Linde[1] this expectation value vanishes at sufficiently high temperature and the electroweak symmetry is restored. Such high temperatures were presumably realized in the very early universe immediately after the big bang. As the universe cooled there must

Electroweak Physics and the Early Universe, Edited by
Romao and Freire, Plenum Press, New York, 1994

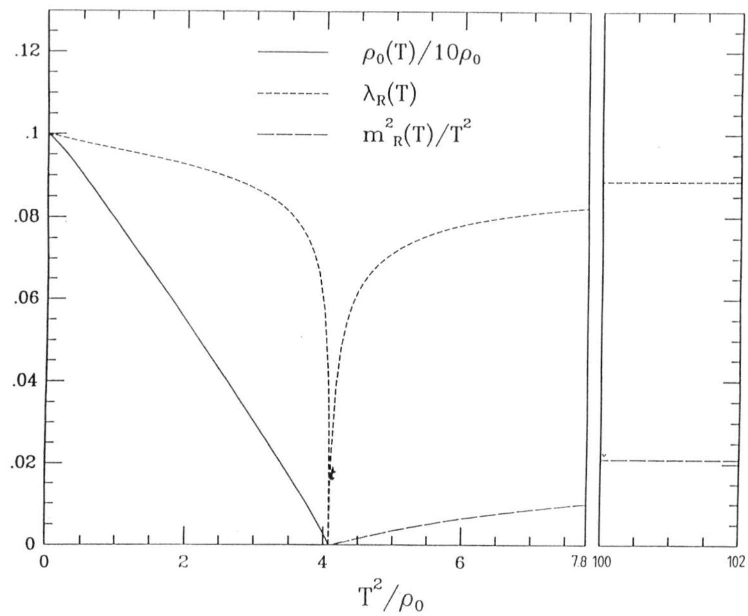

Figure 1. The minimum of the potential, the renormalized coupling and the mass for a wide range of temperatures with $\lambda_R = 0.1$ in a pure O(4) scalar theory.

have been a phase transition from the the symmetric to the spontaneously broken phase of the standard model. This phase transition may have important consequences for our present universe, one example being the possible creation of the excess of matter compared to antimatter (baryon asymmetry)[2]. The physical implications of the high temperature electroweak phase transition depend strongly on its nature (whether it is second or first order) and its details. More specifically, the possibility of generating the baryon asymmetry imposes the requirement of out of equilibrium conditions[3], which is satisfied only if the phase transition is of the first order; the exact amount of produced baryon number is very sensitive to the details of the fluctuations which drive the transition (profile of bubbles, velocity of the wall etc.)[4]; and avoiding the washing out of any generated baryon asymmetry requires a sufficiently strong first order phase transition[5].

Most theoretical studies of the electroweak phase transition use high temperature perturbation theory[6] for a computation of the temperature-dependent effective potential for the Higgs field. This is often supplemented by an appropriate resummation of graphs or a solution of a corresponding gap equation for the mass terms[7]. Near a phase transition, however, the use of perturbation theory becomes questionable. Roughly speaking, one can trust perturbation theory only to the extent that mean field theory gives a qualitatively correct description of the corresponding statistical system. This is known to be not always the case for critical phenomena. A good example for the breakdown of high temperature perturbation theory in the vicinity of the phase transition are scalar field theories[8]. The second order character of the transition and the corresponding critical exponents are not reproduced by perturbation theory.

The deeper reason for the breakdown of perturbation theory lies in the effective three-dimensional character of the high temperature field theory[9]. Field theory at nonvanishing temperature T can be formulated in terms of an Euclidean functional integral where the "time dimension" is compactified on a torus with radius T^{-1} [10]. For phenomena at distances larger than T^{-1} the Euclidean time dimension cannot be resolved. Integrating over modes with momenta $p^2 > (2\pi T)^2$ or, alternatively, over the higher Fourier modes on the torus (the $n \neq 0$ Matsubara frequencies) leads to "dimensional reduction" to an effective three-dimensional theory. This is very similar to dimensional reduction in Kaluza-Klein theories[11] for gravity. The change of the effective dimensionality for distances larger than T^{-1} is manifest in the computation[8] of the temperature-dependent effective potential in scalar theories. The scale dependence of the renormalized couplings is governed by the usual perturbative β-functions only for $p^2 > (2\pi T)^2$. In contrast, for smaller momenta $p^2 < (2\pi T)^2$ the running of the couplings was found to be determined by three-dimensional β-functions instead of the perturbative four-dimensional ones - as proposed in a different setting in ref. 12. The effect of the three-dimensional running is clearly manifest in the temperature-dependence of the couplings shown[8] in fig. 1, especially for the renormalized quartic scalar coupling λ_R, which vanishes for T approaching the critical temperature T_c. As an alternative to integrating out all modes with $p^2 > (2\pi T)^2$ an effective three-dimensional theory for the long distance electroweak physics has been obtained in ref. 13 by integrating out the higher Matsubara frequencies*.

* For an earlier treatment of dimensional reduction in high temperature QCD see ref. 14.

If the three-dimensional running of the couplings becomes important, the physics of the phase transition is dominated by classical statistics even in case of a quantum field theory. A second order phase transition is characterized by an infinite correlation length. The critical exponents which describe the behaviour near the critical temperature are always those of the corresponding classical statistical system. Since the fixpoints of the three-dimensional β-functions are very different from the four-dimensional (perturbative) fixpoints, we conclude that high temperature perturbation theory is completely misleading in the vicinity of a second-order phase transition. This argument extends to sufficiently weak first order transitions. A second related example for the breakdown of perturbation theory is the symmetric phase of the electroweak gauge theory. The gauge bosons are massless in perturbation theory and the three-dimensional running always dominates at large distances[15].

In order to understand the high temperature behaviour of a theory we should understand the qualitative features of the β-functions in three dimensions. These β-functions have nothing to do with the ultraviolet regularization of the field theory - in this respect there is no difference between vanishing and nonvanishing temperature. They are rather related to the infrared behaviour of the theory or the dependence of Green functions on some sort of infrared cutoff. According to Wilson's concept of the renormalization group these β-functions describe the scale dependence of the couplings if one looks at the system on larger and larger distances. For an understanding of systems with approximate scaling in a certain range it is useful to define dimensionless couplings. One divides out an appropriate power of the infrared cutoff k which plays the role of the renormalization scale. For example, the dimensionless quartic scalar coupling λ in the effective three-dimensional theory is related to the four-dimensional coupling λ_4 and the temperature by

$$\lambda = \frac{\bar{\lambda}_3}{k} = \lambda_4 \frac{T}{k} \tag{1}$$

For a pure N-component scalar theory the qualitative behaviour of the function $\beta_\lambda = \frac{\partial \lambda}{\partial t}$, $t = \ln k$, is shown in fig. 2. Arrows denote the flow of λ with decreasing k and we have assumed a massless theory. One clearly sees an infrared stable fixpoint corresponding to the second-order phase transition.

Next we turn to scalar QED which should describe the phase transition of superconductors or of the abelian Higgs model at high temperatures. The supposed form of the functions $\beta_{e^2} = \partial e^2/\partial t$ and $\beta_{\lambda/e^2} = \partial\left(\frac{\lambda}{e^2}\right)/\partial t$ is shown* in fig. 3. One observes a fixpoint for e^2 but no fixpoint for λ/e^2 or λ. Therefore λ decreases until it vanishes and the corresponding phase transition is presumably of the first order. For comparison we also show the β-function for a number M of complex scalar fields exceeding a critical value M_{cr}. Here an ultraviolet and an infrared stable fixpoint appear. For small initial λ (λ/e^2 smaller than the UV fixpoint) the phase transition remains of the first order. The UV fixpoint corresponds to a triple point and for initial λ/e^2 larger than the UV fixpoint one

* The true phase diagrams are multidimensional and the diagrams of Fig. 3 may be interpreted as projections on appropriate trajectories. The fixpoint structure is independent of particular trajectories.

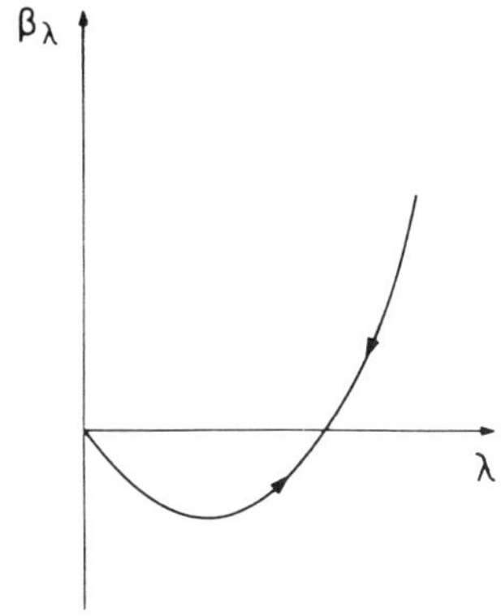

Figure 2. The β-function for λ in a pure N component scalar theory.

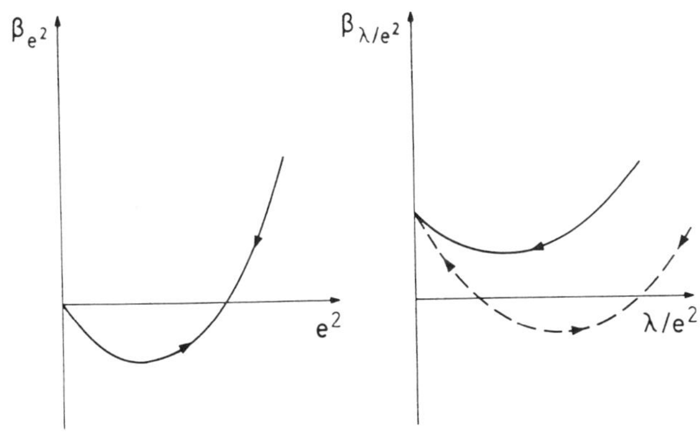

Figure 3. Qualitative forms for the β-functions for e^2 and λ/e^2 in an abelian Higgs model.

has a second-order transition with critical behaviour governed by the IR fixpoint. The cirtical number of charged scalars M_{cr} (above which a second order transition is possible) is not very well known.

A nonabelian gauge theory like the electroweak theory is confining also in three dimensions. We have depicted the running of the gauge coupling g^2 and the ratio λ/g^2 in fig. 4. For sufficiently small initial λ (small physical Higgs boson mass) $\lambda(k)$ reaches zero for k much larger than the three-dimensional confinement scale. One then expects a first-order transition which is analogous to the four-dimensional Coleman-Weinberg scenario[16]. Typical mass scales are of the order k_{cw} where $\lambda(k_{cw}) = 0$. In this case it is possible that high temperature perturbation theory gives reliable results. On the other hand, if the three-dimensional confinement scale $\Lambda^{(3)}_{conf}$ (the value of k for which the gauge coupling diverges or becomes very large) is reached with $\lambda(\Lambda^{(3)}_{conf}) > 0$ the behaviour near the phase transition is described by a strongly interacting electroweak theory. Then strong effective coupling constants appear not only in the symmetric phase, but also in the phase with spontaneous symmetry breaking. The phase transition may either be of the first order or an analytical crossover may replace the phase transition[15]. A second-order transition seems unlikely.* In any case, the "strongly interacting phase transition" will be very different from perturbative expectations.

2. PERTURBATIVE INFRARED DIVERGENCES AND "CUBIC TERMS"

In high temperature perturbation theory the electroweak phase transition is predicted to be of the first order. This follows from a "cubic term" in the effective potential generated by thermal fluctuations. In order to get an idea for which values of the scalar field and the temperature perturbation theory may be applied, we have to estimate the reliability of the cubic term. For a complex two-component Higgs scalar φ the cubic term is proportional $(\varphi^\dagger\varphi)^{3/2}$ and we should understand the origin of such a nonanalytic behaviour of the effective potential. We will see that it is closely linked to the issue of infrared divergences for the quartic scalar coupling[8]. Indeed, a term $\sim (\varphi^\dagger\varphi)^{3/2}$ implies that the quartic scalar coupling diverges for $\varphi \to 0$. This infrared divergence[17] is immediately apparent by inspection of the diagram in fig. 5 for which the $n = 0$ Matsubara frequency gives a contribution

$$\Delta\lambda_4 \sim T \int d^3q \frac{g_4^4}{(q^2 + \frac{1}{2}g_4^2\varphi^\dagger\varphi)^2} \sim \frac{g_4^3 T}{(\varphi^\dagger\varphi)^{1/2}} \tag{2}$$

The φ-dependent gauge boson mass plays here the role of the effective infrared cutoff and

* A second-order transition requires a massless scalar degree of freedom at the critical temperature. The Higgs scalar presumably acquires a mass through strong electroweak interactions both in the symmetric and SSB phase. Chiral condensation phenomena of quarks and leptons similar as for quarks in QCD seem not very plausible for the high temperature electroweak theory.

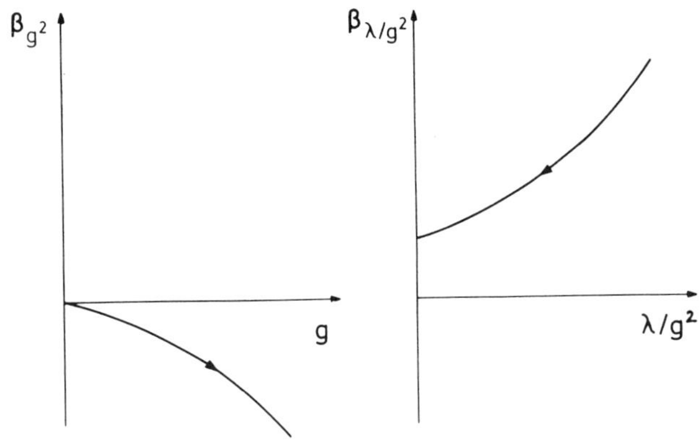

Figure 4. Qualitative forms for the β-functions for g^2 and λ/g^2 in a nonabelian Higgs model.

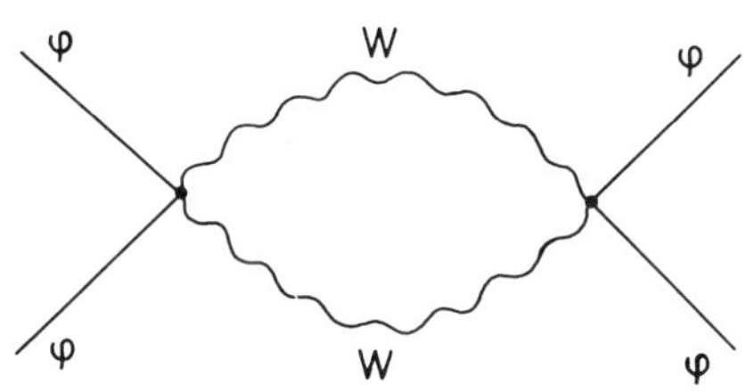

Figure 5. Infrared divergent (for $\varphi \to 0$) diagram contributing to the quartic scalar coupling.

we may associate

$$k = m_W(\varphi) = \frac{1}{\sqrt{2}} g_4 (\varphi^\dagger \varphi)^{1/2} \tag{3}$$

Inserting the correction to the quartic term $\sim \Delta\lambda\ (\varphi^\dagger\varphi)^2$ in the effective potential gives exactly the cubic term mentioned before. The question arises now to which extent the perturbative treatment of the effectively three-dimensional behaviour can be trusted. A similar cubic term appears also in the perturbative treatment of the pure scalar theory and has been shown[8] to be completely misleading in the vicinity of the critical temperature. (For $T = T_c$ the effective scalar potential has not a divergent, but a vanishing quartic coupling and is dominated by the six-point function, $U \sim (\varphi^\dagger\varphi)^3$.) What about gauge theories?

At the scale $k_T = 2\pi T$ the three-dimensional gauge coupling and quartic scalar coupling are given by

$$\bar\lambda_3(k_T) = \lambda_4 T, \quad \bar g_3^2(k_T) = g_4^2 T \tag{4}$$

For $k < k_T$ the running of the couplings becomes effectively three-dimensional. In the limit of small $\bar\lambda_3 \ll \bar g_3^2$ we neglect the contributions from scalar loops and the running of $\bar\lambda_3$ follows* from gauge boson loops (fig. 5)

$$\frac{\partial \bar\lambda_3}{\partial t} = \frac{9}{64\pi} \frac{\bar g_3^4}{k} \tag{5}$$

Let us for a moment neglect the running of the gauge coupling. With the infrared cutoff identified with $m_W(\varphi)$ (3) and defining $\bar\lambda_3$ by an appropriate derivative of the effective potential U

$$\begin{aligned} \bar\lambda_3(k) &= T\frac{\partial^2 U}{\partial \rho^2} \\ \rho = \varphi^\dagger\varphi = \varphi^2&, \quad k^2 = \frac{1}{2} g_4^2 \rho \end{aligned} \tag{6}$$

the flow equation (5) can be turned into a differential equation for $U(\rho)$. Solving

$$\varphi \frac{\partial}{\partial \varphi} \bar\lambda_3 = \frac{9\sqrt{2}}{64\pi} g_4^3 \frac{T^2}{\varphi} \tag{7}$$

for $\varphi < \varphi_T$, $\frac{1}{2} g_4^2 \varphi_T^2 = k_T^2 = (2\pi T)^2$ one obtains

$$\bar\lambda_3(\varphi) = \left(\lambda_4 + \frac{9 g_4^4}{128\pi^2}\right) T - \frac{9\sqrt{2} g_4^3}{64\pi} \frac{T^2}{\varphi} \tag{8}$$

From the definition (6) of $\bar\lambda_3$ follows the second order differential equation

$$\frac{\partial^2 U}{\partial \rho^2} = \lambda_4 + \frac{9 g_4^4}{128\pi^2} - \frac{9\sqrt{2} g_4^3}{64\pi} T \rho^{-\frac{1}{2}} \tag{9}$$

---

* If the contribution of the Debye-screened A_0 mode is omitted, the r.h.s. of eq. (5) should be reduced by a factor 2/3, see sect. 5.

The general solution has two integration constants, one of them being irrelevant here

$$U(\rho) = -\mu^2(T)\rho + \frac{1}{2}(\lambda_4 + \frac{9g_4^4}{128\pi^2})\rho^2 - \frac{3\sqrt{2}}{16\pi}g_4^3 T\rho^{3/2} \tag{10}$$

The temperature-dependent mass term $\mu^2(T)$ appears here as an integration constant and can be determined from high temperature perturbation theory (which should be valid for $\varphi = \varphi_T$)

$$-\mu^2(T) = -\mu_0^2 + \frac{3g_4^2}{16}T^2 \tag{11}$$

The potential (10) is nothing else than the result of high temperature perturbation theory. We conclude that the latter is reliable if the scalar loops can be omitted ($\bar{\lambda}_3 \ll \bar{g}_3^2$) and if the running of the gauge coupling can be neglected. Additional terms from the scalar loops could be included in the evolution equation (5) without changing the qualitative picture. The crucial question concerns the neglection of the running of the gauge coupling. This will determine the range of validity of perturbation theory and we will turn back to this question below.

3. AVERAGE ACTION

A useful tool for describing the running of couplings in arbitrary dimension is the average action[18]. Consider a simple model with a real scalar field χ. The average scalar field is easily defined by

$$\phi_k(x) = \int d^d y f_k(x-y)\chi(y) \tag{12}$$

with f_k decreasing rapidly for $(x-y)^2 > k^{-2}$ and properly normalized. The average is taken over a volume of size $\sim k^{-d}$. The average action $\Gamma_k[\varphi]$ obtains then by functional integration of the "microscopic variables" χ with a constraint forcing $\phi_k(x)$ to equal $\varphi(x)$ up to small fluctuations. It is the effective action for averages of fields and therefore the analogue in continuous space of the block spin action[19] on the lattice. All modes with momenta $q^2 > k^2$ are effectively integrated out. Lowering k permits to explore the theory at longer and longer distances. The average action has the same symmetries as the original action. As usual it may be expanded in derivatives, with average potential $U_k(\rho), \rho = \frac{1}{2}\varphi^2$, kinetic term, etc.

$$\Gamma_k = \int d^d x \left\{U_k(\rho) + \frac{1}{2}Z_k(\rho)\partial_\mu\varphi\partial^\mu\varphi + ...\right\} \tag{13}$$

In a suitable formulation[20] the effective average action becomes the generating functional for 1PI Green functions with an infrared cutoff set by the scale k. It interpolates between the classical action for $k \to \infty$ and the effective action for $k \to 0$. In this version an exact nonperturbative evolution equation describes the dependence of Γ_k on the infrared cutoff k ($t = \ln k$)

$$\frac{\partial}{\partial t}\Gamma_k = \frac{1}{2}\int \frac{d^d q}{(2\pi)^d}\left(\Gamma_k^{(2)} + R_k\right)^{-1}\frac{\partial}{\partial t}R_k \tag{14}$$

Here $R_k(q)$ is a suitable infrared cutoff which may depend on q^2, as, for example, $R_k = q^2 \exp\left(-\frac{q^2}{k^2}\right)\left(1-\exp\left(-\frac{q^2}{k^2}\right)\right)^{-1}$ or $R_k = k^2$. The two-point function $\Gamma_k^{(2)}$ obtains by second functional variation of Γ_k

$$\Gamma_k^{(2)}(q', q) = \frac{\delta^2 \Gamma_k}{\delta\varphi(-q')\delta\varphi(q)} \tag{15}$$

Therefore $(\Gamma_k^{(2)} + R_k)^{-1}$ is the exact propagator in presence of the infrared cutoff R_k and the flow equation (14) takes the form of the scale variation of a renormalization group-improved one-loop expression. We emphasize that the evolution equation is fully nonperturbative and no approximations are made. A simple proof can be found in ref. 20. The exact flow equation (14) can be shown to be equivalent with earlier versions of "exact renormalization group equations"[21] and it encodes the same information as the Schwinger-Dyson equations[22].

An exact nonperturbative evolution equation is not yet sufficient for an investigation of nonperturbative problems like high temperature field theories. It is far too complicated to be solved exactly. For practical use it is crucial to have a formulation that allows to find reliable nonperturbative approximative solutions. Otherwise speaking, one needs a description of Γ_k in terms of only a few k-dependent couplings. The flow equations for these couplings can then be solved numerically or by analytical techniques. It is on the level of such truncations of the effective average action that suitable approximations have to be found. In this respect the formulation of the effective average action[20] offers important advantages: The average action has a simple physical interpretation and eq. (14) is close to perturbation theory if the couplings are small. The formulation is in continuous space and all symmetries - including chiral symmetries or gauge symmetries[15] - can be respected. Since Γ_k has a representation as a functional integral alternative methods (different from solutions of the flow equations) can be used for an estimate of its form. Furthermore, the flow equation (14) is directly sensitive to the relevant infrared physics since the contribution of particles with mass larger than k is suppressed by the propagator on the r.h.s. of eq. (14). The closed form of this equation does not restrict one a priori to given expansions, like in 1PI n-point functions. In addition the momentum integrals in eq. (14) are both infrared and ultraviolet convergent if a suitable cutoff R_k is chosen. Only modes in the vicinity of $q^2 = k^2$ contribute substantially. This feature is crucial for gauge theories where the formulation of a gauge-invariant ultraviolet cutoff is difficult without dimensional regularization.

4. STRONG ELECTROWEAK INTERACTIONS

We are now ready to discuss the running of the three-dimensional gauge coupling. We start from the effective average action for a pure $SU(N_c)$ Yang-Mills theory. It is a gauge-invariant functional of the gauge field A and obeys the exact evolution equation[15] (with Tr including a momentum integration)

$$\frac{\partial}{\partial t}\Gamma_k[A] = \frac{1}{2}\mathrm{Tr}\left\{\frac{\partial R_k[A]}{\partial t}\left(\Gamma_k^{(2)}[A] + GF[A] + R_k[A]\right)^{-1}\right\} - \epsilon_k[A] \tag{16}$$

Here $GF[A]$ is the contribution from a generalized gauge-fixing term in a covariant background gauge

$$GF[A] = \Gamma_k^{gauge(2)}[A, \bar{A}]_{|\bar{A}=A} \tag{17}$$

and $\epsilon_k[A]$ is the ghost contribution[15]. The infrared cutoff R_k is in general formulated in terms of covariant derivatives. We make the simple truncation

$$\begin{aligned} \Gamma_k[A] &= \frac{1}{4}\int d^dx Z_{F,k} F_{\mu\nu}F^{\mu\nu} \\ \Gamma_k^{gauge}[A, \bar{A}] &= \frac{1}{2}\int d^dx Z_{F,k}(D_\mu[\bar{A}](A^\mu - \bar{A}^\mu))^2 \end{aligned} \tag{18}$$

and observe that in our formulation the gauge coupling $\hat{g}$ appearing in $F_{\mu\nu}$ and D_μ is a constant independent of k. The only k-dependent coupling can be associated with the dimensionless renormalized gauge coupling

$$g^2(k) = k^{d-4}\bar{g}_d^2(k) = k^{d-4}Z_{F,k}^{-1}\hat{g}^2. \tag{19}$$

The running of g^2 is related to the anomalous dimension η_F

$$\begin{aligned} \eta_F &= -\frac{\partial}{\partial t}\ln Z_{F,k} \\ \frac{\partial g^2}{\partial t} &= \beta_{g^2} = (d-4)g^2 + \eta_F g^2 \end{aligned} \tag{20}$$

Using configurations with constant magnetic field it was found[15] to obey

$$\frac{\partial g^2}{\partial t} = (d-4)g^2 - \frac{\frac{44}{3}N_c v_d a_d g^4}{1 - \frac{20}{3}N_c v_d b_d g^2} \tag{21}$$

with*

$$\begin{aligned} v_d^{-1} &= 2^{d+1}\pi^{\frac{d}{2}}\Gamma\left(\frac{d}{2}\right) \\ a_d &= \frac{(26-d)(d-2)}{44} n_1^{d-4} \\ b_d &= \frac{(24-d)(d-2)}{40} l_1^{d-2} \end{aligned} \tag{22}$$

Only the momentum integrals ($x \equiv q^2$)

$$\begin{aligned} n_1^d &= -\frac{1}{2}k^{-d}\int_0^\infty dx\, x^{\frac{d}{2}}\frac{\partial}{\partial t}\left(\frac{\partial P}{\partial x}P^{-1}\right) \\ l_1^d &= -\frac{1}{2}k^{2-d}\int_0^\infty dx\, x^{\frac{d}{2}-1}\frac{\partial}{\partial t}P^{-1} \end{aligned} \tag{23}$$

---

* The vanishing of the β-function for $\bar{g}^2$ for $d = 2$ and $d = 26$ is no accident. In lowest order in the ϵ-expansion[23] the denominator in the last term is absent and $v_3 a_3$ is replaced by $v_4 a_4$.

depend on the precise form of the infrared cutoff R_k appearing in

$$P(x) = x + Z_k^{-1} R_k(x) \tag{24}$$

In four dimensions one has $a_4 = 1$, $v_4 = 1/32\pi^2$ and eq. (21) reproduces the one-loop result for β_{g^2} in lowest order g^4. For an appropriate choice of R_k with $b_4 = 1$ an expansion of eq. (21) also gives 93 % of the perturbative g^6 coefficient. For $d \neq 4$ the solution of eq. (21) is

$$\begin{aligned} \frac{g^2(k)}{(1+\delta g^2(k))^\gamma} &= \frac{g^2(k_0)}{(1+\delta g^2(k_0))^\gamma}\left(\frac{k}{k_0}\right)^{d-4} \\ \delta &= \frac{N_c v_d}{3}\left(\frac{44}{4-d}a_d - 20 b_d\right) \\ \frac{1}{\gamma} &= 1 - \frac{5}{11}(4-d)\frac{b_d}{a_d} \end{aligned} \tag{25}$$

Concerning the high temperature field theory we should use the three-dimensional β-function and associate k_0 with the scale $k_T = 2\pi T$, where the three-dimensional running sets in. The "initial value" of the gauge coupling reads $g^2(k_T) = 2\alpha(k_T)$ with α the four-dimensional fine structure constant. For $k < k_T$ the three-dimensional gauge coupling increases with a power behaviour instead of the four-dimensional logarithmic behaviour. The three-dimensional confinement scale $\Lambda^{(3)}_{conf}$ - where g^2 diverges - is proportional to the temperature. For the electroweak theory ($\alpha_w = 1/30$) and the choice (24) $P(x) = x/(1 - \exp - \frac{x}{k^2})$ one finds[15]

$$\Lambda^{(3)}_{conf} = 0.23T \quad (0.1T) \tag{26}$$

Here the number in brackets corresponds to a "lowest order approximation" where b_3 is put to zero in the β-function (21).

For the symmetric phase of the electroweak theory we can neglect the scalar fluctuations in a good approximation and the estimate (26) directly applies. One has to deal with a strongly interacting gauge theory with typical nonperturbative mass scales only somewhat below the temperature scale! Similar to QCD one expects that condensates like $\langle F_{ij}F^{ij}\rangle$ play an important role[15,24]. More generally, the physics of the symmetric phase corresponds to a strongly coupled $SU(2)$ Yang-Mills theory in three dimensions: The relevant excitations are "W-balls" (similar to glue balls) and scalar bound states. All "particles" are massive (except the "photon") and the relevant mass scale is set by $\Lambda^{(3)}_{conf} \sim T$. Also the values of all condensates are given by appropriate powers of the temperature. Since the temperature is the only scale available the energy density must have the same T-dependence as for an ideal gas

$$\rho = cT^4 \tag{27}$$

Only the coefficient c should be different from the value obtained by counting the perturbative degrees of freedom*. We expect that quarks and leptons form $SU(2)$ singlet bound

* A similar remark also applies to high temperature QCD. We expect quantitative modifications of early cosmology due to the difference between c and the ideal gas value.

states similar to the mesons in QCD*. A chiral condensate seems, however, unlikely in the high temperature regime and we do not think that fermions play any important role for the dynamics of the electroweak phase transition. The "photon" (or rather the gauge boson associated to weak hypercharge) decouples from the W-balls. Its effective high temperature coupling to fermion and scalar bound states is renormalized to a very small value. As for the phase with spontaneous symmetry breaking, the fermions and "photon" can be neglected for the symmetric phase. We conclude that the high temperature phase transition of the electroweak theory can be described by an effective three-dimensional Yang-Mills-Higgs system. It is strongly interacting in the symmetric phase. Depending on the value of the mass of the Higgs boson it may also be strongly interacting in the phase with spontaneous symmetry breaking if the temperature is near the critical temperature. A more detailed investigation of this issue will be given in the next section.

5. RENORMALIZATION GROUP-IMPROVED EFFECTIVE POTENTIAL FOR THE ELECTROWEAK PHASE TRANSITION

Before performing a complete study of the average action for the standard model at high temperature, the most important issues can be addressed in a simplified treatment. For large enough $\rho, \rho \geq \varphi_T^2, \varphi_T^2 = 8\pi^2T^2/g_4^2$ (see sect. 2) we believe high temperature perturbation theory and take the one-loop expression[6,7]

$$U_3(\rho_3) = -\mu^2(T)\rho_3 + \frac{1}{2}\left(\bar{\lambda}_3 + \frac{9\bar{g}_3^4}{128\pi^2 T}\right)\rho_3^2 - \frac{1}{12\pi}\left(6m_B^3 + 3m_E^3 + m_1^3 + 3m_2^3\right) \quad (28)$$

with

$$\begin{aligned} m_B^2 &= \frac{1}{2}\bar{g}_3^2\rho_3 \\ m_E^2 &= m_D^2(T) + \frac{1}{2}\bar{g}_3^2\rho_3 \\ m_1^2 &= 3\bar{\lambda}_3\rho_3 - \mu^2(T) \\ m_2^2 &= \bar{\lambda}_3\rho_3 - \mu^2(T) \end{aligned} \quad (29)$$

Here we use already a three-dimensional language with $U_3 = U/T, \rho_3 = \rho/T, \bar{\lambda}_3 = \lambda_4 T, \bar{g}_3^2 = g_4^2 T$. The couplings λ_4 and g_4^2 are evaluated at the scale $k_T = 2\pi T$ and $\mu^2(T), m_D^2(T)$ take the perturbative values. For $\rho < \varphi_T^2$ we can compute a renormalization group-improved potential by using a flow equation similar to (5) with $k \equiv m_B$. In the language of the average action this amounts to an infrared cutoff $P(x) = x + k^2$. This yields in the $\alpha = 0$ gauge the approximative evolution equation[15]

$$\frac{\partial}{\partial t}\bar{\lambda}_3(k) = \frac{3}{64\pi}k^{-1}\left\{2\bar{g}_3^4(k) + \bar{g}_3^4(k)\frac{m_B}{m_E} + 12\bar{\lambda}_3^2(k)\frac{m_B}{m_1} + 4\bar{\lambda}_3^2(k)\frac{m_B}{m_2}\right\} \quad (30)$$

* We use here a language appropriate for the excitations of the three-dimensional Euclidean theory. Interpretation in terms of relativistic particles has to be used with care!

We recover eq. (5) if we neglect $\bar{\lambda}_3$ on the r.h.s. and approximate $m_B = m_E$. For an alternative derivation we could obtain eq. (30) by taking appropriate derivatives of the potential $U_3(\rho_3)$ (28) using the definitions (6) and treating the rations m_B/m_E etc. as independent of k. For the correct interpretation of the flow equation we should also adapt the relation between k and ρ to the situation with running gauge coupling

$$k^2 = m_B^2 = \frac{1}{2}\bar{g}_3^2(k)\rho_3 \tag{31}$$

$$\bar{\lambda}_3 = \frac{\partial^2 U_3}{\partial \rho_3^2} \tag{32}$$

The solution of the flow equation (30) with initial values at k_T specified by (28) and the subsequent solution of the second order differential equation (32) for $U_3(\rho_3)$ should give a good approximation for the temperature dependent effective potential in a range of ρ to be discussed below.

In eq. (30) the factor m_B/m_E accounts properly for the Debye screening of the A_0 mode. We take here $m_B/m_E = 0$ for simplification of the discussion. We also observe that our equation becomes invalid for vanishing or negative $\bar{\lambda}\rho_3 - \mu^2(T)$. This problem is related to the approach to convexity of the effective scalar potential and absent in a more complete treatment[25]. We simplify by taking $\mu^2(T) = 0$ in the mass ratios such that $m_B^2/m_1^2 = \frac{1}{6}\bar{g}_3^2/\bar{\lambda}_3, m_B^2/m_2^2 = \frac{1}{2}\bar{g}_3^2/\bar{\lambda}_3$ and

$$\frac{\partial}{\partial t}\bar{\lambda}_3 = \frac{3}{32\pi k}\left\{\bar{g}_3^4 + (\sqrt{6}+\sqrt{2})\bar{\lambda}_3^{\frac{3}{2}}\bar{g}_3\right\} + 2\eta\bar{\lambda}_3 \tag{33}$$

Here we have included the effects of wave function renormalization and switched to a renormalized coupling $\bar{\lambda}_3$. The dominant contribution to the anomalous dimension η comes from gauge boson fluctuations and reads[15]

$$\eta = -\frac{1}{4\pi}\bar{g}_3^2 k^{-1} \tag{34}$$

For the three-dimensional running of the gauge coupling we use eq. (21) with $b_3 = 0$ and a correction factor τ accounting for the difference between the $\alpha = 0$ gauge needed here and the $\alpha = 1$ gauge for which (21) was computed and for the contribution from scalar loops

$$\frac{\partial}{\partial t}\bar{g}_3^2 = -\frac{23}{24\pi}\tau\bar{g}_3^4 k^{-1} \tag{35}$$

We expect a value of τ near one. We can only use the system of equations (33) and (35) for k larger than the three-dimensional confinement scale. This gives a lower bound[15] on $\rho, \rho > \rho_{np}$: Let us assume that renormalization group-improved perturbation theory breaks down at a scale

$$k_{np} = 1.1\Lambda_{conf}^{(3)} = 1.1T\frac{23\tau\alpha_w}{6+\frac{23\tau}{2\pi}\alpha_w} \approx 0.14\tau T \tag{36}$$

At this scale $\bar{g}_3^2(k_{np})$ has increased by a factor of about ten as compared to $\bar{g}_3^2(k_T)$

$$\frac{\bar{g}_3^2(k_{np})}{\bar{g}_3^2(k_T)} = \frac{11}{1+\frac{23\tau\alpha_W}{12\pi}} \approx 10.8 \tag{37}$$

and the dimensionless gauge coupling $g^2(k) = \bar{g}_3^2(k)/k$ has reached the nonperturbative region

$$\frac{g^2(k_{np})}{4\pi^2} = \frac{60}{23\tau\pi} = 0.83\tau^{-1} \tag{38}$$

From (31)(36) and (37) we infer

$$\rho_{np} = 0.26\frac{\alpha_w\tau^2}{1+\frac{23\tau\alpha_W}{12\pi}}T^2 \approx (0.09\tau T)^2 \tag{39}$$

For values of ρ below ρ_{np} we expect a complete breakdown even for renormalization group-improved perturbation theory due to condensates and similar strong interaction phenomena. We also observe that the nonperturbative improvement of the evolution equation (21) for $\bar{g}^2$ enhances ρ_{np} by a factor of about 4.8. Our best estimate for the onset of strong nonperturbative phenomena is therefore*

$$\rho_{np} = \left(\frac{T}{5}\right)^2 \tag{40}$$

For $\rho > \rho_{np}$ it is instructive to consider the evolution of the ratio $\bar{\lambda}_3/\bar{g}_3^2$ and use $\bar{g}_3^2(k)$ instead of k as a variable

$$\frac{\partial}{\partial \ln \bar{g}_3^2}\left(\frac{\bar{\lambda}_3}{\bar{g}_3^2}\right) = -\frac{9}{92\tau}\left\{1+\frac{92\tau-48}{9}\frac{\bar{\lambda}_3}{\bar{g}_3^2}+(\sqrt{6}+\sqrt{2})\left(\frac{\bar{\lambda}_3}{\bar{g}_3^2}\right)^{3/2}\right\} \tag{41}$$

This corresponds to the β-function for λ/g^2 shown in fig. 4. The r.h.s. is negative and $\bar{\lambda}_3$ is always driven from positive towards negative values as $\bar{g}_3^2$ increases. The initial ratio $\bar{\lambda}_3/\bar{g}_3^2$ at k_T corresponds to the zero temperature ratio between Higgs scalar mass and W-boson mass up to small (four-dimensional) logarithmic corrections**

$$\frac{\bar{\lambda}_3}{\bar{g}_3^2}(k_T) \approx \left(\frac{m_H}{2m_W}\right)^2 \tag{42}$$

---

* This does not mean that high temperature perturbation theory can be trusted for $\rho > \rho_{np}$. Only the renormalization group-improved potential discussed here has a chance to remain valid at such low values of ρ!

** This statement is not valid for very small scalar masses when Coleman-Weinberg symmetry breaking operates.

For not too large values of m_H/m_W we may, for the purpose of a simplified analytical discussion, neglect the last term in eq. (41) and use the approximative solution

$$\begin{aligned}\bar{\lambda}_3(k) &= \left(\frac{\bar{g}^2(k)}{\bar{g}^2(k_T)}\right)^{\frac{12}{23\tau}} \bar{\lambda}_3(k_T) \\ &- \frac{9}{92\tau - 48}\bar{g}^2(k)\left(1 - \left(\frac{\bar{g}^2(k_T)}{\bar{g}^2(k)}\right)^{1-\frac{12}{23\tau}}\right)\end{aligned} \tag{43}$$

The scale k_{cw} (where $\bar{\lambda}_3(k_{cw}) = 0$) can be taken as a characteristic scale for the phase transition. It obeys

$$\frac{\bar{g}^2(k_{cw})}{\bar{g}^2(k_T)} = \left(1 + \frac{92\tau - 48}{9}\frac{\bar{\lambda}_3(k_T)}{\bar{g}_3^2(k_T)}\right)^{\frac{23\tau}{23\tau-12}} \tag{44}$$

Equating k_{cw} with k_{np} (36) we find a critical value for the ratio $\bar{\lambda}_3(k_T)/\bar{g}_3^2(k_T)$ (for $\tau = \frac{24}{23}$)

$$\left(\frac{\bar{\lambda}_3(k_T)}{\bar{g}_3^2(k_T)}\right)_{cr} = 0.43 \tag{45}$$

We conclude that for a Higgs boson mass exceeding the critical value

$$(m_H)_{cr} \approx 1.3 m_W \approx 105 \text{ GeV} \tag{46}$$

the electroweak phase transition is described by a strongly interacting $SU(2)$ gauge theory. Not only the symmetric phase but all phenomena related to the phase transition are dominated by nonperturbative effects! We emphasize that the critical value (46) should not be interpreted as an accurate bound. Even for Higgs masses smaller than 100 GeV the strong nonperturbative effects are very important and may dominate, for example, the whole region of the effective potential between the origin and the local maximum*. Our main conclusion is that for a Higgs boson mass of the order m_W or even somewhat below it is impossible to give a quantitative description of the phase transition without taking the strong nonperturbative effects such as condensates into account.

One last remark concerns the "nonperturbative region" in the effective potential for $\rho < \rho_{np}$ (39). (For m_H taking the critical value this concerns the region inside the turning point $\partial^2 U/\partial\rho^2 = 0$.) Even though the relatively simple renormalization group-improved treatment proposed in this section is not valid here, we can estimate the difference $\Delta U_3 = U_3(\rho_{np}) - U_3(0)$ by a simple scale argument: It has to be proportional to the third power of k_{np}

$$\Delta U_3 = K k_{np}^3 \tag{47}$$

Since there is no small dimensionless quantity in the problem and k_{np} is determined relatively accurately, the constant K should be near one. Restoring the four-dimensional language we therefore estimate the nonperturbative contribution to be roughly

$$\Delta U \approx 3 \cdot 10^{-3} T^4 \tag{48}$$

---

* If we use the improved evolution equation (21) instead of (35), the critical value is lowered further.

We can also offer a speculative picture how the transition could be described as an analytical crossover for very large m_H: As the temperature raises a condensate $\langle F_{ij}F^{ij}\rangle$ (or some other condensate) may start forming at some temperature $\tilde{T}$ for which the absolute minimum of the effective scalar potential still occurs at $\rho_0(\tilde{T}) \neq 0$. For a further increase of the temperature beyond $\tilde{T}$ the magnetic condensate $\langle F_{ij}F^{ij}\rangle$ will increase whereas $\rho_0(T)$ decreases. In the two-dimensional plane spanned by the condensate and ρ the arrow $(\langle F_{ij}F^{ij}\rangle, \rho_0)$ may turn continuously from the ρ-direction for $T = \tilde{T}$ to the condensate direction for very large temperatures. No jump in the particle masses or other quantities would be expected for such a crossover. This picture gives a hint that a "strongly interacting electroweak transition" may need more degrees of freedom than the Higgs scalar for a meaningful description of the vacuum structure!

We conclude that for realistic values of the Higgs boson mass nonperturbative techniques are necessary for a reliable description of the electroweak phase transition. The "strongly interacting electroweak phase transition" is in several aspects close to the high temperature phase transition in QCD. Similar methods for a description of both phenomena have to be developed, in particular for an understanding of the temperature dependence of various condensates. This constitutes an interesting theoretical laboratory, with possible applications ranging from early cosmology to high energy heavy ion collisions.

Acknowledgement

I am grateful to B. Bergerhoff, F. Freire, D. Litim, S. Lola, M. Reuter, and N. Tetradis for collaboration on high temperature gauge theories and to W. Buchmüller and M. Shaposhnikov for fruitful discussions. I thank the organizers of the Sintra workshop for a wonderful meeting.

REFERENCES

1. D. A. Kirzhnitz, JETP Lett. **15**, 529 (1972);
   D. A. Kirzhnits and A. D. Linde, Phys. Lett. **B72**, 471 (1972).
2. V. A. Kuzmin, V. A. Rubakov, and M. E. Shaposhnikov, Phys. Lett. **B155**, 36 (1985);
   M. E. Shaposhnikov, Nucl. Phys. **B287**, 757 (1987); ibid. **299**, 797 (1988).
3. A. D. Sakharov, JETP Lett. **5**, 24 (1967).
4. A. G. Cohen, D. B. Kaplan, and A. E. Nelson, Phys. Lett. **B245**, 561 (1990); Nucl. Phys. **B349** 727 (1991);
   N. Turok and J. Zadrozny, Phys. Rev. Lett. **65**, 2331 (1990); Nucl. Phys. **B358**, 471 (1991);
   M. Dine, P. Huet, and R. Singleton, Nucl. Phys. **B375**, 625 (1992);
   M. Dine, R. G. Leigh, P. Huet, A. D. Linde, and D. A. Linde, Phys. Rev. **D46**, 550 (1992);
   G. R. Farrar and M. E. Shaposhnikov, preprint CERN-TH-6734-93;
   M. B. Gavela, M. Lozano, J. Orloff, and O. Pène, preprints CERN-TH 7262/94, 7263/94.

5. A. I. Bochkarev and M. E. Shaposhnikov, Mod. Phys. Lett. **A2**, 417 (1987);
A. I. Bochkarev, S. V. Kuzmin, and M. E. Shaposhnikov, Phys. Lett. **B244**, 257 (1990).
6. D. A. Kirzhnitz and A. D. Linde, Phys. Lett. **72B**, 471 (1972); JETP **40**, 628 (1974); Ann. Phys. **101**, 195 (1976);
S. Weinberg, Phys. Rev. **D9**, 3357 (1974);
A. D. Linde, Nucl. Phys. **B216**, 421 (1983), Rep. Prog. Phys. **47**, 925 (1984).
7. L. Dolan and R. Jackiw, Phys. Rev. **D9**, 3320 (1974);
G. W. Anderson and L. J. Hall, Phys. Rev. **D45**, 2685 (1992);
M. Carrington, Phys. Rev. **D45**, 2933 (1992);
W. Buchmüller, Z. Fodor, T. Helbig, and D. Walliser, preprint DESY 93-121, to appear in Ann. Phys.;
D. Bödeker, W. Buchmüller, Z. Fodor, and T. Helbig, preprint DESY 93-147, to appear in Nucl. Phys.;
J. R. Espinosa, M. Quiros, and F. Zwirner, Phys. Lett. **B314**, 206 (1993).
8. N. Tetradis and C. Wetterich, Nucl. Phys. **B398**, 659 (1993); preprint HD-THEP-93-36.
9. T. Appelquist and R. Pisarski, Phys. Rev. **D23**, 2305 (1981);
S. Nadkarni, Phys. Rev. **D27**, 917 (1983);
N. P. Landsman, Nucl. Phys. **B322**, 498 (1989).
10. J. Kapusta, Finite temperature field theory, Cambridge University Press (1989).
11. T. Kaluza, Sitzungsber. Preuss. Akad. Wiss. Berlin, Math. Phys. **K1**, 966 (1921);
O. Klein, Z. f. Phys. **37**, 895 (1926).
12. D. O'Connor and C. R. Stephens, Nucl. Phys. **B360**, 297 (1991), preprint DIAS-STP-93-19;
D. O'Connor, C. R. Stephens, and F. Freire, Mod. Phys. Lett. **A8**, 1779 (1993), preprint THU 92/37.
13. A. Jakovác, K. Kajantie, and A. Patkós, Helsinki Preprint HU-TFT-94-01, hepph-9312355;
K. Farakos, K. Kajantie, K. Rummukainen, and M. Shaposhnikov, preprint CERN-TH6973/94.
14. P. Lacock, D. E. Miller, and T. Reisz, Nucl. Phys. **B369**, 501 (1992);
L. Kärkkäinen, P. Lacock, B. Petersson, and T. Reisz, Nucl. Phys. **B395**, 733 (1993).
15. M. Reuter, C. Wetterich, Nucl. Phys. **B391**, 147 (1993), and Nucl. Phys. **B408**, 91 (1993); Heidelberg preprints HD-THEP-93-40/41, to appear in Nucl. Phys. B.
16. S. Coleman and E. Weinberg, Phys. Rev. **D7**, 1888 (1973).
17. A. D. Linde, Phys. Lett. **96B**, 289 (1980);
D. Gross, R. Pisarski, and L. Yaffe, Rev. Mod. Phys. **53**, 43 (1981).
18. C. Wetterich, Nucl. Phys. **B352**, 529 (1991).
19. L. P. Kadanoff, Physica **2**, 263 (1966);
K. G. Wilson, Phys. Rev. **B4**, 3184 (1971);
K. G. Wilson, I. G. Kogut, Phys. Rep. **12**, 75 (1974);
F. Wegner in: *Phase transitions and critical phenomena*, vol. 6, eds. C. Domb and M. S. Green (Academic Press, New York 1976);
G. Mack. T. Kalkreuter, G. Palma, M. Speh, Int. J. Mod. Phys. C, Vol. **3**, No. 1, 121-47 (1992).

20. C. Wetterich, Phys. Lett. **B301**, 90 (1993);
C. Wetterich, Z. Phys. **C60**, 461 (1993).
21. F. Wegner, A. Houghton, Phys. Rev. **A8**, 401 (1973);
K. G. Wilson, I. G. Kogut, Phys. Rep. **12**, 75 (1974);
S. Weinberg, Critical Phenomena for Field Theorists, Erice Subnucl. Phys. 1 (1976);
J. Polchinski, Nucl. Phys. **B231**, 269 (1984);
A. Hasenfratz, P. Hasenfratz, Nucl. Phys. **B270**, 685 (1986).
22. F. J. Dyson, Phys. Rev. **75**, 1736 (1949);
J. Schwinger, Proc. Nat. Acad. Sc. **37**, 452, 455 (1951).
23. P. Ginsparg, Nucl. Phys. **B170** [FS1], 388 (1980);
P. Arnold and L. Yaffe, Preprint UW/PT-93-24 (1993);
W. Buchmüller and Z. Fodor, preprint DESY 94-045.
24. M. Shaposhnikov, Phys. Lett. **B316**, 112 (1993).
25. A. Ringwald and C. Wetterich, Nucl. Phys. **B334**, 506 (1990);
N. Tetradis and C. Wetterich, Nucl. Phys. **B383**, 197 (1992).

HIGH TEMPERATURE PHASE TRANSITIONS WITH THE EFFECTIVE AVERAGE ACTION

N. Tetradis

Theoretical Physics
1 Keble Rd.
Oxford OX1 3NP

INTRODUCTION

According to the original argument of Kirzhnits and Linde [1] the electroweak symmetry is restored at temperatures much larger than the Fermi scale (= 174 GeV). As the universe cooled after the big bang there must have been a phase transition from the symmetric to the spontaneously broken phase of the standard model. This phase transition may have many important consequences for our present universe, one example being the possible creation of the baryon asymmetry [2]. A reliable description of the electroweak phase transition cannot be obtained in the framework of the perturbative evaluation of the effective action [3, 4]. The reason is the appearance of infrared divergences at high orders of the perturbative expansion: corrections involving scalar fields are proportional to powers of $\lambda T/m_s(T)$, which diverge near the critical temperature where $m_s(T_{cr}) = 0$ [4]. If Goldstone modes are present their masses vanish at all temperatures in the spontaneously broken phase. Similarly loops with gauge fields involve powers of $g^2T/m_A(T)$, which are problematic in the whole symmetric phase [5].

THE EFFECTIVE AVERAGE ACTION

In order to reliably compute the long distance behaviour we follow Wilson's philosophy [6] of iteratively integrating out the short distance modes and using an effective action to describe their physical effects. We follow a new approach [7] based on the intuitive picture of an effective action for averages of fields. The field average is the realization in the continumm of the block spin idea of Kadanoff and Wilson [6]. The average is taken over a volume of size k^{-d} and the corresponding effective average action Γ_k depends on the scale k. In a completely equivalent formulation [8], we start by adding an infrared cutoff term $\Delta S_k[\chi]$ in the original action $S[\chi]$. Through a Legendre transformation we obtain the generating functional for the 1PI Green functions $\tilde{\Gamma}_k[\phi]$, where ϕ is the expectation value of the field χ. The use of the infrared cutoff results in the effective integration of only the modes with $q^2 \geq k^2$. Finally, the effective average action $\Gamma_k[\phi]$ is obtained by removing the infrared cutoff. For k equal to the ultraviolet cutoff Λ, Γ_k becomes equal to the classical action S (no integration of modes), while for $k \to 0$ it tends towards the effective action Γ (all the modes are included). For $k > 0$ the effective average action describes a "coarse grained effective action" in the sense used in statistical mechanics. It is the appropriate functional for all processes with an effective physical infrared cutoff of the order of k.

The means for practical calculations is provided by an exact evolution equation $(t = \ln(k/\Lambda))$ [8]

$$\frac{\partial}{\partial t}\Gamma_k[\phi] = \frac{1}{2}\mathrm{Tr}\Big\{(\Gamma_k^{(2)}[\phi] + R_k)^{-1}\frac{\partial}{\partial t}R_k\Big\}. \tag{1}$$

Here $\Gamma_k^{(2)}$ is the second functional derivative of the effective average action with respect to ϕ. We consider an effective average action of the form

$$\Gamma_k = \int d^dx\Big\{U_k(\rho) + \frac{1}{2}\partial^\mu\phi_a Z_k(\rho)\partial_\mu\phi^a + \frac{1}{4}\partial^\mu\rho Y_k(\rho)\partial_\mu\rho\Big\}, \tag{2}$$

where $\rho = \frac{1}{2}\phi_a\phi^a$. We obtain an evolution equation for U_k by expanding the r.h.s. of eq. (2) around a constant field configuration. The evolution of the wave function renormalizations Z_k, Y_k (which determines the anomalous dimensions) requires an expansion around a background with a small momentum dependence [7, 8, 11]. If we neglect the effects of wave function renormalization, the average effective potential U_k satisfies the evolution equation $(t = \ln(k/\Lambda),\ x = q^2)$

$$\frac{\partial}{\partial t}U_k(\rho) = \frac{1}{2}(2\pi)^{-d}\int d^dq\frac{\partial P}{\partial t}\left(\frac{1}{P + U_k'(\rho) + 2U_k''(\rho)\rho} + \frac{N-1}{P + U_k'(\rho)}\right). \tag{3}$$

Primes denote derivatives with respect to $\rho = \frac{1}{2}\phi_a\phi^a$. For $q^2 \gg k^2$ the inverse "average" propagator $P(q^2)$ approaches the standard inverse propagator q^2 exponentially fast, whereas for $q^2 \ll k^2$ the infrared cutoff prevents the propagation. It is easy to recognize the second term in eq. (3) as the contribution from the $N-1$ Goldstone bosons (U' vanishes at the minimum). The first term is related to the radial mode.

RUNNING IN FOUR AND THREE DIMENSIONS

In order to solve eq. (3) we parametrize U_k in terms of the location of its minimum and its successive derivatives with respect to ρ at the minimum. This results in an infinite system of coupled differential equations, which we solve approximately by truncating at a finite number of equations and keeping an equal number of ρ-derivatives. As a first step we truncate after the second derivative of the potential.
In the *spontaneously broken regime* U_k has its minimum at $\rho_0(k) \neq 0$, where $U_k'(\rho_0) = 0$. We define $\bar\lambda(k) = U_k''(\rho_0(k))$, and obtain [7, 9, 11]

$$\frac{d\rho_0}{dt} = -\ v_d k^{d-2}\Big\{3L_1^d(2\bar\lambda\rho_0) + (N-1)L_1^d(0)\Big\} \tag{4}$$

$$\frac{d\bar\lambda}{dt} = -\ v_d k^{d-4}\bar\lambda^2\Big\{9L_2^d(2\bar\lambda\rho_0) + (N-1)L_2^d(0)\Big\}, \tag{5}$$

with $v_d^{-1} = 2^{d+1}\pi^{\frac{d}{2}}\Gamma\left(\frac{d}{2}\right)$. In the *symmetric regime* $(\rho_0(k) = 0)$ we define $m^2 = U_k'(0)$, $\bar\lambda(k) = U_k''(0)$ and find

$$\frac{dm^2}{dt} = (N+2)v_d k^{d-2}\bar\lambda L_1^d(m^2) \tag{6}$$

$$\frac{d\bar\lambda}{dt} = -\ (N+8)v_d k^{d-4}\bar\lambda^2 L_2^d(m^2). \tag{7}$$

The functions $L_n^d(w)$ have been discussed extensively in refs. [7, 11, 12]. They fall off for large values of w/k^2 following a power law. As a result they introduce threshold behaviour for the contributions of massive modes to the evolution equations. When the "running" squared mass $2\bar{\lambda}\rho_0$ or m^2 becomes much larger than the scale k^2 the massive mode decouples. The systems of eqs. (4),(5) and (6),(7) can be solved for given "short distance values" $\rho_0(k = \Lambda)$ and $\bar{\lambda}(k = \Lambda)$. We start at the cutoff Λ and follow the renormalization group flow towards the infrared ($k \to 0$). In this way we obtain the ground state of the theory $\rho_0 = \rho_0(k = 0)$, as well as the renormalized couplings, mass terms etc. Even when one starts in the broken regime at $k = \Lambda$, it is possible that the evolution, as given by eqs. (4),(5), may drive $\rho_0(k)$ to zero at some non-zero k_s. From that point on the theory is in the symmetric regime and we continue the evolution using eqs. (6),(7) with boundary conditions $m^2(k_s) = 0$ and $\bar{\lambda}(k_s)$ given by its value at the end of the running in the broken regime.

At non-zero temperature the zero component of the momentum q_0 has a discrete spectrum: $q_0 \to 2\pi mT$, and the integration over q_0 in eq. (3) is replaced by a summation over m. As before, we can parametrize the temperature dependent potential $U_k(\rho, T)$ in terms of its minimum and derivatives. The evolution equations are given by (4),(5) and (6),(7) with the obvious generalizations $\rho_0(k) \to \rho_0(k, T)$ etc. For $T \ll k$ (low temperature region) the temperature dependent functions $L_n^d(w, T)$ are equal to the zero temperature $L_n^d(w)$. In the opposite limit $T \gg k$ (high temperature region) they are given by the simple expression

$$L_n^d(w, T) = \frac{v_{d-1}}{v_d}\frac{T}{k}L_n^{d-1}(w) \qquad \text{for } T \gg k, \tag{8}$$

with v_d defined as in eqs. (4), (5). The two regions of T/k are connected by a small interval in which the $L_n^d(w, T)$ have a complicated dependence on w and T (intermediate region).

From this point on we concentrate on the four-dimensional theory. For zero temperature eq. (4) results in an evolution for $\rho_0(k)$ proportional to k^2. This is the reflection in our language of the quadratic divergences of standard perturbation theory [7]. In the limit $k \to 0$, $\rho_0(k)$ settles down at the renormalized value for the minimum of the effective potential: $\rho_0 = \rho_0(0)$. Eq. (5) results in the logarithmic running for the quartic coupling in four dimensions. The functions L_n^4 introduce threshold effects in the evolution. As a result the logarithmic running of the coupling stops as $k \to 0$ for $N = 1$, due to the decoupling of the massive radial mode. In the case $N \neq 1$, due to the presence of massless Goldstone modes, the logarithmic running never stops and this results in an infrared free theory. For this reason we define the renormalized coupling at some non-zero k according to: $\lambda_R = \bar{\lambda}\left(\sqrt{2\lambda_R\rho_0}\right)$ for all N.

For non-zero temperature we have to distinguish between the three different regions as k decreases: a) In the low temperature region ($k \gg T$) the evolution is purely four-dimensional. b) In the intermediate region the running is complicated and we can study it only numerically. c) In the high temperature region ($k \ll T$) we can use eq. (8) in order to rewrite the evolution equations in the spontaneously broken regime as

$$\frac{d\rho_0'}{dt} = -\, v_3 k\Big\{3L_1^3(2\bar{\lambda}'\rho_0') + (N-1)L_1^3(0)\Big\} \tag{9}$$

$$\frac{d\bar{\lambda}'}{dt} = -\, v_3 k^{-1}(\bar{\lambda}')^2\Big\{9L_2^3(2\bar{\lambda}'\rho_0') + (N-1)L_2^3(0)\Big\}, \tag{10}$$

with $\rho_0'(k, T) = \rho_0(k, T)/T$ and $\bar{\lambda}'(k, T) = \bar{\lambda}(k, T)T$. Comparison with eqs. (4), (5) shows that the above equations are exactly the ones of the three-dimensional theory for the effective three-dimensional couplings $\rho_0'(k, T)$, $\bar{\lambda}'(k, T)$. In terms of the quantities $\kappa(k, T) = \rho_0'(k, T)/k = \rho_0(k, T)/(kT)$, $\lambda(k, T) = \bar{\lambda}'(k, T)/k = \bar{\lambda}(k, T)T/k$ the evolution equations have a fixed point $(\kappa_{fp}, \lambda_{fp})$ corresponding to the phase transition [7, 9]. There is a critical line separating the spontaneously broken from the symmetric phase. If, as k is lowered, the system enters the high temperature region at some point very close to the critical line the system spends very long "time" $|t| = -\ln(k/\Lambda)$ near the critical point, before deviating either towards the spontaneously broken or the symmetric phase. During this "time" it loses memory of the initial conditions. As a result the dynamics near the phase transition is independent of the "short distance" values of the couplings and has purely three-dimensional character. Solving numerically the evolution equations we determine the non-zero temperature renormalized vacuum $\rho_0(T) = \rho_0(0,T)$ and the renormalized coupling (defined again at non-zero k) $\lambda_R(T) = \bar{\lambda}\left(\sqrt{2\lambda_R(T)\rho_0(T)}, T\right)$. In the symmetric phase the theory is not infrared free and we define: $m_R^2(T) = m^2(0,T)$, $\lambda_R(T) = \bar{\lambda}(0,T)$.

THE PHASE TRANSITION FOR THE $O(N)$-SYMMETRIC ϕ^4 THEORY

For any given "short distance" theory, through the solution of the evolution equations, we can determine the renormalized zero temperature theory in terms of ρ_0 and λ_R. We can also obtain $\rho_0(T)$ and $\lambda_R(T)$ for non-zero temperatures $T < T_{cr}$. For $T \geq T_{cr}$ the symmetry is restored ($\rho_0(T) = 0$) and the non-zero temperature theory is described in terms of $m_R^2(T)$ and $\lambda_R(T)$. We find a *second order phase transition*. The renormalized quartic coupling $\lambda_R(T)$ remains close to its zero temperature value λ_R for a large range of temperatures and drops quickly to zero at $T = T_{cr}$. For $T \gg T_{cr}$ the coupling $\lambda_R(T)$ quickly grows to approximately its zero temperature value λ_R, while $m_R^2(T)$ asymptotically becomes proportional to T^2 as $T \to \infty$. In the symmetric phase the quartic coupling is positive for all temperatures and vanishes only for $T \to T_{cr}$. The critical temperature can be accurately determined and agrees very well with the perturbative expression $T_{cr}^2 = 24\rho_0/(N+2)$, for small λ_R. This is not surprising since perturbation theory breaks down in region of size $|T - T_{cr}|/T_{cr} = \mathcal{O}(\lambda_R)$. For small λ_R this region shrinks to zero and the value of T_{cr} can be reliably determined. Excellent agreement with perturbation theory is also found for $m_R^2(T)$ in the limit $T \to \infty$ (for small λ_R).

The temperature dependence of $\rho_0(T), m_R^2(T)$ for $T \to T_{cr}$ can be characterized by critical exponents: $\rho_0(T) \propto (T_{cr}^2 - T^2)^{2\beta}$, $m_R^2(T) \propto (T^2 - T_{cr}^2)^{2\nu}$. In the limit $T \to T_{cr}$ the critical exponents approach asymptotic values. These are independent of λ_R and therefore fall into universality classes determined only by N. They are equal to the critical exponents of the zero temperature three-dimensional theory. The numerical values for the critical exponents [9] are in rough agreement with precision results in the context of three-dimensional field theory [13]. Improved results are obtained when truncations of the exact evolution equation which preserve higher derivatives of U_k and wave function renormalization effects are used [11]. These are in agreement with the known values at the $1-5\%$ level for exponents $\beta, \nu, \gamma, \delta$, and less accurately for η. The scaling laws are satisfied at the 0.1% level. The critical behaviour of $\lambda_R(T)$ is related to the resolution of the problem of infrared divergences. We find that the quantity $\lambda_R(T)T/m_R(T)$ reaches an asymptotic value depending on N, in the limit $T \to T_{cr}$, which is in good agreement with results of three-dimensional field theory [13]. We conclude that the infrared problem, arising from the divergence of perturbative contributions involving powers of $\bar{\lambda}T/m_R(T)$, disappears when the bare coupling $\bar{\lambda}$ is replaced by the renormalized one λ_R.

All the above results are in exact agreement with an independent study of the high temperature phase transition in the large N limit with other methods (saddle point evaluation of the functional integral for the effective potential, Schwinger-Dyson equations) [10].

OTHER APPLICATIONS AND WORK IN PROGRESS

Work in progress focuses on the improvement of the accuracy for the critical exponents and the study of the equation of state for the critical three-dimensional theory. The zero and high temperature phase structure of a two scalar model with a $Z_2 \times Z_2 \times Z_2$ symmetry is also under investigation. The model exhibits three fixed points, cross-over phenomena and first and second order phase transitions. Another question which has been answered in the context of the effective average action concerns the convexity of the effective potential [12]. From its construction (integration of flucuations with characteristic momenta larger than a given infrared cutoff k), the effective average action is the appropriate quantity for the study of physics at a scale k. In practice one has to identify k with the appropriate physical infrared cutoff scale of the problem (such as the Hubble parameter for inflationary cosmology). For the study of critical bubbles during a first order phase transition the scale k is determined by a quantity such as the correlation length in any of the two minima, or the bubble radius or thickness (a detailed investigation is necessary for specific models). The concept of averaging has been developed for gauge fields and an exact evolution equation has been formulated for gauge theories [14] (see talks by C. Wetterich and M. Reuter). The study of this evolution equation along the lines presented in this talk will lead to a reliable description of the electroweak phase transition.

References

[1] D.A. Kirzhnits and A.D. Linde, Phys. Lett. B **42**, 471 (1972).

[2] V.A. Kuzmin, V.A. Rubakov, and M.E. Shaposhnikov, Phys. Lett. B **155**, 36 (1985); M.E. Shaposhnikov, Nucl. Phys. B **287**, 757 (1987); *ibid* **299**, 797 (1988).

[3] S. Coleman and E. Weinberg, Phys. Rev. D **7**, 1888 (1973); R. Jackiw, Phys. Rev. D **9**, 1686 (1974).

[4] D.A. Kirzhnits and A.D. Linde, JETP **40**, 628 (1974); Ann. Phys. **101**, 195 (1976); L. Dolan and R. Jackiw, Phys. Rev. D **9**, 3320 (1974); S. Weinberg, Phys. Rev. D **9**, 3357 (1974).

[5] A.D. Linde, Phys. Lett. B **96**, 293 (1980).

[6] K.G. Wilson, Phys. Rev. B **4**, 3174;3184 (1971); K.G. Wilson and I.G. Kogut, Phys. Rep. **12**, 75 (1974).

[7] C. Wetterich, Nucl. Phys. B **352**, 529 (1991); Z. Phys. C **57**, 451 (1993).

[8] C. Wetterich, Phys. Lett. B **301**, 90 (1993).

[9] N. Tetradis and C. Wetterich, Nucl. Phys. B. **398**, 659 (1993).

[10] M. Reuter, N. Tetradis and C. Wetterich, Nucl. Phys. B **401**, 567 (1993).

[11] N. Tetradis and C. Wetterich, preprint HD-THEP-93-28, in press Nucl. Phys. B.

[12] N. Tetradis and C. Wetterich, Nucl. Phys. B **383**, 197 (1992).

[13] J. Zinn-Justin, Quantum field theory and critical phenomena, Oxford Science Publications (1989); G. Parisi, Statistical field theory, Addison-Wesley (1988).

[14] M. Reuter and C. Wetterich, Nucl. Phys. B **391**, 147 (1993); *ibid* **408**, 91 (1993), preprints HD-THEP-93-40 and 41.

EXACT RENORMALIZATION GROUP EQUATIONS FOR GAUGE THEORIES

M. Reuter

Deutsches Elektronen-Synchrotron DESY
Notkestrasse 85, D-22603 Hamburg, Germany

ABSTRACT

The construction of a new nonperturbative evolution equation for nonabelian gauge theories is described. The solution of this equation, the scale-dependent "effective average action", interpolates between the classical action and the conventional effective action. It is gauge invariant at all intermediate scales.

1 INTRODUCTION

It is well known that for nonabelian gauge theories conventional perturbation theory suffers from severe infrared divergences. These problems are particularly dramatic in low dimensional quantum field theories, for instance in the three dimensional theories which obtain as the high temperature limit of four dimensional models. However, any reliable description of the electroweak phase transition in the standard model, say, has to take account of the intrinsically nonperturbative phenomena related to the long-wavelength modes of the theory. Here we are going to describe a method which can be applied to problems of this type even in situations where perturbation theory breaks down. It involves the scale-dependent effective average action Γ_k.The mass scale k has the interpretation of a variable infrared cutoff [1]. The functional Γ_k interpolates between the classical action S for large values of k, and the conventional effective action for k approaching zero: $\Gamma_{k\to\infty} = S, \Gamma_{k\to 0} = \Gamma$. The effective average action Γ_k describes the physics at the scale k in the sense that it has the modes with momenta $q^2 > k^2$ integrated out already whereas those with momenta $q^2 < k^2$ have been left untouched. In many important cases where perturbation theory is inapplicable the limit $k \to 0$ exists and can be computed. For a simple theory involving only a scalar field φ the scale dependence of Γ_k is governed by the renormalization group equation

$$\frac{\partial}{\partial t}\Gamma_k[\varphi] = \frac{1}{2}\mathrm{Tr}\left[\frac{\partial}{\partial t}R_k \, \left(\Gamma_k^{(2)}[\varphi] + R_k\right)^{-1}\right] \tag{1}$$

Here $t \equiv \ln k$, and $\Gamma_k^{(2)}$ denotes the matrix of the second functional derivatives of Γ_k. The operator $R_k \equiv R_k(-\partial^2)$ or, in momentum space, $R_k \equiv R_k(q^2)$ describes the details of how the small momentum modes are cut off and it is to some extent arbitrary. It has to vanish for $q^2 \gg k^2$ and to become a mass-like term proportional to k^2 for small momenta $q^2 \ll k^2$.

If one tries to generalize (1) for gauge theories one immediately faces a difficulty if one requires $\Gamma_k = \Gamma_k[A_\mu]$ to be a gauge invariant functional of its argument. Because of the gauge degrees of freedom, $\Gamma_k^{(2)}[A]$ has zero modes and, as a consequence, the trace on the RHS of eq.(1) does not exist. One way of solving this problem is to give up the gauge invariance of Γ_k and to evolve a completely gauge fixed effective action. The disadvantage of this method is that it is very hard to solve the resulting flow equation beyond perturbation theory: as the restrictions due to gauge invariance do not apply anymore, any truncation of the flow equation has to retain many gauge-variant terms. For this reason we shall now describe a different method [2, 3] which employs background field techniques in order to maintain gauge invariance at any scale k. In this approach very simple truncations of the flow equation are sufficient in order to analyze its nonperturbative contents.

2 THE GAUGE INVARIANT EVOLUTION EQUATION

We consider the d-dimensional theory of a scalar multiplet χ^a and a nonabelian gauge field $\mathcal{A}$ defined by the classical action $S[\chi, \mathcal{A}]$. After having decomposed $\mathcal{A}(x)$ as a background field $\bar{A}(x)$ plus a fluctuation $a(x)$, $\mathcal{A} = \bar{A} + a$, our starting point is the following scale-dependent modification of the familiar generating functional in the background gauge [4, 5]:

$$\begin{aligned}\exp W_k[J, K; \bar{A}] \;&=\; E_k[\bar{A}] \int D\chi Da \exp -\Big\{\hat{S}[\chi, a, \bar{A}] + \Delta_k S_{\mathrm{S}}[\chi; \bar{A}] + \Delta_k S_{\mathrm{G}}[a; \bar{A}] \\ &\quad - \int d^dx \Big(J_a(x)\chi^a(x) + K_z^\mu(x) a_\mu^z(x)\Big)\Big\}\end{aligned} \tag{2}$$

Here

$$\hat{S}[\chi, a, \bar{A}] = S[\chi, \bar{A} + a] + \frac{1}{2\alpha}\int d^dx \left(D_\mu[\bar{A}]a^\mu\right)^2 - \ln\det\Big(-D_\mu[\bar{A}]D^\mu[\bar{A} + a]\Big) \tag{3}$$

contains the background gauge fixing term and the associated Faddeev-Popov determinant. The novel feature of (2) are the infrared cutoffs $\Delta_k S_{\mathrm{S}}$, $\Delta_k S_{\mathrm{G}}$ and E_k for the scalar, the gauge field and the Faddeev-Popov ghosts, respectively:

$$\begin{aligned}\Delta_k S_{\mathrm{S}} &= \frac{1}{2}\int d^dx \; \chi_a(x) \; R_k(\mathcal{D}_{\mathrm{S}}[\bar{A}])^a{}_b \; \chi^b(x) \\ \Delta_k S_{\mathrm{G}} &= \frac{1}{2}\int d^dx \; a_y^\nu \; R_k(\mathcal{D}_{\mathrm{T}}[\bar{A}])_{\nu z}^{y\mu} \; a_\mu^z \\ &\quad + \frac{1}{2}(\frac{1}{\alpha_k} - 1)\int d^dx \; (D^\nu[\bar{A}]a_\nu)_y (\mathcal{D}_{\mathrm{S}}^{-1}[\bar{A}] R_k(\mathcal{D}_{\mathrm{S}}[\bar{A}]))^y{}_z (D^\mu[\bar{A}]a_\mu)^z \\ E_k[\bar{A}] &= \det\Big[1 + \mathcal{D}_{\mathrm{S}}^{-1}[\bar{A}] R_k(\mathcal{D}_{\mathrm{S}}[\bar{A}])\Big]\end{aligned} \tag{4}$$

In order to maintain gauge invariance the cutoffs are defined in terms of the covariant operators

$$
\begin{aligned}
(\mathcal{D}_{\rm S}[\bar{A}])^{y}{}_{z} &= -(D_\mu[\bar{A}]D^\mu[\bar{A}])^{y}{}_{z} \\
(\mathcal{D}_{\rm T}[\bar{A}])^{y\mu}_{\nu z} &= (\mathcal{D}_{\rm S}[\bar{A}])^{y}{}_{z}\,\delta^\mu_\nu + 2i\bar{g}\,(T_w)^{y}{}_{z}\,F^{w\mu}_{\nu}[\bar{A}]
\end{aligned} \tag{5}
$$

where $\bar{g}$ denotes the bare gauge coupling and T_w are the generators of the gauge group. Acting on a_μ, the covariant derivatives are taken in the adjoint representation (w, y and z are adjoint group indices) and when acting on χ they are in the corresponding representation of the matter field. It is important to note that, being the difference of two gauge fields, the fluctuation a_μ transforms homogeneously under gauge transformations. Therefore the functionals (4) are invariant under gauge transformations of $\bar{A}_\mu$ combined with simultaneous homogeneous transformations of χ and a_μ. As a consequence, the functional $W_k[J, K; \bar{A}]$ is invariant under gauge transformations of $\bar{A}$ combined with homogeneous transformations of the sources J and K.

The function R_k appearing in (4) has the same form as the one in (1), but the argument $-\partial^2$ is replaced by the covariant operators (5) now. The use of $\mathcal{D}_{\rm T}$ in the cutoff term for the transverse gauge fluctuations is motivated by the fact that an expansion of the classical Yang-Mills action about $\bar{A}$ leads exactly to $\mathcal{D}_{\rm T}$ if one includes the gauge fixing term with $\alpha = 1$. Combining the pieces quadratic in a_μ coming from $\hat{S}$ and $\Delta_k S_{\rm G}$ one finds the inverse propagator $\bar{P}(\mathcal{D}_{\rm T}) = \mathcal{D}_{\rm T} + R_k(\mathcal{D}_{\rm T})$. This confirms that R_k acts as a mass-like infrared cutoff for the modes with small eigenvalues of $\mathcal{D}_{\rm T}$. If one combines E_k with the Faddeev-Popov determinant one finds a similarly modified propagator for the ghosts.

Next we perform the Legendre transformation of $W_k[J, K; \bar{A}]$ *at fixed* $\bar{A}$. This leads us to the functional $\tilde{\Gamma}_k[\varphi, \bar{a}; \bar{A}]$ where φ and $\bar{a}$ are the fields conjugate to the sources J and K, respectively. The effective average action Γ_k is defined in terms of $\tilde{\Gamma}_k$ in the following way:

$$
\Gamma_k[\varphi, \bar{a}; \bar{A}] = \tilde{\Gamma}_k[\varphi, \bar{a}; \bar{A}] - \Delta_k S_{\rm S}[\varphi; \bar{A}] - \Delta_k S_{\rm G}[\bar{a}; \bar{A}] \tag{6}
$$

It can be shown [2] that this functional, with the extra terms included, has all the desired features of a scale dependent effective action. Let us combine $\bar{a}$ with the background field $\bar{A}$ into the new field $A^z_\mu = \bar{A}^z_\mu + \bar{a}^z_\mu$. The gauge field A has the standard inhomogeneous transformation properties. Replacing the variable $\bar{a}$ by A the effective action Γ_k becomes now a functional of φ, A and $\bar{A}$: $\Gamma_k = \Gamma_k[\varphi, A, \bar{A}]$. By construction $\Gamma_k[\varphi, A, \bar{A}]$ is manifestly gauge-invariant under the simultaneous transformations of φ, A and $\bar{A}$. Finally one can work out the behaviour of Γ_k under a change of the scale k. One finds a remarkably simple exact renormalization group equation:

$$
\begin{aligned}
\frac{\partial}{\partial t}\Gamma_k[\varphi, A, \bar{A}] &= \frac{1}{2}{\rm Tr}\left[\left(\frac{\partial}{\partial t}R_k[\bar{A}]\right)\left(\Gamma_k^{(2)}[\varphi, A, \bar{A}] + R_k[\bar{A}]\right)^{-1}\right] \\
&\quad -{\rm Tr}\left[\left(\frac{\partial}{\partial t}R_k(\mathcal{D}_{\rm S}[\bar{A}])\right)\left(\mathcal{D}_{\rm S}[\bar{A}] + R_k(\mathcal{D}_{\rm S}[\bar{A}])\right)^{-1}\right]
\end{aligned} \tag{7}
$$

In this equation $\Gamma_k^{(2)}[\varphi, A, \bar{A}] = \Gamma_k^{(2)}[\varphi, \bar{a}; \bar{A}]_{|\bar{a}=A-\bar{A}}$ denotes the matrix of the second functional derivatives of $\Gamma_k[\varphi, A, \bar{A}]$ with respect to φ and A at fixed $\bar{A}$. The first trace on the RHS of (7) extends over the Hilbert space of the matter fields and gauge fluctuations. (For the operator R_k we used a compact matrix notation. Its precise definition can be read off from (4).) The second trace reflects the effect of the Faddeev-Popov ghosts.

Following the lines of the conventional background method [4] one might be tempted to equate the two gauge field arguments of Γ_k, and to work with the functional $\bar{\Gamma}_k[\varphi, A] \equiv$

$\Gamma_k[\varphi, A, A]$. However, it is important to note that the evolution equation (7) cannot be rewritten in terms of $\bar{\Gamma}_k[\varphi, A]$ alone, since $\Gamma_k^{(2)}$ does not involve derivatives with respect to $\bar{A}$. In fact, let us introduce the decomposition

$$\Gamma_k[\varphi, A, \bar{A}] = \bar{\Gamma}_k[\varphi, A] + \Gamma_k^{\text{gauge}}[\varphi, A, \bar{A}] \tag{8}$$

This leads to

$$\Gamma_k^{(2)}[\varphi, A, \bar{A}] = \bar{\Gamma}_k^{(2)}[\varphi, A] + \Gamma_k^{\text{gauge}(2)}[\varphi, A, \bar{A}]_{|\bar{A}} \tag{9}$$

where the second functional derivative $\Gamma_k^{\text{gauge}(2)}$ is performed at fixed $\bar{A}$. The interpretation of (8) and (9) is as follows. Because $\bar{\Gamma}_k[\varphi, A]$ is a gauge invariant functional, $\bar{\Gamma}_k^{(2)}[\varphi, A]$ has the usual zero modes. They are gauge fixed by the generalized gauge fixing term Γ_k^{gauge} whose Hessian $\Gamma_k^{\text{gauge}(2)}$ has no gauge zero modes.

In practice, any solution of (7) will require some kind of a truncation. One possibility is to write down an ansatz for $\Gamma_k[\varphi, A, \bar{A}]$ which contains two different gauge fields. In ref.[2] this strategy has been applied to pure Yang-Mills theory. If one wants to work with one gauge field only, one may use the following iterative procedure [3]. One starts from the decomposition $\Gamma_k^{\text{gauge}}[\varphi, A, \bar{A}] = S_{\text{gf}}[A, \bar{A}] + \hat{\Gamma}_k^{\text{gauge}}[\varphi, A, \bar{A}]$ where $S_{\text{gf}}[A, \bar{A}]$ is the classical background gauge fixing term from (3). Its presence guarantees that the (rather poor) approximation $\hat{\Gamma}_k^{\text{gauge}} \equiv 0$ would not lead to any infinities. We can do better than that, however. We define the final effective average action, depending on *one* gauge field only, as

$$\Gamma_k[\varphi, A] = \bar{\Gamma}_k[\varphi, A] + C_k[A] = \Gamma_k[\varphi, A, A] + C_k[A] \tag{10}$$

Here $C_k[A]$ is a gauge invariant functional to be chosen in such a way that its second functional derivative absorbs the dominant part of $\hat{\Gamma}_k^{\text{gauge}(2)}$. It obeys $C_0 = C_\infty = 0$. In [3] we have described a canonical choice for $C_k[A]$. If we assume for a moment that $Q_k \equiv \hat{\Gamma}_k^{\text{gauge}(2)} - C_k^{(2)}$ is known, it is easy to write down an evolution equation for $\Gamma_k[\varphi, A]$. It is similar to (7) with $\Gamma_k^{(2)}[\varphi, A, \bar{A}]$ replaced by $\Gamma_k^{(2)}[\varphi, A] + S_{\text{gf}}^{(2)} + Q_k[\varphi, A]$. In the context of the abelian Higgs model we have shown [3] that one obtains already rather accurate results if one solves the (truncated) renormalization group equation for $\Gamma_k[\varphi, A]$ with the approximation $Q_k = 0$. The power of this approach resides in the fact that the accuracy of these results can be systematically improved in an iterative way. From the solution for $\Gamma_k[\varphi, A]$ obtained with $Q_k = 0$ one can compute an improved $Q_k \neq 0$ which is then used on the RHS of the evolution equation in order to determine an improved $\Gamma_k[\varphi, A]$, $\cdots$. For further details of this method we have to refer to [3] where also the abelian Higgs model is worked out as an example and the implications for the electroweak phase transition are discussed.

REFERENCES

[1] C. Wetterich and N. Tetradis, these proceedings, and references therein

[2] M. Reuter and C. Wetterich, Preprint DESY 93-152, to appear in Nucl. Phys. B

[3] M. Reuter and C. Wetterich, Preprint DESY 94-017;
Nucl. Phys. **B391** (1993) 147; Nucl. Phys. **B408** (1993) 91

[4] L. F. Abbott, Nucl. Phys. **B185** (1981) 189

[5] W. Dittrich and M. Reuter, *Selected Topics in Gauge Theories*, Lecture Notes in Physics, Vol.244, Springer, Berlin, 1986

ENVIRONMENTALLY FRIENDLY RENORMALIZATION IN FINITE-TEMPERATURE FIELD THEORY

M.A. van Eijck[1], Denjoe O'Connor[2] and C.R.Stephens[2]
[1]Institute for Theoretical Physics, University of Amsterdam
Valckenierstraat 65, 1018 XE Amsterdam, The Netherlands
[2]DIAS, 10 Burlington Road, Dublin 4, Ireland

INTRODUCTION

Phase transitions in relativistic quantum field theory are a topic of great theoretical and practical interest. Some of the more interesting ones are the deconfinement phase transition in QCD and the electroweak phase transition, the latter having generated much attention recently due to its possible role in explaining the baryon asymmetry of the universe.

It is well known that a quantitative description of finite temperature phase transitions presents grave difficulties, not least because of the infrared divergences (IR) that have been known since the work of Dolan and Jackiw and Weinberg[1]. It is first important to understand the physical origin of such difficulties in order that they may be overcome. In past work[2,3] we have emphasized that the origin of such difficulties lies in the fact that the effective degrees of freedom (EDOF) of a system can change qualitatively as a function of temperature. For instance, above the QCD deconfinement phase transition quarks are free whereas below it they exist only as bound states in mesons and hadrons. As a function of temperature then there is a drastic change in the nature of the EDOF from quarks to baryons. The IR problems intrinsic to a description of the phase transition in $\lambda\phi^4$ theory have an analogous origin. A perturbation expansion in terms of the zero temperature coupling in this case breaks down near the critical temperature because the EDOF there are three dimensional rather than four dimensional. A popular approach to this problem has been to *assume* dimensional reduction and proceed to describe the critical region as an effective three dimensional theory using ε expansion methods for instance[5,6]. However, whether a

Electroweak Physics and the Early Universe, Edited by
Romao and Freire, Plenum Press, New York, 1994

particular field theory will dimensionally reduce or not is something one should derive not assume. In other words the ideal would be to develop a quantitatively reliable method based solely on information about the zero temperature theory.

We have shown previously[2,3,4] that environmentally friendly renormalization provides a very powerful tool for the qualitative and quantitative description of finite temperature phase transitions. The general gist of the approach is the following: the EDOF (fluctuations) in a system are very often sensitive to the "environment" in which they live. Temperature is just one example of an environmental variable which drastically affects the EDOF. Given that one is interested in how the physical behaviour of systems change as a function of "scale" (momentum, mass, temperature, magnetic field etc.), and that the renormalization group (RG) is an extremely powerful tool for investigating physics as a function of scale, then one should implement a RG which is environmentally friendly in the sense that viewed as a coarse graining procedure it coarse grains as closely as possible the true environment dependent EDOF. This means that the RG should depend on any relevant environmental parameters in the problem. In particular in the context of finite temperature field theory it should be explicitly temperature dependent.

Although one can fruitfully ground one's intuition in a coarse graining point of view it is important to point out that the environmentally friendly RG's we employ are not in practice of the Wilsonian type such as the "average effective action"[7]. Rather, they exploit the reparametrization invariance of field theories, in that one may parameterize the results of one's experiments in terms of different sets of renormalized parameters. Although non-perturbatively speaking physical results cannot depend on the particular renormalization scheme chosen, when one resorts to an approximation scheme such as perturbation theory one finds that the reliability of the approximation scheme may be quite crucially dependent on the parameters chosen. More specifically one should always try to implement an environmentally friendly reparametrization. We believe that an approach based on a reparametrization point of view, although less intuitive, is more powerful than that based on coarse graining. The latter is always concerned with integrating out degrees of freedom as a function of an effective cutoff. The RG flow is always with respect to this cutoff. In contrast, as has been emphasized previously[3], and as will also be seen herein, the reparametrization point of view, in principle, allows for the RG flow to be with respect to any parameter in the problem. One can even implement several RG's at the same time as we will see below, where RG's based on an arbitrary fiducial momentum scale κ and on an arbitrary fiducial temperature scale τ are used together. An important feature of environmentally friendly RG's is that the fixed points of such RG's will coincide with all the points of conformal invariance of a particular theory.

ENVIRONMENTALLY FRIENDLY RENORMALIZATION

We will briefly review in this section environmentally friendly renormalization in the context of finite temperature field theory; more specifically we will consider two environmentally friendly RG's. In the first an RG is generated within which an arbitrary, fiducial mass is the running parameter and in the second an arbitrary, fiducial temperature. We will refer to the latter as "running the environment". As discussed in the introduction, in our formalism a particular RG is associated with reparametrizations of the "couplings" (masses, interaction strengths etc.) of a field theory. Here we exhibit a reparametrization from bare to renormalized parameters that yields perturbatively valid answers for any value of the temperature, and in particular in the critical regime $T \sim T_c$.

For the most part we will consider $\lambda\phi^4$ theory in this paper, defined by the Euclidean action

$$S[\phi_B] = \int_0^{\frac{1}{T}} \int d^4x \left(\frac{1}{2}(\nabla\phi_B)^2 + \frac{1}{2}m_B^2\phi_B^2 + \frac{\lambda_B}{4!}\phi_B^4 + j_B(x)\phi_B + \frac{1}{2}t_B(x)\phi_B^2 \right)$$

which proffers a paradigm for considerations of the Higgs sector of the standard model, though we will consider gauge theories briefly in a later section. $j_B(x)$ and $t_B(x)$ are "sources" which generate correlators of ϕ_B and ϕ_B^2 respectively. Note that these sources could represent real effects. Here a constant t_B will be a mass parameter. We will work initially in the symmetric phase ($j_B = 0$, $T > T_c$) and consider reparametrizations obtained from a set of normalization conditions. For the first case, running a fiducial mass, we choose the following

$$\Gamma^{(2)}(p = 0, m(T) = m_{min}, \lambda, T, \kappa) = m_{min}^2 \tag{1}$$

$$\frac{\partial\Gamma^{(2)}}{\partial\vec{p}^2}(p_0 = 0, \vec{p}, m(T) = \kappa, \lambda, T, \kappa)|_{\vec{p}=0} = 1 \tag{2}$$

$$\Gamma^{(4)}(p = 0, m(T) = \kappa, \lambda, T, \kappa) = \lambda \tag{3}$$

$$\frac{\partial\Gamma^{(2)}}{\partial t}(p = 0, m(T) = \kappa, \lambda, T, \kappa) = 1 \tag{4}$$

where t is the renormalized mass parameter. Note that we are here parameterizing the system in terms of an an arbitrary, fiducial screening length, κ^{-1}, not the actual screening length, m^{-1}, which would lead to a Callan-Symanzik type equation. In Eq. (1) m_{min} represents the value of the physical mass when $t = 0$ and is an RG invariant point. If $m_{min} = 0$ the theory exhibits a second order phase transition. Eq.'s (2-4) fix the multiplicative renormalization constants Z_ϕ, Z_{ϕ^2} and Z_λ which renormalize ϕ, the composite operator ϕ^2 and λ respectively. Using these conditions, and implementing a [2,1] Padé resummation of the two loop Wilson functions one finds[3]

$$\beta(h, \frac{T}{\kappa}) = -\varepsilon(\frac{T}{\kappa})h + \frac{h^2}{1 + 4\left(\frac{(5N+22)}{(N+8)^2}f_1(\frac{T}{\kappa}) - \frac{(N+2)}{(N+8)^2}f_2(\frac{T}{\kappa})\right)h} \tag{5}$$

$$\gamma_{\phi^2}(h, \frac{T}{\kappa}) = \frac{(N+2)}{(N+8)}\frac{h}{1 + \frac{6}{(N+8)}(f_1(\frac{T}{\kappa}) - \frac{1}{3}f_2(\frac{T}{\kappa}))h} \tag{6}$$

and

$$\gamma_\phi(h, \frac{T}{\kappa}) = 2\frac{(N+2)}{(N+8)^2}f_2(\frac{T}{\kappa})h^2 \tag{7}$$

where $\varepsilon(\frac{T}{\kappa}) = 1 + \kappa\frac{d}{d\kappa}\ln(\sum_n m^{-3})$

$$f_1(\frac{T}{\kappa}) = 2\frac{\sum_{n_1,n_2}\left(\frac{1}{m_1^3}(\frac{1}{M} - \frac{1}{2m_2}) + \frac{1}{m_1 M^2}(\frac{1}{m_1} + \frac{2}{m_2})\right)}{(\sum_n \frac{1}{m^3})^2} \quad \text{and} \quad f_2(\frac{T}{\kappa}) = 4\frac{\sum_{n_1,n_2}\frac{1}{M^3 m_1}}{(\sum_n \frac{1}{m^3})^2}$$

with

$$m_i = (1 + \frac{4\pi^2 n_i^2 \kappa^2}{T^2})^{\frac{1}{2}}, \quad m_{12} = (1 + \frac{4\pi^2\kappa^2}{T^2}(n_1 + n_2)^2)^{\frac{1}{2}}, \quad M = m_1 + m_2 + m_{12}$$

γ_ϕ and γ_{ϕ^2} are the anomalous dimensions of ϕ and ϕ^2 respectively. In Eq.'s (5-7) the coupling h, or floating coupling[2,3], is defined via the relation $h = a_2(\frac{T}{\kappa})\lambda(\kappa)$, where a_2 is the coefficient of λ^2 in $\beta(\lambda)$. After the characteristic equations are solved we can use our freedom in choosing the arbitrary fiducial mass κ to be the physical finite temperature mass $m(T)$. An important feature of Eq. (5) is that it exhibits more than one fixed point. As $\frac{T}{m(T)} \rightarrow 0$, one obtains the Gaussian fixed point as expected in four dimensions. As $T \rightarrow T_c$, i.e. $m(T) \rightarrow 0$ one finds a non-trivial fixed point, $h = 1.732$ for $N = 1$, for instance. The value of the fixed point and the corresponding critical exponents are in exact agreement with corresponding two loop Padé resummed results[8] in three dimensional critical phenomena. Observe that the coupling $\lambda \rightarrow 0$ as the critical temperature is approached. As emphasized in O' Connor and Stephens[2,3] (and later also by Tetradis and Wetterich[7]) this however does not mean that the theory is non-interacting in this limit but only that λ is an inappropriate measure of the effective interactions, the floating coupling h being a more transparent measure.

Notice that we have derived the fact that the theory is three dimensional near the phase transition instead of assuming it, as we are able to examine the complete crossover between four and three dimensional behaviour, as $\frac{T}{m(T)}$ varies between 0 and ∞, in a perturbatively controllable fashion. Here we are probing different values of $m(T)$ by varying the mass parameter t for fixed temperature, below we will use a complementary approach where T is varied for fixed t.

One can think of $d_{eff} = 4 - \varepsilon(\frac{T}{m(T)})$, which interpolates between four and three when $\frac{T}{m(T)}$ varies between 0 and ∞, as a measure of the effective dimension of the system. Near $T = 0$ the EDOF of the problem are four dimensional and near $T = T_c$, three dimensional. The power of our approach is that we have implemented a reparametrization which is temperature dependent in such a fashion that it tracks the evolving nature of the EDOF between the four and three dimensional limits. In the above we considered only the symmetric phase. An environmentally friendly κ RG can also be implemented for the broken phase[9].

We see then how environmentally friendly methods can yield a perturbatively reliable description of the finite temperature behaviour from $T = 0$ to $T = \infty$. The description that we have obtained though is in terms of the finite temperature screening length and the finite temperature coupling rather than the parameters of the zero temperature theory, which are, after all, the ones we would have experimental access to. Additionally there are certain quantities, such as the critical temperature itself, which are very difficult to obtain from this RG. We consider therefore a complementary approach wherein we renormalize at a fiducial temperature τ. Such a RG was first introduced by Matsumoto et al.[10]. Here we will show how such a RG can be utilized and exploited in an environmentally friendly context. Once again the reparametrization we will use is defined by a set of normalization conditions which we take to be

$$\Gamma^{(2)}(p=0, M, \lambda, \bar{\phi}, T=\tau) = M^2 \tag{8}$$

$$\Gamma^{(4)}(p=0, M, \lambda, \bar{\phi}, T=\tau) = \lambda \tag{9}$$

$$\frac{d}{d\vec{p}^{\,2}}\Gamma(p_0=0, \vec{p}, M, \lambda, \bar{\phi}, T=\tau)\bigg|_{\vec{p}=0} = 1 \tag{10}$$

where M^{-1} is the finite temperature screening length at the fiducial temperature τ and $\bar{\phi}$ represents the minimum of the effective potential at that temperature, i.e. it satisfies the equation of state

$$\Gamma^{(1)}(p=0, M, \lambda, \bar{\phi}, T=\tau) = 0. \tag{11}$$

The beta functions, obtained by differentiating these renormalization conditions with respect to τ for fixed bare parameters, depend on derivatives $d\bar{\phi}/d\tau$. These can be eliminated using the equation of state. The differential equations describing an infinitesimal change in the normalization point are

$$\tau\frac{dM^2}{d\tau} = \beta_M(\lambda, M, \tau) \qquad \tau\frac{d\lambda}{d\tau} = \beta_\lambda(\lambda, M, \tau) \tag{12}$$

The functions β_M and β_λ take different functional forms above and below the critical temperature. To one loop for $T > T_c$

$$\beta_M = \frac{\lambda}{2}\tau\frac{\partial\bigcirc}{\partial\tau} \qquad \beta_\lambda = -\frac{3}{2}\lambda^2\tau\frac{d\bigcirc}{d\tau} \tag{13}$$

and below T_c we find

$$\begin{aligned}\beta_M &= -\lambda\left(\tau\frac{\partial\bigcirc}{\partial\tau} + \frac{3}{2}M^2\tau\frac{\partial\bigcirc}{\partial\tau} + \frac{1}{2}M^4\tau\frac{\partial\bigcirc}{\partial\tau}\right)\\ \beta_\lambda &= -\frac{3}{2}\lambda^2\tau\frac{d}{d\tau}\left(\bigcirc - \frac{34}{3}M^2\bigcirc + 18M^4\bigcirc\right)\end{aligned} \tag{14}$$

We have presented our results in diagramatic form as this renders the structure of these expressions more readily apparent and easily adoptable to other situations. Our diagrammatic notation is that the dots on a diagram represent the location of insertions in that diagram. The insertions could arise from different sources. All the one loop diagrams can be obtained from the basic diagram $\bigcirc$ which using dimensional regularization is given by

$$\bigcirc = -\frac{\Gamma(-\frac{d}{2})M^d}{(4\pi)^{d/2}} - \frac{2\tau^d}{(4\pi)^{(d-1)/2}\Gamma(\frac{d+1}{2})}\int_0^\infty \frac{dq}{\sqrt{q^2+\frac{M^2}{\tau^2}}}\frac{1}{e^{\sqrt{q^2+\frac{M^2}{\tau^2}}}-1} \tag{15}$$

The derivative with respect to M^2 of a diagram with k dots gives $-k$ times a diagram with $k+1$ dots. Note that M is retained in a non-perturbative manner while λ is retained only to lowest non-trivial order. This accounts for the asymmetry between the partial derivatives in β_M but total derivatives in β_λ. Such a procedure, which can be consistently carried out to higher order, automatically uses the fully dressed mass, M, and in particular sums all daisy insertions on all the internal lines of the β function diagrams. Such an artifice is crucial in that without it the tadpole insertions, or "shift", which constitute the largest temperature effects would ensure a breakdown in perturbation theory. In using such a procedure we differ substantially from other authors[11] who all neglect the term $\tau\frac{dM^2}{d\tau}$ in β_λ in Eq.'s (17) and (18). As demonstrated below this is inconsistent in the vicinity of the critical point.

We will now examine the structure of our flow equations near the critical point. The asymptotic expansion for small M of (15) is

$$\bigcirc = \tau^d \Big(-\frac{2\Gamma(d/2)\zeta(d)}{\pi^{d/2}} + \frac{\Gamma(\frac{d-2}{2})\zeta(d-2)}{2\pi^{d/2}}\frac{M^2}{\tau^2} - \frac{\Gamma(\frac{1-d}{2})}{(4\pi)^{(d-1)/2}}\left(\frac{M}{\tau}\right)^{(d-1)} - \frac{\Gamma(\frac{d-4}{2})\zeta(d-4)}{(4\pi)^2\pi^{(d-4)/2}}\left(\frac{M}{\tau}\right)^4 + \dots \Big) \tag{16}$$

for any dimension d between three and four. Thus in the neighbourhood of the critical point for the four dimensional theory, we find

$$\beta_M = \lambda\left(\frac{\tau^2}{12} - \frac{M\tau}{8\pi} + \dots\right)$$
$$\beta_\lambda = -\frac{3\tau\lambda^2}{16\pi M}\left(1 - \frac{1}{2M^2}\tau\frac{dM^2}{d\tau} + \dots\right) \tag{17}$$

for $T > T_c$ and

$$\beta_M = -\lambda\left(\frac{\tau^2}{6} - \frac{29M\tau}{64\pi} + \dots\right)$$
$$\beta_\lambda = -\frac{5\lambda^2\tau}{64\pi M}\left(1 - \frac{1}{2M^2}\tau\frac{dM^2}{d\tau} + \dots\right) \tag{18}$$

for $T < T_c$. The differential equation for λ is easy to solve since it is a total derivative. We find that as the critical point is approached $\lambda \to 0$ which is consistent with what we found using an RG where a fiducial mass was the running parameter. As explained above this is merely a manifestation of the fact that in the critical region a more appropriate coupling would be the effective three dimensional coupling $\frac{\lambda T}{M(T)}$. Explicitly

$$\lambda = \frac{16\pi M}{3\tau} + \dots \quad \text{for } T > T_c \text{ and} \quad \lambda = \frac{64\pi M}{5\tau} + \dots \quad \text{for } T < T_c \tag{19}$$

The solution of the differential equation for λ can be substituted into that for M to find, for example, for $T > T_c$

$$\frac{1}{M^2}\tau\frac{dM^2}{d\tau} = \frac{4\pi\tau}{9M} - \frac{2}{3} + \dots \tag{20}$$

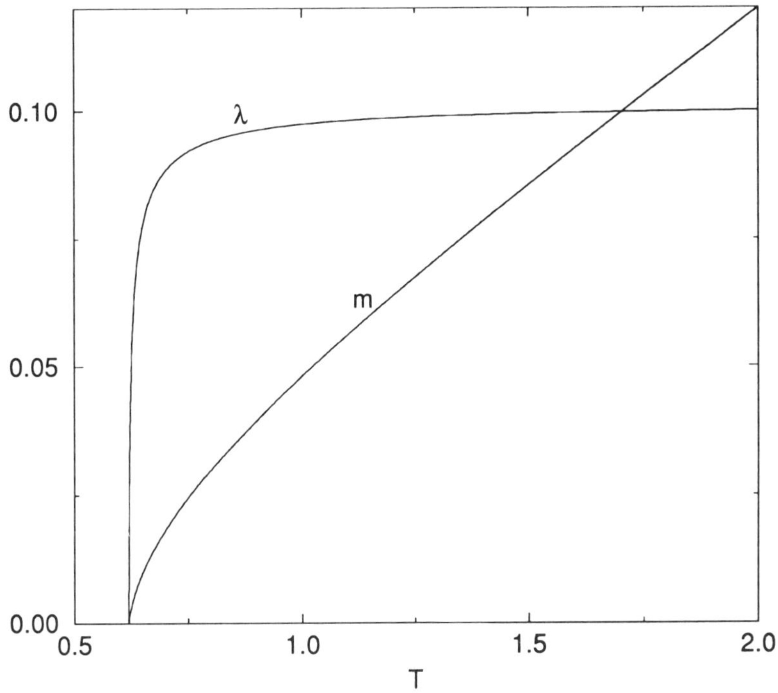

Fig. 1. Solutions of the coupled differential equations (12) for the mass M and coupling λ as a function of temperature T.

The fact that this quantity diverges as the critical region is approached gives ample justification as to why it was necessary to use total derivatives and not partial derivatives in Eq.'s (13) and (14).

The solutions to the equations (12) for $T > T_c$ are plotted in Fig. 1 where we have chosen $\tau = T$, the physical temperature. As emphasized in Freire and Stephens[3] such a choice is essential if a perturbatively reliable description is to be obtained for all possible temperatures. The critical temperature is clearly visible and is a solution of $M(T_c, M(T = T_{in}), \lambda(T = T_{in})) = 0$. The critical temperature being a non-universal quantity depends on the "non-universal" initial conditions of the RG flow. We emphasize however that T_c nevertheless is obtained from the RG flow. If we use Eq.'s (12) for $T < T_c$ we can choose the initial temperature $T_{in} = 0$ and thereby relate T_c to the zero temperature parameters. As M goes continuously to zero for $T \to T_c$ we see that the phase transition is second order. The effective dimension $d_{eff}(\frac{T}{M(T)})$ provides a quantitative measure of the extent to which dimensional reduction has occurred. The latter in this setting is manifestly *derived* from the solutions of Eq. (12) for $T < T_c$ based only on a knowledge of the zero temperature parameters. Two loop results for $T \geq T_c$ have been obtained and along with a more detailed discussion of the results in the broken phase will be presented elsewhere.

Let us now make some brief comments about the effective potential. The effective potential is convex, as it must be on physical grounds, the region between the minima being physically inaccessible for any constant background field. The minima of the effective potential are obtained from the equation of state (11). We find to one loop that

$$\bar{\phi} = \pm\sqrt{\frac{3}{\lambda(T)}}M(T)\left[1 + 9\lambda(T)M^2(T)(\bigcirc - \frac{3}{2}\lambda(T)M^2(T)\bigcirc) + \ldots\right] \tag{21}$$

where $M(T)$ and $\lambda(T)$, solutions of equations (12), are functions of the zero temperature parameters and T.

The two environmentally friendly RG's we have shown here are complementary. Although they both in principle give a complete description of the finite temperature behaviour some questions are more easily addressed using one versus the other, in particular the τ RG allows for a completely RG determination of the critical temperature. Together the two RG's offer perturbatively reliable answers for basically any physical quantity.

NON-ABELIAN GAUGE FIELDS

We now discuss briefly some aspects of finite temperature non-abelian gauge fields, further details may be found in other work[4]. Here we will concentrate on the gauge coupling in the magnetic sector. We use as a renormalization condition that the static (i.e. zero energy), spatial three-gluon vertex equals the tree-level vertex in the symmetric momentum configuration

$$\Gamma^{abc}_{ijk}(p^0_i = 0, \vec{p}_i, g_{\kappa,\tau}, T = \tau)\big|_{\substack{\text{symm.}\\ \kappa}} = g_{\kappa,\tau} f^{abc}\left[\eta_{ij}(p_1 - p_2)_k + cycl.\right]. \tag{22}$$

where η_{ij} is the metric and f_{abc} are the structure constants of the gauge group.

In contradistiction to the previously discussed cases this chosen normalization condition depends on two flow parameters simultaneously, the momentum scale κ, and the temperature scale τ. Therefore we can perform a RG analysis with respect to both parameters, i.e. we can run more than one environmental parameter at the same time. The τ RG is needed to draw conclusions about the temperature dependence of the coupling. This cannot be done using the κ-scheme alone without assuming something about the temperature dependence of the initial value of the coupling used in solving the differential equation.

For the calculation we use the Landau gauge Background Field Feynman rules[12] resulting from the Vilkovisky-de Witt effective action in order to get rid of ambiguities arising from gauge dependence. In terms of the coupling $\alpha_{\kappa,\tau} := g^2_{\kappa,\tau}/4\pi^2$ the β function equations are

$$\kappa\frac{d\alpha_{\kappa,\tau}}{d\kappa}\Big|_\tau = \beta_{vac} + \beta_{th}, \qquad \tau\frac{d\alpha_{\kappa,\tau}}{d\tau}\Big|_\kappa = -\beta_{th}, \tag{23}$$

where the vacuum contribution is, as usual,

$$\beta_{vac} = \alpha^2_{\kappa,\tau}\left(-\frac{11}{6}N_c + \frac{1}{3}N_f\right) \tag{24}$$

and where, in terms of the IR and UV convergent integrals

$$F_n^\eta = \int_0^\infty dx \frac{x^n}{e^{\kappa x/2\tau} - \eta} \left[\log \left| \frac{x+1}{x-1} \right| - 2 \sum_{k=0}^{\frac{n}{2}-1} \frac{x^{2k+1}}{2k+1} \right]$$

and

$$G_n^\eta = \int_0^\infty dx \frac{1}{e^{\kappa x/2\tau} - \eta} \mathrm{P} \frac{x}{(x^2-1)^n}$$

the thermal contribution is given by

$$\beta_{th} = \alpha_{\kappa,\tau}^2 \left[\left(\frac{21}{16} F_0^1 + \frac{3}{4} F_2^1 - \frac{3}{2} G_0^1 - \frac{25}{8} G_1^1 - G_2^1 \right) N_c + \left(\frac{1}{4} F_0^{-1} + \frac{3}{4} F_2^{-1} - \frac{3}{2} G_0^{-1} - G_1^{-1} \right) N_f \right]. \tag{25}$$

Because the two beta functions (23) are not exactly each other's opposite the RG improved coupling is not just a function of the ratio κ/τ. There is another dimensionful scale (such as Λ_{QCD}) that comes from an initial condition for these differential equations. The solution of the set of coupled differential equations can be written in the form

$$\alpha_{\kappa,\tau} = \frac{1}{\left(\frac{11}{6} N_c - \frac{1}{3} N_f\right) \ln \frac{\kappa}{\Lambda_{QCD}} - f\left(\frac{\kappa}{\tau}\right)} \tag{26}$$

where the function f satisfies $\beta_{th} = \alpha_{\kappa,\tau}^2 \kappa df/d\kappa$ with the initial condition $\lim_{\tau \downarrow 0} f = 0$ so that we can identify Λ_{QCD} with the usual zero-temperature QCD scale. Actually this function f can be found in terms of the functions F and G:

$$f = \left(\frac{21}{16} F_0^1 + \frac{1}{4} F_2^1 + \frac{7}{8} G_1^1 \right) N_c + \left(\frac{1}{4} F_0^{-1} + \frac{1}{4} F_2^{-1} \right) N_f. \tag{27}$$

Fig. 2 is a contour plot of the effective coupling as a function of both momentum and temperature scale. The results are best trusted in places where the coupling is small. For physical reasons we have to restrict ourselves in any case to the region where the coupling is positive, which is below the uppermost line in the graph.

The high-temperature behaviour (i.e. for $\tau \gg \kappa$) is determined by

$$f \longrightarrow N_c \frac{21\pi^2}{16} \frac{\tau}{\kappa} + \left(\frac{11}{6} N_c - \frac{1}{3} N_f \right) \ln \frac{\kappa}{\tau} + O(1). \tag{28}$$

As emphasized in van Eijck et al.[4] for increasing temperatures at fixed momentum scale the coupling approaches an IR Landau pole, an indication that we are entering a strong-coupling regime, whereas the opposite sign would lead to asymptotic freedom in this limit. Buchmüller and Fodor[6] have recently come to a similar conclusion based on an assumed dimensional reduction in the electroweak model. Stimulated by the original belief[13] that high-temperature QCD would be asymptotically free as in the high-momentum situation, it was suggested[14] that this strong coupling in the IR would be an artifact of the one-loop calculation and that a higher-loop calculation or a resummation could change it. We however believe that this will not happen, as the sign appears quite naturally if one realises that this limit $\tau/\kappa \to \infty$ is an IR limit where confinement takes place. Unless at higher loop order the magnetic mass increases quickly enough with temperature in order to act as an efficient IR cutoff, we cannot get around this problem without actually solving confinement. We believe this to be an important consideration for phase transitions which involve non-abelian gauge fields.

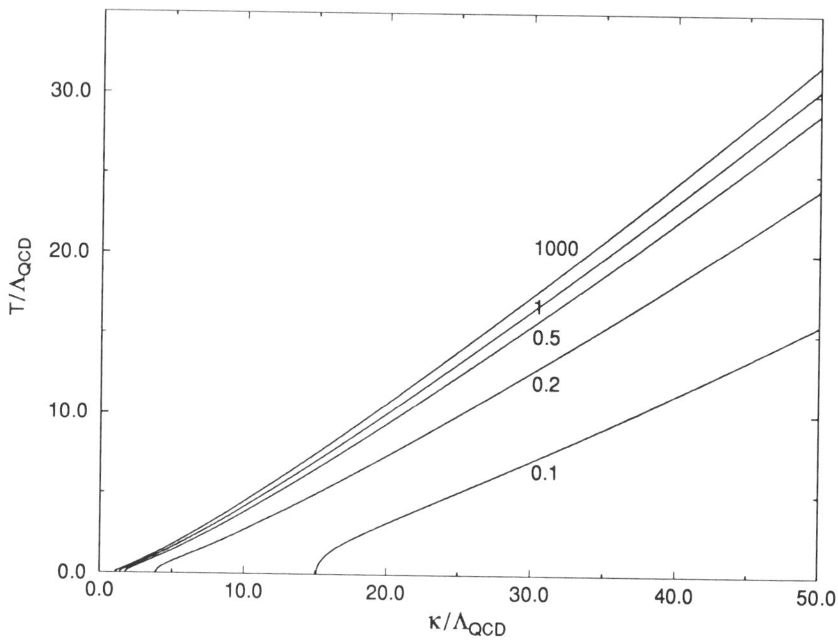

Fig. 2. Contour plot of the running coupling $\alpha_{\kappa,\tau}(\kappa,\tau)$ for QCD with three colours and six fermion flavours. The fermions have been taken massless. Only below the curve $\alpha_{\kappa,\tau} = \infty$ (close to $\alpha_{\kappa,\tau} = 1000$) is the coupling positive and finite.

In the regime $\tau \gg \kappa$ the beta functions behave as in a three-dimensional theory so that we designate this as the region where dimensional reduction occurs. Here it is natural, as for $\lambda\phi^4$, to use a different dimensionless coupling, for instance, $u = \alpha_{\kappa,\tau}\frac{\tau}{\kappa}$. However in this case such a reparametrization cannot remove the pole and will not give a different behaviour.

If we allow the momentum scale to change with temperature simultaneously, the high-temperature limit can be taken in many ways. In the region $\tau \gg \kappa$ the shape of the contours is given by $\tau \sim \kappa \ln \frac{\kappa}{\Lambda_{QCD}}$. This characterizes exactly along which paths in the (τ,κ)-plane the coupling increases or decreases. For example at a fixed ratio τ/κ

(no matter what this ratio is) we eventually find a coupling that decreases like $1/\ln \kappa$, much in the same way as at zero temperature. This is a natural contour to consider for a weak-coupling regime[15] where one could treat the quark-gluon plasma as a perfect gas, as then the thermal average of the momentum of massless quanta at temperature T is proportional to the temperature. However at low momenta the assumption of weak coupling breaks down. Furthermore, instead of considering quantities at the average momentum it is more appropriate to use thermal averages of the quantities themselves as a weighted integral over all momenta[16]. But once again one runs into problems at the low-momentum end as long as we cannot treat the strong-coupling regime.

DISCUSSION AND CONCLUSIONS

There have been several approaches to trying to ameliorate the IR problems inherent in a quantitative discussion of the electroweak phase transition. One of the most popular has been associated with the resumming of infinite sets of Feynman diagrams, and in particular ring diagrams. This can be done either in the Higgs sector alone or in the full electroweak model where both scalar loops and gauge loops must be summed. Dimensional reduction is then motivated by considerations of the ring-resummed effective potential. Given that dimensional reduction occurs an effective three dimensional theory has been treated using critical phenomena techniques such as ε-expansions. Dimensional reduction occurs when $\frac{T}{M(T)} \gg 1$ (i.e. in the vicinity of a weakly first order or second order phase transition). In this regime the ring-resummed effective potential is a bad approximation to the true effective potential. Consequently it is not particularly consistent to motivate dimensional reduction by considerations of the ring-resummed effective potential. Naturally a more desireable approach would be to derive whether dimensional reduction occurs or not using a formalism that treats in a perturbatively reliable manner not only the zero temperature regime and the "deep IR" regime (where dimensional reduction occurs) but also the entire crossover between these qualitatively different regimes. Environmentally friendly renormalization, as briefly reviewed in this contribution, is precisely such a formalism.

We have explained how environmentally friendly renormalization gives a complete description of the finite temperature behaviour of $\lambda\phi^4$ theory which is closely related to the Higgs sector of the standard model. We have seen also that there is a significant difference when treating the magnetic sector of non-abelian gauge fields. The implication in this case is that the RG we have employed is not sufficiently environmentally friendly and is manifest in the breakdown in perturbation theory as one runs into the IR. There are several possibilities for making it more environmentally friendly. If the suggestion above that the crossover is to confined degrees of freedom is true then an RG must be implemented that explicitly takes account of this fact, i.e. a reparametrization must be found that captures this crossover. If such a reparametrization could be found then it would provide a completely perturbative description of the crossover between asymptotic freedom and confinement. If in contrast the problem is resolved by the appearance of a magnetic mass in higher orders in perturbation theory then the β functions should depend on this mass. The above considerations would apply to the electroweak model in the symmetric phase. In the broken phase the Higgs mechanism gives a mass and therefore an IR cutoff to the gauge bosons. The extent to which this can act as an efficient IR cutoff as the phase transition is approached is not clear without a more detailed analysis. Certainly we believe that environmentally friendly renormalization is an extremely useful tool in the analysis of the electroweak phase transition, and more generally in the context of phase transitions in particle physics and cosmology.

ACKNOWLEDGEMENTS

DOC would like to thank NWO for financial support during a visit to the University of Amsterdam.

REFERENCES

1. L. Dolan and R. Jackiw, *Phys. Rev.* **D9** 3320 (1974); S. Weinberg *ibid* 3357.
2. Denjoe O' Connor and C.R. Stephens, *Nucl. Phys.* **B360** 297 (1991).
3. Denjoe O'Connor, C.R. Stephens and F. Freire, *Mod. Phys. Lett* **A25** 1779 (1993); *Class. Quan. Grav.* **23** (1993) S243; F. Freire and C.R. Stephens, *Z. Phys.* **C60** 127 (1993); "Environmentally Friendly Renormalization"; Preprint THU-93/14, DIAS-STP-93-19; (to be published in Int. Jou. Mod. Phys. **A**).
4. M.A. van Eijck, C.R. Stephens and C.W. van Weert, *Mod. Phys. Lett.* **A9** (1994) 309.
5. P. Ginsparg, *Nucl.Phys.* **B170** **[FS1]** 388 (1980); M. Gleiser and E.W. Kolb, *Phys. Rev.* **D48** 1560 (1993).
6. W. Buchmüller and Z. Fodor, DESY-preprint DESY 94-045 (1994).
7. N. Tetradis and C. Wetterich, *Nucl. Phys.* **B398** 659 (1993).
8. G.A. Baker, B.G. Nickel and D.I. Meiron, *Phys. Rev.* **B17** 1365 (1978).
9. F. Freire, Denjoe O' Connor and C.R. Stephens, *J. Stat. Phys.* **74** 219 (1994).
10. H. Matsumoto, Y. Nakano and H. Umezawa, *Phys. Rev.* **D29** 1116 (1984).
11. P.Elmfors, "Temperature Renormalization Group and Resummation", Nordita preprint NORDITA-92/63; K. Funakobo and M. Sakamoto, *Phys. Lett.* **B186** 205 (1987); Y. Fujimoto, K. Ideura, Y. Nakano and H. Yoneyama, *Phys. Lett.* **B167** 406 (1986).
12. A. Rebhan, *Z. Phys.* **C30** 309 (1986).
13. J.C. Collins and M.J. Perry, *Phys. Rev. Lett.* **34** 1353 (1975).
14. N.P. Landsman, *Phys. Lett.* **B232** 240 (1989).
15. K. Enquist and K. Kajantie, *Mod. Phys. Lett.* **A2** 479 (1987).
16. K. Enquist and K. Kainulainen, *Z. Phys.* **C53** 87 (1992).

SPHALERON-LIKE TRANSITIONS AT HIGH TEMPERATURES

Jan Ambjørn

The Niels Bohr Institute
Blegdamsvej 17
DK-2100 Copenhagen Ø, Denmark

Abstract

We study sphaleron-like transitions and the associated spectrum of the Dirac Hamiltonian (including the coupling to the Higgs fields) at temperatures prevailing in the early universe. The temperature fluctuations are created by numerical lattice simulations of the classical field equations. It is shown that there are frequent changes in the Chern-Simons number by one unit and that such changes are accompanied by eigenvalues crossing zero for the Dirac Hamiltonian. When the eigenvalues cross zero there is a change of sign of the generalized chirality $\tilde{\Gamma}_5 = (-1)^{2T+1}\gamma_5$ which labels these modes.

1 Introduction

One of the most remarkable quantum effects in the electroweak theory is the anomaly of the baryonic current [1]. This leads to a non-conservation of baryon- and lepton-number. It is of little importance in the present day universe where the processes which violate baryon- and lepton-number conservation are exponentially suppressed. There is evidence that this suppression is not present at the high temperatures prevailing in the early universe and it has played an increasing role in attempts to explain the baryon asymmetry observed in the universe today.

In this article I will describe an attempt to "observe" the high temperature fluctuations and check whether or not they correspond to the "baryon-changing" configurations, i.e. whether they change Chern-Simons number and whether such a change is related to a level shifting of the corresponding Dirac operator. While the connection between a change in Chern-Simons number and the spectral flow of the *massless* Dirac operator has been clarified long ago, the connection is not well established in the case where the Higgs field is included. Since the Higgs field *is* a significant part of the standard model one cannot ignore.

Unfortunately almost any interesting question related to the high temperature phase of the electroweak theory is difficult to address by analytical tools, due to the infrared singularities of the associated high temperature perturbation theory. A generic non-perturbative way to study the high temperature fluctuations of the gauge and Higgs fields is by means of lattice simulations. If we want to address questions related to the anomaly we encounter the difficulty that topology is involved, an inherent continuum concept, and it is important to verify that configurations generated on the lattice during the simulations will qualify as representatives of continuum configurations.

The anomaly is most easily understood when the gauge fields are are treated as background fields. This means that for questions related to the anomaly we can treat the gauge (and Higgs) fields as classical fields. In addition it is generally believed that high temperature fluctuations are essentially classical fluctuations. From these two point of views it makes good sense to use classical dynamics for the gauge and the Higgs fields if we want to study the anomaly at high temperature. In [2] a real time evolution of classical gauge and Higgs fields was used as an approximation to the high temperature fluctuations present in the electroweak theory in the early universe. The baryon number violation was studied by observing the change in the Chern-Simons number during the time evolution. Since the temperature fluctuations are not small in these simulations it is important to ask if these fluctuations might spoil the picture of level crossing, which somehow is the justification of viewing the change in Chern-Simons number as responsible for a change in fermion number.

In [3] it was shown that the gauge field fluctuations which lead to a change in Chern-Simons numbers at the same time cause the lowest eigenvalue of the *massless* Dirac Hamiltonian to cross zero, in agreement with the level crossing picture where such a crossing results in the conversion of particles to anti-particles (or vice versa). In addition the measurement of the spatial extension of energy lumps associated with the fluctuations corroborated the interpretation of these as representing continuum physics. However, in the full electroweak theory an important ingredient is the Higgs field. It couples to the fermions and is responsible for particle masses in the broken phase. At first sight it is maybe surprising that the level crossing picture is still true in this case since the eigenstates of the full Dirac Hamiltonian are massive when one includes the coupling to the Higgs field. This is in particular true if we consider the ground state configuration in the broken phase : $\phi = \phi_0$, $A_\mu = 0$. The situation is different in the presence of a sphaleron. For a sphaleron configuration there is precisely one normalizable eigenmode and it has zero energy [4, 5]. As a consequence a gauge-Higgs configuration which changes continuously from one vacuum configuration to a neighboring one, passing through a sphaleron configuration on the way, will trigger an adiabatic change of the lowest positive eigenvalue such that it passes through zero and ends up as the "highest" negative eigenvalue [5].

In many ways the situation around a sphaleron configuration is similar to the situation around the simplest non-abelian monopole. In that case there exists an index theorem for open three-dimensional space [6] which tells us that the difference n between the number of normalizable eigenfunctions of positive chirality and the normalizable eigenfunctions of negative chirality, of the Dirac Hamiltonian with eigenvalue zero, is related to the charge of the monopole Q by

$$Q = \frac{4\pi n}{e}. \tag{1}$$

n can be expressed in terms of the Higgs field:

$$n = \frac{1}{4\pi} \int_{S_\infty} d^2 x_i \varepsilon_{ijk} \varepsilon^{abc} \tilde{\phi}^a \partial_j \tilde{\phi}^b \partial_k \tilde{\phi}^c, \tag{2}$$

where $\tilde{\phi}^a = \phi^a/|\phi|$ for a Higgs field in the adjoint representation. Eq. (2) reflects that $|\phi|$ is assumed to be different from zero at spatial infinity and consequently that the map $x \in S^2_\infty \to \tilde{\phi}(x) \in S^2$ has winding number n. In the case of the sphaleron there is not a similar topological interpretation since the Higgs field is a complex doublet, i.e. it has four real components and we will have a map $x \in S^2_\infty \to \tilde{\phi}(x) \in S^3$. This map is always contractible to the trivial map ($\pi_2(S^3) = 0$). However, for a continuous family of static fields $\phi(x;\tau), A_i(x;\tau)$, $\tau \in [0,1]$, which interpolates between two neighboring vacua when τ changes from 0 to 1, there has to be at least one value τ_0 of τ such that the Higgs field $\phi(x,\tau_0)$ has a zero. In the case of the electroweak theory and the so-called minimal path which connects two neighboring vacua the sphaleron is precisely such a configuration. If one follows a minimal path it is known that at the sphaleron there will be a normalizable eigenmode of the Dirac operator and that it has zero eigenvalue. It is intuitively reasonable that a zero of the Higgs field is a necessary condition for having a zero eigenvalue of the full Dirac operator. However, there is to my knowledge no topological theorem which tells us that the full Dirac operator has eigenvalue zero for some τ_0. This is in contrast to the case of the massless Dirac operator, which only couples to the gauge field, and where it follows from the Atiyah-Patodi-Singer index theorem that the number of eigenvalues of the Dirac Hamiltonian which crosses zero is directly related to the change in Chern-Simons number [7, 8, 9].

Recently a number of articles have addressed the question of level crossing for the full Dirac Hamiltonian in the electroweak theory [10, 11, 12]. The proof of the validity of the level crossing picture has either been of numerical nature, for a specific chosen path in (ϕ, A_i) configuration space, or it has used some special symmetry properties of the sphaleron solutions and is only valid for some range of the coupling constants. In this article we make an attempt to investigate the validity of the level crossing picture in the case of fluctuations of the gauge *and especially the Higgs fields* which are typical for the high temperatures of the early universe.

There is one more reason to focus on the Higgs field. The role of this field in the high temperature fluctuations, and in particular its relation to the masses of the fermions and therefore to the level shifting, is not very clear. If the symmetry is restored the classical expectation value of the Higgs field is equal to zero, even if we have imposed complete gauge fixing. However, the field still couples to the fermions and the statement that $\langle\varphi\rangle_T = 0$ refers to the thermal average $\langle\cdot\rangle_T$. When we discuss temperature induced transitions of gauge and Higgs fields between neighboring vacua connected by large gauge transformations we deal with dynamical processes caused by thermal fluctuations, and it is strictly speaking misleading to refer to thermal averages. In fact we want in the discussion of the anomaly and level shifting to use an approximation where the gauge and Higgs fields are treated as background fields *and even above the symmetry restoration temperature we can talk about "sphaleron-like" transitions.* In the simulations we perform the expectation value $\langle|\varphi|^2\rangle$ will typically be quite small compared to the classical vacuum expectation value $\langle|\varphi|^2\rangle_0$ at zero temperature, due to temperature fluctuations. This indicates that the system is in the symmetric phase rather than in the broken phase. It is nevertheless perfectly possible (for a finite volume) to talk about the fields as located in the neighborhood of a definite vacuum and the transition rate between such vacua. The picture becomes especially clear if one talks about the family of all fields which interpolate between the neighborhoods of two different vacua, and then focuses on those configurations in which the high energy thermal fluctuations have been stripped off. By such a procedure it becomes clear that the low frequency modes of the gauge fields can be viewed as interpolating between the different vacua and that the Chern-Simons number will change by one unit during the interpolation [3]. In this article we will use the spectral flow of the Dirac operator as a tool to include the Higgs field in the analysis.

2 The model

Let us consider the electroweak theory where for simplicity we ignore the hyper-charge sector. It is of no importance for the electroweak anomaly. We also restrict ourselves to two families. The generalization to the twelve existing doublets is trivial.

The bosonic part of the Lagrangian is given by

$$\mathcal{L}_b = \frac{1}{4} F^a_{\mu\nu} F^a_{\mu\nu} + |D_\mu \varphi|^2 - \lambda(|\varphi|^2 - v_c^2)^2 \tag{3}$$

where $F^a_{\mu\nu}$ denotes the electroweak field strength.

The fermionic part of the Lagrangian is given by

$$\begin{aligned} \mathcal{L}_f &= \bar{\psi}_L \not{D} \psi_L + \bar{\chi}_R \not{\partial} \chi_R + \bar{\psi}'_L \not{D} \psi'_L + \bar{\chi}'_R \not{\partial} \chi'_R \\ &\quad - \bar{\psi}_L (h_u \tilde{\varphi} \chi_{uR} + h_d \varphi \chi_{dR}) - (h_u \bar{\chi}_{uR} \tilde{\varphi}^\dagger + h_d \bar{\chi}_{dR} \varphi^\dagger) \psi_L \\ &\quad - \bar{\psi}'_L (h'_u \tilde{\varphi} \chi'_{uR} + h'_d \varphi \chi'_{dR}) - (h'_u \bar{\chi}'_{uR} \tilde{\varphi}^\dagger + h'_d \bar{\chi}'_{dR} \varphi^\dagger) \psi'_L \end{aligned} \tag{4}$$

where the left handed doublets are

$$\psi_L = \begin{pmatrix} \psi_u \\ \psi_d \end{pmatrix}_L \qquad \psi'_L = \begin{pmatrix} \psi'_u \\ \psi'_d \end{pmatrix}_L \tag{5}$$

and the four right handed singlets are $\chi_{uR}, \chi_{dR}, \chi'_{uR}, \chi'_{dR}$. With an abuse of notation we collect them as vectors, like the doublets, even if they of course behave like singlets under electroweak $SU(2)$ transformations:

$$\chi_R = \begin{pmatrix} \chi_u \\ \chi_d \end{pmatrix}_R \quad \chi'_R = \begin{pmatrix} \chi'_u \\ \chi'_d \end{pmatrix}_R \tag{6}$$

The Higgs doublet φ and its conjugated field $\tilde{\varphi}$ (which transforms as φ under electroweak transformations) are defined as usual:

$$\varphi = \begin{pmatrix} \phi_+ \\ \phi_0 \end{pmatrix}, \qquad \tilde{\varphi} = \begin{pmatrix} \phi_0^* \\ -\phi_+^* \end{pmatrix} = i\tau_2 \varphi^* \tag{7}$$

There are four independent Yukawa couplings h_u, h_d, h'_u and h'_d. We have only two independent couplings in the corresponding lattice model and for simplicity we therefore choose $h'_u = h_u$ and $h'_d = h_d$. $h_u \approx h_d$ mimics the the quark sector (as the notation indicates), while $h_d = 0$ describes the lepton sector.

It is well known that the chiral theory defined by (4) can be given a formulation as a vector-like theory of Dirac fermions [14]. This is due to the fact that the representations of $SU(2)$ are real. We can therefore introduce the charge-conjugated fields

$$\psi'^c_R = \varepsilon C \psi'_L = \begin{pmatrix} \psi'^c_{dR} \\ -\psi'^c_{uR} \end{pmatrix} \tag{8}$$

where $\varepsilon = i\tau_2$ acts on the isotopic indices and C is the charge-conjugation matrix, acting on spinor indices: $\psi^c \equiv C\psi$. In a similar way we define

$$\chi'^c_L = C \chi'_R = \begin{pmatrix} \chi'^c_{dL} \\ -\chi'^c_{uL} \end{pmatrix}. \tag{9}$$

If we use the chiral representation of the γ matrices a Dirac fermion can be written as

$$f_D = \begin{pmatrix} f_L \\ f_R \end{pmatrix}. \tag{10}$$

and we can now introduce the two Dirac spinors:

$$\psi_D = \begin{pmatrix} \psi_L \\ \psi_R'^c \end{pmatrix}, \qquad \chi_D = \begin{pmatrix} \chi_R \\ \chi_L'^c \end{pmatrix}. \tag{11}$$

This allows us finally to introduce the *eight*-component spinor Ψ by:

$$\Psi = \begin{pmatrix} \psi_D \\ \chi_D \end{pmatrix}, \tag{12}$$

and using the matrix

$$M = \Phi \begin{pmatrix} h_u & 0 \\ 0 & h_d \end{pmatrix}, \qquad \Phi = \begin{pmatrix} \phi_0^* & \phi_+ \\ -\phi_+^* & \phi_0 \end{pmatrix}, \tag{13}$$

we can write the Lagrangian (4) as follows

$$-\mathcal{L}_f = \bar{\Psi} \begin{pmatrix} \not{D} & -M \\ -M^\dagger & \not{\partial} \end{pmatrix} \Psi \tag{14}$$

such that each entry of the matrix in eq. (14) is a 2×2 matrix in isospinor indices and a 4×4 matrix in spinor indices.

Note that (14) resembles the ordinary Dirac equation but that *(i)* the mass is spatially varying, depending on φ, *(ii)* the mass term may multiply the 1st and 2nd (and 3rd and fourth) components differently if $h_u \neq h_d$ and, most importantly, *(iii)* only the first 4 components (ψ_D) of the eight component spinor Ψ have gauge interactions. The last four (χ_D) are gauge singlets.

The notation introduced here follows [11], which also introduced the generalized chirality matrix $\tilde{\Gamma}_5$ acting on the eight component fermion field Ψ:

$$\tilde{\Gamma}_5 = \begin{pmatrix} \gamma_5 & 0 \\ 0 & -\gamma_5 \end{pmatrix}. \tag{15}$$

The matrix $\tilde{\Gamma}_5$ anticommutes with the matrix in (14) and also with the matrix

$$\Gamma_0 = \begin{pmatrix} \gamma_0 & 0 \\ 0 & \gamma_0 \end{pmatrix} \tag{16}$$

which enters in the definition of $\bar{\Psi}$. Consequently the Lagrangian (14) is invariant under the generalized chiral transformation [11]

$$\Psi \rightarrow e^{i\theta\tilde{\Gamma}_5/2}\Psi, \qquad \bar{\Psi} = e^{i\theta\tilde{\Gamma}_5/2}\bar{\Psi}. \tag{17}$$

The symmetry expresses nothing but the fermion number conservation of the electroweak theory. In fact the current associated with the symmetry is:

$$J_\mu^5 = \bar{\Psi}\gamma_\mu\tilde{\Gamma}_5\Psi = \bar{\psi}_D\gamma_\mu\gamma_5\psi_D - \bar{\chi}_D\gamma_\mu\gamma_5\chi_D, \tag{18}$$

or, expressed in terms of the original fields of the model:

$$J_\mu^5 = \bar{\psi}_L\gamma_\mu\psi_L + \bar{\psi}'_L\gamma_\mu\psi'_L + \bar{\chi}_R\gamma_\mu\chi_R + \bar{\chi}'_R\gamma_\mu\chi'_R \tag{19}$$

i.e. the fermionic current. As remarked in the introduction this current is known to be anomalous:

$$\partial_\mu J^5_\mu = \frac{g_w^2}{32\pi^2} F^a_{\mu\nu} \tilde{F}^{a\mu\nu} \tag{20}$$

and leads to the famous non-conservation of baryon number in the electroweak theory. The reason we introduce the generalized chirality dictated by $\tilde{\Gamma}_5$ is that the analogy to ordinary abelian symmetry and the level shifting picture will be more clear, as emphasized recently by several authors [11, 12], who proved that under certain assumptions a transition from one vacuum to another, related by a large gauge transformation, will result in a level shift. It is our goal to try to verify this level shifting for realistic high temperature configurations which do not necessarily satisfy the assumptions used in the analytical proofs of the level shifting and, in addition, to provide further evidence that we really observe continuum-like configurations during the simulations.

3 Lattice Formulation

The translation of the continuum formulas above to the lattice is rather trivial for the bosonic sector and non-trivial for the fermionic sector due to the well known problems with fermion doubling. I refer to [13] for references to original papers and for a discussion of technical aspects. Here it is sufficient to say the following: Only the bosonic part is given dynamics. The fermionic part is only used to generate spectral flow for a given bosonic background.

3.1 The bosons

The basic approximation for the bosonic fields is the use of classical physics to generate the high temperature thermal fluctuations. In this way we use the classical equations of motion in ordinary Minkowski space to generate the time sequence of configurations. For this time sequence we can study transitions (in *real* time from one topological sector to another, and the correspondent spectral flow of eigenvalues of the Dirac operator. In order that the classical equations of motion can describe the system at a finite temperature it is necessary that the system is large enough and that the initial conditions for the fields and their momenta reflect this temperature. The last requirement will be fulfilled if we first heating the system by a standard metropolis algorithm applied to the classical Hamiltonian. In this way we will get the correct initial thermal distribution of fields and momenta if we pick one such generated configuration as our starting configuration for the classical equations of motion (of course we repeat the computer experiments many times and in this way we will in addition select many different initial configurations).

In order to solve the classical equations of motion we discretize space in a cubic lattice. Lattice gauge theory allows us to maintain a natural concept of gauge invariance. More precisely we do the following: we choose the standard Wilson action for Euclidean lattice gauge theory combined with the standard Higgs action on a lattice and make the obvious change of sign of the spatial part relative to the temporal part in order to get a Minkowskian signature. After that we make the lattice spacing in the temporal direction much smaller than in the spatial directions by scaling the lattice spacing "a" to "$a\Delta t$". In this way the action will be split into a kinetic and a potential part:

$$S = \sum_t (E_{kin} - E_{pot}) \tag{21}$$

where the summation is over all time slices. From S one can derive the classical equations of motion which connect time slice t and $t + \Delta t$. The equations of motion will automatically satisfy the Gauss constraint by this procedure if we work in temporal gauge, as will be the case. So in the following we will always assume $A_0 = 0$.

If we define an energy functional H by

$$H = E_{kin} + E_{pot} \tag{22}$$

it will not be strictly conserved by the equations of motion since time is still discrete. However, in the limit $\Delta t \to 0$, H reduces to the correct Hamiltonian of the system and the energy will be conserved by the equations of motion. For $\Delta t \leq 0.05$ we have, for the range of coupling constants used in the simulations, found very good conservation of the energy.

For a detailed discussion of the equation of motion derived from (21) we refer to [2]; let us only here give explicitly the form of $\sum_t E_{kin}$ and $\sum_t E_{pot}$ used in the simulations:

$$\begin{aligned}\sum_t E_{kin} &= \frac{\beta_G}{(\Delta t)^2}\sum_{\Box(t)}(1 - \frac{1}{2}\mathrm{Tr}\, U_{\Box(t)}) + \\ &\quad \frac{\beta_H}{2(\Delta t)^2}\sum_{x,t}\mathrm{Tr}\left(\Phi_x^\dagger \Phi_x - \Phi_x^\dagger U_{x,x+\hat{0}}\Phi_{x+\hat{0}}\right)\end{aligned} \tag{23}$$

$$\begin{aligned}\sum_t E_{pot} &= \beta_G \sum_{\Box(s)}(1 - \frac{1}{2}\mathrm{Tr}\, U_{\Box(s)}) + \frac{\beta_H}{2}\sum_{x,t,i}\mathrm{Tr}\left(\Phi_x^\dagger \Phi_x - \Phi_x^\dagger U_{x,x+\hat{i}}\Phi_{x+\hat{i}}\right) \\ &\quad + \beta_R \sum_{x,t}(\frac{1}{2}\mathrm{Tr}\,\Phi_{x,t}^\dagger \Phi_{x,t} - v^2)^2\end{aligned} \tag{24}$$

where $U_{x,x+\mu}$ is the lattice gauge field on the link connecting x and $x + \hat{\mu}$, $U_\Box$ denotes the plaquette action for the gauge field and $\Box(t)$ stands for a plaquette in the $\hat{0} - \hat{i}$ plane while $\Box(s)$ refers to a plaquette in a $\hat{i} - \hat{j}$ plane. The $SU(2)$ Higgs doublet φ is represented as a matrix

$$\Phi \equiv \begin{pmatrix} \phi_0^* & \phi_+ \\ -\phi_+^* & \phi_0 \end{pmatrix}. \tag{25}$$

The summation over spatial points is finite and limited by the lattice 3-volume (periodic boundary conditions) while it is infinite in the time direction.

The Metropolis updating used to thermalize the system uses as weight

$$\exp(-H_\xi), \qquad H_\xi = H + \xi G^2 \tag{26}$$

where H is the lattice hamiltonian given by (22), while G denotes the lattice version of Gauss constraint (see [2] for details). By choosing ξ sufficiently large one can (at the expense of slowing down the thermalization time) control the violation of Gauss constraint to be arbitrarily small in the Metropolis updating.

If we write the classical continuum Hamiltonian as

$$H = \int d^3x \left[\frac{1}{2}E_i^a E_i^a + \frac{1}{4}F_{ij}^a F_{ij}^a + |\pi|^2 + |D_i\varphi|^2 + \lambda_c(|\varphi|^2 - v_c^2)^2\right] \tag{27}$$

the tree value connection between the lattice parameters in (23), (24) and the continuum coupling constants in (27) can trivially be worked out. We do not need them here so for details I refer to the original articles [2, 13].

3.2 The fermions

Let us begin with the continuum formulation as given by (14). If we choose a chiral representation of the Dirac matrices we may according to the Lagrangian (14) write the equations of motion (suppressing the primed fields, which are not essential to the formulation) as

$$\begin{pmatrix} i\sigma_i D_i & M \\ M^\dagger & -i\sigma_i\partial_i \end{pmatrix} \begin{pmatrix} \psi_L \\ \chi_R \end{pmatrix} = E \begin{pmatrix} \psi_L \\ \chi_R \end{pmatrix}, \tag{28}$$

where each entry is a 2×2 matrix in spinor space and a 2×2 matrix in isospace. The resulting system is very similar to the Dirac equation for *four*-component Dirac spinors (recall the comments following (14)).

In order put the fermions on the ($3d$ spatial) lattice we use the so called staggered fermions. There will in this connection be a problem with a faithful representation of the Dirac algebra for the matrices α_i and β since a representation these four matrices in the staggered fermion formalism requires a four-dimensional lattice, while we are only using a three-dimensional lattice. The solution is to introduce an additional "time slice" of link variables, identical to the first *except* that on the second time slice there will be no gauge fields, and coupled to the original one as if it was displaced by one unit in a fictitious Euclidean time direction. In this way we get two components ψ and χ (as in (28)) living on separate time slices coupled via the Higgs interaction. The system of lattice equations then reads

$$\tfrac{i}{2}\textstyle\sum_i^3 \eta_i(x)[U_i(x)\psi(x+\hat{\imath}) - U_i^\dagger(x-\hat{\imath})\psi(x-\hat{\imath})] + \eta_4(x)M(x)\chi(x) = E\psi(x), \tag{29}$$

$$\tfrac{i}{2}\textstyle\sum_i^3 \eta_i(x)[\chi(x+\hat{\imath}) - \chi(x-\hat{\imath})] + \eta_4(x)M^\dagger(x)\psi(x) = E\chi(x), \tag{30}$$

where ψ and χ are now single spin component gauge doublet fields and $\eta_\mu(x) = (-1)^{x_1+x_{\mu-1}}$, $\mu = 1, 2, 3, 4$. The following symmetries exist:

$$(\psi, \chi, E) \Leftrightarrow (\eta_4\psi, -\eta_4\chi, -E), \tag{31}$$

and, if $h_u = h_d$,

$$(\psi, \chi, E) \Leftrightarrow (\tau_2\psi^*, -\tau_2\chi^*, -E). \tag{32}$$

These symmetries ensure that each lattice mode is doubly degenerate if $h_u = h_d$; if not, there will be an additional fine structure in the spectrum.

On our three-dimensional lattice with staggered fermions there is no straight forward definition of the γ_5 matrix. A suitable definition with matches the continuum one given by (15) is as follows:

$$\left\langle\Psi|\tilde{\Gamma}_5|\Psi\right\rangle \equiv \langle\psi|\Gamma_5|\psi\rangle - \langle\chi|\Gamma_5|\chi\rangle, \qquad \Psi \equiv \begin{pmatrix} \psi \\ \chi \end{pmatrix} \tag{33}$$

where

$$\langle\phi_n|\Gamma_5|\phi_n\rangle = \frac{i}{2^3}\sum_x \sum_{ijk=\pm 1} (-1)^{x_2}\phi_n^\dagger(x)\mathcal{U}_{111}\phi_n(x + i\hat{1} + j\hat{2} + k\hat{3}). \tag{34}$$

Here $\phi_n(x)$ is one of the components ψ_n or χ_n of the nth eigenmode Ψ_n of the (lattice) Dirac equation (29)-(30) in the notation (34), and $\mathcal{U}_{111}$ denotes the average of the SU(2) link products over the six equivalent paths joining the two sites, thus ensuring the expectation value is gauge invariant. Notice there is an important difference with the continuum operator. Since Γ_5 is a three-link operator, $\Gamma_5^2 \neq 1$.

For all calculations of the eigenvalue spectrum presented in this paper, anti-periodic boundary conditions in the three spatial directions have been imposed on the fermion fields. The choice of boundary conditions is usually not critical in Monte Carlo simulations [21]; however, equation (30) describes almost free fermions as $h_u, h_d \to 0$, and if periodic boundary conditions are used there will be mixing between the zero mode associated with the sphaleron transition and near zero constant modes. If anti-periodic boundary conditions are used the lowest plane wave modes have energies of $O(\pi/N)$ and this danger is removed.

Because of the symmetry of the spectrum about $E = 0$, the system described by the lattice Hamiltonian (29)-(30) is not truly chiral: if it were there would be spectral asymmetry, and true spectral flow would result when a mode crossed zero. Of course, the reason the model is not chiral is due to the species replication inherent to the lattice approach. For every positive energy level which crosses zero to become negative, there will be a corresponding negative energy level crossing zero to become positive. The symmetries (31),(32) ensure that at each zero crossing *four* SU(2) doublets are involved: two cross one way and two the other. Thus there is no true baryogenesis on the lattice, since for each unit of baryon number created there is another one destroyed. Of course, in the continuum limit we expect the species to become independent (in the quenched approximation), and thus we should be able to focus simply on the crossing itself as the event of interest.

4 Calibration

As mentioned above there seem not to be any analytical results which tell us in detail about the level crossing for chiral fermions coupled to the gauge-Higgs system encountered in the standard model. In order to know what to expect we have investigated this question numerically on the lattice by taking families of smooth configurations[1] which interpolate between different classical vacua and for these configurations we have followed the flow of eigenvalues and the change of chiralities. We have investigated the following scenarios:

(1) A family of gauge-Higgs field configurations which interpolate between two vacua and where the zero of the Higgs field appears at the same time as the Chern-Simons number $N_{cs}(t)$ of the gauge field is equal to 1/2.

(2) A family of field configurations where the gauge- and Higgs fields still interpolate between two vacua, but where the zero of the Higgs field is displaced in time relative to the time t where $N_{cs} = 1/2$.

(3) A family of field configurations where only the gauge field interpolates between two neighboring gauge vacua, while the Higgs field remains constant.

(4) A family of field configurations where only the Higgs field interpolates between two neighboring Higgs vacua, while the gauge field stays constant (for instance zero).

The field configurations are constructed as follows: Let us first define a gauge transformation $V(x)$ with winding number one:

$$V(x) = (-1)\exp\left(\frac{2\pi i \sigma_i v_i}{L|v(x)|}\max\{|v_j(x)|, j = 1,2,3\}\right), \tag{35}$$

[1] By "smooth" configurations we mean lattice configurations which are approximations to continuum configurations.

where x denotes a lattice point, L the linear size of the lattice, $v_i = x_i - \frac{1}{2}L$ and σ_i are the Pauli matrices. The trivial vacuum is given by

$$U_i^{(0)}(x) = \hat{1}, \qquad \phi^{(0)}(x) = \begin{pmatrix} 0 \\ 1 \end{pmatrix}, \tag{36}$$

while a vacuum configuration with winding number one is given by

$$U_i^{(1)}(x) = V(x)V^{-1}(x+\hat{e}_i) = \exp(i\sigma_i u_i(x)) \tag{37}$$

$$\phi^{(1)}(x) = V(x)\phi^{(0)}. \tag{38}$$

where $\hat{e}_i$ is a unit lattice vector in direction i and $u_i(x)$ will be close to zero for large lattices.

A continuous family of configurations which interpolate between the two vacua is simply given by

$$U_i(x,t) = \exp(if(t)\sigma_i u_i(x)), \tag{39}$$

$$\phi(x,t) = (1-g(t))\phi^{(0)} + f(t)\phi^{(1)}(x), \tag{40}$$

where $f(t)$ and $g(t)$ are monotonic, continuous functions which interpolate between 0 and 1 when t changes from 0 to the final time which we denote T.

Note that there exists precisely one space-time point (x,t) where $\phi(x,t) = 0$ and the time t is determined by $1 - g(t) = f(t)$. We denote this time t_h. If $g(t) = f(t)$ it will coincide with the time $t_{1/2}$ where $N_{cs}(t) = 1/2$, but if $g \neq f$ it will in general be different from $t_{1/2}$.

We will choose $f(t) = t/T$ in the actual numerical calculations. This means that $t_{1/2} = T/2$. In case (1) and (4) mentioned above we choose $g(t) = f(t)$. In case (2) we choose $g(t) = \min\{3t/T, 1\}$, which means that the zero of the Higgs field will be located at $t = T/4$, ($t_h = T/4$) i.e. displaced by $T/4$ relative to $t_{1/2}$.

Let us now describe the spectral flow and the change in chirality for the four families of field configurations mentioned above. The Chern-Simons number changes smoothly from 0 to -1 as a function of the time t. The lowest eigenvalue cross zero at $t = 1/2$ and the chirality jumps abruptly from -1 to 1 at $t = 1/2$ for all values of the Yukawa coupling. Everything behaves as expected from the continuum considerations.

In case (2) where the zero of the Higgs field is displaced from $t_{1/2}$ the time t_0 at which the lowest eigenvalue crosses zero changes continuously with increasing Yukawa coupling from $t_{1/2}$, corresponding to the crossing at $N_{cs} = 1/2$ for zero Yukawa coupling, towards t_h, which is the time where the Higgs field has a zero. The flip in chirality always occurs at t_0, as expected.

In case (3) where the Higgs field is constant we see that the level crossing is still present for moderate values of the Yukawa coupling. This is in sharp contrast to the situation where we insert an explicit mass term in the Dirac equation. In that case we know from the continuum equation that there is a mass gap, separating the negative and the positive eigenvalues. When the Yukawa coupling is increased t_0 is gradually displaced towards T, and for a finite value of the Yukawa coupling t_0 reaches T and there is no longer any level crossing and no chirality flip.

The behavior of the lowest eigenvalue in the situation where the gauge field is zero (i.e. $U_i(x)$ is given by (36)) and Higgs field is given by (40) with $g(t) = f(t) = t/T$ is in a certain sense dual to the one in case (3). For small Yukawa coupling there is no level crossing, but eventually a zero and a corresponding chirality flip takes place at T and with increasing Yukawa coupling t_0 moves towards t_h.

The results in this section are relevant when we want to interpret the data obtained in real lattice simulations. In particular we see that increasing the Yukawa coupling can displace the time t_0 of zero level crossing. It is somewhat surprising that there is no need for a zero in the Higgs field in order to observe level crossing, provided the Yukawa coupling is sufficiently small. However, our three dimensional lattice has periodic boundary conditions and since it is finite, it imitates a compact space of finite volume. This means that the "continuum" configuration imitated by (39) for the gauge field and (36) for the Higgs field will have finite energy, and we expect spectral flow of the continuum Dirac operator to be continuous with respect to the coupling constants in the lagrangian. Since we have level crossing for zero Yukawa coupling by the Atiyah-Singer index theorem and since the chiral symmetry is not broken for finite Yukawa coupling, we must have level crossing for small Yukawa couplings as well, even if the Higgs field is nowhere zero. This is indeed what we observe. For sufficiently large Yukawa coupling the level crossing can disappear without violating any continuity requirements. It is more difficult to decide whether these continuity arguments remain valid if the three dimensional space is non-compact. The very simple configurations we have used will have infinite energy in the non-compact case and might not belong to a reasonable class of field configurations where one can define a spectral flow of the Dirac operator $H(A_i(t), \phi(t))$. Similar remarks are valid in case (4). On the other hand the results obtained in case (2) are probably true also for a non-compact continuum space, even if if the actual configuration used here has infinite energy in the continuum limit for $t \neq 0, T$. The reason is that it is easy to find finite energy continuum field configurations with the characteristics used in case (2). An analytic proof of the behavior observed in case (2) has recently been given if space is one dimensional [27]. It would be interesting to extend it to three dimensions.

At high temperature we might certainly encounter situations where the phase of the Higgs and the gauge field are placed in different gauge sectors as in case (3) or (4), since the finite temperature provides the needed infinite energy. Consequently cases (3) and (4) are relevant for our interpretation of data not only because the spatial lattice used in the simulations is finite, but also because the situations might be generic at the high temperatures prevailing in the early universe above the electroweak phase transition.

5 The measurements

5.1 The procedure

Equipped with an initial configuration at a given temperature we can now solve the classical equations of motion and at the same time follow the evolution of the Chern-Simons number by measuring the integral

$$Q(t) \equiv N_{CS}(t) - N_{CS}(0) = \frac{1}{32\pi^2} \int_0^t dt \int d^3x \; F^a_{\mu\nu} F^{a\mu\nu}. \tag{41}$$

This quantity has a simple implementation on the lattice and will directly give us the change in Chern-Simons number, which is a gauge invariant quantity. This procedure has been applied successfully before (see [2] for details). The typical measurement of $Q(t)$ will result in a curve which stays for some time around integer values of $Q(t)$, while the transitions between these values take place rather rapidly. Superimposed on these movements of $Q(t)$ will be short wavelength thermal fluctuations, which occasionally can mask somewhat the picture of $Q(t)$ jumping between integer values. If desired, it is possible to strip off these thermal fluctuations in the following way. By solving the classical equations of motions we get a time sequence of configurations. For each of these configurations we can apply the simplest relaxation equation:

$$\frac{\partial \varphi}{\partial t} = -\frac{\delta H}{\delta \varphi}, \qquad \frac{\partial A}{\partial t} = -\frac{\delta H}{\delta A} \tag{42}$$

where H is the Hamiltonian (22). This technique is well known from the lattice study of monopoles and instantons [21, 23, 24, 25, 26]. If one "cools" the configurations too much one forces them to a vacuum configuration, which is not what we want. However, only a few cooling steps will reduce the energy of the gauge configurations by a factor 80 without breaking the link to the original configuration since the part of the field configurations which survives this relaxation is precisely the long wavelength fluctuations which we imagine are responsible for the large scale changes of $Q(t)$. Indeed, the picture of the transitions between different integer values of $Q(t)$ is considerably sharpened by such a cooling.

In order to check for level crossing during the time-development of the gauge field configurations we calculate the eigenvalues of the Dirac Hamiltonian for each time step, using the staggered fermion formalism discussed in detail in sections 2 and 3. Since the Dirac operator $H_D(A(t), \varphi(t))$ will depend implicitly on the time t due to the time dependence of the gauge and Higgs field we will get a spectral flow of the eigenvalues and we can compare this with the expectation that a crossing of zero should be related to a change in Chern-Simons number and that diving and rising eigenvalues have different generalized chiralities.

5.2 Results for the massless Dirac operator

In [2] the Chern-Simons number was measured by the procedure outlined above. At high temperatures there were rapid transitions where the Chern-Simons number jumped by one unit. In [3] the spectral flow of the *massless* Dirac operator, i.e. the Dirac operator where there is no coupling to the Higgs field, was investigated for the same field configurations. It was shown that there was indeed a crossing of zero of the lowest eigenvalue associated with a sphaleron-like transition where the Chern-Simons number changed by one unit. The generalized chirality operator was introduced in [13] and it was shown that one had an abrupt change in the chirality from -1 to 1 (or opposite) at the same time as the lowest eigenvalue crossed zero.

In addition a number of more complicated situations had a natural interpretation when one combined the information of chirality with that of a change in Chern-Simons number and the flow of eigenvalues. As an example I can mention the situation where there seemingly was a fluctuation in Chern-Simons number , but it was not very large and the Chern-Simons number returned to its original value. One could at the same time observe the diving of two eigenvalues but the change in chirality had opposite sign. This is in accordance with the idea that *two* sphalerons of opposite sign are present on the lattice. In the same way a number of other situations could be understood in a consistent way.

We conclude that the gauge fields seem to qualify as continuum configurations. The change in Chern-Simons number is reflected closely in the spectral flow of the eigenvalues of the massless Dirac operator, which is implemented by means of staggered fermions.

5.3 Spectral flow of the full Dirac operator

In order to investigate the role of the Higgs field in the sphaleron-like transitions we have also measured the spectral flow of the full Dirac operator and the change of generalized chirality $\tilde{\Gamma}_5$ in the case where $h_u = h_d$ (and we will denote the common coupling constant by h_y).

For small values of h_y (i.e. typical $h_y \leq 0.1$) we see a picture consistent with the massless case. However, for larger values of h_y the situation becomes problematic: Both the eigenvalues and $\langle \tilde{\Gamma}_5 \rangle$ will depend on the relaxation of the configurations in a crucial way. In case there is no relaxation we will see signals close to noise. In case we use as much as 8 relaxation steps for each configuration (i.e. we strip off as much as 96% of the thermal energy) we see a signal which can be as level crossing as long as $h_y \leq 0.5$. For larger values there is no reasonable signal. All this is in contrast to the massless case where only the gauge fields are tested and in contrast to the smooth toy configurations we constructed by hand. For the toy configurations one can have h_y as large as two and still see a clear level crossing and a corresponding flip in the generalized chirality. It seems as if the radial part of the Higgs field introduce a lot of spurious transitions due to too wild fluctuations.

6 Discussion

The purpose of this work has been to establish the possibility to investigate in detail by numerical methods a number of questions related to sphaleron-like transitions in the early universe. By simulations we have generated conditions which should have some resemblance with the conditions near the electroweak phase transition: a small expectation value of $|\varphi|^2$ (compared to the tree-value determined from the Lagrange function), but the temperature still not so high that it completely masks the transitions between neighboring vacua. We are here in the lucky situation that we can use our finite lattice volume in a constructive way. Had we been working at infinite or very large volume V it is clear that the temperature fluctuations in globally defined quantities like the Chern-Simons number would be very large since they essentially grow with the square root of the volume. By fine tuning the volume and the coupling constants we can work at high temperature and corresponding small expectation values of $|\varphi|^2$, but still have an acceptably small change in the Chern-Simons number . The risk is that finite volume artifacts will be present, and certainly we would have to check this aspect carefully if we attempted to measure actual transition rates. Then one would have to make sure than the transition rate grew correctly with volume. In this study we have concentrated on the qualitative aspects of the individual transitions: is there any trace of topology left in the high temperature phase above the electroweak transition and can it be associated with level crossing in the usual way?

We have seen that the gauge fields indeed behave like continuum configurations with respect to change of Chern-Simons number and level crossing and chirality of the massless fermions. As mentioned in the introduction it is not a priori clear how this picture survives the coupling to the Higgs sector of the electroweak theory. However, for not too large values of the Yukawa coupling h_y ($h_y \leq 0.5$) we have observed the diving of a single eigenvalue to zero and an associated flip in the generalized chirality $\tilde{\Gamma}_5$ when the Chern-Simons number changes by one unit. The natural interpretation of this is that energy eigenvalues of eigenmodes of opposite generalized chirality cross at zero energy. The crossing is not linked in any precise way to the point where $\Delta N_{CS} =$

1/2, but as discussed in sec. 4 one would not expect this for general gauge-Higgs configurations[2]. The observed picture of level crossing and especially chirality flip is enhanced by stripping off the high energy thermal fluctuations of the gauge and the Higgs fields. This is an indication that the low frequency part of the Higgs field indeed has some topological content, even if it is clear that the Higgs configurations from the outset are rather "wild". It also shows that we presumably have a "renormalization" of the chirality due to fluctuations. The break down of any sensible eigenvalue flow and chirality flip for $h_y \geq 1$ is probably due to the fluctuations of the radial part of the Higgs field. In sec. 4 where the radial part of the Higgs field was smooth we could choose a considerably larger h_y before we encounter any problems with eigenvalues and chirality. When we use the relaxation algorithm on the configuration the situation does not improve much. On the other hand we know from direct measurements that the phase of the Higgs field and the gauge field are almost totally correlated after 8 relaxation steps and that we get very sensible results for smaller values of h_y. Since our relaxation algorithm works rather slowly on the radial part of the Higgs field, it is natural to expect the fluctuations of the radial part to develop into lattice artifacts for some threshold value of h_y. This threshold value should depend somewhat on the relaxation and we have observed this dependence.

Finally we wanted to use the measured flow of eigenvalues and the chirality flip to extract information about the topology of the Higgs field. The results obtained in sec. 4 made this procedure more ambiguous than desired. Before relaxation we can by direct measurement see that the phase of the Higgs field and the gauge field are rather decorrelated. After relaxation they are highly correlated. The basic picture of level crossing and chirality flip is not changed by relaxation for $h_y \leq 0.5$ although it becomes much clearer. Unfortunately we have seen in sec. 4 that we can have level crossing and chirality flip even if the phase of the gauge and the Higgs field develop into different vacua. This means the the cooling itself could induce a change of the relative phase of the gauge and the Higgs field (from being in different vacuum sectors to being in the same vacuum sector) without a drastic change in the level crossing and chirality flip. To determine in more detail the topology associated with the Higgs field would therefore require additional independent measurements.

Acknowledgement

It is a pleasure to thank the organizers of the conference for providing such perfect and simulating environment. In addition I thank my collaborators K. Farakos, S. Hands, G. Koutsoumbas and G. Thorleifsson for numerous discussions and advises. Part of this work was carried out at the Danish computer center UNI-C and made possible by a grant from the Danish Research Council for Natural Science and Technology.

References

[1] G. 't Hooft, Phys. Rev. Lett. 37 (1976) 8; Phys. Rev. D14 (1976) 3432.

[2] J. Ambjørn, T. Askgaard, H. Porter and M.E. Shaposhnikov, Phys. Lett. B244

[2]Even for Yukawa coupling zero we are not aware of any analytic considerations which tell us that the crossing has to take place at the point where $\Delta N_{CS} = 1/2$. Furthermore our configurations do not interpolate between two vacuum configurations due to the finite temperature. The best we can say is that we start out in some neighborhood of a classical vacuum and end up near another classical vacuum.

(1990) 479; Nucl. Phys. B353 (1990) 346.

[3] J. Ambjørn and K. Farakos, Phys. Lett. B294 (1992) 248; Nucl. Phys. B (Proc. Suppl) 30 (1993) 711.

[4] A. Ringwald, Phys.Lett. B213 (1988) 61.

[5] J. Boguta and J. Kunz, Phys.Lett. B154 (1985) 407.

[6] C. Callias, Commun.Math.Phys. 62 (1978) 213.

[7] C.G. Callan, R. Dashen and D.J. Gross, Phys. Rev. D17 (1978) 2717.

[8] N. H. Christ, Phys.Rev. D21 (1980) 1591.

[9] J. Kiskis, Phys.Rev. D18 (1978) 3590.

[10] J. Kunz and Y. Brihaye, *Fermions in the background of sphaleron barrier*, preprint, January 1993.

[11] A.A. Anselm and A.A. Johansen *Can electroweak θ-term be observable ?*, preprint LNPI-1778.

[12] M. Axenides, A. Johansen and H.B. Nielsen, *The level crossing phenomenon with Yukawa Interactions*, preprint NBI-HE-93-46.

[13] J. Ambjørn, K. Farakos, S. Hands, G. Koutsoumbas and G. Thorleifsson, *Level crossing for hot sphalerons*, NBI-HE-93-62, to appear in Nucl.Phys.B.

[14] I-Hsiu Lee and R.E. Shrock, Nucl.Phys. B305 (1988) 305.

[15] T. Reisz, Z.Phys. C53 (1992) 169; A. Irbäck, P. Lacock, D. Miller, B. Petersson and T. Reisz, Nucl.Phys. B363 (1991) 34; P. Lacock, D. Miller and T. Reisz, Nucl.Phys. B369 (1992) 501.

[16] C. Burden and A.N. Burkitt, Europhys. Lett. 3 (1987) 545.

[17] C.P. van den Doel and J. Smit, Nucl. Phys. B228 (1983) 122.

[18] H. Kluberg-Stern, A. Morel, O. Napoly and B. Petersson, Nucl. Phys. B220 [FS8] (1983) 447.

[19] M.F.L. Golterman and J. Smit, Nucl. Phys. B245 (1984) 61.

[20] J. Smit and J.C. Vink, Phys. Lett. B194 (1987) 433; Nucl. Phys. B298 (1988) 557.

[21] S.J. Hands and M. Teper, Nucl. Phys. B347 (1990) 819.

[22] D. Daniel and S.N. Sheard, Nucl. Phys. B302 (1988) 471.

[23] S.J. Hands, Nucl. Phys. B329 (1990) 205.

[24] I.M. Barbour and M. Teper, Phys.Lett. B175 (1986) 445.

[25] E. Ilgenfritz, M. Laursen, M. Müller-Preussker, G. Schierholz and H. Schiller, Nucl.Phys. B268 (1986) 693.

[26] M. Laursen and G. Schierholz, Z.Phys. C38 (1988) 501.

[27] M. Axenides, A. Johansen and H.B. Nielsen, *On sphaleron deformations induced by Yukawa interactions*, preprint NBI-HE-93-75.

FLUCTUATION CORRECTIONS TO THE TRANSITION RATE OF THE ELECTROWEAK SPHALERON*

Jürgen Baacke

Institut für Physik
Universität Dortmund
D-44221 Dortmund

Abstract

I discuss various aspects of the numerical computation of the fluctuation determinant of the electroweak sphaleron. This talk is based on work done together with S. Junker, the results of this work are compared with those of previous computations and with analytical estimates.

INTRODUCTION

The electroweak theory has been found [1, 2] to possess a topologically nontrivial solution which describes a saddlepoint between two topologically distinct vacua.

The recent interest in this solution has been centered around its possible rôle in generating baryon number violating processes in the early universe or even at accelerator energies [3] - [5]. The rate of sphaleron transitions in the range of temperatures $M_W(T) \ll T \ll M_W(T)/\alpha_W$ has been derived by Arnold and McLerran [4]. It is given by

$$\Gamma = \frac{\omega_-}{2\pi} T^{-3} \mathcal{N}_{tr} \mathcal{N}_{rot} 8\pi^2 e^{-E_{sph}/T} \kappa \tag{1}$$

Here ω_- is the absolute value of the eigenvalue of the unstable mode, the prefactors $\mathcal{N}_{tr/rot}$ refer to the normalization of the translation and rotation zero modes. E_{sph} is the sphaleron energy.

The quantity κ takes account of the quantum fluctuations of the sphaleron and is the object being considered here. It is given by

$$\kappa = \mathrm{Im} \left(\frac{\det \mathcal{M}_{FP}^S \det \mathcal{M}_{gf}^0}{\det \mathcal{M}_{FP}^0 \det' \mathcal{M}_{gf}^S} \right)^{1/2} \tag{2}$$

where the symbols $\mathcal{M}$ denote the small fluctuation operators of the gauge fixed action (gf) and the Fadeev-Popov action (FP) evaluated around the sphaleron (S) and the

Electroweak Physics and the Early Universe, Edited by
Romao and Freire, Plenum Press, New York, 1994

vacuum (0), respectively. If the fluctuation operators are diagonalized the determinants are formally given by the product of their squared eigenfrequencies λ_α^2. The determinant $\det' \mathcal{M}^S_{gf}$ of the gauge fixed action is to be evaluated without the zero modes and taking the imaginary part is equivalent to replacing the eigenvalue of unstable mode $\lambda_-^2 = -\omega_-^2$ by its absolute value.

The exact calculation of determinants of generalized Laplaceans as κ constitutes a considerable numerical problem. At first the formal expression has to be defined in a regularized form. Then the divergent parts and the zero mode contributions have be taken out and the resulting finite quantity has to be evaluated numerically. I will describe below various regularized definitions and the numerical approach related to these definitions.

For the sphaleron exact computations of the fluctuation determinant have been presented in Refs. [6] and [7, 8]. Approximations to the fluctuation determinant based on the heat kernel expansion have been developed by Diakonov et al. [9] and been evaluated by Carson and McLerran [10]. More recently Hellmund, Kripfganz and Schmidt [11] have reconsidered this approximation. I will comment on this expansion after presenting the numerical results.

BASIC RELATIONS

The action of a pure SU(2) gauge theory with minimal Higgs sector is given by

$$S = \frac{1}{g^2}\int d^4x\mathcal{L}(x) \tag{3}$$

with the Lagrangean density

$$\mathcal{L} = -\frac{1}{4}F^a_{\mu\nu}F^{\mu\nu a} + 2\left(\mathcal{D}_\mu\Phi\right)^\dagger\left(\mathcal{D}^\mu\Phi\right) - \frac{1}{2}\xi^2\left(\Phi^\dagger\Phi - 1\right)^2, \tag{4}$$

Here the original space and field variables have been rescaled as

$$x\mu \to x_\mu/M_W, \qquad W^a_\mu \to \frac{M_W}{g}W^a_\mu \qquad \text{and} \qquad \Phi \to \frac{v}{\sqrt{2}}\Phi. \tag{5}$$

and we have introduced the ratio of Higgs and W masses ξ via

$$\xi^2 \equiv \frac{m_H^2}{m_W^2} = \frac{8\lambda}{g^2}. \tag{6}$$

As a consequence of the rescaling the action does not depend on the vacuum expectation value of the Higgs field and the couplings appear only in the mass ratio ξ. As explained in [4] at high temperatures the fields can be considered as essentially static and the theory reduces to a three dimensional theory if the vacuum expectation value is rescaled to the temperature dependent one [12, 4]

$$\begin{aligned} v(T) &= v(0)\sqrt{1 - T^2/T_c^2}, &(7)\\ \text{where} \quad T_c^2 &= 2v(0)^2 \Big/ \left[1 + \frac{3g^2}{8\lambda}\right] &(8)\end{aligned}$$

is the critical temperature for symmetry restoration. While the masses become then functions of the temperature, $M_W(T)$ and $M_H(T)$, this dependence cancels in their ratio ξ.

One obtains for the three dimensional Euclidean high temperature action

$$S_E = \frac{1}{g_3^2}\int d^3x\mathcal{L}_E(x), \tag{9}$$

with the Lagrangean density

$$\mathcal{L}_E(x) = \frac{1}{4}F^a_{\mu\nu}F_{\mu\nu a} + 2\left(\mathcal{D}_\mu\Phi\right)^\dagger\left(\mathcal{D}_\mu\Phi\right) + \frac{1}{2}\xi^2\left(\Phi^\dagger\Phi - 1\right)^2. \tag{10}$$

The effective coupling constant of the three dimensional theory g_3 which describes the interaction at high temperatures is given by

$$g_3^2(T) \equiv \frac{g^2T}{M_W(T)}. \tag{11}$$

For temperatures

$$T \ll M_W(T)/\alpha_w \tag{12}$$

this coupling is much smaller than 1 and an expansion with respect to g_3 should be reliable.

The sphaleron [1, 2] is a saddle point solution of the corresponding equations of motion. With the Ansatz ("singular gauge")

$$\begin{aligned} W_{S}{}^a_0 &= 0 \\ W_{S}{}^a_j &= \frac{f_A(r)-1}{r}\varepsilon_{jam}\hat{x}_m \\ \Phi_S &= H_0(r)\begin{pmatrix}0\\1\end{pmatrix} \end{aligned} \tag{13}$$

the energy functional takes the form

$$E_S = \frac{M_W(T)}{\alpha_w}\int dr\,\mathcal{H}(r), \tag{14}$$

whith the Hamiltonian density

$$\mathcal{H}(r) = f'^2_A + \frac{1}{2r^2}(f_A^2-1)^2 + 2r^2H'^2_0 + (f_A-1)^2H_0^2 + \frac{1}{2}\xi^2r^2(H_0^2-1)^2. \tag{15}$$

The sphaleron configuration is obtained by solving the corresponding equations of motion for the profiles $f_A(r)$ and $H_0(r)$ subject to the boundary conditions

$$\begin{aligned} f_A(0) &= -1 \\ f_A(\infty) &= 1 \\ H_0(0) &= 0 \\ H_0(\infty) &= 1. \end{aligned} \tag{16}$$

DEFINITION OF THE FLUCTUATION OPERATOR

The preexponential factor κ is defined to be the determinant of the fluctuation operator. This operator is defined as the second functional derivative of the action with respect to the fluctuations of the field around the classical sphaleron solution. In the present context the fields have been expanded as

$$\begin{aligned} W_0^a &= a_0^a \\ W_i^a &= W_{S}{}^a_i + a_i^a \end{aligned} \tag{17}$$

$$\begin{aligned} \Phi &= (H_0+h)U(\varphi)\begin{pmatrix}0\\1\end{pmatrix} \\ &= \Phi_S + \Phi^{(1)} \end{aligned} \tag{18}$$

where $U(\phi)$ is given by †

$$U(\varphi) = \exp(i\tau^a \varphi^a / H_0). \tag{19}$$

A second ingredient to the fluctuation expansion consists in fixing a gauge for the fluctuations. We used, as it was done in [6], the background gauge condition

$$\mathcal{F}_a = \left(\tilde{\mathcal{D}}_\mu a^\mu\right)^a - i\left[\Phi_{cl}^\dagger \tau^a \Phi^{(1)} - \Phi^{(1)\dagger}\tau^a \Phi_{cl}\right] = 0 \tag{20}$$

for $a = 1, 2, 3$. The tilde indicates that only the classical fields are used in the covariant derivative. The gauge fixing involves also adding the corresponding Fadeev Popov action.

Inserting the field expansion into the action one obtains then the fluctuation operators for gauge and Fadeev Popov fields in the general form

$$\mathcal{M} = \mathcal{M}^0 + \mathcal{V} \tag{21}$$

Here all operators refer to coupled channels; the vacuum part $\mathcal{M}^0$ has only diagonal elements of the form $-\Delta + m^2$ with $m = M_W$ or $m = M_H$ and $\mathcal{V}$ is a potential or vertex operator induced by the classical field configuration.

Since classical solutions in quantum field theory are obtained as extrema of the energy functional, they have usually a high degree of symmetry. The sphaleron has a generalized form of spherical symmetry, a symmetry with respect to K spin rotations, where $\vec{K} = \vec{J} + \vec{I}$ is defined as the sum of total angular momentum and isospin. It is therefore convenient to expand the fluctuations into K spin partial waves. Such a partial wave analysis has been performed in [6] , in the work by Junker and myself we based our analysis on previous work [13] where it was used to investigate the sphaleron stability. The basic algebra and the resulting equations are given in detail in [7].

After the fluctuation expansion and its K spin partial wave decomposition the fluctuation operator $\mathcal{M}$ reduces to corresponding partial wave operators which we denote by $\mathbf{M}_K = \mathbf{M}_K^0 + \mathbf{V}_K$. These are again matrix valued coupled channel operators; the vacuum part $\mathbf{M}_K^0$ is a diagonal matrix of radial Schrödinger type operators $-d^2/dr^2 - (2/r)d/dr - l(l+1)/r^2 + m^2$ where l can take the values $K, K \pm 1$ and $K \pm 2$ and m can be the W boson or Higgs mass. The logarithm of the fluctuation determinant decomposes accordingly:

$$\begin{aligned} \mathcal{J} &\equiv \ln\left\{\frac{\det \mathcal{M}}{\det \mathcal{M}^0}\right\} = \\ \sum_K d_K \mathbf{J}_K &= \sum_K d_K \ln\left\{\frac{\det \mathbf{M}}{\det \mathbf{M}^0}\right\} \end{aligned} \tag{22}$$

where d_K denotes the degeneracy of the partial wave, here $d_K = (2K+1)$.

DEFINITION AND COMPUTATION OF THE FLUCTUATION DETERMINANT

The fluctuation determinant has only been defined formally up to now, what we need is a definition that can be converted into a numerical receipe for its evaluation. While we attempt here an exact evaluation it is useful to consider the perturbative expansion of the fluctuation operator because of its close relationship to renormalized

†This exponential parametrization is - contrary to our statement in [7] - equivalent to the linear one of [13] since in the expansion of the fields only the terms of first order in the small fields are relevant. Second order terms arising in expanding the fields do not contribute to the second order action since they appear multiplied with the first order functional derivative of the action at the classical field configuration which vanishes due to the classical equations of motion.

perturbation theory. Since the fluctuation operator takes the form $\mathcal{M} = \mathcal{M}^0 + \mathcal{V}$ we may expand the logarithm of the fluctuation determinant as

$$\mathcal{J} = \ln\left\{\frac{\det \mathcal{M}}{\det \mathcal{M}^0}\right\} = \sum \frac{(-1)^{k+1}}{k} \mathrm{Tr}\left(\mathcal{G}^0\mathcal{V}\right)^k \tag{23}$$

which is a sum of one loop graphs with the Feynman propagator given by the Green function $\mathcal{G}^0$ of the free fluctuation operator, in momentum space $\mathcal{G}^0 = (p^2 + m^2)^{-1}$. This expansion shows that we have to expect divergent contributions from the graphs of low order in $\mathcal{V}$, depending on the dimension of space. In the context of the sphaleron solution the divergent graphs are tadpole graphs with external Higgs fields [4, 6]. So in general a regularization is necessary. As long as one considers just the contribution of one partial wave this aspect is relevant only in subtracting the vacuum contribution. This is not generally necessary, it is possible to exclude it from the outset (see below). However the necessity of handling perturbative divergences becomes important at the latest if the contributions of partial waves are summed up.

A regularized definition of the fluctuation determinant which is widely used in the mathematical literature and is often applied to cases where if the eigenvalues of the fluctuation operator are known analytically is the ζ function definition

$$\mathcal{J} = -\zeta'_{\mathcal{M}}(0) + \zeta'_{\mathcal{M}^0}(0) \tag{24}$$

where the generalized zeta function of an operator $\mathcal{M}$ is defined in terms of its eigenvalues λ_α^2 as

$$\zeta_{\mathcal{M}}(z) = \sum_\alpha \frac{1}{\lambda_\alpha^{2z}} \tag{25}$$

This definition is certainly less appropriate as a basis of numerical computations. Also the formulation is very remote from the one of renormalized perturbation theory.

Another way to define the fluctuation determinant, used in the context of the electroweak theory by Carson et al. [6] and more recently by Diakonov et al. [14] is the Schwinger proper time definition

$$\begin{aligned}\mathcal{J} &= \sum_\alpha (\ln(\lambda_\alpha^2) - \ln(\lambda_\alpha^{(0)2})) \\ &= \lim_{\epsilon\to 0} \int_\epsilon^\infty \frac{d\zeta}{\zeta} \sum_\alpha (\exp(-\zeta\lambda_\alpha^2) - \exp(-\zeta\lambda_\alpha^{(0)2}))\end{aligned} \tag{26}$$

As mentioned above here the eigenvalues λ_α^2 of the operator $\mathcal{M}$ have to be determined numerically. In order to do that, a finite space boundary of radius R has to be introduced in order to obtain a discrete spectrum. At the end of course one has to perform the limit $R \to \infty$. This method is closely related to the heat kernel expansion and its relation to renormalized perturbation theory has become standard.

Another practical definition of the fluctuation determinant, the one used by us, is based on the relation

$$\sum_\alpha (\ln(\lambda_\alpha^2) - \ln(\lambda_\alpha^2 + \Lambda^2)) = 2\int_0^\Lambda d\nu\nu \sum_\alpha \frac{1}{\nu^2 + \lambda_\alpha^2} \tag{27}$$

where we have introduced a Pauli Villars cutoff Λ from the outset. The limit $\Lambda \to \infty$ exists after subtracting from the integrand the divergent tadpole graphs. It is not necessary to determine the eigenvalues explicitly since the expression can be related to the Euclidean Green function via

$$\sum_\alpha \frac{1}{\nu^2 + \lambda_\alpha^2} = \int d^3x \sum_\alpha \frac{\phi_\alpha^*(\vec{x})\phi_\alpha(\vec{x})}{\nu^2 + \lambda_\alpha^2} \tag{28}$$

$$= \int d^3x \mathcal{G}_E(\vec{x}, \vec{x}, \nu) \tag{29}$$

Here ϕ_α are the eigenmodes and the Euclidean Green function is a solution of the equation

$$(\mathcal{M} + \nu^2)\mathcal{G}_E(\vec{x}, \vec{x}', \nu) = \delta^3(\vec{x} - \vec{x}') \tag{30}$$

This equation can - after partial wave reduction - be solved numerically without any reference to the eigenvalues of the fluctuation operator. This constitutes an advantage with respect to the previous method. The Euclidean Green function has its poles on the imaginary ν axis while the computation is done for real ν.

The partial wave Green functions can be found numerically [15] by determining the solutions f_l^-, regular at $r = 0$, and f_l^+, regular as $r \to \infty$, of the corresponding radial differential equations

$$(\mathbf{M}_l + \nu^2) f_l^{\pm}(r) = 0 \tag{31}$$

(We describe the case of a single channel and use l instead of K). The partial wave Green function is then obtained in the usual way as

$$G_l(r, r', \nu) = f_l^-(r_<) f_l^+(r_>)/[r^2 W(f_l^-, f_l^+)] \tag{32}$$

where W is the Wronskian of f_l^+ and f_l^- ($r^2 W(r) = const.$). The detailed procedure for a multichannel system as we have it here is described extensively in Ref. [15]. We just mention that it is very convenient to decompose the functions f_l via

$$f_l^{\pm} = b_l^{\pm}(1 + h^{\pm}) \tag{33}$$

where $b_l^+ = i_l$ and $b_l^+ = k_l$, the modified spherical Bessel functions are the solutions for the free operator $\mathbf{M}_l^0 + \nu^2$ with the corresponding boundary conditions. One solves then numerically the differential equations one obtains for the functions $h^{\pm}$. One has then

$$G_l(r, r, \nu) - G_l^0(r, r, \nu) = (h^+ + h^- + h^- h^+) i_l k_l / (r^2 W) \tag{34}$$

so that the vacuum term is not subtracted but *omitted*.

The final equation for the fluctuation determinant is then

$$\mathcal{J} = 2 \int_0^\infty d\nu \nu F(\nu) \tag{35}$$

where

$$\begin{aligned} F(\nu) &= \int d^3x (\mathcal{G}_E^S(\vec{x}, \vec{x}, \nu) - \mathcal{G}_E^0(\vec{x}, \vec{x}, \nu)) \\ &= \sum_l d_l \int dr r^2 G_l(r, r, \nu) \end{aligned} \tag{36}$$

In general the Green function is matrix valued and a trace has to be taken. d_l is degeneracy, $2l + 1$ in the special case.

There is finally a very fast method - based on a theorem on functional determinants [16] - which has been developed recently by V. Kiselev and the author [17]. Applied to the partial wave fluctuation operators the theorem takes the form

$$\frac{\det \mathbf{M}_l}{\det \mathbf{M}_l^0} = \lim_{r \to \infty} \frac{\psi(r)}{\psi^0(r)} \tag{37}$$

where the wave functions ψ and ψ^0 are solutions of

$$\mathbf{M}^{(0)} \psi^{(0)}(r) = 0 \tag{38}$$

which are regular at $r = 0$ and satisfy

$$\lim_{r \to 0} \frac{\psi(r)}{\psi^0(r)} = 1 \tag{39}$$

For a more precise formulation of the theorem we refer the reader to [16]. For details of this method which takes over many advantages from the previous one we refer to [17], [18] and the talk of V. Kiselev at this meeting. This method turns out - however - to have a problem when applied to topological solutions like instantons or sphalerons. The requirement that the ratio of the solutions tends to 1 as $r \to 0$ can obviously not be satisfied if the potential $\mathbf{V}_l$ modifies the centrifugal barrier as it turns out to be the case for topological solutions. Attempts to solve this problem e. g. by a suitable regularization have not been successful up to now.

ZERO MODES, UNSTABLE MODE AND RENORMALIZATION

Equations (35,36) are still formal, since the integral over ν is divergent both at $\nu = 0$ and at $\nu = \infty$ due to zero modes and ultraviolet divergences respectively.

Renormalization requires two steps
- the divergent tadpole graphs have to be removed from the numerical calculation and M_W and M_H have to be replaced by their temperature dependent counterparts $M_W(T)$ and $M_W(T)$ as discussed in Ref. [4].
- while the zero temperature part of the tadpole contribution is absorbed into the zero temperature renormalization there is a finite contribution which has to be restituted. This is discussed in [12],[4] and a bit more explicitly in [8].

The first step was performed by subtracting the tadpole contribution - which is known analytically - from $F(\nu)$; the difference $F_{ren}(\nu)$ behaves then as ν^{-3} as $\nu \to \infty$ and the numerical ν integral becomes ultraviolet convergent.

The second step was forgotten in [7] and this lead to a large discrepancy w. r. t. the results of [6]. The finite contribution that has to be restituted is [8]

$$(\ln \kappa)_{tad} = \frac{1}{2}(9 + \frac{3}{2}\xi^2(1+\xi)) \int drr^2(H_0^2 - 1) \tag{40}$$

As one may see from Eqs. (28),(29) and (36) F_{ren} behaves as $6\nu^{-2}$ near zero due to the presence of 6 zero modes, leading to a logarithmic divergence of the ν integration. It can be realized that Eq.(27) has been constructed in such a way that the contribution of each eigenmode is related to the evaluation of the integral of $F_{ren}(\nu)$ at its lower limit; the procedure for removing the zero mode contributions consists therefore in evaluating the integral with a sufficiently small lower limit ϵ and then adding $6\ln(\epsilon)$. The resulting expression stays finite as $\epsilon \to 0$ and this limit is the final answer:

$$\ln \kappa = \lim_{\epsilon \to 0}(\int_\epsilon^\infty d\nu \nu F_{ren}(\nu) + 6 \ln \epsilon). \tag{41}$$

It should be noted that ν and therefore ϵ have the dimension of energy, so the scale used for these variables matters. We have performed our calculations in units of M_W while in Ref. [6] the scale was gv ; we had to take this into account for comparing our results. In the complete rate (Eq.(1)) the dimensional scale reappears in the scale of the translation and rotation mode prefactors so that the final result is unique as it should.

It remains to consider the rôle of the unstable mode. It leads to a pole $1/(\nu^2 - \omega_-^2)$ in $F(\nu)$ where ω_- is the absolute value of the corresponding imaginary eigenvalue λ_-. Since this eigenvalue should be included in the determinant with its absolute value, the correct prescription for evaluating the integral is the principal value one. In praxi we have subtracted from $F_{ren}(\nu)$ an expression

$$F_{pole} = \frac{1}{\nu^2 - \omega_-^2} - \frac{1}{\nu^2 + \sigma^2} \tag{42}$$

with some suitable value for σ. This subtraction makes the integrand regular at $\nu = \omega_-$ without spoiling its ultraviolet convergence. We have then added the analytic principal value integral of this expression back to the numerical integral.

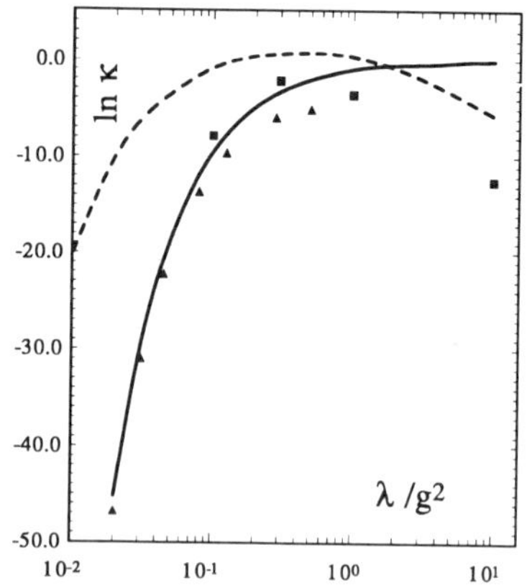

Figure 1. **The Fluctuation Determinant**: We plot the logarithm of the fluctuation determinant κ for various values of the ratio $\lambda/g^2 = \xi^2/8$. The triangles are our results, the squares those of Ref. [9]. The dashed line is the estimate of Ref. [7] and the solid line the estimate based on the effective action of Ref. [22].

Our analytical and numerical procedure contains various implicit checks:
(i) The asymptotic behaviour of $F(\nu)$ and $F_{ren}(\nu)$ follows from analyzing the perturbation expansion of the Euclidean Green function to be as ν^{-1} and ν^{-3} respectively. Note that for $F(\nu)$ not only the power behaviour but also its absolute normalization are determined.
(ii) The behaviour of the partial wave contributions at large values of K spin can be derived from a perturbative analysis to be as K^{-2}.
(iii) The zero modes and unstable mode poles occur with the correct positions and residues.

Furthermore the differential equations for all partial waves have been checked to a certain extent by verifying their gauge invariance before gauge fixing [13]. The decoupling of the time components of the vector field after adding the gauge fixing term represents a further check.

For details of the numerical computations we refer the reader to [7]. Our results are plotted in Fig. 1 together with previous results and estimates. The unit used for κ is gv^6 as in Ref. [6] while we used the scale unit M_W throughout (see Eq. (5)).

COMMENTS AND CONCLUSIONS

As we have seen after correcting some errors the results of Refs. [6] and [7, 8] have become more or less compatible. They become furthermore consistent with an estimate obtained by inserting the sphaleron configuration into the effective "Φ^3" potential generated by the gauge field loops, an approximation that should be good if $M_H \ll M_W$. It is given by

$$(\ln \kappa)_{\Phi^3} = 3 \int dr r^2 (H_0^3 - 1) \tag{43}$$

However the approximation of Refs. [9] and [10] still disagrees substantially with the numerical results. This feature seems to be understood [11, 8]. The approximation was based on the heat kernel expansion, and this expansion was carried out, as usual, up to a fixed power of the heat kernel "time" variable, thus generating higher and higher derivative terms. The "zero-derivative" effective potential is however generated by summing the *nonderivative* terms to *all* powers in heat kernel time and this does generate a good approximation namely Eq. (43). Similar expansions [18]-[21] have been used recently, including also the two-derivative "Z factors" in various contexts. In the case of fermion loops they have also been compared [18] to exact results and found to approximate them well for large fermion masses.

In conclusion the numerical computations of the fluctuation corrections to the sphaleron transition rate and analytical approximations to these corrections are found now to be consistent which each other. Though the agreement between the two exact computations is not perfect the precision reached should be sufficient for cosmological applications.

REFERENCES

1. N. Manton, *Phys. Rev.* D28:2019 (1983)
2. F. Klinkhamer and N. Manton, *Phys. Rev.* D30:2212 (1984)
3. V. Kuzmin, V. Rubakov and M. Shaposhnikov, *Phys. Lett.* 155B:36 (1985)
4. P. Arnold and L. McLerran, *Phys. Rev.* D36:581 (1987) ; *ibid.* D37:20 (1988)
5. A. Ringwald, *Phys. Lett.* B201:510 (1988)
6. L. Carson, X. Li, L. McLerran and R.-T. Wang, *Phys. Rev.* D42:2127 (1990)
7. J. Baacke and S. Junker, *Phys. Rev.* D49:2055 (1994)
8. J. Baacke and S. Junker, Dortmund preprint DO-TH-93-13EA, to be published in *Phys. Rev.* D
9. D. I. Diakonov, V. Yu. Petrov and A. V. Yung, *Phys. Lett.* 130B:385 (1983) ; *Yad. Fiz.* 39:240 (1984) [*Sov. J. Nucl. Phys.*39:150 (1984)]
10. L. Carson and L. McLerran, *Phys. Rev.* D41:647 (1990)
11. M. Hellmund, J. Kripfganz and M. G. Schmidt, Heidelberg preprint HD-THEP-93-23 rev (1993)
12. G. Dolan and R. Jackiw, *Phys. Rev.* D9:3320 (1974)
13. J. Baacke and H. Lange, *Mod. Phys. Lett.* A7:1455 (1992)
14. D.Diakonov, M. Polyakov, P. Sieber, J. Schaldach and K. Goeke, Bochum preprint RUB-TPII-62/93
15. J. Baacke, *Z. Physik* C47:619 (1990) ; *ibid.* C53:407 (1992)
16. S. Coleman, 'The Uses of Instantons' in "The Whys of Subnuclear Physics", A. Zichichi ed., PlenumPress, New York 1979
17. J. Baacke and V. G. Kiselev, *Phys. Rev.* D48:5648 (1993)
18. J. Baacke, H. So and A. Sürig, Dortmund preprint DO-TH-94-06, to be publ. in *Z. Phys.*
19. I. Moss,D. Toms and A. Wright, *Phys. Rev.* D48:1671 (1992)
20. D. Bödeker, W. Buchmüller, Z. Fodor and T. Helbig, DESY 93-147 (1993)
21. J. Kripfganz, A. Laser and M. G. Schmidt, Heidelberg preprint HD-THEP-94-13
22. M. E. Shaposhnikov, *Nucl. Phys.* B287:757 (1987) ; *ibid.* B299:797 (1988)

HOT ABELIAN HIGGS MODEL: REAL-TIME PROPERTIES

A. Krasnitz

Interdisziplinäres Projektzentrum für Supercomputing
ETH-Zentrum, CH-8092 Zürich, Switzerland

Abelian Higgs model (AHM) in 1+1 dimensions is one of the simplest yet nontrivial field-theoretic analogs of the Standard Model (SM) as far as fermion number nonconservation at high temperature is concerned. Clearly, there is little hope of quantitatively understanding this phenomenon in SM if a similar goal cannot be achieved for AHM, which lends itself much easier to both analytical and numerical studies. This explains considerable amount of work done on this aspect of AHM in the recent years [1, 2, 3, 4, 5].

Prior to the reported work, the status of knowledge on thermally induced fermion-number violation in AHM has been as follows. Much like in the SM, the dynamics of Chern-Simons number N_{CS} (and thus of the fermion number) in AHM can reliably be studied in the classical approximation at temperatures well above the gauge-boson mass scale. A relative simplicity of AHM allows to give an accurate estimate of the N_{CS} diffusion rate at temperatures T low compared to the energy $E_{\rm sph}$ of a sphaleron saddle-point configuration separating vacua labeled by consecutive integer values of N_{CS}. Not surprisingly, the rate Γ exhibits thermal-activation behavior [2]

$$\Gamma = f(\xi)T^{2/3}\left(\frac{E_{\rm sph}}{T}\right)^{\frac{7}{6}}\exp\left(-\frac{E_{\rm sph}}{T}\right), \tag{1}$$

where $\xi = 2m_W^2/m_H^2$. Unlike the SM case, in AHM both the sphaleron energy and the f prefactor are known analytically. Numerical real-time simulations of AHM at low temperatures, including those of the current work, indeed agree with (1).

Surprising as it may sound, at the opposite end of thermal scale, $T \gg E_{\rm sph}$, less was known about the rate in AHM than in SM. In SM, the only dimensionful quantity relevant at such high temperatures is the temperature itself, hence, on dimensional grounds, the rate is expected to behave as T^4. In AHM, however, the gauge coupling constant is dimensionful, and therefore, while a power-law behavior was generally believed to be the case [6, 7], the corresponding power could not be established from dimensional considerations. Nevertheless, a simple scaling argument given in the following indicates that this power is 2/3.

As usually is the case, the intermediate-temperature regime (T comparable to E_{sph}) is the least accessible to analytical estimates. Numerical work [4, 5] revealed a surprising feature: at temperatures 5 to 7 times less than E_{sph} the rate practically stops growing and remains roughly constant until the sphaleron scale is reached. Our simulations show that this phenomenon persists at smaller lattice spacings and is therefore not a lattice artifact.

We shall now elaborate on the points just made. Due to space limitations we will concentrate on the physics and not discuss our numerical methods in any detail. Those will be described elsewhere. We begin by giving a scaling argument in favor of $T^{2/3}$ behavior of the rate at high temperatures. Consider the lattice-regularized classical AHM whose temporal-gauge Hamiltonian can be written in suitably chosen units as

$$H(v,a) = \frac{1}{2a}\sum_{j=1}^{N}[\xi E_{j,j+1}^2 + |\pi_j|^2 + \frac{a^2}{2}(|\phi_j|^2 - v^2)^2 + |\phi_{j+1} - e^{iaA_{j,j+1}}\phi_j|^2], \tag{2}$$

where π_j and $E_{j,j+1}$ are canonically conjugate momenta of the complex scalar (Higgs) field ϕ_j, and the spatial component of the gauge field $A_{j,j+1}$, residing respectively on sites and links of an N-site chain. The lattice spacing is a. Periodic boundary conditions are assumed. The reader can verify by direct substitution that if $\phi_j(t)$, $\pi_j(t)$, $A_{j,j+1}(t)$, $E_{j,j+1}(t)$ are solutions of the equations of motion following from $H(v,a)$ with $H(v,a) = \mathcal{E}$, then $c\phi_j(ct)$, $c\pi_j(ct)$, $cA_{j,j+1}(ct)$, $cE_{j,j+1}(ct)$ play the same role for $H(cv, a/c)$ with $H(cv,a/c) = c^3\mathcal{E}$. Consider now diffusion of Chern-Simons number defined as

$$N_{\text{CS}}(t) = \frac{a}{2\pi}\sum_{j=1}^{N} A_{j,j+1}(t). \tag{3}$$

The corresponding diffusion rate is

$$\Gamma(a,v,T) = \lim_{N,t\to\infty}\frac{1}{aNt}\langle(N_{\text{CS}}(t) - N_{\text{CS}}(0))^2\rangle_T. \tag{4}$$

By $\langle\rangle_T$ we mean integration over all the initial values of dynamical variables with the weight

$$\exp\left(-\frac{H(v,a)}{T}\right)\prod_{j=1}^{N}\delta\left(C_j(a)\right), \tag{5}$$

where the delta-function factors impose a set of Gauss' constraints

$$C_j(a) \equiv \frac{1}{a}(E_{j,j+1} - E_{j-1,j}) - \text{Im}(\pi_j\phi_j) = 0. \tag{6}$$

Since $H(v,a)$ is a periodic function of the A variables, the latter need only be integrated in the range $|aA| \leq \pi$. Multiplying all the integration variables (*i.e.*, all the initial values of fields and momenta) in (4) by c and using the scaling property of $H(v,a)$ under this transformation we obtain a simple scaling rule for the rate:

$$\Gamma\left(\frac{a}{c}, cv, c^3T\right) = c^2\Gamma(a,v,T). \tag{7}$$

For this rule to be satisfied the rate must have a functional form

$$\Gamma = T^{\frac{2}{3}}g\left(a^3T, \frac{v^3}{T}\right), \tag{8}$$

where $g(x,y)$ is an arbitrary function of its two arguments. In particular, since $E_{\text{sph}} \propto v^3$, (1), derived for the continuum theory, shows that $g(0,y) \propto y^{7/6}\exp(-y)$ for $y \gg 1$. As

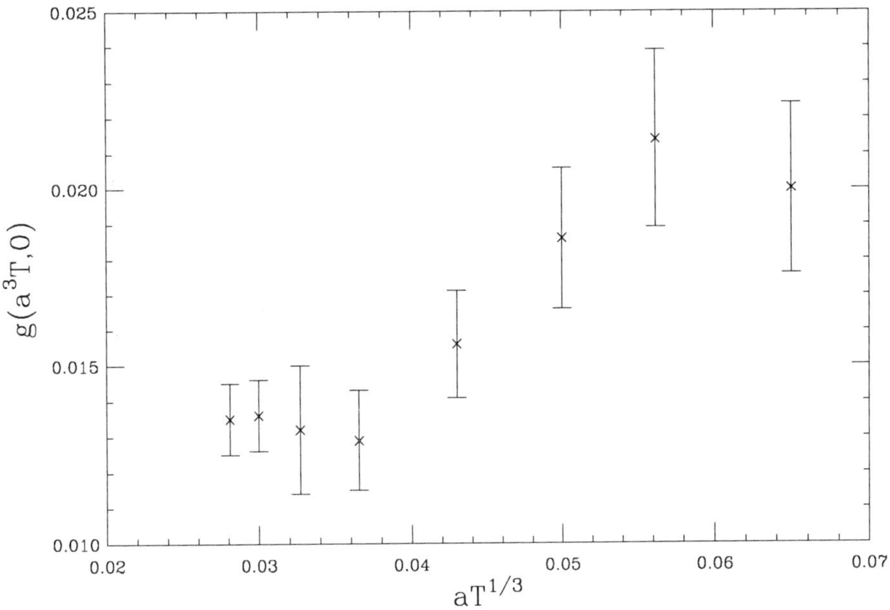

Figure 1. $g(a^3T, 0)$ measured at small values of $aT^{1/3}$. $\xi = 10$

for the temperatures far above the sphaleron scale, the continuum limit of the rate, if it exists, is $g(0,0)T^{2/3}$ as claimed. Both result (1) and numerical evidence strongly suggest that the rate does indeed have a finite continuum limit for any nonzero v^3/T. It is also plausible that the rate becomes insensitive to the details of the Higgs potential at energy scales far below T, *i.e.*, Γ becomes independent of v if $v^3 \ll T$. We therefore expect $0 < g(0,0) < \infty$.

We verified the scaling rule (7) by a numerical simulation. Since in this part of the work we are primarily interested in the high-temperature regime, we simplify matters by setting $v = 0$. The rate measurement then gives $\Gamma(a = 1/2, T = 1)/\Gamma(a = 1, T = 1/8) = 4.\pm0.4$, in agreement with (7). Next, we performed a series of rate measurements closer to the continuum limit of the $v = 0$ theory in order to estimate $g(0,0)$. The measurements, done at lattice spacings 1/16 and 1/8, span a range of values of a scaling-invariant parameter $x = aT^{1/3}$ between 0.028 and 0.065. The results, presented in Figure 1, indicate that $g(x, 0)$ approaches a finite value as $x \to 0$.

Next, we studied the intermediate-temperature properties of the rate. As in previous work on the subject, we fixed $v = 1$, resulting in $E_{\mathbf{sph}} \approx 0.94$. The inverse-temperature range of interest is then $1 < \beta < 5$, where an extremely weak temperature dependence of the rate was found in previous work. By simulating the theory closer to continuum limit ($a = 1/16$ compared to $a \geq 1/8$ in the earlier work), we verified that this feature is not a lattice artifact: as Figure 2 shows, the plateau region becomes wider and more pronounced as the lattice spacing shrinks. For higher temperatures, $\beta \leq 1$, the rate growth resumes but shows a strong lattice spacing dependence. It would be interesting to check by simulations at smaller ($a < 1/16$) lattice spacings whether the rate values begin to approach the $T^{2/3}$ asymptote.

Let us summarize the properties of N_{CS} diffusion rate in AHM as they emerge from this and previous works. At low ($T \ll E_{\mathbf{sph}}$) the rate follows the activation-theory prediction (1) and is exponentially suppressed by a Boltzmann factor involving the sphaleron barrier height. At extremely high temperatures ($T \gg E_{\mathbf{sph}}$) the rate,

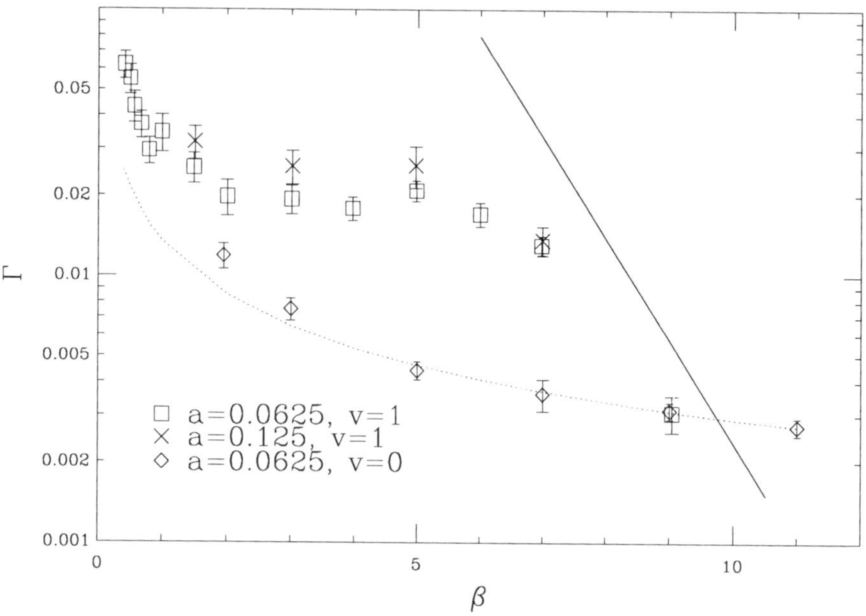

Figure 2. Measured rate dependence on the inverse temperature β. The solid line is the activation-theory prediction (1). The dashed line shows the $T^{2/3}$ law for the high-temperature theory, with $g(0,0)$ estimated at $\beta = 11$, $a = 1/16$. ξ=10.

if it has a finite continuum limit, should grow as $T^{2/3}$. Our simulations do point to a finite continuum limit of the high-temperature rate. The crossover between the two extreme regimes involves a temperature interval characterized by a very slow or no rate growth. The scaling-invariant prefactor of $T^{2/3}$ is not known in detail at the moment. Determining its dependence on a^3T, v^3/T, and the ξ parameter at intermediate and high temperatures should be a goal of future numerical and analytical studies of AHM.

This work owes a great deal to continuous advice and help from Ph. de Forcrand, R. Potting, and A. Bochkarev. I acknowledge enlightening discussions with A. Kovner, J. Smit, A. Wipf, W.H. Tang, E. Mottola, and S. Habib.

References

[1] D. Yu. Grigoriev, V. A. Rubakov, and M. E. Shaposhnikov, Nucl. Phys. B **326** 737 (1989).

[2] A. I. Bochkarev and G. G. Tsitsishvili, Phys. Rev. D **40** 1378 (1989).

[3] A. I. Bochkarev and Ph. de Forcrand, Phys. Rev. D **44** 519 (1991).

[4] A. Krasnitz and R. Potting, Phys. Lett. B **318** 492 (1993); Nucl. Phys. B (Proc. Suppl.) **34** 613 (1994).

[5] J. Smit and W.H. Tang, Nucl. Phys. B (Proc. Suppl.) **34** 616 (1994).

[6] E. Mottola and A. Wipf, Phys. Rev. D **39** 588 (1989).

[7] M. Dine, O. Lechtenfeld, B. Sakita, W. Fischler, and J. Polchinski, Nucl. Phys. B **342** 381 (1990).

NON-EQUILIBRIUM EVOLUTION OF DISORIENTED CHIRAL CONDENSATES

Emil Mottola

Theoretical Division, Los Alamos National Laboratory
Los Alamos, New Mexico 87545 USA

ABSTRACT

The nonequilibrium time evolution of the hot hadronic plasma produced in a relativistic heavy ion collision is studied in the $O(N)$ linear sigma model to leading order in the large N expansion. The chiral phase transition to the broken symmetry phase as the plasma expands and cools is described by the evolution of the order parameter $< \sigma >$. In order to create a large domain of disoriented chiral condensate (DCC) long wavelength fluctuations must become and remain unstable for times longer than the characteristic strong interaction time scale of a few fm/c. Numerical evidence is presented that such instabilities survive for proper times that are at most $3\,fm/c$, and hence do not give rise to a substantial signal of DCC in the coherence of the final state pions. The large N development of time-dependent field equations described in this work is quite general and should be applicable to the electroweak phase transition, bubble wall propagation, and baryogenesis problems as well.

1. INTRODUCTION

The possibility of producing large correlated regions of quark condensate $< \bar{q}_i q_j >$ pointing along the wrong direction in isospin space was proposed to explain rare events where there is a deficit or excess of neutral pions observed in cosmic ray experiments[1-2]. The idea that such a disoriented chiral condensate (DCC) may be produced in relativistic heavy ion collisions has motivated several experimental proposals[3] and a number of theoretical papers[4-11]. It has been recognized that the possibility of forming large regions of DCC relies on the existence of a substantial region in which the hot plasma formed after the collision evolves *out of equilibrium*. In fact, if thermal equilibrium is approximately preserved by the dynamics, the typical correlation length would be determined by the pion mass and no significant correlation of final state pions can be expected. Hence, the main question any investigation of DCC should answer is whether during the evolution that follows the collision there are instabilities in the fluctuations around the mean condensate field. If so, the fluctuations will grow in amplitude and large correlations can be imprinted on the final state pions. If not, correlations in the

Electroweak Physics and the Early Universe, Edited by
Romao and Freire, Plenum Press, New York, 1994

final state are expected to remain small and no dramatic DCC signal will be observed.

It should be clear from the outset that the answer to this question will depend on the assumptions one is willing to make about the initial state of the plasma immediately after the collision. For example, the typical way to introduce instability into the problem has been by making a rapid "quench." In a quench, the hot plasma created after the collision is subject to a sudden external action (the expansion) that is assumed to modify the effective potential, turning it "upside down," without changing its instantaneous state. This sudden approximation is clearly an idealization which is appropriate only if the expansion is much more rapid than the typical interaction time in the plasma. However, the time scale for interactions and instabilities in the plasma is determined by the interaction strength and cross sections. Using the linear sigma model as the effective field theory to describe the chiral phase transition, the quartic coupling must be rather large ($\lambda_r \approx 10$) in order to model low energy $\pi - \pi$ scattering data, even approximately, as we shall see below. In this moderately strong coupling regime interaction time scales are *not* long compared to the plasma expansion time scale, the quench approximation is not justified, and instabilities persist only for quite short periods of proper time. This makes it difficult to obtain large sizes for the correlated domains.

In the approach described here we do not put in instabilities by hand but begin instead in a region of stability at an initial value of the proper time τ_0 from the collision event in the symmetric phase, above the critical temperature $T_c^2 = 12 f_\pi^2/N$. The system cools via its expansion into the vacuum and this results in the transition to the chirally broken phase.

It is important to realize that intuitions developed in cosmological phase transitions, such as the electroweak transition discussed elsewhere in these proceedings where the coupling is extremely weak, may not apply at all to the chiral phase transition of the strongly interacting plasma. In a weakly coupled system the domains can grow for long periods of time and a totally different picture can emerge. Nevertheless, the formalism to study the chiral phase transition employed here is quite general and should work in the electroweak case equally well.

2. THE LINEAR σ-MODEL IN THE LARGE N APPROXIMATION

In order to describe the chiral phase transition of hot hadronic matter we use the $O(N)$ linear sigma model, which describes the low energy phenomenology of pions, with the correct chiral symmetry properties. We will choose the parameters of the model to give reasonable values for three experimentally determined quantities : the mass of the pion m_π, the pion decay constant f_π and the s wave $\pi - \pi$ phase shifts above threshold. For $N = 4$ the chiral phase transition in this model takes place at a temperature of order $T_c = \sqrt{3} f_\pi$ in qualitative agreement with the lattice estimates. The mesons are organized in an $O(4)$ vector $\Phi = (\sigma, \vec{\pi})$ and the action is

$$S = \int d^4x \{ \frac{1}{2} \partial\Phi \cdot \partial\Phi - \frac{1}{4} \lambda (\Phi \cdot \Phi - v^2)^2 + H\sigma \}. \tag{1}$$

We will describe the evolution of the mean value $\bar{\Phi} \equiv < \Phi >$ and the two point correlation functions including the effects of quantum and thermal fluctuations. Perturbation theory is useless for our purpose[12], and a scheme which is non-perturbative in λ must be adopted. In the literature, the mean field Hartree (or Gaussian) ansatz and the large N expansion of the $O(N)$ sigma model have been considered[13,14]. We will adopt the latter since it presents several advantages. On the one hand, the expansion is systematic and allows us to study higher order corrections, as well as being directly generalizable to gauge theories. Work is in progress in this direction and results will be presented elsewhere [15]. On the other hand, the naive Gaussian approximation is known to violate the Goldstone theorem, and leads to an unphysical (and not necessarily small) mass for the pions in the $H = 0$ limit.

The large N effective equations can be obtained in a variety of ways. A very convenient method is to use an alternative equivalent action, which is a functional of the original fields Φ and of the auxiliary constrained field $\chi = \lambda(\Phi^2/2N - v^2)$ [16].

$$\tilde{S}[\Phi, \chi] = \int d^4x\{-\frac{1}{2}\phi_i(\Box + \chi)\phi_i + \frac{\chi^2}{4\lambda} + \frac{1}{2}\chi v^2 + H\sigma\}. \tag{2}$$

As this action is quadratic in Φ we can perform the functional integral over those fields and are left with the integration over χ which, to leading order in $1/N$ can be calculated by the stationary phase method. The effective action, which is a functional of the mean fields of Φ and χ, is obtained by a Legendre transformation of the generating functional of the connected Green's functions. To the lowest order (in large N expansion) the effective action is given by

$$\Gamma[\Phi, \chi] = \int d^4x\{-\frac{1}{2}\phi_i(\Box + \chi)\phi_i + \frac{\chi^2}{4\lambda} + \frac{1}{2}\chi v^2 + H\sigma + \frac{i}{2}N\text{tr}\ \text{ln}G_0^{-1}\}, \tag{3}$$

where

$$G_0^{-1}(x, y) = i[\Box + \chi(x)]\ \delta^4(x - y), \tag{4}$$

and where for notational convenience we drop the overbars and denote the expectation values as Φ and χ. By varying this effective action we derive the equations of motion for the fields Φ and the equation of constraint (gap equation) for the composite field χ, which also plays the role of the effective mass for the mean values Φ_i. We find

$$\chi(x) = \lambda\big(-v^2 + \Phi^2(x) + NG_0(x, x)\big). \tag{5}$$

The function $G_0(x, x)$ that appears in (5) is the coincidence limit of the propagator $G_0(x, y)$ that inverts the operator G_0^{-1} defined in (4).

The initial value problem associated with equations (4)– (5) will be solved in the next section. First I would like to address the issue of the values of the parameters appearing in the above equations so that they describe low energy pion physics. The measurable quantities we fix are the pion mass, $m_\pi = 135\ MeV$ and the pion decay constant $f_\pi = 92.5\ MeV$. The physical masses can be related to the parameters of the theory by computing the inverse propagators of the pion and sigma fields. The pion mass is determined by the the vacuum expectation value of χ,

$$\chi_v = m_\pi^2 \equiv m^2\ , \tag{6}$$

while the vacuum expectation value of σ determines f_π:

$$\sigma_v = f_\pi. \tag{7}$$

One more parameter in the model must be fixed, namely the value of the quartic self-coupling λ. This can be determined by the energy of the sigma resonance, or better, by comparing to the experimental data on the s-wave phase shifts of $\pi - \pi$ scattering in the isosinglet channel. This leads to a value of the renormalized coupling λ_r of around 10^{17}. Because of the existence of the Landau pole singularity in the large N propagators, and the associated issue of the triviality of the theory as the ultraviolet cutoff is removed to infinity, we must regard the sigma model as an effective field theory with a UV cutoff, Λ. Since the cutoff must be at least as high as the physics we wish to describe, there is a maximum value of the coupling[18],

$$\lambda_r^{max} \approx \frac{2\pi^2}{\ln(\frac{2\Lambda}{m})} \tag{8}$$

as Λ tends to infinity. For $\Lambda = 1Gev$, $\lambda_r^{max} = 13$. Correspondingly, there is also a maximum value of the σ mass of approximately $450MeV$. This is in rough agreement with the (poorly determined) experimental number. Demanding that the *sigma* resonance be above the two pion threshold of $270Mev$ requires $\lambda_r > 7$ Hence the renormalized quartic coupling is constrained to lie between 7 and 13 with a preferred value of around 10. This moderately strong coupling will make it difficult to sustain large regions of DCC for very long times as we shall see.

3. COOLING BY EXPANSION

Experimentally, a flat plateau in the distribution of produced particles per unit rapidity is observed in the central rapidity region of relativistic heavy ion collisions. This suggests the existence of an approximate Lorentz boost invariance. Thus, the simplest picture of a collision, due to Landau[19,20], is one in which the excited nuclei are highly contracted pancakes receding away from the collision point at approximately the speed of light. Of course, this is an approximate picture that is not valid for large values of the spatial rapidity and for transverse distances of the order of the nucleus size. The natural coordinates are then the proper time τ and the spatial rapidity η,

$$\tau \equiv (t^2 - x^2)^{1/2}, \qquad \eta \equiv \frac{1}{2}\log(\frac{t-x}{t+x}). \tag{9}$$

The observed symmetry will be respected by the model if one imposes initial values on a $\tau =$constant hypersurface (and not at constant laboratory time t).

According to this simple model of a collision, the plasma evolves in a highly inhomogeneous way when viewed from the laboratory frame. In fact, by analyzing a constant t surface we realize that the field configurations strongly depend on the spatial coordinate x. Near the light cone $|x| = t$ the system is "hot" (corresponding to small values of the proper time τ). On the other hand, for small values of x (that correspond to larger values of τ) the system is "colder". This type of configuration, hot in the outside and cold in the inside, is schematically known as "Baked Alaska". Fixing boost invariant initial conditions at an initial proper time τ_0 we introduce the expansion and "cooling" of the plasma in a natural way. Therefore, we do not need to introduce any ad hoc cooling mechanisms by hand. It appears as a result of the evolution, which is fully out of equilibrium[10]. In this way, we can really test the validity of the quench approximation that has been used in earlier work. In the coordinates (9) the Minkowski line element becomes

$$ds^2 = d\tau^2 - \tau^2 d\eta^2 - dx_\perp^2. \tag{10}$$

In these coordinates the equations (4)–(5) become

$$\tau^{-1}\partial_\tau\, \tau\partial_\tau\, \Phi_i(\tau) + \chi(\tau)\, \Phi_i(\tau) = H\delta_{i1} \tag{11}$$

$$\chi(\tau) = \lambda(-v^2 + \Phi_i^2(\tau) + N < \phi^2(x,\tau) >), \tag{12}$$

where the quantum field $\phi(x,\tau)$ satisfies the Klein Gordon equation

$$\left(\tau^{-1}\partial_\tau\,\tau\partial_\tau\; -\tau^{-2}\partial_\eta^2-\partial_\perp^2+\chi(x)\right)\phi(x,\tau)=0. \tag{13}$$

and I have assumed that the mean fields are functions only of the proper time variable τ (boost invariance). The similarity to cosmologically expanding spacetimes is apparent.

The quantum field $\phi(x,\tau)$ defined here is an auxiliary field which allows us to calculate the Wightman function $G_0(x,y)$ by taking the expectation value:

$$G_0(x,y;\tau)\equiv<\phi(x,\tau)\;\phi(y,\tau)>. \tag{14}$$

When the expectation value of the pion field is zero, then this field corresponds to one component of the pion field.

We expand this field in an orthonormal basis

$$\phi(\eta,x_\perp,\tau)\equiv\frac{1}{\tau^{1/2}}\int[d^3\mathrm{k}]\left(\exp(i\mathrm{k}\cdot\mathbf{x})f_\mathrm{k}(\tau)\;a_\mathrm{k}\;+h.c.\right), \tag{15}$$

where $\mathrm{k}\cdot\mathbf{x}\equiv k_\eta\eta+\vec{k}_\perp\vec{x}_\perp$, $[d^3\mathrm{k}]\equiv dk_\eta d^2k_\perp/(2\pi)^3$ and the mode functions $f_\mathrm{k}(\tau)$ evolve according to:

$$\ddot{f}_\mathrm{k}+(\frac{k_\eta^2}{\tau^2}+\vec{k}_\perp^2+\chi(\tau)+\frac{1}{4\tau^2})f_\mathrm{k}=0. \tag{16}$$

A dot here denotes the derivative with respect to the proper time τ. The expectation value $<\phi^2(x,\tau)>$ can be expressed in terms of the mode functions f_k and of the distribution functions

$$n_\mathrm{k}\equiv<a_\mathrm{k}^\dagger a_\mathrm{k}>,\quad g_\mathrm{k}\equiv<a_\mathrm{k}a_\mathrm{k}>, \tag{17}$$

which entirely characterize the initial state of the quantum field. For simplicity, we will assume that the initial state is described by a density matrix which is diagonal in the number basis (like a thermal state). In such a case, the only nonvanishing distribution is n_k. Thus, replacing the above expressions in (12) we have:

$$\chi(\tau)=\lambda\left(-v^2+\Phi_i^2(\tau)+\frac{1}{\tau}N\int[d^3\mathrm{k}]|f_\mathrm{k}(\tau)|^2\;(1+2\;n_\mathrm{k})\right). \tag{18}$$

We assume that the initial density matrix at τ_0 is one of local thermal equilibrium in the comoving frame

$$n_\mathrm{k}(\tau_0)\equiv n_\mathrm{k}=\frac{1}{e^{\beta_0E_\mathrm{k}^0}-1}, \tag{19}$$

where $\beta_0=1/T_0$, $E_\mathrm{k}^0=\sqrt{k_\eta^2/\tau^2+k_\perp^2+\chi(\tau_0)}$. This thermal assumption is for convenience only, and could easily be relaxed with perhaps more realistic initial data in future studies. The mode functions are otherwise chosen to be those corresponding to the instantaneous adiabatic vacuum at the initial time.

Although we are dealing with a cut-off theory the issue of renormalization can still be discussed formally in the large N theory. All quadratic cut-off dependence may be absorbed in the bare mass λv^2 and logarithmic dependence on the cut-off absorbed into the coupling constant λ_r. After a few simple manipulations (that involve adding and substracting the appropriate terms in (18)) we can write the equation for χ as

$$\begin{aligned}\chi(\tau)=\chi(\tau_0)\;&+\;\frac{\lambda_r}{Z}(\Phi_i^2(\tau)-\Phi_i^2(\tau_0))+\frac{N\lambda_r}{\tau Z}\int[d^3\mathrm{k}]\Big\{|f_\mathrm{k}(\tau)|^2(1+2\;n_\mathrm{k})-\\ &-\;\frac{1}{2\tilde{\omega}_\mathrm{k}(\tau)}\Big\}-\frac{N\lambda_r}{\tau_0 Z}\int[d^3\mathrm{k}]\frac{1}{\omega_\mathrm{k}(\tau_0)}n_\mathrm{k},\end{aligned} \tag{20}$$

where $\tilde{\omega}_{\mathbf{k}}(\tau) \equiv (k_\eta^2/\tau^2 + \vec{k}_\perp^2 + \chi(\tau_0))^{1/2}$ and the renormalized coupling λ_r and Z are defined as

$$\lambda = \lambda_r/Z, \qquad Z = 1 - \lambda_r \delta\lambda, \qquad \delta\lambda \equiv \frac{N}{4\tau} \int^\Lambda [d^3\mathrm{k}] \frac{1}{\tilde{\omega}_{\mathbf{k}}^3(\tau)}. \tag{21}$$

The initial value $\chi(\tau_0)$ comes from solving the gap equation (18) at the initial time using the fact that in the initial equilibrium state at τ_0, $\Phi_{eq}(T_0) = H\delta_{i1}/\chi_{eq}(T_0)$. Notice that the set of coupled equations, (11), (16), (20), are entirely written in terms of the renormalized quantities λ_r and χ_0. Changing the value of the cutoff will change both the value of the integrals and the value of Z in Eq. (20) leaving χ cutoff independent.

We see that there are unstable modes f_k whenever $(k_\eta^2 + 1/4)/\tau^2 + k_\perp^2 + \chi$ is negative so that only the long wave length modes can become unstable. χ can classically become negative only when

$$\chi(\tau_0) + \lambda\Phi_i^2(\tau) - \lambda\Phi_i^2(\tau_0) < 0, \tag{22}$$

so that the most unstable case has $\Phi(\tau) = 0$. Therefore, at the classical level the bare coupling must satisfy

$$\lambda > \frac{\chi(\tau_0)}{\Phi_i^2(\tau_0)} \tag{23}$$

for there to be any instabilities. Once instabilities grow, they cause exponential growth of the modes in the mode sum which contributes a positive quantity to the equation for χ. Thus, the stronger the coupling the quicker χ returns to a positive value. The initial value of χ is restricted in this way to be positive. Starting with a negative χ_0 is equivalent to the quench approximation

In studying possible initial conditions, let us consider first the case where the initial value $\sigma(\tau_0)$ is the thermal equilibrium one corresponding to a temperatuare of $200MeV$. We varied the value of the initial proper time derivative of the sigma field expectation value and found that there is a narrow range of initial values that lead to the growth of instabilities, namely

$$0.25fm^{-2} < |\dot{\sigma}| < 1.3fm^{-2}.$$

Surprisingly when $|\dot{\sigma}| > 1.3fm^{-2}$ instabilities no longer occur.

Figures 1-4 summarize the results of the numerical simulation for the evolution of the system (11),(16) and (20). We display the auxiliary field χ in units of fm^{-2} , the classical fields Φ in units of fm^{-1} and the proper time in units of fm^{-1}.

In Fig. 1 and Fig. 2 the proper time evolutions of the auxiliary field χ field are presented for two different values of f_π, where the initial conditions were fixed at $\tau = 1$. In the case of $f_\pi = 92.5MeV$ we see that the instability lasts for less than $3fm$. As discussed earlier, the regime of exponential growth of the unstable modes occurs whenever $\chi < 0$. We notice that when we rotate the expectation value from the σ to the π^1 direction initially, the typical time that instability exsists is of the same magnitude. We also notice that if we choose the time derivative to be zero and start from an equilibrium configuration, then the expansion alone is insufficient to generate instabilities. We find that in order to generate instabilities we require fluctuations in the classical kinetic terms such as $\dot{\sigma}$ or $\dot{\pi}_i$, and as discussed earlier, these initial conditions must be in a very narrow range to produce instabilites. Thus the rapid quench conditions assumed by other authors comprises only a small region of the phase space of initial fluctuations expected in an initial thermal distribution. As expected, at late proper times, the auxiliary field approaches its equilibrium value of m_π^2.

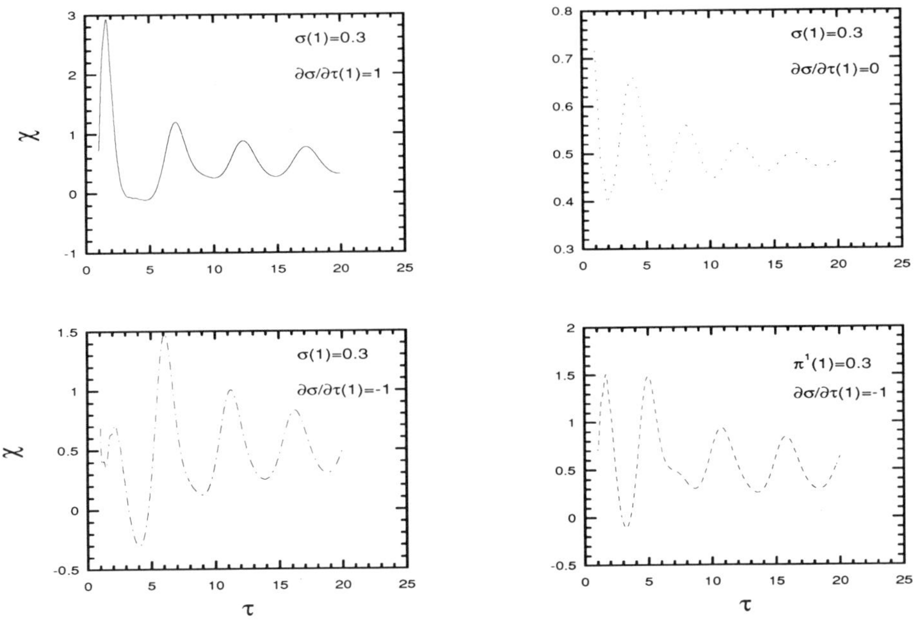

Figure 1. Proper time evolution of the χ field for four different initial conditions with f_π=92.5MeV. Note that instability exists and fluctuations can grow only when χ becomes negative, which happens only for a short interval of proper time for some initial conditions.

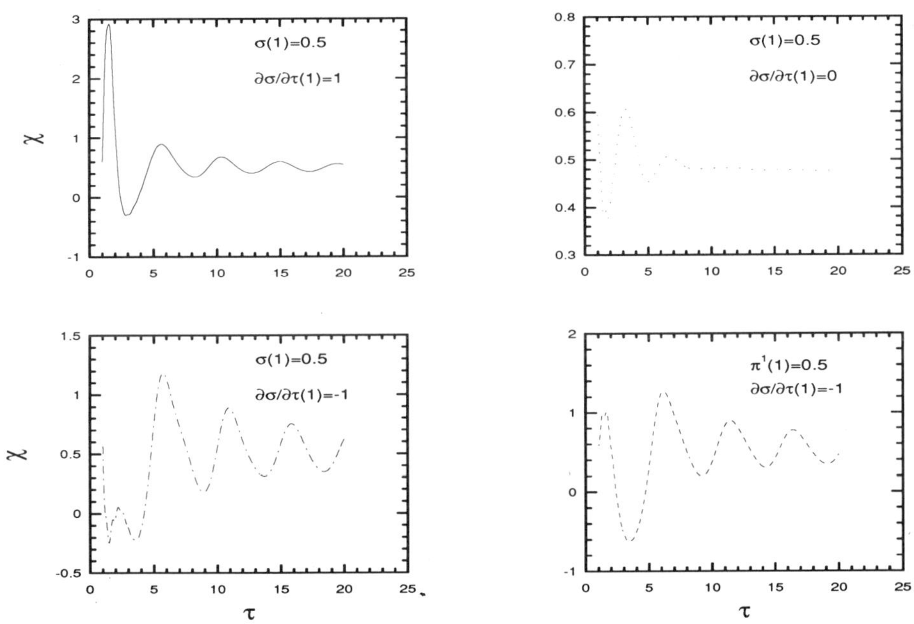

Figure 2. Same as Fig. 1, but for f_π=125MeV.

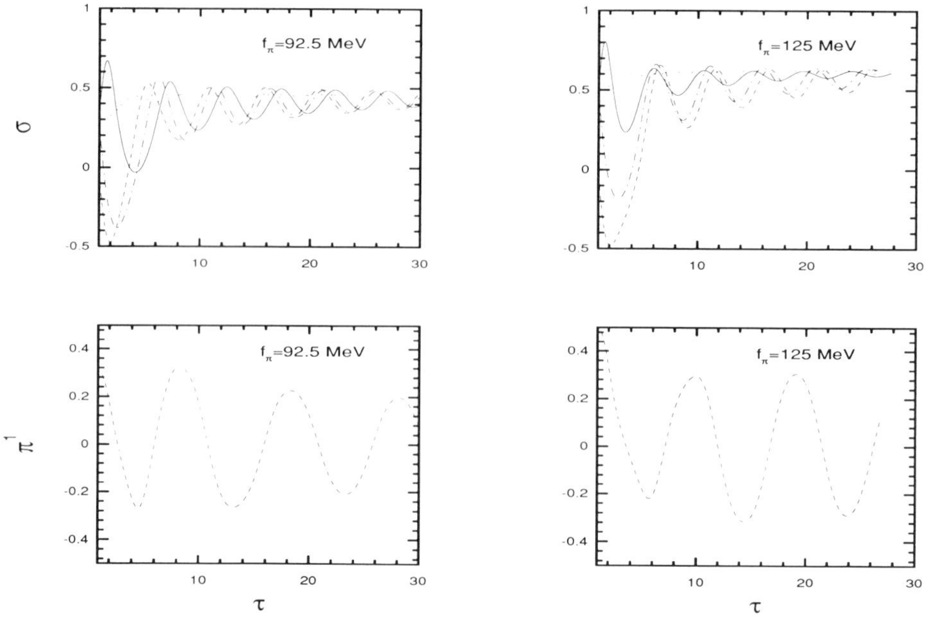

Figure 3. Proper time evolution of the σ and π fields for the following initial conditions: Solid line is for $\sigma(1) = \sigma_T$, $\pi^i(1)$=0 and $\dot{\sigma}(1)$=1. Dotted line is for $\sigma(1)=\sigma_T$, $\pi^i(1)$=0 and $\dot{\sigma}(1)$=0. Dashed- dotted line is for $\sigma(1)=\sigma_T$, $\pi^i(1)$=0 and $\dot{\sigma}(1)=-1$. Dashed line is for $\sigma(1)$=0, $\pi^i(1)=\sigma_T$ and $\dot{\sigma}(1)=-1$. At $T=200MeV$, $\sigma_T=0.3fm^{-1}$ for $f_\pi=92.5MeV$ and $\sigma_T=0.5fm^{-1}$ for $f_\pi=92.5MeV$.

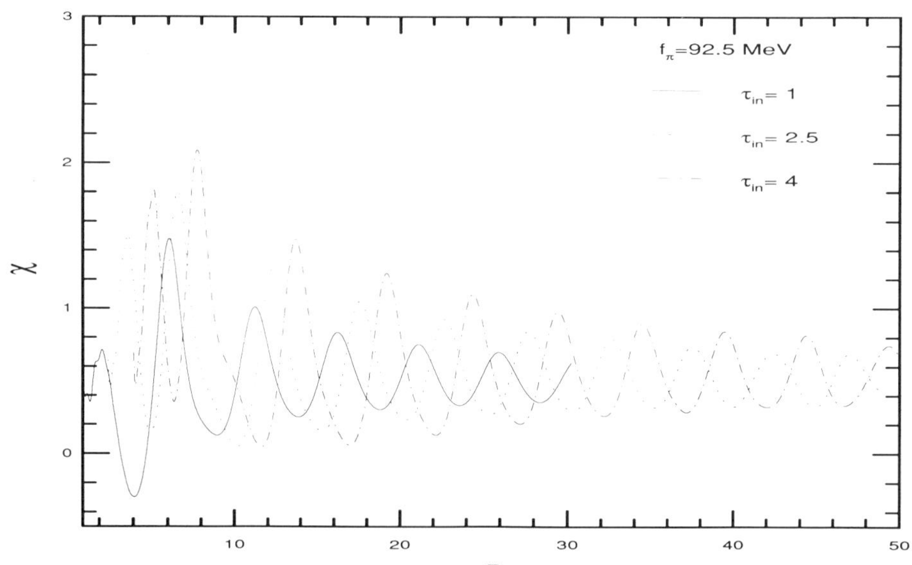

Figure 4. Proper time evolution of the χ field for three different initial proper times τ_0=1,2.5,4 with $f_\pi=92.5MeV$, $\sigma(\tau_{in})=\sigma_T$, $\pi^i(\tau_{in})$=0 and $\dot{\sigma}(\tau_{in})=-1$.

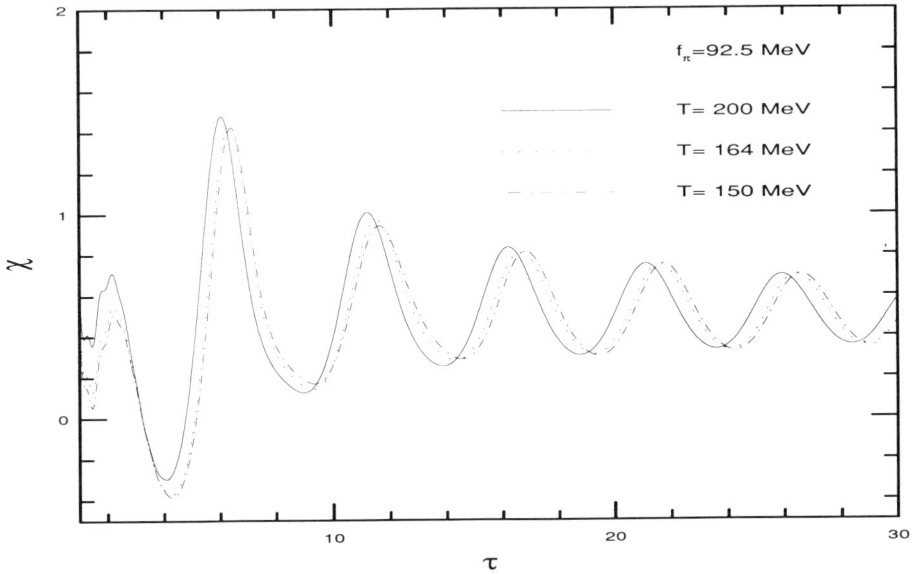

Figure 5. Proper time evolution of the χ field for three different initial thermal distributions with T=200,164,150MeV for the case of $\sigma(1)=\sigma_T$, $\pi^i(1)=0$ and $\dot{\sigma}(1)=-1$.

A crude estimate for the size of a disoriented chiral condensate may be obtained by multiplying the typical life time of the instabilies by the speed of light. With this estimate the size of these regions are of the order of a few fermi. In Fig. 3 we plot the proper time evolution of the classical fields σ and π for the same two choices for f_π as in Fig. 1 and Fig. 2. We see that the sigma field asymptotes to its vacuum value f_π and the π field gradually converges to its equilibrium value of zero.

In Fig. 4 we show the effect of starting the initial value problem at later times, namely $\tau_0 = 2.5, 4\, fm$, and compare them to the case previously studied with $\tau_0 = 1 fm$, which had a modest region of instability growth. We see that as we increase τ_0 we decrease the possibility of instability growth, and with these late initial times it is not possible to produce instabilities even with kinetic energy fluctuations.

In Fig. 5 we study the effect of the initial temperature on our time evolution problem. For $f_\pi = 92.5$ the critical temperature is $160MeV$ (in the absence of explicit symmetry breaking term). We see that in the vicinity of the critical temperature the effects of varying the initial temperature is minor.

Next, we are interested in seeing how our results differ from the case where the system evolves in local thermal equilibrium. This is described by two correlation lengths, the inverse of the effective pion mass associated with $\chi(\tau)$, and the inverse of the proper time evolving effective temperature $T(\tau) = T(\tau_0)(\tau_0/\tau)^3$ corresponding to a boost invariant hydrodynamical models. We see from Fig. 6 that in the case of $\sigma(1) = \sigma_T$, $\pi^i(1) = 0$ and $\dot{\sigma}(1) = -1$, where maximum instabilitiy exists, complex structures are formed as contrasted to the local thermal equilibrium evolution. The interpolating phase space distribution $n(k_\eta, \mathbf{k}_\perp, \tau)$ obtained numerically, clearly exhibits a larger correlation length in the transverse direction than the $n_T(k_\eta, \mathbf{k}_\perp, \tau)$ equilibrium one, and has correlation in rapidity of the order of 1-2 units of rapidity. We notice that in both directions there are structures that does not lend themselves to a simple interpretation. On the other hand, the local thermal equilibrium evolution is quite regular apart from

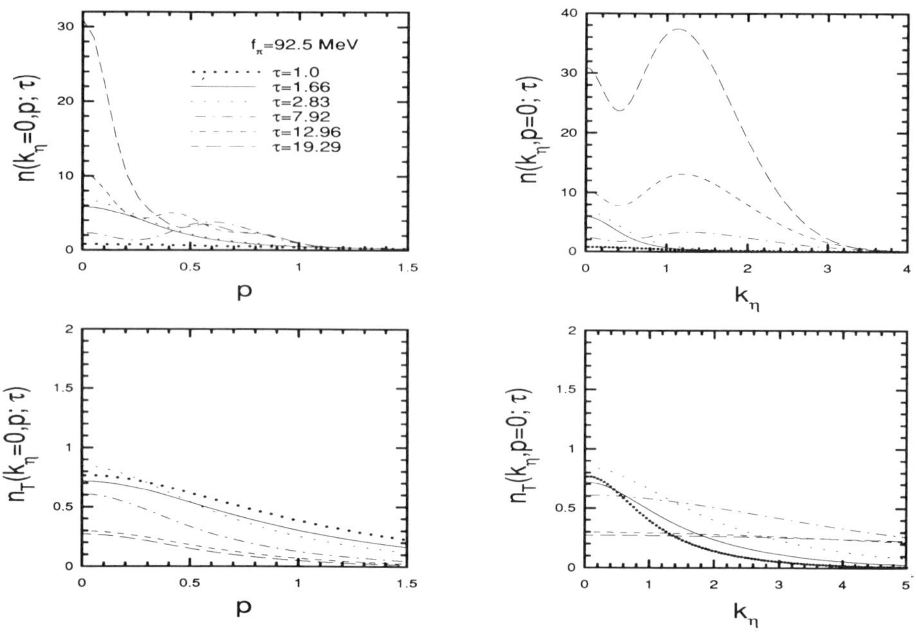

Figure 6. Slices of k_η=0 and $p\equiv|\mathbf{k}_\perp|$=0 of the proper time evolution of the phase space particle number density $n(k_\eta,\mathbf{k}_\perp,\tau)$ for $\sigma(1)$=σ_T, $\pi^i(1)$=0 and $\dot{\sigma}(1)$=−1 compared with the corresponding local thermal equilibrium densities $n_T(k_\eta,\mathbf{k}_\perp,\tau)$.

the normalization of the distributions that are changing with time due to oscillation in the quantity $\chi(\tau)$ which is damped to its equilibrium value once the system expands sufficiently.

In Fig. 7 we plot the time evolution of the number operator for the case in which we did not have any fluctuations in the kinetic energy. In this case with no instabilities we do not see complicated structures, and the transverse distribution is similar to the equilibrium case.

The pair density function $g(k_\eta, \mathbf{k}_\perp, \tau)$ is another interesting quantity with information about the correlations of pions in the final state. In the case where there are no instabilities ($\dot{\sigma}(1) = 0$) we see in Fig. 8 that although the transverse distribution is relatively simple, the distribution in the η direction (whose fourier transform gives the rapidity distribution) has many length scales. When we have instabilities ($\dot{\sigma}(1) = -1$) then both distributions are complicated and possess several length scales, as seen in the lower part of Fig. 8.

4. CONCLUSIONS

In this talk I have presented results of numerical simulations in the regime of the chiral phase transition in the linear sigma model for a wide variety of initial conditions starting above the critical temperature for an expanding plasma of pions and sigmas. Assuming that this model gives a reasonable discription in this temperature range, we found initial conditions where instabilities grew in the scenario required for the formation of disoriented chiral condensates. However low energy phenomenology also requires a moderately large renormalized coupling constant. The constraints on our model which are imposed in order to approximate the low energy physics of pion interactions require a large renormalized coupling constant. This had the effect of rapidly

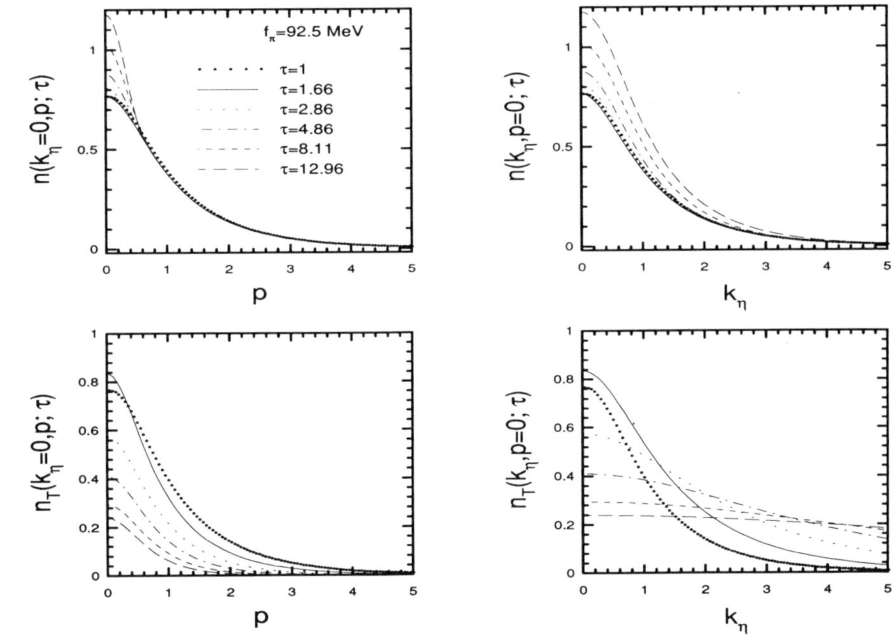

Figure 7. Same as Fig. 6, but for $\sigma(1)=\sigma_T$, $\pi^i(1)=0$ and $\dot{\sigma}(1)=0$.

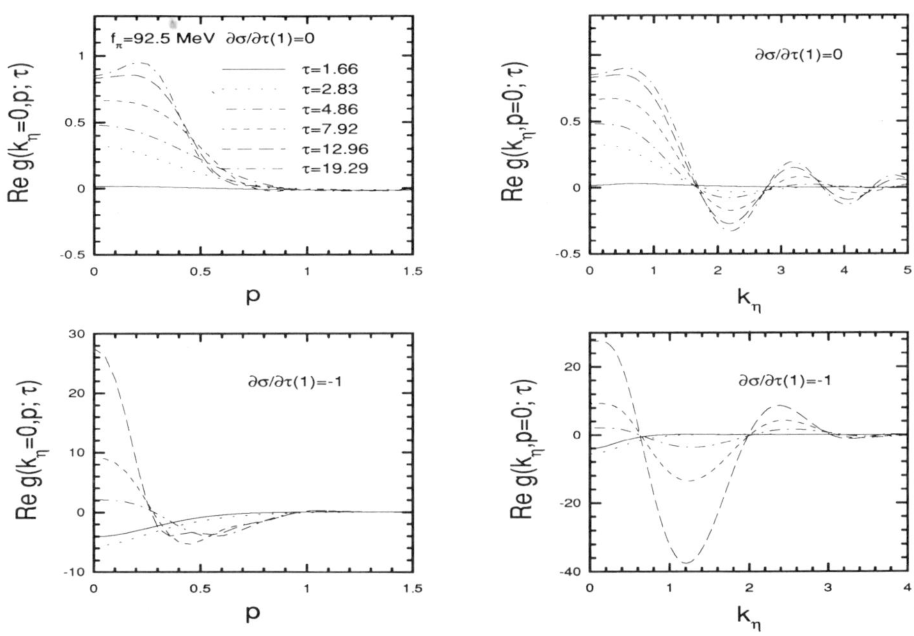

Figure 8. Slices of $k_\eta=0$ and $p\equiv|\mathbf{k}_\perp|=0$ of the proper time evolution of the real part of the phase space pair density $g(k_\eta,\mathbf{k}_\perp,\tau)$ for $\sigma(1)=\sigma_T$, $\pi^i(1)=0$ and $\dot{\sigma}(1)=0,-1$.

damping the instabilities, and we found no evidence in this model for large domains of disoriented chiral condensates. We did, however, see rather large departures in the phase space number density from one which would result from an evolution in local thermal equilibrium. These departures show a narrowing of the moementum space distribution which could be interpreted as a larger spatial correlation length. In our simulations we assumed a reasonable mechanism for cooling– namely the expansion of the plasma following its production in a collision. We ignored scattering, which occurs at the next order in the $1/N$ expansion, and would almost certainly tend to decrease the correlations of any DCC seen in this lowest order calculation.

In the future we plan to include these scattering effects which will introduce another time scale, namely the equilibration time scale, into the problem. The extension of this work to other field theories is fairly straightforward. In particular the Higgs field interacting with the hot, dense medium of quarks, W's and Z's at the electroweak phase transition may be treated by a modified large N expansion, where N is the number of quark flavors. This should provide a systematic technique to study bubble wall propagation, scattering and damping effects in the hot EW plasma, and eventually put the current models of EW baryogenesis discussed in these proceedings on a firmer theoretical footing.

Acknowledgements

This work was done in collaboration with Juan Pablo Paz, Fred Cooper and Yuval Kluger. I would like to thank Los Alamos post-docs Alex Kovner and Salman Habib as well for stimulating criticisms and discussions.

REFERENCES

1. A. Anselm, Phys. Lett. **B217**, 169 (1989); A. Anselm and M. Ryskin, Phys. Lett. **B226**, 482 (1991).
2. C.M.G. Lattes, Y. Fujimoto and S. Hasegawa, Phys. Rep. **65**, 151 (1980).
3. J.D. Bjorken, Int. Jour. of Mod. Phys.**A7**, 4189 (1992); Acta Phys. Pol. **B23** 561 (1992).
4. K. Rajagopal and F. Wilczek, Nucl. Phys. **B379**, 395 (1993).
5. P. Bedaque and A. Das, Mod. Phys. Lett. **A8**, 3151 (1993).
6. S. Yu. Khlebnikov, Mod. Phys. Lett. A **8**, 1901 (1993).
7. I. Kogan, Phys. ReV. D **48**, 3971 (1993); I. Kogan, Princeton Preprint PUPT-1424 (1993).
8. S. Gavin, A. Gocksch and R.D. Pisarski, Brookhaven Preprint BNL-GGP-1 (1993).
9. Z. Huang and X.N. Wang, LBL Preprint LBL-34931 (1993).
10. S. Gavin and B. Mueller, Brookhaven Preprint BNL-GM-1 (1993).
11. D. Boyanovsky, H.J. De Vega and R. Holman, Pittsburgh Univ. Preprint PITT-94-01 (1994).
12. D. Boyanovsky, D.S. Lee and A. Singh, Phys. Rev. **D48**, 800 (1993).
13. R. Jackiw and A. Kerman, Phys. Lett. **A71** 158 (1979); F. Cooper, S.Y. Pi and P. Stancioff Phys. Rev. D**34**,3831 (1986); S.Y. Pi and M. Samiullah, Phys. Rev. **D36** 3128 (1987); F. Cooper and E. Mottola, Phys. Rev. **D36**, 3114 (1987).
14. H. E. Stanley, Phys. Rev. **176** 718 (1968); K. Wilson, Phys. Rev. **D7**

2911 (1973); S. Coleman, R. Jackiw and H. D. Politzer, Phys. Rev. **D10** 2491 (1974); C. Bender , F. Cooper and G. Guralnik, Phys. Rev. Lett. **45**, 501 (1980); G. A.Baker Jr., F. Cooper, L. P. Benofy and D. Preston, Nucl. Phys. B 210 [FS6], 273 (1982).

15. F. Cooper, S. Habib, Y. Kluger, E. Mottola, J.P. Paz, and Paul R. Anderson, Los Alamos preprint LA-UR-94-783 (Phys. Rev. D in press); F. Cooper, Y. Kluger, E. Mottola, and J.P. Paz, LA-UR-94-1569, Bulletin Board: hep-ph@xxx.lanl.gov - 9404357.
16. R. Root, Phys. Rev. **D11** 831 (1975); C. M. Bender, F. Cooper and G. S. Guralnik, Ann. Phys. **109** 165 (1977); C. M. Bender and F. Cooper, Ann. Phys. **160** 323 (1985).
17. J. L. Basdevant and B. W. Lee Phys. Rev. **D2**1680 (1970).
18. W. Bardeen and Moshe Moshe Phys. Rev.**D28** 1372 (1983).
19. L. D. Landau, Izv. Akad. Nauk SSSR **17** 51 (1953); L. D. Landau and S. Z. Belenkij, Usp. Phys. Nauk. **56**, 309 (1955) (Nuovo Cimento Suppl. **3**, 15 (1956)).
20. E. V. Shuryak, Phys. Lett. **34B**, 509 (1971); F. Cooper, G. Frye and E. Schonberg, Phys. Rev. **D11** 192 (1975); J.D. Bjorken, Phys. Rev. **D27**, 140 (1983).

NON-EQUILIBRIUM QUANTUM FIELD EVOLUTION IN FRW COSMOLOGIES

Hector J. de Vega

Laboratoire de Physique Théorique et Hautes Energies
Université Pierre et Marie Curie (Paris VI)
Laboratoire Associé au CNRS UA280
Tour 16, 1er. étage, 4, Place Jussieu
75252 Paris, Cedex 05, France

We derive the effective equations for the out of equilibrium time evolution of the order parameter and the fluctuations of a scalar field theory in spatially flat FRW cosmologies. After setting the problem in general we propose a non-perturbative, self-consistent Hartree approximation. The method consists of evolving an initial functional thermal density matrix in time and is suitable for studying phase transitions out of equilibrium. The renormalization aspects are studied in detail and we find that the counterterms depend on the initial state. We investigate the high temperature expansion and show that it breaks down at long times. The infinite time limit is computed for de Sitter spacetime. We obtain the time evolution of the initial Boltzmann distribution functions, and argue that in the Hartree approximation, the time evolved state is a "squeezed" state. We illustrate the departure from thermal equilibrium by studying the case of a free field in de Sitter and radiation dominated cosmologies. It is found that a suitably defined non-equilibrium entropy per mode increases linearly with comoving time in a de Sitter cosmology, whereas it is *not* a monotonically increasing function in the radiation dominated case. This work has been done in collaboration with D. Boyanovsky and R. Holman.

1 Introduction

The usual assumption in inflation [1] is that the dynamics of the spatial zero mode of the (so-called) inflaton field is governed by some approximation by the effective potential which incorporates the effects of quantum fluctuations of the field. Thus the equation of motion is usually an ordinary non-linear differential equation of the form:

$$\ddot{\phi} + 3\frac{\dot{a}}{a}\dot{\phi} + V'_{\text{eff}}(\phi) = 0 \tag{1}$$

where $a(t)$ is the expansion factor in the metric:

$$ds^2 = dt^2 - a^2(t)\, d\vec{x}^2 \tag{2}$$

Electroweak Physics and the Early Universe, Edited by
Romao and Freire, Plenum Press, New York, 1994

The problem here is that the effective potential is really only suited for analyzing *static* situations; it is the effective action evaluated for a field configuration that is constant in time [2]. Thus, it is inconsistent to use the effective potential in a *dynamical* situation. Notice that such inconsistency appears for **any** inflationary scenario (old, new, chaotic, ...).

More generally, the standard methods of high temperature field theory are based on an equilibrium formalism[3, 4]; there is no time evolution in such a situation. Such techniques preclude us from treating non-equilibrium situations such as surely exist for very weakly coupled theories in the early universe.

More precisely, the effective potential develops an imaginary part around points ϕ_0 where $V''(\phi_0) < 0$ [ϕ_0 : false vacuum]. This fact leads to long-wavelength modes with imaginary frequencies $\omega(k) = \sqrt{k^2 + V''(\phi_0)}$ for $0 \leq k < \sqrt{|V''(\phi_0)|}$. When the field ϕ starts localized around ϕ_0 , these **long-wavelength** unstable modes push the system to evolve on time away from ϕ_0. The smaller is the non-linear coupling, the better is this small fluctuations approximations and the more important are these unstabilities.

In collaboration with Daniel Boyanovsky and Rich Holman we try to rectify this situation by addressing three issues[5, 6, 7, 8]: a) obtaining the evolution equations for the order parameter including the *quantum fluctuations*, b) studying departures from thermal equilibrium if the initial state is specified as a thermal ensemble, c) understanding the renormalization aspects and the validity of the high temperature expansion.

Our ultimate goal is to study the dynamics of phase transitions in the early universe, in particular, the formation and evolution of correlated domains and symmetry breaking in an expanding universe. From some of our previous studies on the dynamics of phase transitions[5, 6] in Minkowski space, we have learned that the familiar picture of "rolling" is drastically modified when the fluctuations are taken into account. As the phase transition proceeds fluctuations become large and correlated regions (domains) begin to grow. This enhancement of the fluctuations modifies substantially the evolution equation of the order parameter. Thus the time dependence of the order parameter is not enough to understand the dynamical aspects of the phase transition; it must be studied in conjunction with that of the fluctuations.

Our approach is to use the functional Schrödinger formulation, wherein we specify the initial wavefunctional $\Psi[\Phi(\vec{\cdot}); t]$ (or more generally a density matrix $\rho[\Phi(\vec{\cdot}), \tilde{\Phi}(\vec{\cdot}); t]$), and then use the Schrödinger equation to evolve this state in time. We can then use this state to compute all of the expectation values required in the construction of the effective equations of motion for the order parameter of the theory, as well as that for the fluctuations.

One advantage of this approach is that it is truly a dynamical one; we set up initial conditions at some time t_o by specifying the initial state and then we follow the evolution of the order parameter $\phi(t) \equiv \langle\Phi(\vec{x})\rangle$ and of the fluctuations as this state evolves in time. Another advantage is that it allows for departures from equilibrium. Thus, issues concerning the restoration of symmetries in the early universe can be addressed in a much more general setting.

Our analysis applies quite generally to any arbitrary spatially flat FRW cosmology. We also determine the time evolution of the initial (Boltzmann) distribution functions, relate the time evolution to "squeezed states" and perform a numerical integration in the case of free fields for de Sitter and radiation dominated cosmologies. We expect to provide a quantitative analysis of the evolution of the order parameter and the dynamics of phase transitions for interacting fields in a forthcoming article [8].

The initial state we pick for the field $\Phi(\vec{x}, t)$ is that corresponding to a thermal density matrix centered at $\phi(t)$. It is then useful to try to understand the high temperature limit of our calculations. We are able to compute both the leading and subleading terms in the high T expansion of $\langle \phi^2(t) \rangle$. ¿From this we show that the high T expansion cannot be valid for all time, but breaks down in the large time limit. We then compute the limit of long times ($t \to \infty$). during the phase transition.

The time evolution of the Boltzmann distribution functions (initially the thermal equilibrium distribution functions) is obtained in ref.[7]. It is pointed out that to one-loop order and also in the Hartree approximation, the time evolved density matrix describes quantum "squeezed" states and time evolution corresponds to a Bogoliubov transformation.

To illustrate the departure of equilibrium, we have studied numerically in ref.[7] the case of a free massive scalar field in de Sitter and radiation dominated cosmologies. It was found that a suitably defined coarse-grained non-equilibrium entropy (per $\vec{k}$ mode) grows linearly with time in the de Sitter case but it is not a monotonically increasing function of time in the radiation dominated case. This result may cast some doubt on the applicability of this definition of the non-equilibrium entropy. There still remain some (open) fundamental questions regarding the connection of this entropy and the thermodynamic entropy of the universe, in particular whether the amount of entropy produced is consistent with the current bounds.

This work sets the stage for a numerical study of the dynamics of phase transitions in cosmology fully incorporating the non-equilibrium aspects in the evolution of the order parameter and which at the same time can account for the dynamics of the fluctuations which will necesarily become very important during the phase transition.

We have applied similar methods to investigate the formation of disordered chiral condensates in high energy collisions [9].

2 Evolution Equations

We consider the inflaton scalar field $\Phi(\vec{x}, t)$ in the spatially flat FRW cosmology (2) with action given by

$$S = \int d^4x a^3(t) \left[\frac{1}{2}\dot{\Phi}^2(\vec{x}, t) - \frac{1}{2}\frac{(\vec{\nabla}\Phi(\vec{x}, t))^2}{a(t)^2} - V(\Phi(\vec{x}, t)) \right] \tag{3}$$

$$V(\Phi) = \frac{1}{2}[m^2 + \xi \mathcal{R}] \; \Phi^2(\vec{x}, t) + \frac{\lambda}{4!} \; \Phi^4(\vec{x}, t) \tag{4}$$

$$\mathcal{R} = 6\left(\frac{\ddot{a}}{a} + \frac{\dot{a}^2}{a^2}\right) \tag{5}$$

with $\mathcal{R}$ the Ricci scalar and ξ a non-minimal coupling constant.

In the Schrödinger representation (at an arbitrary fixed time t), the Hamiltonian becomes

$$H(t) = \int d^3x \left\{ -\frac{\hbar^2}{2a^3(t)} \frac{\delta^2}{\delta\Phi(\vec{x})^2} + \frac{a(t)}{2}(\vec{\nabla}\Phi)^2 + a^3(t) \; V(\Phi) \right\} \tag{6}$$

Since we consider a 'thermal ensemble', we work with a (functional) density matrix $\hat{\rho}$ with matrix elements in the Schrödinger representation $\rho[\Phi(\vec{\cdot}), \tilde{\Phi}(\vec{\cdot}); t]$. We will *assume* that the density matrix obeys the functional Liouville equation

$$i\hbar \frac{\partial \hat{\rho}}{\partial t} = [H(t), \hat{\rho}] \tag{7}$$

Normalizing the density matrix such that $Tr\hat{\rho} = 1$, we define as 'order parameter'

$$\phi(t) = \frac{1}{\Omega}\int d^3x \langle \Phi(\vec{x},t)\rangle = \frac{1}{\Omega}\int d^3x \; Tr\left[\hat{\rho}(t)\Phi(\vec{x})\right] \tag{8}$$

where Ω is the comoving volume.

The evolution equations for the order parameter are as follows:

$$\frac{d^2\phi(t)}{dt^2} + \frac{3}{a(t)}\frac{da(t)}{dt} + \frac{1}{\Omega}\int d^3x \; \langle \frac{\delta V(\Phi)}{\delta \Phi(\vec{x})} \rangle = 0 \tag{9}$$

It is now convenient to write the field in the Schrödinger picture as

$$\begin{aligned} \Phi(\vec{x}) &= \phi(t) + \eta(\vec{x},t) && (10) \\ \langle \eta(\vec{x},t)\rangle &= 0 && (11) \end{aligned}$$

Expanding the right hand side of (9) we find the effective equation of motion for the order parameter:

$$\frac{d^2\phi(t)}{dt^2} + 3\,\frac{\dot{a}(t)}{a(t)}\frac{d\phi(t)}{dt} + V'(\phi(t)) + \frac{V'''(\phi(t))}{2\Omega}\int d^3x = 0 \tag{12}$$

where primes stand for derivatives with respect to ϕ.

This equation of motion is clearly *very* different from the one obtained by using the effective potential. It may be easily seen (by writing the effective action as the classical action plus the logarithm of the determinant of the quadratic fluctuation operator) that this is the equation of motion obtained by the variation of the one-loop effective action.

The *static* effective potential is clearly not the appropriate quantity to use to describe scalar field dynamics in an expanding universe. Although there may be some time regime in which the time evolution is slow and fluctuations rather small, this will certainly *not* be the case at the onset of a phase transition. As the phase transition takes place, fluctuations become dominant and grow in time signaling the onset of long range correlations[5, 6].

3 Hartree equations

Motivated by our previous studies in Minkowski space[5, 6] which showed that the growth of correlation and enhancement of fluctuations during a phase transition may not be described perturbatively, we proceeded to obtaining the equations of motion in a Hartree approximation[7]. This approximation is non-perturbative in the sense that it sums up infinitely many diagrams of the cactus or bubble-type[3]. The Hartree approximation becomes exact in the $N \to \infty$ limit of an $O(N)$ vector theory. It at least provides a consistent non-perturbative framework in which correlations and fluctuations can be studied. Corrections to the Hartree approximation can be systematically implemented in a consistent fashion.

The Hartree self-consistent approximation is implemented as follows. We decompose the field as in (10), using a potential as in (4). We find that the Hamiltonian becomes

$$\begin{aligned} H &= \int d^3x \left\{ -\frac{\hbar^2}{2a^3(t)}\frac{\delta^2}{\delta\eta^2} + \frac{a(t)}{2}\left(\vec{\nabla}\eta\right)^2 + a^3(t)\left(V(\phi) + V'(\phi)\,\eta + \frac{1}{2!}V''(\phi)\,\eta^2 \right.\right. \\ &+ \left.\left. \frac{1}{3!}\lambda\,\phi\,\eta^3 + \frac{1}{4!}\lambda\,\eta^4\right)\right\} \qquad (13) \end{aligned}$$

The Hartree approximation is obtained by assuming the factorization

$$\eta^3(\vec{x},t) \rightarrow 3\langle\eta^2(\vec{x},t)\rangle\eta(\vec{x},t) \tag{14}$$

$$\eta^4(\vec{x},t) \rightarrow 6\langle\eta^2(\vec{x},t)\rangle\eta^2(\vec{x},t) - 3\langle\eta^2(\vec{x},t)\rangle^2 \tag{15}$$

where $\langle\cdots\rangle$ is the average using the time evolved density matrix. This average will be determined self-consistently (see below). Translational invariance shows that $\langle\eta^2(\vec{x},t)\rangle$ can only be a function of time. This approximation makes the Hamiltonian quadratic at the expense of a self-consistent condition. In the time independent (Minkowski) case this approximation sums up all the daisy (or cactus) diagrams and leads to the self-consistent gap equation[3]. In this approximation the Hamiltonian becomes

$$\begin{aligned} H &= \Omega\, a^3(t)\, \mathcal{V}(\phi) + \\ &\int d^3x \left\{ -\frac{\hbar^2}{2\,a^3(t)}\frac{\delta^2}{\delta\eta^2} + \frac{a(t)}{2}\left(\vec{\nabla}\eta\right)^2 + a^3(t)\left(\mathcal{V}^{(1)}(\phi)\eta + \frac{1}{2}\mathcal{V}^{(2)}(\phi)\eta^2\right)\right\} \end{aligned} \tag{16}$$

$$\mathcal{V}(\phi) = V(\phi) - \frac{1}{8}\lambda\langle\eta^2\rangle^2\,,\ \mathcal{V}^{(1)}(\phi) = V'(\phi) + \frac{1}{2}\lambda\,\phi\,\langle\eta^2\rangle\,,\ \mathcal{V}^{(2)}(\phi) = V''(\phi) + \frac{1}{2}\lambda\langle\,\eta^2\rangle\,.$$

It is convenient to introduce the discrete Fourier transform of the fields in the comoving frame as

$$\eta(\vec{x},t) = \frac{1}{\sqrt{\Omega}}\sum_{\vec{k}}\eta_{\vec{k}}(t)\, e^{-i\vec{k}\cdot\vec{x}} \tag{17}$$

The Hamiltonian (16) then takes the form

$$\begin{aligned} H &= \Omega\, a^3(t)\, V(\phi(t)) + \\ &\frac{1}{2}\sum_{\vec{k}}\left\{-\frac{\hbar^2}{a^3(t)}\frac{\delta^2}{\delta\eta_{\vec{k}}\delta\eta_{-\vec{k}}} + 2\,a^3(t)\,\mathcal{V}'_{\vec{k}}(\phi(t))\,\eta_{-\vec{k}} + \omega^2_{\vec{k}}(t)\,\eta_{\vec{k}}\,\eta_{-\vec{k}}\right\} \end{aligned} \tag{18}$$

where the time dependent frequencies ($\omega^2_{\vec{k}}(t)$) and the linear term in η have the values

$$\omega^2_{\vec{k}}(t) = a(t)\,\vec{k}^2 + a^3(t)\,\mathcal{V}^{(2)}(\phi(t)) \tag{19}$$

$$\mathcal{V}^{(1)}_{\vec{k}}(\phi(t)) = \mathcal{V}^{(1)}(\phi(t))\,\sqrt{\Omega}\,\delta_{\vec{k},0} \tag{20}$$

We propose the following Gaussian ansatz for the functional density matrix elements in the Schrödinger representation

$$\begin{aligned} \rho[\Phi,\tilde{\Phi},t] &= \prod_{\vec{k}}\mathcal{N}_k(t)\exp\left\{-\left[\frac{A_k(t)}{2\hbar}\eta_k(t)\eta_{-k}(t) + \frac{A^*_k(t)}{2\hbar}\tilde{\eta}_k(t)\tilde{\eta}_{-k}(t) + \frac{B_k(t)}{\hbar}\eta_k(t)\tilde{\eta}_{-k}(t)\right]\right. \\ &\left. + \frac{i}{\hbar}\,\pi_k(t)\,\left(\eta_{-k}(t) - \tilde{\eta}_{-k}(t)\right)\right\} \end{aligned} \tag{21}$$

$$\eta_k(t) = \Phi_k - \phi(t)\sqrt{\Omega}\,\delta_{\vec{k},0}\quad .\quad \tilde{\eta}_k(t) = \tilde{\Phi}_k - \phi(t)\sqrt{\Omega}\,\delta_{\vec{k},0}$$

where $\phi(t) = \langle\Phi(\vec{x})\rangle$ and $\pi_k(t)$ is the Fourier transform of $\langle\Pi(\vec{x})\rangle$. This form of the density matrix is dictated by the hermiticity condition $\rho^\dagger[\Phi,\tilde{\Phi},t] = \rho^*[\tilde{\Phi},\Phi,t]$; as a result of this, $B_k(t)$ is real. The kernel $B_k(t)$ determines the amount of "mixing" in the density matrix, since if $B_k = 0$, the density matrix corresponds to a pure state because it is a wave functional times its complex conjugate.

In order to solve for the time evolution of the density matrix (7) we need to specify the density matrix at some initial time t_o. It is at this point that we have to *assume* some physically motivated initial condition. We believe that this is a subtle point that has not received proper consideration in the literature. A system in thermal equilibrium has time-independent ensemble averages (as the evolution Hamiltonian commutes with the density matrix) and there is no memory of any initial state. However, in a time dependent background, the density matrix will evolve in time, departing from the equilibrium state and correlation functions or expectation values may depend on details of the initial state.

We will *assume* that at early times the initial density matrix is *thermal* for the modes that diagonalize the Hamiltonian at t_o (we call these the *adiabatic* modes). The effective temperature for these modes is $k_B T_o = 1/\beta_o$. It is only in this initial state that the notion of "temperature" is meaningful. As the system departs from equilibrium one cannot define a thermodynamic temperature. Thus in this case the "temperature" refers to the temperature defined in the initial state.

Inserting the Gaussian Ansatz (21) into the Liouville equation yields upon using the hamiltonian (16) the following equations for the functions $A_k(t)$, $B_k(t)$ and $\mathcal{N}_k(t)$:

$$i\frac{\dot{\mathcal{N}}_k}{\mathcal{N}_k} = \frac{1}{2a^3(t)}(A_k - A_k^*) \tag{22}$$

$$i\dot{A}_k = \left[\frac{A_k^2 - B_k^2}{a(t)^3} - \omega_k^2(t)\right] \tag{23}$$

$$i\dot{B}_k = \frac{B_k}{a^3(t)}(A_k - A_k^*) \tag{24}$$

The equation for $B_k(t)$ reflects the fact that a pure state $B_k = 0$ remains pure under time evolution.

Writing A_k in terms of its real and imaginary components $A_k(t) = A_{Rk}(t) + iA_{Ik}(t)$ (and because B_k is real) we find that

$$\frac{B_k(t)}{A_{Rk}(t)} = \frac{B_k(t_o)}{A_{Rk}(t_o)} \tag{25}$$

and that the time evolution is unitary (as it should be), that is

$$\frac{\mathcal{N}_k(t)}{\sqrt{(A_{Rk}(t) + B_k(t))}} = \text{constant} \tag{26}$$

For the chosen thermal initial conditions at $t = t_0$, we have

$$A_k(t_o) = A_k^*(t_o) = \mathcal{W}_k(t_o)\ a^3(t_o)\ \coth\left[\beta_o\hbar\mathcal{W}_k(t_o)\right] \tag{27}$$

$$B_k(t_o) = -\frac{\mathcal{W}_k(t_o)a^3(t_o)}{\sinh\left[\beta_o\hbar\mathcal{W}_k(t_o)\right]}$$

$$\mathcal{N}_k(t_o) = \left[\frac{\mathcal{W}_k(t_o)a^3(t_o)}{\pi\hbar}\tanh\left[\frac{\beta_o\hbar\mathcal{W}_k(t_o)}{2}\right]\right]^{\frac{1}{2}} \tag{28}$$

where

$$\mathcal{W}_k(t_o) = \frac{\left[\vec{k}^2 + m^2(T_o)\right]^{\frac{1}{2}}}{a(t_o)} \quad , \quad \frac{m^2(T_o)}{a^2(t_o)} = \mathcal{V}^{(2)}(\phi(t_o)) \tag{29}$$

The initial density matrix is normalized such that $Tr\rho(t_o) = 1$. Since time evolution is unitary such a normalization will be constant in time. For $T_o = 0$ the density matrix describes a pure state since $B_k = 0$.

It is convenient to introduce the complex function:

$$\mathcal{A}_k(t) = \tanh\left[\beta_o \hbar W_k(t_o)\right] \mathrm{Re}A_k(t) + i\ \mathrm{Im}A_k(t) \tag{30}$$

Then $\mathcal{A}_k(t)$ obeys a Riccati equation

$$i\dot{\mathcal{A}}_k(t) = \frac{1}{a^3(t)}\left[{\mathcal{A}_k}^2(t) - \omega_k^2(t)a^3(t)\right] \tag{31}$$

with the initial conditions:

$$\mathcal{A}_k(t_o) = \mathcal{W}_k(t_o)a^3(t_o) \tag{32}$$

Eq.(31) can be easily linearized by introducing the functions $\varphi_k^H(t)$ as

$$\mathcal{A}_k(t) = -ia^3(t)\frac{\dot{\varphi}_k^H(t)}{\varphi_k^H(t)} \tag{33}$$

The equal time two-point function thus becomes

$$\langle \eta^2(\vec{x},t)\rangle = \frac{\hbar}{2}\int \frac{d^3k}{(2\pi)^3} \mid \varphi_k^H(t) \mid^2 \coth\left[\beta_o \hbar \mathcal{W}_k(t_o)/2\right], \tag{34}$$

which leads to the following set of self-consistent time dependent Hartree equations:

$$\ddot{\phi} + 3\frac{\dot{a}}{a}\dot{\phi} + V'(\phi) + \lambda\,\phi\,\frac{\hbar}{2}\int \frac{d^3k}{(2\pi)^3}\frac{\mid \varphi_k^H(t) \mid^2}{2}\coth\left[\beta_o \hbar \mathcal{W}_k(t_o)/2\right] = 0 \tag{35}$$

$$\left[\frac{d^2}{dt^2} + 3\frac{\dot{a}(t)}{a(t)}\frac{d}{dt} + \frac{\vec{k}^2}{a^2(t)} + V''(\phi(t)) + \right.$$
$$\left.\lambda\frac{\hbar}{2}\int \frac{d^3k}{(2\pi)^3}\frac{\mid \varphi_k^H(t) \mid^2}{2}\coth\left[\beta_o \hbar \mathcal{W}_k(t_o)/2\right]\right]\varphi_k^H(t) = 0 \tag{36}$$

$$\varphi_k^H(t_o) = \frac{1}{\sqrt{a^3(t_o)\mathcal{W}_k(t_o)}} \quad , \quad \dot{\varphi}_k^H(t)\mid_{t_o} = i\sqrt{\frac{\mathcal{W}_k(t_o)}{a^3(t_o)}} \tag{37}$$

That is, we are faced with two complicated coupled non-linear and non-local equations for $\phi(t)$ and $\varphi_k(t)$.

The Hartree self-consistent equations (35) need regularization and renormalization since the momentum integrals of the mode functions (34) diverge.

Because the Bose-Einstein distribution functions are exponentially suppressed at large momenta, the finite temperature contribution will always be convergent and we need only address the zero temperature contribution.

A WKB analysis of eq.(35) provides the large k behaviour of the mode functions $\varphi_k(t)$ [7]:

$$\frac{\mid \varphi_k^H(t) \mid^2}{2} \overset{k\to\infty}{=} \frac{1}{2a^2(t)k} + \frac{1}{4k^3}\left[\frac{\dot{a}^2(t_o)}{a^2(t)} - \left(-\frac{\mathcal{R}}{6} + V''(\phi) + \frac{\lambda}{2}\langle\eta^2(\vec{x},t)\rangle\right)\right] + \mathcal{O}(1/k^4) + \cdots \tag{38}$$

Introducing an upper momentum cut-off Λ we obtain

$$\langle \eta^2(\vec{x},t)\rangle = \frac{\hbar}{8\pi^2}\frac{\Lambda^2}{a^2(t)} + \frac{\hbar}{8\pi^2}\ln\left(\frac{\Lambda}{K}\right)\left[\frac{\dot{a}^2(t_o)}{a^2(t)} - \left(-\frac{\mathcal{R}}{6} + V''(\phi(t)) + \frac{\lambda}{2}\langle\eta^2(\vec{x},t)\rangle\right)\right] + \text{finite} \tag{39}$$

where we have introduced a renormalization point K, and the finite part depends on time, temperature and K.

There are several physically important features of the divergent structure obtained above. First, the quadratically divergent term reflects the fact that the physical momentum cut-off is being red-shifted by the expansion. (This term will not appear in dimensional regularization).

Secondly, the logarithmic divergence contains a term that reflects the initial condition (the derivative of the expansion factor at the initial time t_o). The initial condition breaks any remnant symmetry. For example, in de Sitter space there is still invariance under the de Sitter group, but this is also broken by the initial condition at an arbitrary time t_o.Thus this term is not forbidden, and its appearance does not come as a surprise. As a consequence of this term, we need a time dependent term in the bare mass proportional to $1/a^2(t)$.

We are now in a position to present the renormalization prescription within the Hartree approximation. The differential equation for the mode functions (36) *must be finite.* Thus the renormalization conditions are obtained from

$$m_B^2(t) + \frac{\lambda_B}{2}\phi^2(t) + \xi_B\mathcal{R} + \frac{\lambda_B}{2}\langle\eta^2\rangle_B = m_R^2 + \frac{\lambda_R}{2}\phi^2(t) + \xi_R\mathcal{R} + \frac{\lambda_R}{2}\langle\eta^2\rangle_R \tag{40}$$

where the subscripts B, R refer to bare and renormalized quantities respectively and $\langle\eta^2\rangle_B$ is read from (39)

$$\langle\eta^2\rangle_B = \frac{\hbar}{8\pi^2}\frac{\Lambda^2}{a^2(t)} \tag{41}$$

$$+\frac{\hbar}{8\pi^2}\ln\left(\frac{\Lambda}{K}\right)\left[\frac{\dot{a}^2(t_o)}{a^2(t)} - \left(-\frac{\mathcal{R}}{6} + m_R^2 + \frac{\lambda_R}{2}\phi^2(t) + \xi_R\mathcal{R} + \frac{\lambda_R}{2}\langle\eta^2\rangle_R\right)\right] + \text{ finite}$$

Using the renormalization conditions (40) we obtain

$$m_B^2(t) + \frac{\lambda_B\hbar}{16\pi^2}\frac{\Lambda^2}{a^2(t)} + \frac{\lambda_B\hbar}{16\pi^2}\ln\left(\frac{\Lambda}{K}\right)\frac{\dot{a}^2(t_o)}{a^2(t)} = m_R^2\left[1 + \frac{\lambda_B\hbar}{16\pi^2}\ln\left(\frac{\Lambda}{K}\right)\right] \tag{42}$$

$$\lambda_B = \frac{\lambda_R}{1 - \frac{\lambda_R\hbar}{16\pi^2}\ln\left(\frac{\Lambda}{K}\right)} \tag{43}$$

$$\xi_B = \xi_R + \frac{\lambda_B\hbar}{16\pi^2}\ln\left(\frac{\Lambda}{K}\right)\left(\xi_R - \frac{1}{6}\right) \tag{44}$$

$$\langle\eta^2\rangle_R = I_R + J \tag{45}$$

where

$$I_R = \hbar\int\frac{d^3k}{(2\pi)^3}\left\{\frac{|\varphi_k^H(t)|^2}{2} - \frac{1}{2ka^2(t)} + \frac{\theta(k-K)}{4k^3}\left[-\frac{\mathcal{R}}{6} - \frac{\dot{a}^2(t_o)}{a^2(t)} + m_R^2 + \frac{\lambda_R}{2}\phi^2(t) + \xi_R\,\mathcal{R} + \frac{\lambda_R}{2}\langle\eta^2\rangle_R\right]\right\} \tag{46}$$

$$J = \hbar\int\frac{d^3k}{(2\pi)^3}\frac{|\varphi_k^H(t)|^2}{\exp\beta_o\hbar\mathcal{W}_k(t_o) - 1} \tag{47}$$

The conformal coupling $\xi = 1/6$ is a *fixed point* under renormalization[10]. In dimensional regularization the terms involving Λ^2 are absent and $\ln\Lambda$ is replaced by a simple pole at the physical dimension. Even in such a regularization scheme, however, a time dependent bare mass is needed. The presence of this new renormalization allows us to introduce a new renormalized mass term of the form

$$\frac{\Sigma}{a^2(t)}$$

This counterterm may be interpreted as a squared mass red-shifted by the expansion of the universe. However, we shall set $\Sigma = 0$ for simplicity.

Notice that there is a weak cut-off dependence on the effective equation of motion for the order parameter.

For *fixed* λ_R, as the cutoff $\Lambda \to \infty$

$$\lambda_B \approx -\frac{(4\pi)^2}{\ln\left(\frac{\Lambda}{K}\right)} \; , \quad \xi_B = \frac{1}{6} + O\left(\frac{1}{\ln \Lambda}\right) \; , \quad m_B^2(t) = \frac{1}{a^2(t)}\left[\frac{\Lambda^2}{\ln\left(\frac{\Lambda}{K}\right)} + \dot{a}^2(t_0)\right] + O\left(\frac{1}{\ln \Lambda}\right) \tag{48}$$

This approach to 0^- of the bare coupling as the cutoff is removed translates into an instability in the bare theory. This is a consequence of the fact that the N-component Φ^4 theory for $N \to \infty$ is asymptotically free (see ref.[11]), which is not relieved in curved space-time. Clearly this theory is sensible only as a low-energy cut-off effective theory, and it is in this restricted sense that we will ignore the weak cut-off dependence and neglect the term proportional to the bare coupling in (49).

The renormalized self-consistent Hartree equations thus become after letting $\Lambda = \infty$:

$$\ddot{\phi} + 3\,\frac{\dot{a}}{a}\,\dot{\phi} + m_R^2\phi + \xi_R\,\mathcal{R}\,\phi + \frac{\lambda_R}{2}\,\phi^3 + \frac{\lambda_R\,\phi}{2}\,\langle\eta^2\rangle_R = 0 \tag{49}$$

$$\left[\frac{d^2}{dt^2} + 3\,\frac{\dot{a}(t)}{a(t)}\frac{}{d}dt + \frac{\vec{k}^2}{a^2(t)} + m_R^2 + \xi_R\,\mathcal{R} + \frac{\lambda_R}{2}\,\phi^2 + \frac{\lambda_R}{2}\,\langle\eta^2\rangle_R\right]\varphi_k^H(t) = 0$$

where $\langle\eta^2\rangle_R$ is given by equations (46), (47).

4 High Temperature Limit

One of the payoffs of understanding the large-k behavior of the mode functions (as obtained in the previous section via the WKB method) is that it permits the evaluation of the high temperature limit. We shall perform our analysis of the high temperature expansion for the Hartree approximation.

The finite temperature contribution is determined by the integral

$$J = \hbar \int \frac{d^3k}{(2\pi)^3}\frac{\mid \varphi_k^H(t) \mid^2}{e^{\beta_o\hbar W_k(t_o)} - 1} \tag{50}$$

For large temperature, only momenta $k \geq T_o$ contribute. Thus the leading contribution is determined by the first term in eq.(38)). We find

$$J = \frac{1}{12\hbar}\left[\frac{k_B T_o a(t_o)}{a(t)}\right]^2 [1 + \mathcal{O}(1/T_o) + \cdots] \tag{51}$$

Thus we see that the leading high temperature behavior results in an effective time dependent temperature

$$T_{eff}(t) = T_o\left[\frac{a(t_o)}{a(t)}\right]$$

This expression corresponds to what would be obtained for an *adiabatic* (isentropic) expansion for radiation (massless particles) evolving in the cosmological background.

This behavior only appears at *leading* order in the high temperature expansion. There are subleading terms that we computed subsequently [7]. To avoid cluttering of notation, we will set $k_B = \hbar = 1$ in what follows.

We define

$$m^2(T_o) \equiv m_R^2 + \xi_R \mathcal{R}(t_o) + \frac{\lambda_R}{2}\phi^2(t_o) + \frac{\lambda_R}{2}\langle \eta^2(t_o)\rangle_R \tag{52}$$

and we will assume that $m^2(T_o) \ll T_o^2$. Since we are interested in the description of a phase transition, we will write

$$m_R^2 + \xi_R \mathcal{R}(t_o) + \frac{\lambda_R}{2}\phi^2(t_o) = -\frac{\lambda_R T_c^2}{24}; \quad T_c^2 > 0 \tag{53}$$

Thus, to leading order in T_o

$$m^2(T_o) = \frac{\lambda_R}{24}(T_o^2 - T_c^2) \tag{54}$$

Our high temperature expansion will assume *fixed* $m(T_o)$ and $m(T_o)/T_o \ll 1$.

It becomes convenient to define the variable

$$x^2 = \frac{k^2}{T_o^2 a^2(t_o)} + \frac{m^2(T_o)}{T_o^2}$$

$$\mid \varphi_k^H(t) \mid^2 = \left|\varphi^H(a(t_o)\sqrt{x^2T_o^2 - m^2(T_o)}; t)\right|^2$$

Recall from our WKB analysis that the leading behavior for $k \to \infty$ is (see equation (38)

$$\frac{\mid \varphi_k^H(t) \mid^2}{2} \to \frac{1}{2a^2(t)k}$$

adding and subtracting this leading term in the integral J and performing the above change of variables, we have

$$\begin{aligned}
J &= J_1 + J_2 \\
J_1 &= \left[\frac{a(t_o)}{a(t))}\right]^3 \left(\frac{T_o}{\pi}\right)^2 \int_{\frac{m(T_o)}{T_o}}^{\infty} \frac{x\,dx}{e^x - 1}\left[a^3(t)\sqrt{x^2T_o^2 - m^2(T_o)}\frac{\mid \varphi_k^H(t) \mid^2}{2} - \frac{a(t)}{2a(t_o)}\right] \\
J_2 &= \frac{T_o^2}{2\pi^2}\left[\frac{a(t_o)}{a(t)}\right]^2 \int_{\frac{m(T_o)}{T_o}}^{\infty} \frac{x\,dx}{e^x - 1} \\
&= \frac{1}{2}T_o^2\left[\frac{a(t_o)}{a(t)}\right]^2\left[\frac{1}{12} - \frac{m(T_o)}{2\pi^2 T_o} + \frac{m^2(T_o)}{8\pi^2T_o^2} + \mathcal{O}\left(\frac{m^3(T_o)}{T_o^3}\right) + \cdots\right]
\end{aligned} \tag{55}$$

We now must study the high temperature expansion of J_1. We restrict ourselves to the determination of the linear and logarithmic dependence on T_o. For this purpose, it becomes convenient to introduce yet another change of variables $x = \frac{m(T_o)}{T_o}z$ and use the fact that in the limit $T_o \gg m(T_o)$,

$$\frac{z}{e^{\frac{m(T_o)}{T_o}z} - 1} \approx \frac{T_o}{m(T_o)}[1 - \frac{m(T_o)}{2T_o}z + \cdots]$$

This yields the following linear and logarithmic terms in T_o:

$$J_{1lin} = \left[\frac{a(t_o)}{a(t)}\right]^3 \frac{T_o m(T_o)}{\pi^2}\int_1^{\infty} dz\left\{a^3(t)m(T_o)\sqrt{z^2-1}\frac{\mid \varphi_k^H(t) \mid^2}{2} - \frac{a(t)}{2a(t_o)}\right\} \tag{56}$$

Note that the above integral is finite.

The logarithmic contribution is obtained by keeping the $\mathcal{O}(1/k^3)$ in the large momentum expansion of $\mid \varphi_k^H(t) \mid^2$ given by equation (38) (in terms of the new variable z). We obtain after some straightforward algebra:

$$J_{1log} = -\frac{\ln\left[\frac{m(T_o)}{T_o}\right]}{8\pi^2}\left\{-\frac{\mathcal{R}}{6} - \frac{\dot{a}^2(t_o)}{a^2(t)} + \left[\frac{a(t_o)}{a(t)}\right]^2\left[m^2(T_o) + \frac{\lambda_R T_c^2}{24}\right] - \frac{\lambda_R T_c^2}{24}\right\} \tag{57}$$

That is, in the limit $T_o >> m(T_o)$, $J_1 = J_{1lin} + J_{1log} + O((T_o)^0)$.

Comparing the $\mathcal{O}(T_o^2, T_o, \ln T_o)$ contributions it becomes clear that they have very different time dependences through the scale factor $a(t)$. Thus the high temperature expansion as presented will not remain accurate at large times since the term quadratic in T_o may become of the same order or smaller than the linear or logarithmic terms. The high temperature expansion and the long time limit are thus not interchangeable, and any high temperature expansion is thus bound to be valid only within some time regime that depends on the initial value of the temperature and the initial conditions.

As an illustration, we calculated J_{1lin} explicitly in the case of de Sitter space [7].

5 Large Time Limit

The expansion factor $a(t)$ tends to infinity in the limit $t \to \infty$. As we have just seen, the high temperature expansion breaks down in this limit. Physical momenta are given by

$$l = \frac{k}{a(t)} \tag{58}$$

For large $a(t)$, only comoving momenta $k = l\, a(t) \to \infty$ will be relevant. Thus, we can again use the WKB method to evaluate the mode functions $\varphi_k^H(t)$ in this regime. Let us consider for example the de Sitter universe ($a(t) = a_o e^{Ht}$). We find from eq.(35)

$$\left|\varphi^H_{la(t)}(t)\right|^2 \overset{a(t)\to\infty}{=} \frac{H^2+l^2}{2l^2\sqrt{l^2+\frac{m(T_o)^2}{a(t)^2}}}\left[1+\frac{1}{2l^2}\left(\frac{m(T_o)^2}{a(t)^2}+\frac{\dot{a}(t_0)^2}{a(t)^2}\right)+O\left(\frac{1}{[a(t)l]^4}\right)\right] \tag{59}$$

for $m/H << 1$. Using the asymptotic behaviour (59) in eqs.(45)-(47) leads to [8]

$$\langle\eta^2\rangle_R = \frac{1}{(2\pi)^2}\left\{H^2 Z(\frac{K}{m(T_0)}) + Y(T_0)\left[\frac{a(t_0)}{a(t)}\right]^2 + O(a(t)^{-4})\right\} \tag{60}$$

where

$$Z(\frac{K}{m(T_0)}) = \frac{1}{2}(x-1)\left(1+\frac{\dot{a}(t_0)^2}{m^2}\right) + \log\left(\frac{2}{1+x}\right) \text{ with } x \equiv \sqrt{1+\frac{m^2}{K^2}} \tag{61}$$

$$Y(T) = \int_0^\infty \frac{k\,dk}{e^{\frac{\sqrt{k^2+m(T)^2}}{T}}-1} + \frac{m^4}{4K^2}\frac{1}{(1+x)^2} + \frac{\dot{a}(t_0)^2}{2}\log\left(\frac{2}{1+x}\right) \tag{62}$$

It should be noticed that the integral in $Y(T)$ is the mean value of $\frac{1}{k}$ for a free Bose gas at temperature T_0. Notice that both $Y(T)$ and $Z(\frac{K}{m(T_0)})$ depend on the renormalization point K and that $Z(\frac{K}{m(T_0)})$ is **positive** for all K .

Assuming that $\phi(t) \to \phi_\infty$ (a constant) for $t \to \infty$, the Hartree equations (49) yield for $t \to \infty$,

$$\phi_\infty^2 = -\frac{2}{\lambda_r}\left[m_r^2 + 12H^2\xi_r\right] - H^2\, Z(\frac{K}{m(T_0)}) \tag{63}$$

Since $Z(\frac{K}{m(T_0)}) > 0$ and we consider $\lambda_r << 1$, this equation has a solution provided

$$m_r^2 + 12H^2\xi_r < 0 \tag{64}$$

The physical interpretation is as follows, in the symmetry breaking case where eq.(64) holds, the order parameter $\phi(t)$ for large times t (when the universe cools down fast) tends to the non-zero value ϕ_∞ given by eq.(63) independent of the initial value $\phi(t_0)$.

The equal time two-point field correlator is given in the Hartree approximation by

$$\Delta(\vec{x},t) = \langle \eta(\vec{x},t)\eta(0,t)\rangle = \frac{1}{2}\int \frac{d^3k}{(2\pi)^3}\, e^{i\vec{k}.\vec{x}} \mid \varphi_k^H(t) \mid^2 \coth\left[\beta_o\hbar\mathcal{W}_k(t_o)/2\right], \tag{65}$$

For long times [$a(t) >> 1$] we find using the previous WKB results for $\mid \varphi_k^H(t) \mid^2$ (38),

$$\Delta(\vec{x},t) \overset{a(t)\to\infty}{=} \left(\frac{H}{2\pi}\right)^2 \frac{A(r)}{r} + \frac{B(r)}{[2\pi r a(t)]^2}\,, \quad \text{where} \tag{66}$$

$$A(r) = \frac{\pi}{2m} - \int_r^\infty dr\; K_0(mr) \quad , \quad B(r) = mrK_1(mr)\,, \quad r \equiv |\vec{x}| \tag{67}$$

We find in terms of the physical (redshifted) length $R \equiv r\,a(t)$ when $t \to \infty$ and R is kept fixed:

$$\Delta(\vec{x},t) \overset{a(t)\to\infty}{=} \frac{1}{(2\pi R)^2} + (\frac{H}{2\pi})^2\left[Ht - \log(2mR) + 1 - \gamma\right] + O(\log a(t)/[a(t)]^2) \tag{68}$$

Notice that $m = m(T_0)$ fixes the scale of the correlations.

References

[1] E. W. Kolb and M.S. Turner, "The Early Universe", Addison Wesley (Frontiers in Physics) (1990).

A. Linde, Particle Physics and Inflationary Cosmology, Harwood Academic Publishers (1990), and references therein.

[2] For a review of the effective potential and equilibrium methods as applied in cosmology see for example: R. H. Brandenberger, Rev. of Mod. Phys. 57, 1 (1985); Int. J. Mod. Phys. 2A, 77 (1987) and ref.[1].

[3] L. Dolan and R. Jackiw, Phys. Rev. D9, 3320 (1974).

[4] S. Weinberg, Phys. Rev. D9, 3357 (1974).

[5] D. Boyanovsky and H. J. de Vega, Phys. Rev. D47, 2343 (1993).

[6] D. Boyanovsky, D-S Lee and A. Singh, Phys. Rev. D 48, 800 (1993).

[7] D. Boyanovsky, H.J. de Vega and R. Holman, 'Non-Equilibrium evolution of scalar fields in FRW Cosmologies I' , LPTHE preprint 93-52, hep-ph/9310319. To appear in Phys. Rev. D.

[8] D. Boyanovsky, H.J. de Vega and R. Holman, in progress (1994).

[9] D. Boyanovsky, H.J. de Vega and R. Holman, Can Disordered Chiral Condensates Form? A Dynamical Perspective. LPTHE-Paris preprint 94-01, hep-ph/9401308. January 1994.

[10] N. D. Birrell and P. C. W. Davies, "Quantum fields in curved space", (Cambridge 1982).

[11] S. Coleman, R. Jackiw and H. D. Politzer, Phys. Rev. D10, 2491 (1974);

L. F. Abbott, J. S. Kang, and H. J. Schnitzer, Phys. Rev. D13, 2212 (1976);

J. Avan and H. J. de Vega, Phys. Rev. D29, 2891 and 2904 (1984).

A REPORT ON QUANTUM SPINODAL DECOMPOSITION

Da-Shin Lee

Department of Physics and Astronomy
University of Pittsburgh
Pittsburgh, P. A. 15260, U.S.A.

The dynamics of phase separation via spinodal decomposition has been well understood within the context of classical nonequilibrium statistical mechanics [1]. Following a "critical quench", the dynamics is primarily associated with the onset and growth of unstable long wavelength modes. The new inflationary Universe scenario, in which spinodal decomposition is conjectured to have taken place during phase transitions in the early Universe, gives the impetus for a detailed study of such processes in quantum field theories[2, 3]. More recently, this subject has assumed even greater importance due to the suggestion by Wilczek and Rajagopal that the region of misaligned chiral vacuum might form following a quench in heavy ion collisions. The long wavelength pion modes may be amplified leading to emission of coherent pions during the chiral phase transition[4, 5].

In this article, we introduce and develop the necessary techniques from nonequilibrium quantum statistical mechanics to study the processes of formation and growth of correlated domain via spinodal decomposition in weakly coupled quantum scalar field theories. The situation under consideration is very different from the classical description of spinodal decomposition in statistical mechanics. The classical approach to spinodal decomposition is based on a "coarse grained" time dependent Landau-Ginzburg equation, which is first order in time and purely dissipative. Thermal fluctuations are usually introduced as a Langevin noise typically uncorrelated, that obeys the fluctuation dissipation relation. In our case, we study a *quantum field theory*, the Heisenberg field equations are second order, non-dissipative, and both quantum and thermal fluctuations are present in the initial state (density matrix)[6].

We model a "critical quench" by a time dependent Hamiltonian

$$H(t) = \int_{\Omega} d^3x \left\{ \frac{1}{2}\Pi^2(x) + \frac{1}{2}(\vec{\nabla}\Phi(x))^2 + \frac{1}{2}m^2(t)\Phi^2(x) + \frac{\lambda}{4!}\Phi^4(x) \right\} \tag{1}$$

$$m^2(t) = m_i^2\Theta(-t) - m_f^2\Theta(t), \quad m_i^2 = \mu^2[\frac{T_i^2}{T_c^2} - 1], \quad m_f^2 = \mu^2[1 - \frac{T_f^2}{T_c^2}] \tag{2}$$

with $\mu^2 > 0$; $T_i > T_c$; $T_f \ll T_c$. Furthermore, we assume that for all times $t < 0$ the system is in thermal equilibrium at the initial temperature T_i, thus described by the density matrix $\hat{\rho}_i = e^{-\beta_i H_i}, H_i = H(t < 0)$. In the Schroedinger picture,the density matrix evolves in time via the quantum Liouville equation. For the calculation of *real time* correlation functions, we introduce the generating functional based upon the method of closed-time-path[7]. In particular, the relevant quantity of interest is the *equal time* correlation function. We find it to be

$$\langle\Phi_{\vec{k}}(t)\Phi_{-\vec{k}}(t)\rangle = \frac{1}{2\omega_<(k)}\mathcal{U}_k^+(t)\mathcal{U}_k^-(t)\coth[\beta_i\omega_<(k)/2] \tag{3}$$

with the mode functions obeying

$$\left[\frac{d^2}{dt^2} + \vec{k}^2 + m^2(t)\right]\mathcal{U}_k^\pm = 0, \quad \mathcal{U}_k^\pm(t<0) = e^{\mp i\omega_<(k)t} \tag{4}$$

where $m^2(t)$ is given by (2) and $\omega_<(k) = \left[\vec{k}^2 + m_i^2\right]^{\frac{1}{2}}$. For $t > 0$, they consist of stable modes ($k^2 > m_f^2$) and unstable modes ($k^2 < m_f^2$). The unstable modes are responsible for the growth of correlations. We are thus led to study *only* the contributions of unstable modes to the correlation function. Recalling that the zero temperature correlation length is $\xi(0) = 1/\sqrt{2}\mu$, we find that for $T_f \approx 0$ the amplitude of the fluctuation inside a "domain" $\langle\Phi^2(t)\rangle$, and the "size" of a domain $\xi_D(t)$ grow as

$$\langle\Phi^2(t)\rangle \approx \frac{\exp[\sqrt{2}t/\xi(0)]}{(\sqrt{2}t/\xi(0))^{\frac{3}{2}}}, \qquad \xi_D(t) \approx (8\sqrt{2})^{\frac{1}{2}}\xi(0)\sqrt{\frac{t}{\xi(0)}} \tag{5}$$

We indicate that the *dynamics* of the phase transition cannot be studied within perturbation theory because of exponential growth of unstable modes. We use a non-perturbative self-consistent Hartree approximation to study the time evolution of domain growth. It is implemented by replacement $m^2(t) \to m^2(t) + \frac{\lambda}{2}\langle\Phi^2(t)\rangle$ (where we use spatial translational invariance). After renormalization, the Hartree equations read

$$[\langle\Phi^2(t)\rangle - \langle\Phi^2(0)\rangle] = \int \frac{d^3k}{(2\pi)^3}\frac{1}{2\omega_<(k)}[\mathcal{U}_k^+(t)\mathcal{U}_k^-(t) - 1]\coth[\beta_i\omega_<(k)/2] \tag{6}$$

$$\left[\frac{d^2}{dt^2} + \vec{k}^2 + m_R^2(t) + \frac{\lambda_R}{2}\left(\langle\Phi^2(t)\rangle - \langle\Phi^2(0)\rangle\right)\right]\mathcal{U}_k^\pm(t) = 0 \tag{7}$$

$$m_R^2(t) = \mu_R^2\left[\frac{T_i^2}{T_c^2} - 1\right]\Theta(-t) - \mu_R^2\left[1 - \frac{T_f^2}{T_c^2}\right]\Theta(t) \tag{8}$$

with $T_i > T_c$, $T_f \ll T_c$ and the boundary conditions (4) on the homogeneous equations for $t < 0$. This set of Hartree equations is extremely complicated for an exact solution, however it has the correct physics in it. Consider the equations for $t > 0$, at very early times, when (the renormalized) $\langle\Phi^2(t)\rangle - \langle\Phi^2(0)\rangle \approx 0$ the mode functions are the same as in the zeroth order approximation, and the unstable modes grow exponentially. By computing self-consistently (6) with these zeroth-order unstable modes, we see that the fluctuation operator begins to grow exponentially. As

$(\langle\Phi^2(t)\rangle - \langle\Phi^2(0)\rangle)$ grows larger, its contribution to the Hartree equation is compensating for the negative mass term, thus weakening the unstabilities, only longer wavelengths are becoming unstable. Thus when $\frac{\lambda_R}{2m_f^2}(\langle\Phi^2(t)\rangle - \langle\Phi^2(0)\rangle) \approx 1$ the instabilities shut-off, this equality determines the "spinodal time" t_s. Figure (1) shows

$$\frac{\lambda}{2m_{f,R}^2}(\langle\Phi^2(\tau)\rangle - \langle\Phi^2(0)\rangle) = 3(\mathcal{D}^{HF}(0,\tau) - \mathcal{D}^{HF}(0,0))$$

(solid line) and also for comparison, the zeroth-order counterpart $3(\mathcal{D}^{(0)}(0,\tau) - \mathcal{D}^{(0)}(0,0))$ (dashed line) for $\lambda_R = 10^{-12}$; $T_i/T_c = 2$. (The value of the initial temperature does not have any particular physical significance and is chosen as a representative value to obtain the behavior of the correlations). We clearly see what we expected, whereas the zeroth order correlation grows indefinitely, the Hartree correlation function is bound in time and oscillatory. At $\tau \approx 10.52$, $3(\mathcal{D}^{(HF)}(0,\tau) - \mathcal{D}^{(HF)}(0,\tau)) = 1$, fluctuations are sampling field configurations near the classical spinodal, fluctuations continue to grow, however, because the derivatives are still fairly large, but after this time, the modes begin probing the stable region in which there is no exponential growth. Then $\frac{\lambda}{2m_{f,R}^2}(\langle\Phi^2(\tau)\rangle - \Phi^2(0)\rangle)$, becomes small again because of the coupling $g \approx 10^{-7}$, and the correction term becomes small. When it becomes smaller than one, the instabilities set in again, modes begin to grow and the process repeats. This gives rise to an oscillatory behavior around $\frac{\lambda_R}{2m_{f,R}^2}(\langle\Phi^2(\tau)\rangle - \Phi^2(0)\rangle = 1$. We see that up to the spinodal time $\tau_s \approx 10.52$ at which $\frac{\lambda}{2m_{f,R}}(\langle\Phi^2(\tau_s)\rangle - \Phi^2(0)\rangle) = 1$, the zeroth order correlation $3\mathcal{D}^{(0)}(0,\tau)$ is very close to the Hartree result, in fact at τ_s, the difference is less than 15%, in particular for these values of the coupling and initial temperature, the zeroth order correlation function leads to $t_s \approx 10.15$, and we may use the zeroth order correlations to provide an analytic estimate for t_s, as well as the form of the correlation functions and the size of the domains.

The fact that the zeroth-order correlation remains very close to the Hartree-corrected correlations up to times comparable to the spinodal is clearly a consequence of the very small coupling.

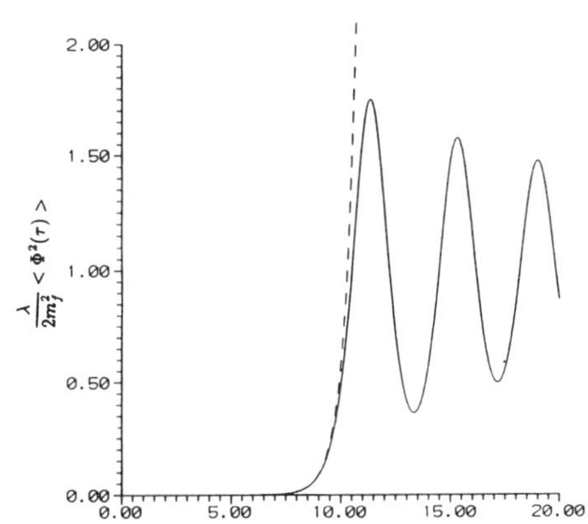

Figure 1. Hartree (solid line) and zero order (dashed line) results for $\frac{\lambda}{2m_f^2}(\langle\Phi^2(\tau)\rangle - \langle\Phi^2(0)\rangle) = \mathcal{D}(0,\tau)$, for $\lambda = 10^{-12}$, $\frac{T_i}{T_c} = 2$.

We conclude that the dynamics of the process of domain formation and growth (spinodal decomposition) is non-perturbative. A self-consistent Hartree approximation is used to study the long time evolution. We find in weakly coupled theories that the size of domains grow at long times as $\xi_D(t) \approx \sqrt{t\xi(0)}$. The domains, however, will grow up to a maximum time at which the fluctuations begin sampling the stable region. This maximum "spinodal time" is approximately given for weakly coupled theories by

$$t_s \approx -\sqrt{2}\xi(0) \ln \left[\left(\frac{3\lambda}{4\pi^3} \right)^{\frac{1}{2}} \left(\frac{(\frac{T_i}{2T_c})^3}{[\frac{T_i^2}{T_c^2} - 1]} \right) \right] \tag{9}$$

When the self-couplings are strong, the phase transition proceeds rapidly, and domains will not have time to grow substantially, and their sizes will be of the order of the zero temperature correlation length.

Acknowledgements

D.-S. Lee would like to thank the organizers for their great hospitality. He would also like to thank D. Boyanovsky, R. Holman, H. de Vega and R. Rivers for useful discussions and illuminating comments. This work was supported in part by NSF grant # PHY-9302534 as well as a Mellon PreDoctoral Fellowship Award.

References

[1] J. S. Langer in "Fluctuations, Instabilities and Phase Transitions",(T. Riste Ed.) page 19, Plenum N.Y. (1975); See also: J. S. Langer in "Solids Far From Equilibrium", Ed. C. Godreche, Cambridge Univ. Press (1992), page 297.

[2] For an introductory review on the subject, see for example L. Abbott and S.-Y. Pi, Inflationary Cosmology, (World Scientific, 1986); See also E. W. Kolb and M. S. Turner, "The Early Universe", Addison Wesley (Frontiers in Physics) (1990).

[3] D. Boyanovsky and H. J. de Vega, Phys. Rev. D47, 2343 (1993); D. Boyanovsky, D.-S. Lee and A. Singh, Phys. Rev. D48, 800 (1993) ; D. Boyanovsky, Phys. Rev. E48, 767 (1993); D. Boyanovsky, H. J. de Vega, and R. Holman, Phys. Rev. D49, 2769 (1994).

[4] K. Rajagopal and F. Wilczek, Nucl. Phys. B379, 395 (1993); Nucl. Phys. B404, 577 (1993).

[5] S. Gavin, A. Gocksch and R. D. Pisarski, Phys. Rev. Lett72 ,2143 (1994); D. Boyanovsky, H. J. de Vega, and R. Holman, "Can Disordered Chiral Condensates Form? A Dynamical Perspective" (PITT 94-01); F. Cooper, Y. Kluger, E. Mottola and J. P. Paz," Noequilibrium Quantum Dynamics of Disoriented Chiral Condensate", HEP-PH/9404357.

[6] A. Guth and S-Y. Pi, Phys. Rev. D32, 1899 (1985); E. J. Weinberg and A. Wu, Phys. Rev. D36, 2474 (1987).

[7] R. Mills, "Propagators for Many Particle Systems" (Gordon and Breach, N. Y. 1969).

NONPERTURBATIVE EVALUATION OF THE SPHALERON TRANSITION RATE

A. Bochkarev

Department of Physics and Astronomy
University of Pittsburgh
Pittsburgh, PA 15260

"You haven't measured it, you've only calculated it..."
Jan Smit.

We review nonperturbative calculations of the rate of sphaleron transitions on the lattice in 1 + 1-dimensional field theories and introduce a way to perform gauge-invariant Gibbs averages in classical non-Abelian Higgs theories.

Effective potential of the Chern-Simons variable N in the Standard Model (SM) is bounded at infinity. For this reason in hot electroweak plasma one expects Brownian motion of the system between topologically distinct vacua labelled by integer vallues of N :

$$\Delta(t) \; = \ll (N(t) \; - \; N(0))^2 \gg \; \rightarrow \; \Gamma t \; , \quad as \;\; t \rightarrow \infty, \tag{1}$$

where $\ll \cdot \gg$ stands for the Gibbs average. Diffusion rate Γ is the rate of sphaleron transitions. The parameter $\Delta(t)$ at large times receives contributions only from the relevant fluctuations interpolating between different classical vacua, which makes it possible in practice to calculate Γ numerically on the lattice. Γ is a nonperturbative definition of the rate of anomalous baryon number nonconservation at high temperatures in the SM [1], which can be used in the semiclassical temperature domain and beyond it. At high temperatures one neglects quantum corrections and solves classical equations of motion to generate static solutions like kinks [2] or sphalerons [3].

The correlator (1) was first evaluated in 1+1-dimensional Abelian Higgs model in [4] and the Brownian behaviour of the Chern-Simons number was found. Two methods to evalute $\Delta(t)$ are known so far. The first one is based on the microscopical classical *real-time* Langevin equation, derived in [4] in such a way that the white noise would respect gauge invariance. The Lagrangian of the model is:

$$\mathcal{L} \; = \; -\frac{1}{4\xi} F_{\mu\nu}^2 \; + \; |D_\mu \phi|^2 \; + \; (|\phi|^2 \; - \; 1/2)^2 \; , \tag{2}$$

with $D_\mu \; = \partial - iA_\mu$. In this Abelian gauge theory one can resolve constraints in the Coulomb ($\partial_1 A_1 \; = \; 0$) gauge using polar coordinates $\phi \; = \; \rho e^{i\alpha}$. The only (gauge-invariant!) degree of freedom left of the gauge field after the constraints have been resolved is the Chern-Simons variable $N = 1/2\pi \int dx A_1$. The momentum p_α canonically conjugated to the gauge-depenent degree of freedom α is the the density of electric charge. The white noise cannot be introduced in the equation for $\dot{p}_\alpha$, because it would violate the conservation of the electric charge. Therefore the noise term must be introduced only to the gauge-invariant variables like ρ. The solution $N(\tau)$ of the resulting first-order Lagevin equation is obtained numerically and the diffusion parameter (1) is evaluated as

$$\Delta(t) \;=\; \frac{1}{\tau_o}\int_o^{\tau_o} d\tau\ \left(N(t+\tau)\;-\;N(\tau)\right)^2, \quad as\ \tau_o \to \infty, \tag{3}$$

In this way the diffusion (1) was observed [4] with the expected thermal activation behaviour of the diffusion rate $\Gamma \sim \exp(-E_{sp}/T)$ at $T < E_{sp}$, where $E_{sp} \simeq 1$ is the sphaleron mass.

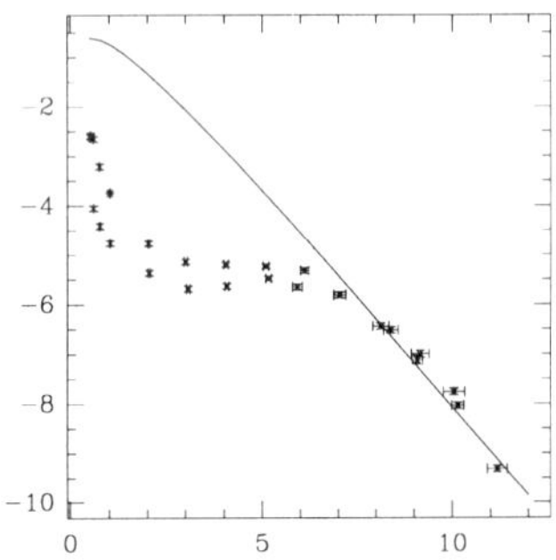

Figure 1. $\ln(\Gamma)$ vs β for lattice spacings $a = (0.16, 0.32)$ in the Abelian Higgs model [5].

The second method is to calculate the Gibbs average (1) directly according to the definition of the classical Green function:

$$\Delta(t) \;=\; \mathcal{Z}^{-1}\int \mathcal{D}p\,\mathcal{D}\phi\ \exp\left[-H(p,\phi)/T\right]\ \Delta_{p\phi}(t) \tag{4}$$

where $\Delta_{p\phi}(t)$ is evaluated on the microcanonical trajectories $p(t), \phi(t)$ - solutions of the Hamiltonian equations of motion with the initial conditions $p(t) = p,\ \phi(t) = \phi$. The Metropolis algorithm generating initial configurations must respect the Gauss law. For this purpose the gauge-invariant Langevin equation of [4] was used in [5] to generate the initial $\{\phi, p\}$. The temperature dependence $\Gamma(T)$ of these canonical measurements of the diffusion rate [5] is shown on Fig.1. At semiclassical temperatures $\beta \equiv T^{-1} > 7$ the data agree well with the analytical predictions of [6], while at high temperatures $\beta < 3$ the rate exhibits the expected T^2-law. In [7] it was found numerically advantageous to use $A_o = 0$ - gauge in Cartezian coordinates during the Hamiltonian evolution.

The gauge-invariant Langevin equation of [4] was introduced in the Abelian theory, where one can resolve the gauge constraints explicitly. To generalize it to the non-Abelian case we suggest to use the unitary gauge. In SU(2)-gauge theory with a scalar doublet φ the unitary gauge implies $\varphi = \sqrt{\varphi^\dagger\varphi} \equiv \rho$, and the Hamiltonian

$$\mathcal{H} \;=\; \frac{1}{2\xi}E^2 \;+\; \frac{1}{2\xi}H^2 \;+\; \frac{1}{4}p^2 \;+\; (\nabla\rho)^2 \;+\; \rho^2 A_i^2 \;+\; V(\rho) \;+\; \frac{1}{4\xi^2\rho^2}\left(D_iE_i\right)^2 \tag{5}$$

can be expressed entirely in terms of canonically conjugated variables (ρ, p) and (A_i^a, E_i^a). The last term (Coulomb energy $\rho^2 A_o^2$) takes into account the Gauss law $D_iE_i \;=\; -2\rho^2 A_o$ to get rid of A_o. The Gauss law is automatically satisfied in, say, Metropolis samplings with the Hamiltonian (5). In this way the gauge-invariant Gibbs distribution in classical non-Abelian gauge theories can be performed by Monte-Carlo lattice simulations. Field configurations generated in the unitary gauge do not contain sphalerons. The latter appear during the canonical evolution according to the equations of motion, which are gauge-invariant.

Langevin and canonical times are two alternative definitions of physical time in the processes at $T \neq 0$. To find a relation between them the sine-Gordon model was studied [8]:

$$\mathcal{H} = \int dx \left\{ \frac{1}{2}(\dot{\phi})^2 + \frac{1}{2}(\Delta\phi)^2 + 1 - cos(\phi) \right\} \tag{6}$$

The diffusing variable now is the zero mode $\bar{\phi} = \frac{1}{2\pi L}\int dx\, \phi(x)$. The model has many degenerate classical vacua and stable kink solutions interpolating between them. At $T \neq 0$ the Brownian motion of $\bar{\phi}$ has been observed. At semiclassical temperatures it is due to creation of kink-antikink pairs, so the thermal activation-type of the temperature dependence $\Gamma \sim exp[-E_K/T]$ has been seen in both Langevin and direct Gibbs measurements.

The complete second-order Langevin equation has friction coefficient γ as an input parameter:

$$\ddot{\phi}_x + \frac{\partial \mathcal{H}}{\partial \phi_x} = -\gamma\dot{\phi}_x + \eta_x(t), \tag{7}$$

with the white noise normalized as $< \eta_x(t)\, \eta_x(0) >= 2T\gamma\delta(x)\delta(t)$. Eq. (7) interpolates between the Hamiltonian equation of motion for $\gamma \to 0$ and the first-order Langevin equation (no $\ddot{\phi}_x$-term in (7)), valid at large γ or t. The absolute value of the rate (the preexponential factor) depends on γ. For moderate to large friction the simulations of (7) show (Fig.2): $\Gamma(\gamma) \sim \gamma^{-1}$; which would be a prediction of the first-order Langevin equation. That is because Γ is extracted from the correlator at large times t, for which the first-order Langevin equation is usually derived.

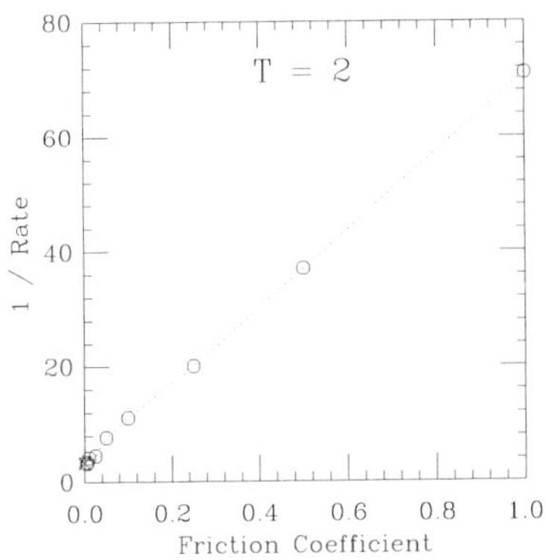

Figure 2. Γ^{-1} vs friction coefficient γ at $T = 2$ in the sine-Gordon model [8].

In the limit $\gamma \to 0$ the rate remains nonzero and has been found [8] to coincide exactly with the corresponding independent direct Gibbs measurement according to (4). These data demonstrate a fundamental conclusion that canonical measurements can be obtained as the $\gamma \to 0$-limit of the second-order Langevin simulations.

In the high-temperature domain $T > E_K$ at moderate to large friction the second-order Langevin simulations [8] show: $\Gamma = 2T/\gamma$, which is obviously the prediction of the *free* first-order Langevin equation. These simulations are insensitive to the dynamics of the system, because at high temperatures the noise-term $\eta_x(t)$ entirely dominates over the regular Hamiltonian force.

The high-temperature domain of canonical measurements was studied in [9]. It corresponds to the second-order Langevin simulations with asymptotically small γ, so that the noise-term does not dominate over the regular force at any fixed T. This set up represents a large isolated system like the early Universe with no external dumping. In addition to the expected temperature behaviour $\Gamma = T^2$ a strong lattice spacing dependence of the diffusion rate was discovered (Fig.3): $\Gamma \sim a^{-2}$. This is in contrast with the moderate to large friction simulations which do not exhibit any lattice-spacing dependence for $a \leq 1$ [8]. The rate, however, remains remarkably constant for the fixed ratio T/a even as one goes into the

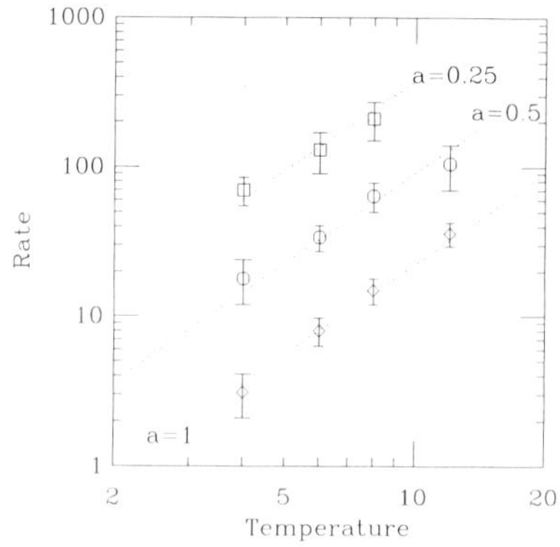

Figure 3. $\Gamma(T)$ for $a = (1, 0.5, 0.25)$ at high T in the sine-Gordon model [9].

semiclassical temperature domain [9]. Presumably, damping with the friction γ screens the high-frequency ultraviolet modes $\omega \sim a^{-1} \geq \gamma$, which are essential beyond the semiclassical domain of temperatures. The lattice spacing dependence of the diffusion rate may be different between the sine-Gordon and Abelian Higgs models, because the dimensionality of the rate is different. The latter is fixed by the requirement that the diffusing variable is dimensionless, since it has integer dimensionless values in the classical vacua. Study of the lattice spacing dependence of the diffusion rate requires rather long simulations. As a decreases, the number of degrees of freedom grows, and it takes more and more time for the system to thermalize itself in the course of the canonical evolution, which always takes place before the Brownian motion sets in. This study is, however, important particularily in higher space-time dimensions, since it is a connection between the simulated classical system and the underlying quantum field theory at finite temperature [2], [10].

Acknowledgements

This work was supported by the U.S. NSF grant PHY90-24764 and the grant of the Pittsburgh Supercomputer Center.

References

[1] V. Kuzmin, V. Rubakov, E. Shaposhnikov, *Phys. Lett.* B155:36(1985).

[2] D. Grigoriev, V. Rubakov, *Nucl. Phys.* B299:67(1988).

[3] D. Grigoriev, V. Rubakov, M. Shaposhnikov, *Nucl. Phys.* B326:737(1989).

[4] A. Bochkarev, Ph. de Forcrand, *Phys. Rev.* D44:519(1991).

[5] J. Smit, W.H. Tang, *Nucl. Phys. (Proc. Suppl.)* B34:616(1994).

[6] A. Bochkarev, G. Tsitsishvili, *Phys. Rev.* D40:R1378(1989).

[7] A. Krasnitz, R. Potting, *Phys. Lett.* B318:492(1993).

[8] A. Bochkarev, Ph. de Forcrand, *Phys. Rev.* D47:3476(1993).

[9] A. Bochkarev, *Nucl. Phys. (Proc. Suppl.)* B34:619(1994).

[10] A. Bochkarev, *Classical Limit of the Quantum Field Theory at Finite Temperature*, Preprint PITT-93-10 (1993).

BUBBLE NUCLEATION IN THEORIES WITH SYMMETRY BREAKING BY RADIATIVE CORRECTIONS

Erick J. Weinberg

Physics Department
Columbia University
New York, NY 10027

INTRODUCTION

In any first order phase transition a quantity of fundamental importance is the rate at which bubbles of the equilibrium ("true vacuum") phase nucleate within the metastable ("false vacuum") phase. The bubble nucleation rate per unit volume, Γ, can be calculated by a method, due to Coleman,[1] which is based on finding a "bounce" solution to the classical Euclidean field equations. Thus, for a theory with a single scalar field, one seeks a solution of

$$\partial_\mu \partial_\mu \phi \equiv \Box \phi = \frac{\partial V}{\partial \phi} \tag{1}$$

for which the ϕ is near its true vacuum value near the origin, but approaches its false vacuum value as any of the x_μ tend to $\pm\infty$.

A difficulty arises when one deals with theories where the symmetry breaking is a result of radiative corrections.[2] In such cases the vacuum structure is not determined by $V(\phi)$, but instead can be found only by examining the effective potential, $V_{\rm eff}(\phi)$. The bounce equation (1) is clearly inappropriate — in fact, if $V(\phi)$ has only a single minimum there will not even be a bounce solution. An obvious solution is to replace $V(\phi)$ by $V_{\rm eff}(\phi)$ in the bounce equation.[3]

Although plausible, and clearly a step in the right direction, this procedure raises some questions. The one-loop radiative corrections generate an effective action which contains not only $V_{\rm eff}$, but also terms involving derivatives (of all orders) of the fields. Can these terms be neglected when dealing with configurations, such as the bounce solution, which are not constant in (Euclidean) space-time? Even if this can be done in a first approximation, what are the nature and magnitude of the corrections which these terms generate? There are also questions relating to $V_{\rm eff}$ itself. First, the effective potential obtained by perturbative calculations differs considerably from that defined by a Legendre transform (the latter must be convex, while the former is not). The latter clearly does not lead to an appropriate bounce, but how precisely does the formalism pick out the former? Further, the perturbative effective potential is known to be complex for certain values of the fields. How is the imaginary part of $V_{\rm eff}$ to be handled?

In this talk I will describe a systematic calculational scheme[4] which gives an answer to these questions. The general idea is to use the path integral approach of Callan and Coleman,[5] but to integrate out certain fields at the outset. This leads to a modified effective action that gives a correct description of the vacuum structure of the theory and has a bounce solution that can provide the basis for a tunneling calculation. To leading approximation, one obtains the same result for Γ as would have been obtained by replacing V by $V_{\rm eff}$ in the standard procedure. The next-to-leading terms, which can be expressed in terms of the effective action, give calculable and significant corrections. The corrections beyond these, although well-defined, do not have a simple expression in terms of the effective action. In particular, the potentially complex terms in the effective potential do not appear directly, but only as part of more complicated functional determinants that can easily be shown to be real.

Throughout this talk I will confine myself to the case of bubble nucleation by quantum mechanical tunneling at zero temperature. Similar issues arise in connection with finite temperature bubble nucleation, even in theories where radiative corrections have little effect on the zero-temperature vacuum structure. It should be possible to deal with these by appropriate extensions of the methods described here.

A SIMPLE MODEL

Although radiative symmetry breaking is of greatest interest in the context of gauge theories, I will begin by using a somewhat simpler model to illustrate the method. This avoids some technical complications associated with gauge theories, which I discuss later. The model has two scalar fields, ϕ and A, and is governed by the Lagrangian

$$\mathcal{L} = \frac{1}{2}\left(\partial_\mu \phi\right)^2 + \frac{1}{2}\left(\partial_\mu A\right)^2 - V(\phi, A) \tag{2}$$

with

$$V = \frac{1}{2}\mu^2\phi^2 + \frac{\lambda}{4!}\phi^4 + \frac{1}{2}m^2A^2 + \frac{f}{4!}A^4 + \frac{1}{2}g^2\phi^2A^2. \tag{3}$$

Both μ^2 and m^2 are positive, so the tree-level potential has only a single symmetric minimum. We would like A-loop corrections to induce symmetry breaking with $\langle\phi\rangle = \sigma$ and $\langle A\rangle = 0$. This is done by choosing parameters so that $\mu = O(g^2\sigma)$, $m = O(g\sigma)$, $f = O(g^2)$, and $\lambda = O(g^4)$.

The one-particle-irreducible Green's functions of this theory are generated by the effective action $S_{\rm eff}$, which can be expanded in the familiar derivative expansion

$$S_{\rm eff}(\phi, A) = \int d^4x \left[V_{\rm eff}(\phi, A) + \frac{1}{2}Z_\phi(\phi, A)(\partial_\mu\phi)^2 + \frac{1}{2}Z_A(\phi, A)(\partial_\mu A)^2 + \cdots\right] \tag{4}$$

where the dots represent terms containing four or more derivatives. The leading contribution to the effective potential is of order g^4. It is given by the sum of the tree-level potential, all graphs with a single A-loop, and the corresponding counterterms and is equal to

$$\begin{aligned}(V_{\rm eff})_{g^4}(\phi) = {} & \frac{1}{64\pi^2}(m^2 + g^2\phi^2)^2 \left[\ln\left(\frac{m^2+g^2\phi^2}{m^2+g^2\sigma^2}\right) - \frac{1}{2}\right] \\ & - \frac{1}{4}\left(\frac{\mu^2}{\sigma^2} + y_1\right)\left(\phi^2 - \sigma^2\right)^2\end{aligned} \tag{5}$$

Figure 1. The two-loop graphs which contribute to the effective potential at order g^6. Solid and dashed lines represent ϕ and A propagators, respectively.

where y_1 is a constant of order g^2m^2/σ^2. The order g^6 contributions to V_{eff} arise from the two graphs shown in Fig. 1, together with several counterterm graphs. In these graphs the propagators are not simply those of the tree-level theory, but take into account the interaction with a background ϕ field. Thus, the A propagator is given by

$$G_A(k^2) = i(k^2 - m^2 - g^2\phi^2)^{-1}. \tag{6}$$

The ϕ propagator is a bit more complicated. Because the tree-level and one A-loop contributions are of the same magnitude, the effects of both must be included, giving

$$G_\phi(k^2) = i\left[k^2 - \mu^2 - (V_{\text{eff}})''_{g^4}\right]^{-1}. \tag{7}$$

(The inclusion of one-loop effects in this propagator means that when calculating higher order corrections one must be careful to avoid double-counting of graphs; this will not be relevant for the calculations discussed here.) At order g^8 there are both three-loop graphs containing A propagators and graphs with a single ϕ loop. The latter graphs bring in terms proportional to $\ln(V_{\text{eff}})''_{g^4}$ that become complex for certain values of ϕ.

This complex effective potential deserves some comment. The effective potential is often defined formally through a Legendre transform of the generating functional of connected Green's functions. The properties of Legendre transforms then imply that it is everywhere convex, a property that the perturbatively calculated effective potential does not enjoy in theories with multiple stable or metastable vacua. To understand this, recall that the effective potential $V_{\text{eff}}^{\text{Leg}}(\hat\phi)$ obtained by Legendre transform is equal to the minimum value of the energy density among all states $|\Psi\rangle$ such that $\langle\Psi|\phi(x)|\Psi\rangle = \hat\phi$. For values of $\hat\phi$ that lie between two vacua, the energy is minimized by states $|\Psi\rangle$ which are superpositions of the two vacua; these lead to a flat effective potential. The perturbative effective potential is most easily understood[6] by decomposing $\phi(\mathbf{x},t)$ into a spatially uniform mode $\phi_0(t)$ and a part $\tilde\phi(\mathbf{x},t)$ whose spatial integral vanishes. In the infinite volume limit the former mode is essentially classical, and one can discuss states in which the wave functional is of the form $\delta(\phi_0 - \hat\phi)\tilde\Psi[\tilde\psi(\mathbf{x}]$. The real part of the perturbative effective potential is the minimum expectation value of the energy density among states of this form subject to the additional condition that $\tilde\Psi[\tilde\psi(\mathbf{x}]$ be concentrated near $\tilde\psi = 0$. For values of $\hat\phi$ between two vacua these latter states are not eigenfunctions of the Hamiltonian and eventually decay to states whose wavefunctionals are concentrated on configurations with large fluctuations in $\tilde\psi(\mathbf{x})$. Being unstable, they have complex energies with the imaginary part of the energy related to their decay rate. In the region where the tree-level potential is concave the decay is essentially classical and is reflected in a perturbative imaginary part to the effective potential. Beyond this region decay proceeds by quantum tunneling and leads a nonperturbative imaginary part.

THE CALLAN-COLEMAN FORMALISM

In theories where the vacuum structure can be read off from the tree-level Lagrangian, Γ can be calculated using a formula that was derived by Callan and Coleman[5] using path integral techniques. The starting point is the quantity

$$G(T) = \langle \phi(\mathbf{x}) = \phi_{\rm fv} |\, e^{-HT} \,| \phi(\mathbf{x}) = \phi_{\rm fv} \rangle = \int [d\phi] e^{-[S(\phi)+S_{\rm ct}(\phi)]} \tag{8}$$

where the Euclidean action

$$S(\phi) = \int d^3x \int_{-T/2}^{T/2} dx_4 \left[\frac{1}{2}(\partial_\mu \phi)^2 + V(\phi) \right] \tag{9}$$

and $S_{\rm ct}$ contains the counterterms needed to make the theory finite.

The path integral is over all configurations such that ϕ takes its false vacuum value $\phi_{\rm fv}$ at $x_4 = \pm T/2$ and at spatial infinity. In the limit $T \to \infty$, Eq. (8) is dominated by the lowest energy state with a non-vanishing contribution (i.e., the false vacuum) and is of the form

$$G(T) \approx e^{-\mathcal{E} T \Omega} \tag{10}$$

where Ω is the volume of space and $\mathcal{E}$ may be interpreted as the energy density of the false vacuum state. Because this is an unstable state, $\mathcal{E}$ is complex with its imaginary part giving the decay rate, which in this case is simply the bubble nucleation rate. Dividing by Ω gives the nucleation rate per unit volume,

$$\Gamma = -2\, \mathrm{Im}\, \mathcal{E}. \tag{11}$$

The path integral may be approximated as the sum of the contributions about all of the stationary (or quasi-stationary) points of the Euclidean action $S(\phi)$: the pure false vacuum, the bounce solution ϕ_b with all possible locations in Euclidean space-time, and all multibounce configurations. In each case the contribution to the path integral is obtained by expanding the field about the classical solution $\bar\phi(x)$:

$$\phi(x) = \bar\phi(x) + \eta(x) \tag{12}$$

and then integrating over η. To leading approximation one keeps only the terms in the action which are quadratic in η. Expanding these in terms of the normal modes of $S''(\bar\phi) = -\Box + V''(\bar\phi)$ gives a product of Gaussian integrals. The evaluation of these integrals is completely straightforward in the case of the false vacuum, but about the bounce solution is complicated by the fact that $S''(\phi_b)$ has four zero and one negative eigenvalues. The zero modes are treated by introducing collective coordinates; integrating over these gives a factor of ΩT. The negative mode is handled by deforming the contour of integration; aside from a factor of $1/2$, this gives a contribution whose imaginary part is just that which would have been obtained from a naive application of the Gaussian integration formula. Finally, the contributions from the multibounce configurations are simply related to that from the single bounce. Summing over all of these stationary points gives

$$G = G_0 + \Omega T G_0 G_b + \frac{1}{2}(\Omega T)^2 G_0 G_b^2 + \ldots = G_0 e^{\Omega T G_b}. \tag{13}$$

This leads to

$$\Gamma = 2\,\mathrm{Im}G_b = e^{-B} K J e^{-[S_{\mathrm{ct}}(\phi_{\mathrm{b}}) - S_{\mathrm{ct}}(\phi_{\mathrm{fv}})]}(1 + \cdots) \tag{14}$$

where

$$B = S(\phi_{\mathrm{b}}) - S(\phi_{\mathrm{fv}}) \tag{15}$$

and

$$K = \left| \frac{\det'[-\Box + V''(\phi_{\mathrm{b}})]}{\det[-\Box + V''(\phi_{\mathrm{fv}})]} \right|^{-1/2}. \tag{16}$$

Here det′ indicates that the translational zero-frequency modes are to be omitted when evaluating the determinant. The determinants are divergent, but these divergences are cancelled by the terms containing S_{ct}. Finally, J is a Jacobean factor associated with the introduction of the collective coordinates; it can be shown to equal $B^2/4\pi^2$.

Scaling arguments can be used to estimate the magnitude of the various terms in the expression for Γ. Let us assume that we can identify a small coupling λ and a dimensionful quantity σ such that the potential can be written in the form

$$V(\phi) = \lambda\sigma^4 U(\psi) \tag{17}$$

where U involves no small couplings and the dimensionless field $\psi = \phi/\sigma$. The minima of the potential must then be located either at $\phi = 0$ or at values of ϕ of order σ.

By defining a dimensionless variable $s = \sqrt{\lambda}\,\sigma x$, we may write the field equations as

$$\Box_s \psi = \frac{\partial U}{\partial \psi}. \tag{18}$$

From the assumptions made above, this equation involves no small parameters and so has a bounce solution in which ψ is of order unity and differs from the false vacuum within a region of a spatial extent (measured in terms of s) which is also of order unity. In terms of the original variables, the bounce has ϕ of order σ and extends over a range of x of order $1/(\sqrt{\lambda}\,\sigma)$.

With the same change of variables, the action becomes

$$S = \frac{1}{\lambda} \int d^4 s \left[\frac{1}{2} \left(\frac{\partial \psi}{\partial s^\mu} \right)^2 + U(\psi) \right]. \tag{19}$$

Since the integrand contains no small parameters, while the volume of the bounce restricts the integration to a region of order unity, the bounce action B is of order λ^{-1}.

Similarly, the determinant factor K becomes

$$\begin{aligned} K &= \frac{1}{2} \left| \frac{\det'[(\lambda\sigma^2)(-\Box_s + U''(\psi_{\mathrm{b}}))]}{\det[(\lambda\sigma^2)(-\Box_s + U''(\psi_{\mathrm{fv}}))]} \right|^{-1/2} \\ &= \frac{1}{2}\lambda^2\sigma^4 \left| \frac{\det'[-\Box_s + U''(\psi_{\mathrm{b}})]}{\det[-\Box_s + U''(\psi_{\mathrm{fv}})]} \right|^{-1/2} \end{aligned} \tag{20}$$

where the explicit factor of $\lambda^2\sigma^4$ on the second line arises because the det$'$ factor involves four fewer modes than the det factor. With this factor extracted, the ratio of determinants is formally of order unity, although divergent. Finally, the Jacobean factor J is proportional to $B^2 \sim \lambda^{-2}$. Putting all of these factors together, we see that the nucleation rate is of the form

$$\Gamma = a\sigma^4 \, e^{-b/\lambda} \tag{21}$$

with a and b both of order unity. In practice b can be calculated rather accurately, since it is not difficult to numerically solve the differential equation for the bounce solution. On the other hand, an accurate determination of a, which involves a functional determinant, is exceedingly difficult.

A MODIFIED FORMALISM

If we were to apply this formalism to our scalar field model, the first step would be to find a bounce solution to the Euclidean field equations implied by the Lagrangian (2). The problem is that, because the tree level potential has only the symmetric minimum, there is no bounce solution. We can circumvent this problem by integrating out the A field from the start, and writing

$$G(T) = \int [d\phi][dA] e^{-S(\phi,A)} \equiv \int [d\phi] e^{-W(\phi)}. \tag{22}$$

W may be thought of as a kind of effective action, although it is not the same as the more usual S_{eff} of Eq. (4). Graphically, the two actions differ in that W receives contributions only from graphs with only internal A-lines and external ϕ-lines. Also, W is divergent while S_{eff}, since it generates renormalized Green's functions, must be finite. Performing the path integral over A fields, one obtains

$$\begin{aligned} W(\phi) = \int d^4x \left[\frac{1}{2} (\partial_\mu \phi)^2 + \frac{1}{2}\mu^2\phi^2 + \frac{\lambda}{4}\phi^4 + \frac{1}{2} \langle x | \ln \left[-\Box + M^2(\phi) \right] | x \rangle \right] \\ + \text{counterterms} + \cdots \end{aligned} \tag{23}$$

where $M^2(\phi) = m^2 + g^2\phi^2$ and the dots represent the two-loop and higher order corrections.

Although W, in contrast with the tree-level action, does reflect the true vacuum structure of the theory, we cannot simply proceed by solving its field equations to obtain a bounce. There are two difficulties here. First, we only have a perturbative expansion for W. Second, and more importantly, W is nonlocal and will therefore lead to quite complicated field equations.

Although nonlocal, W can be approximated by a local functional if ϕ is sufficiently slowly varying. (For the terms shown explicitly in Eq. (23), the requirement is that the change in ϕ over a distance of order $M(\phi)^{-1}$ be small.) This local functional takes the form of a derivative expansion

$$W(\phi) = \int d^4x \left[\hat{V}(\phi) + \frac{1}{2}\hat{Z}(\phi)(\partial_\mu\phi)^2 + \cdots \right]. \tag{24}$$

Although this is similar in form to the expansion for S_{eff}, the functions entering the two expansions are not the same. In line with the remarks made previously, $\hat{V}(\phi)$

and $\hat{Z}(\phi)$ differ from $V_{\rm eff}(\phi,0)$ and $Z_\phi(\phi,0)$ by the omission of graphs with internal ϕ-lines. In particular, to $O(g^4)$ we have

$$\hat{V}_{g^4} = (V_{\rm eff})_{g^4} \tag{25}$$

but at the next order

$$\hat{V}_{g^6} \neq (V_{\rm eff})_{g^6} \tag{26}$$

because the second graph in Fig. 1 contributes to $V_{\rm eff}$ but not to $\hat{V}$.

Returning to the bubble nucleation problem, let us define an action

$$W_0(\phi) = \int d^4x \left[\frac{1}{2}(\partial_\mu\phi)^2 + \hat{V}_{g^4}(\phi)\right] \tag{27}$$

that does display the correct vacuum structure and that, at least for slowly varying ϕ, is a good approximation to W. We can now attempt to evaluate the path integral over ϕ in Eq. (22) by expanding about the stationary points $\bar{\phi}$ of W_0. These include the homogeneous false vacuum and a bounce solution obeying

$$\Box\phi = \frac{\partial\hat{V}_{g^4}}{\partial\phi}. \tag{28}$$

Defining $\eta(x) = \phi(x) - \bar{\phi}(x)$, we obtain

$$W(\phi) = W(\bar{\phi}) + \int d^4z\, W'(\bar{\phi};z)\eta(z) + \frac{1}{2}\int d^4z\, d^4z'\, W''(\bar{\phi};z,z')\eta(z)\eta(z') + O(\eta^3) \tag{29}$$

where primes denote variational derivatives. Note that $W'(\bar{\phi};z)$ does not vanish, since $\bar{\phi}$ is a stationary point of W_0 but not of W. Note also that we cannot use the derivative expansion of W inside the path integral, since the integral includes rapidly varying field configurations. Inserting Eq. (29) into the path integral and then proceeding as in the standard case, we eventually obtain

$$\Gamma = e^{-C_1}\, e^{C_2} \left|\frac{\det' W''(\phi_{\rm b})}{\det W''(\phi_{\rm fv})}\right|^{-1/2} J(1+\cdots) \tag{30}$$

where

$$C_1 = W(\phi_{\rm b}) - W(\phi_{\rm fv}) \tag{31}$$

and

$$C_2 = W'(\phi_{\rm b})[W''(\phi_{\rm b})]^{-1}W'(\phi_{\rm b}) - (\phi_{\rm b} \to \phi_{\rm fv}). \tag{32}$$

Let us now examine the various terms in this expression, beginning with C_1. Scaling arguments of the type described previously show that the bounce solution to Eq. (28) has a characteristic spatial size of order $1/(g^2\sigma)$ and is therefore slowly varying relative to the scale $M(\phi)$ entering the one-loop approximation to W, Eq. (23). This both allows us to use the derivative expansion (24) to evaluate $W(\phi_b)$ and also suppresses the higher-derivative terms in this expansion. Doing the derivative expansion, one finds that the leading contribution to C_1 is

$$\hat{B}_0 = \int d^4x \left[\frac{1}{2}(\partial_\mu\phi_b)^2 + \hat{V}_{g^4}(\phi_b) - (\phi_{\rm b} \to \phi_{\rm fv})\right] \sim \frac{1}{g^4}. \tag{33}$$

Recalling that $\hat{V}_{g^4} = (V_{\rm eff})_{g^4}$, we see that this is just the result one would expect from simply replacing V by $V_{\rm eff}$ in the standard formalism. The next-to-leading contribution is

$$\hat{B}_1 = \int d^4x \left[\frac{1}{2} \hat{Z}_{g^2}(\phi_b)(\partial_\mu \phi_b)^2 + \hat{V}_{g^6}(\phi_b) - (\phi_{\rm b} \to \phi_{\rm fv}) \right] \sim \frac{1}{g^2}. \tag{34}$$

Because $\hat{V}_{g^6} \neq (V_{\rm eff})_{g^6}$ (although $\hat{Z}_{g^2}$ and $(Z_\phi)_{g^2}$ are equal in our model), this term is not simply the next correction to $S_{\rm eff}$. The term beyond this, which is of order unity, involves the $O(g^8)$ contributions to $\hat{V}$ and the $O(g^4)$ contributions to $\hat{Z}$, as well as the leading four-derivative terms.

For C_2 we need both W' and W''. The derivative expansion can be used for the former; because $\bar{\phi}$ is a stationary point of W_0, the leading contribution here is from $\delta W' \equiv W' - W_0'$. Matters are less simple for $W''(\bar{\phi}; z, z')$. Although the derivative expansion can be used when $|z - z'|$ is large, the behavior for small $|z - z'|$ is sensitive to the high momentum modes and so the derivative expansion fails no matter how slowly varying $\bar{\phi}$ is. However, the relation

$$W'' = W_0'' \left[1 + (W_0'')^{-1} \delta W''\right] \tag{35}$$

can be used to obtain formal expansions for $(W'')^{-1}$ and $\det W''$ as power series in $(W_0'')^{-1}\delta W''$. The actual utility of these expansions depends on the size of the contribution from the region of small $|z - z'|$. In the calculation of C_2 this contribution is subdominant and $(W'')^{-1}$ can be approximated by $(W_0'')^{-1}$; one finds that C_2 is of order unity. For the determinant factor, on the other hand, more terms must be retained:

$$\begin{aligned} \det{}'[W''] &= \det{}'[W_0''] \det[I + (W_0'')^{-1}\delta W''] \\ &= \det{}'[W_0''] \exp\left\{ \mathrm{tr} \ln[I + (W_0'')^{-1}\delta W''] \right\} \\ &= \det{}'[W_0''] \exp\left\{ \mathrm{tr}(W_0'')^{-1}\delta W'' + \frac{1}{2} \mathrm{tr} \left[(W_0'')^{-1}\delta W''\right]^2 + \cdots \right\}. \end{aligned} \tag{36}$$

In the last line, the first term in the exponent is of order $1/g^2$ while the second is of order unity. In fact, the former term is given by the second graph of Fig. 1 which, it will be recalled, contributes to $V_{\rm eff}$ but not to $\hat{V}$. It combines with the terms in $\hat{B}_1$ to give the full $O(g^6)$ contribution to the effective potential. Combining all other factors together in a prefactor A of order unity, we obtain

$$\Gamma = A\sigma^4 e^{-(B_0 + B_1)} \tag{37}$$

where

$$B_0 = \hat{B}_0 = \int d^4x \left\{ \left[(V_{\rm eff})_{g^4}(\phi_{\rm b}) + \frac{1}{2}(\partial_\mu \phi_{\rm b})^2 \right] - (\phi_{\rm b} \to \phi_{\rm fv}) \right\} \sim \frac{1}{g^4} \tag{38}$$

$$B_1 = \int d^4x \left\{ \left[(V_{\rm eff})_{g^6}(\phi_{\rm b}) + \frac{1}{2}(Z_\phi)_{g^2}(\phi_{\rm b})(\partial_\mu \phi_{\rm b})^2 \right] - (\phi_{\rm b} \to \phi_{\rm fv}) \right\} \sim \frac{1}{g^2}. \tag{39}$$

After seeing how the expansion of the determinant combines with the exponent factors to reconstruct $S_{\rm eff}$ to leading and next-to-leading orders, it is natural to speculate that this process might continue to higher orders. It does not. In particular, it is at the next order that we would encounter the potentially complex contributions to

V_{eff} which arise from graphs with a single ϕ loop. These would appear in a derivative expansion of the determinant factor

$$\left| \frac{\det' W_0''(\phi_{\text{b}})}{\det W_0''(\phi_{\text{fv}})} \right|^{-1/2} . \tag{40}$$

However, a derivative expansion of $\det W_0''$ is valid only for fields which vary slowly relative to $(V_{\text{eff}})''_{g^4}$. The bounce solution does not satisfy this condition, and so the complex terms in V_{eff} do not appear. (One can make a derivative expansion of the determinant in the denominator, but $(V_{\text{eff}})_{g^8}(\phi_{\text{fv}})$ is real.)

SCALAR QUANTUM ELECTRODYNAMICS

These methods can be applied to gauge theories with only minor modifications. To be specific, consider the case of scalar electrodynamics with the quartic scalar self-coupling taken to be $O(e^4)$ so that the vacuum structure is determined by the one-loop corrections to the effective potential.[2] Directly following the approach used for the scalar field example, one would integrate out the photon field at the start to obtain an effective action for the complex scalar field. The classical equations following from this action have a bounce solution of the desired type which gives the leading approximation to Γ. However, the calculation of the next order terms turns out to be much more complicated than in the scalar case. These complications can be avoided by an alternative approach. Since the bounce solution can be chosen to be entirely real, it is possible to integrate out both the photon field and the imaginary part of the scalar field to give an action $W(\phi)$ which depends only on a single real scalar field. The analysis then proceeds very much as before. The $O(e^4)$ terms in V_{eff} and $\hat{V}$ are identical and, together with the dominant gradient term, lead to a contribution to the exponent proportional to e^{-4}. At order e^6 there are two graphs contributing to V_{eff}, neither of which appears in $\hat{V}$. Both are recovered from the expansion of $\det W''$ and combine with $(Z_\phi)_{e^2} = (\hat{Z})_{e^2}$ to give a contribution to the exponent which is $O(e^{-2})$. V_{eff} becomes complex at order e^8 but, as before, the failure of the derivative expansion of $\det W_0''$ about the bounce solution prevents the offending terms from entering the nucleation rate calculation. In fact, the appearance of infrared divergences at small ϕ means that the derivative expansion cannot be carried out beyond the four-derivative terms. Instead, one must extract the potential and two-derivative terms, leaving a remainder which gives an $O(1)$ contribution to Γ that can be absorbed in the prefactor.

The issue of gauge-dependence also arises. The nucleation rate is a measurable physical quantity and so should be independent of gauge. However, the formulas above express it in terms of the effective potential and other gauge-dependent[7–10] quantities. The Nielsen identities[8] indicate how this conflict can be resolved. In a class of gauges with photon propagator

$$D_{\mu\nu}(k^2) = \frac{-g_{\mu\nu} - \frac{k_\mu k_\nu}{k^2}}{k^2} + \xi \frac{k_\mu k_\nu}{k^4} \tag{41}$$

these identities take the form

$$\xi \frac{\partial S_{\text{eff}}}{\partial \xi} = \int d^4x C[\phi(x)] \frac{\delta S_{\text{eff}}}{\delta \phi} . \tag{42}$$

(We do not need the explicit form of the functional $C[\phi(x)]$, but only the fact that it is $O(e^2)$.) For a uniform ϕ field this reduces to

$$\xi \frac{\partial V_{\text{eff}}}{\partial \xi} = C(\phi) \frac{\partial V_{\text{eff}}}{\partial \phi} . \tag{43}$$

Now let us expand this in powers of e^2 to yield a series of identities. The first of these states that $(V_{\text{eff}})_{e^4}$ is gauge-independent, as is easily verified. The next is

$$\xi \frac{\partial (V_{\text{eff}})_{e^6}}{\partial \xi} = C_{e^2}(\phi) \frac{\partial (V_{\text{eff}})_{e^4}}{\partial \phi} . \tag{44}$$

Further identities are obtained by making derivative expansions of both sides of Eq. (42) and expanding the various terms in powers of e^2. The first new identity obtained in this manner is

$$\xi \frac{\partial (Z_\phi)_{e^2}}{\partial \xi} = \frac{\partial C_{e^2}}{\partial \phi} . \tag{45}$$

Let us now apply these identities to Γ. The leading $(O(1/e^4))$ term in the exponent involves $(V_{\text{eff}})_{e^4}$, and is manifestly gauge-independent. The next term in the exponent, of order $1/e^2$, contains $(V_{\text{eff}})_{e^6}$ and $(Z_\phi)_{e^2}$. Using the identities we have just obtained, we find that the gauge dependence of this term is given by

$$\begin{aligned} \xi \frac{\partial}{\partial \xi} \int d^4x \left[(Z_\phi)_{e^2} (\partial_\mu \phi)^2 + (V_{\text{eff}})_{e^6} \right] &= \int d^4x \left[\frac{\partial C_{e^2}}{\partial \phi} (\partial_\mu \phi)^2 + C_{e^2} \frac{\partial (V_{\text{eff}})_{e^4}}{\partial \phi} \right] \\ &= \int d^4x \left[(\partial_\mu C_{e^2})(\partial_\mu \phi) + C_{e^2} \frac{\partial (V_{\text{eff}})_{e^4}}{\partial \phi} \right] \\ &= \int d^4x C_{e^2} \left[-\Box\phi + \frac{\partial (V_{\text{eff}})_{e^4}}{\partial \phi} \right] . \end{aligned} \tag{46}$$

Because ϕ_b is a solution of the Euclidean field equations, the last line vanishes.

CONCLUSION

In this talk I have described how the decay rate of a metastable vacuum can be calculated in a theory whose vacuum structure is determined by radiative corrections. As in the standard case, the result may be written as a dimensionful prefactor times the exponential of an action involving a bounce solution. To leading approximation this exponent is just the tree-level action supplemented by the dominant one-loop contribution to the effective potential. The first correction to the exponent arises from the next-to-leading contributions to the effective potential and the leading correction to the tree-level kinetic part of the effective action. Although smaller than the leading terms, these give an addition to the exponent which is larger than order unity and is thus more important than the prefactor. It does not appear that this correction need have any particular sign, but rather that it might increase the nucleation rate in some theories and reduce the rate in others.

All further corrections may be absorbed into the prefactor. Although some of these can be identified with particular terms in the effective action, this is not true of all higher corrections. Specifically, the graphs which give rise to complex terms in the effective potential cannot, when calculated in the background of the bounce, be expanded in a derivative expansion. Consequently, the imaginary part of the effective potential does not explicitly enter the bubble nucleation calculation and the problems of interpretation which it would entail are avoided.

Acknowledgements

This work was supported in part by the U. S. Department of Energy.

REFERENCES

1. S. Coleman, Phys. Rev. D**15**, 2929 (1977).
2. S. Coleman and E. J. Weinberg, Phys. Rev. D**7**, 1888 (1973).
3. See, e.g., P. Frampton, Phys. Rev. Lett. **37**, 1378 (1976) and Phys. Rev. D**15**, 2722 (1977); A. Linde, Phys. Lett. **70B**, 306 (1977).
4. E. J. Weinberg, Phys. Rev. D**47**, 4614 (1993).
5. C. G. Callan and S. Coleman, Phys. Rev. D**16**, 1762 (1977).
6. E. J. Weinberg and A. Wu, Phys. Rev. D**36**, 2474 (1987).
7. R. Jackiw, Phys. Rev. D**9**, 1686 (1974).
8. N. K. Nielsen, Nucl. Phys. **B101**, 173 (1975).
9. R. Fukuda and T. Kugo, Phys. Rev. D**13**, 3469 (1976).
10. I. J. R. Aitchison and C. M. Fraser, Ann. Phys. **156**, 1 (1984).

ONE-LOOP CORRECTIONS TO THE BUBBLE NUCLEATION RATE AT FINITE TEMPERATURE

V. G. Kiselev [1]

Institut für Physik, Universität Dortmund
D - 44221 Dortmund , Germany

ABSTRACT

I present an evaluation of the 1-loop prefactor in the lifetime of a metastable state which decays at finite temperature by bubble nucleation. Such a state is considered in one-component φ^4-model in three space dimensions. The calculation serves as a prototype application of a fast numerical method for evaluating the functional determinants that appear in semiclassical approximations.

The decay of metastable states by bubble nucleation appears in a large variety of physical contexts. It has received the attention of particle physicists and cosmologists due to its possible rôle in the evolution of the Universe. The semiclassical approach to bubble formation has been developed by Langer [1, 2] and Coleman and Callan [3, 4]. The leading factor in the transition rate is determined by the classical euclidean trajectory. Quantum corrections may however modify the rate in an significant way. Their evaluation for a realistic model in three space dimensions is an enterprise that can easily reach the limits of practical computability. It is therefore very useful to have method that leads to a fast numerical algorithm.

I present here in brief such a method. The details can be found in a paper written with J. Baacke [5]. We use as a simple model the four-dimensional φ^4-theory at finite temperature T given by the Euclidean action

$$S(\varphi) = \int_0^{1/T} d\tau \int d^3x \left(\frac{1}{2}(\partial_\mu\varphi)^2 + U(\varphi)\right). \tag{1}$$

The one-component ϕ^4-potential reads

$$U(\varphi) = \frac{1}{2}m^2\varphi^2 - \eta\varphi^3 + \frac{1}{8}\lambda\varphi^4 \tag{2}$$

We assume the global minimum to be at $\varphi = \varphi_+ > 0$. There is a local minimum $\varphi_- = 0$.

---

[1]On leave of absence from Institute of Physics, 220602 Minsk, Byelorussia;
e-mail: kiselev@het.physik.uni-dortmund.de

Any state built on the local minimum is metastable. Its decay rate per unit volume $\gamma = \Gamma/V$ at sufficiently high temperature is dominated by the energy $E = ST$ of a fluctuation which looks like a bubble $\phi(x)$ of the ϕ_+-phase. This bubble is in unstable equilibrium between collapse and unbounded expansion. The tree level approximation determines the order of magnitude of the decay rate as $\gamma \approx \exp\{-E(\phi(x))/T\}$.

Fluctuations around the critical bubble contribute a pre-exponential factor to the decay rate which is known to take the form [1, 2]

$$\gamma = \frac{\omega_-}{2\pi}\left(\frac{E(\phi)}{2\pi T}\right)^{3/2} |\mathcal{D}|^{-1/2} \exp\left\{-\frac{E(\phi)}{T} - S_{ct}\right\} \tag{3}$$

to 1-loop accuracy. The coefficient $\mathcal{D}$ here reads

$$\mathcal{D}(T) = \frac{det'(-(\partial/\partial\tau)^2 - \Delta + U''(\phi))}{det(-(\partial/\partial\tau)^2 - \Delta + U''(0))} \tag{4}$$

The prime in the determinant implies omitting of three zero modes in it. The temperature dependence of $\mathcal{D}$ arises from imposing periodic boundary conditions in the time direction with a period $1/T$. $\mathcal{D}(T)$ as introduced in Eq. (4) is ill-defined because of ultraviolet divergences. As discussed in [4] they are absorbed by the counterterm action S_{ct} which has been introduced in the exponent. It will be specified below.

It is straighforward to compute the tree level exponent $-E(\phi)/T$ in the decay rate (3). We parametrize the value of E in the same way as in Ref. [6]:

$$E = \frac{4.851\ m^3}{\eta^2} f(\alpha) \tag{5}$$

where

$$\alpha = \frac{\lambda m^2}{4\eta^2}. \tag{6}$$

The limit $\alpha \to 1$ corresponds to the thin wall approximation. The function $f(\alpha)$ is plotted in Fig. 1.

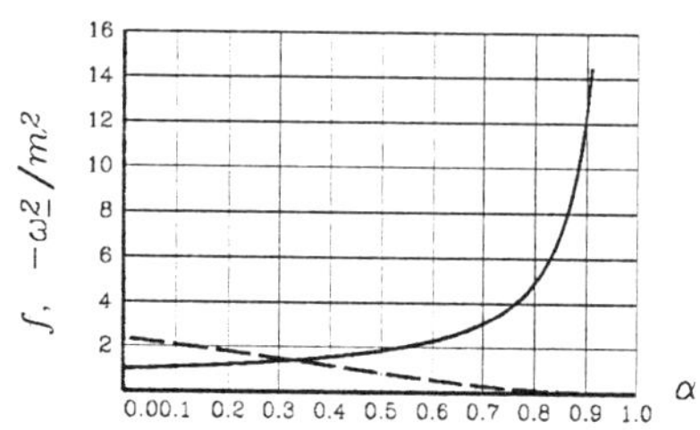

Figure 1. The values of $f(\alpha)$ (solid line) and $-\omega_-^2/m^2$ (dashed line) vs. α.

Let us discuss a method of computing the ratio of functional determinants (4), which is based on earlier papers [7].

The explicit form of the operator in the nominator (4) is

$$-\frac{\partial^2}{\partial\tau^2} - \Delta + m^2 + V(r) \tag{7}$$

with the periodical boundary conditions for the eigenfunctions $\Psi(\tau,\vec{x}) = \Psi(\tau + 1/T, \vec{x})$. The space-dependent part V (7) of $U''(\phi)$ (4) reads

$$V(r) = U''(\Phi) - m^2 = -6\eta\phi(r) + \frac{3}{2}\lambda\phi^2(r) \quad . \tag{8}$$

The free operator in the denominator (4) takes the same form as (7), but with $V(r) = 0$.

The time-independence and spherical symmetry of the background bubble field yield a classification of ω^2_{n,l,n_r}, the eigenvalues of (7), with respect to the number n of their Matsubara frequencies $\nu = 2\pi nT$, radial quantum number n_r, and angular momentum l. One can formally write the ratio of determinants (4) as

$$\mathcal{D}(T) = \prod_{n=-\infty}^{\infty} \prod_{l=0}^{\infty} \prod_{n_r=0}^{\infty} \left[\frac{\omega_{n,l,n_r}}{\omega^{(0)}_{n,l,n_r}} \right]^{2l+1} \tag{9}$$

with $\omega^{(0)}_{n,l,n_r}$ standing for the free eigenvalues.

The outline of our calculation is as follows. We compute first for each partial wave the product over n_r, i.e. the ratio of the determinants of the radial operators

$$J_l(\nu) = \frac{det H_{\nu,l}}{det H^{(0)}_{\nu,l}}; \tag{10}$$

here

$$H_{\nu,l} = -\frac{d^2}{dr^2} - \frac{2}{r}\frac{d}{dr} + \nu^2 + m^2 + V(r) \tag{11}$$

and $H^{(0)}_{\nu,l}$ has the same form, but without $V(r)$.

In order to find $J_l(\nu)$ (10) we make use of a known theorem [8, 9] whose statement is

$$\frac{det H_{\nu,l}}{det H^{(0)}_{\nu,l}} = \lim_{r\to\infty} \frac{\psi_{\nu,l}(r)}{\psi^{(0)}_{\nu,l}(r)} \quad . \tag{12}$$

Here $\psi_{\nu,l}$ and $\psi^{(0)}_{\nu,l}$ are solutions to equations

$$H_{\nu,l}\psi_{\nu,l} = 0 \quad , \qquad H^{(0)}_{\nu,l}\psi^{(0)}_{\nu,l} = 0 \tag{13}$$

and have same regular behavior at $r = 0$. More exactly, the boundary conditions at $r = 0$ must be chosen in such a way that the right-hand side of Eq.(12) tends to 1 at $\nu \to \infty$.

As the next step we calculate the product over l:

$$F(\nu) = \sum_{l=0}^{\infty} (2l+1) \ln J_l(\nu) \tag{14}$$

The function $F(\nu)$ is the sum of all three-dimensional one-loop one-particle-irreducible diagrams.

In terms of this function the final result reads

$$\ln \mathcal{D}(T) = \sum_{n=-\infty}^{\infty} F(2\pi nT) \quad . \tag{15}$$

This expression is formal because of the ultraviolet divergences in it. We first evaluate $\ln \mathcal{D}$ without the divergent terms of the first and the second order in $V(r)$.

$$[\ln \mathcal{D}(T)]^{\overline{(3)}} = \ln \mathcal{D}(T) - [\ln \mathcal{D}(T)]^{(1)} - [\ln \mathcal{D}(T)]^{(2)} \, . \tag{16}$$

The terms $[\ln \mathcal{D}(T)]^{(1)}$ and $[\ln \mathcal{D}(T)]^{(2)}$ are the standard tadpole and fish diagrams. We calculate their renormalised values analytically and add them to our numerical result.

While renormalization requires just standard techniques it is not straightforward here to select a specific renormalization precription because the scheme depends strongly on the physical context in which the first-order phase transition is considered. Though we consider the value (16) as our main result, we apply, for the sake of completeness, a specific scheme of choosing the counterterms.

To this end we fix the zero-temerature mass and the vacuum expectaion value of the scalar field. This implies the following conditions on the temperature-dependent effective potential $U_{eff}(\varphi;T)$ and the 1-loop Euclidean propagator in the true vacuum $D_+(p^2)$:

$$
\begin{aligned}
D_+(-m_+^2) &= 0 \\
U'_{eff}(\phi_+;0) &= 0 \\
U'_{eff}(0;0) &= 0 \\
U_{eff}(\phi_+;0) &= U(\phi_+)
\end{aligned}
\qquad (17)
$$

where m_+^2 stands for the particle mass in the true vacuum. The last two conditions set the false vacuum to be at $\varphi = 0$ and fix the density of energy stored in it. They are more specific for our toy model. The results of the application of this renormalization scheme are plotted in Figs. 2 - 4.

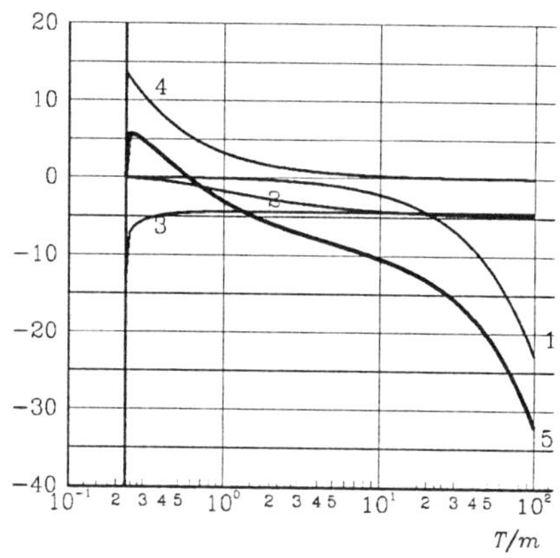

Figure 2. The value of S_{1-loop} at $\alpha = 0.1$ vs. T/m: The curves 1 - 2 represent the temperature dependent parts of $[\ln \mathcal{D}(T)]^{(1)}$ and $[\ln \mathcal{D}(T)]^{(2)}$. The line 3 shows the value of $\overline{[\ln \mathcal{D}(T)]^{(3)}}$. The curve 4 displays the contribution of the renormalized zero-temperature $[\ln \mathcal{D}(T)]^{(1)}$ and $[\ln \mathcal{D}(T)]^{(2)}$ to the bubble energy. This contribution depends on temperature via the factor $1/T$ only. Curve 5 is the full result S_{1-loop}. The temperature range is bounded from the left by $T_{min} = 0.2301m$ where the result has a logarithmic singularity.

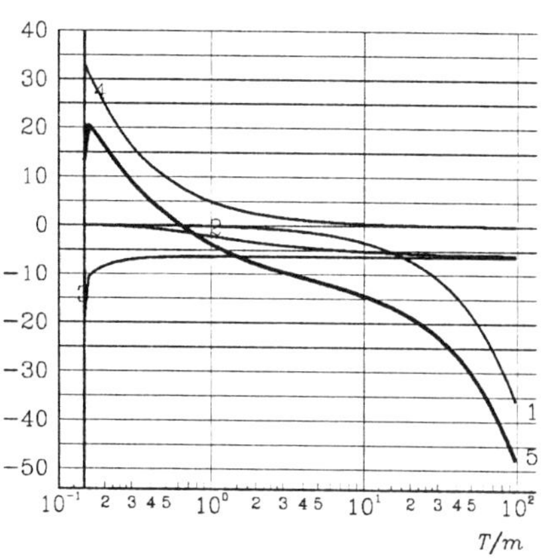

Figure 3. The same as in Fig. 3 at $\alpha = 0.5$; here $T_{min} = 0.148m$.

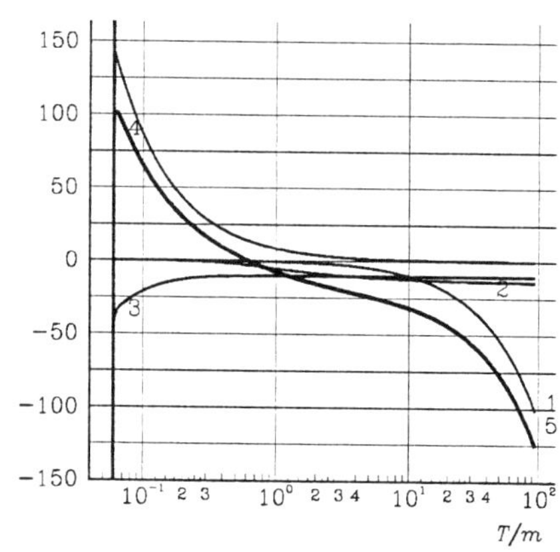

Figure 4. The same as in Fig. 3 at $\alpha = 0.8$; here $T_{min} = 0.0598m$.

The model considered here is only semi-realistic. Nevertheless we would like to add some remarks on our results. We find that the correction to the bubble nucleation rate (3) coming from diagrams of the third order and higher favors the transition (i.e. $\ln \mathcal{D}^{(3)} < 0$). The sign of the full 1-loop contribution to the effective action S_{1-loop} depends on the renormalization scheme. With our choice (17) it enhances $\gamma(T)$ at high temperature due to the tadpole diagram $[\ln \mathcal{D}(T)]^{(1)} \sim T$ as $T \gg m$. This contribution becomes too big $T \gg m$ and then the naive application of the 1-loop approximation becomes inconsistent.

Another feature of our results is a weak logarithmic singularity at $T = T_{min}$, the temperature at which one more fluctuation mode of the critical bubble becomes unstable. Numerically it is unimportant. Moreover, in a finite region of temperature $T_{min} < T < T_{tunn}$ quantum tunneling has to be taken into consideration.

One more restriction comes from the fact that the contribution of the counterterms becomes logarithmically large as $\alpha \to 0$. It is due to the high difference of the mass scales of states built on the false and true vacua.

To sum up, we present a method for calculating functional determinants. The method has been tested under realistic conditions. In particular we have shown that within this formalism the problems connected with ultraviolet divergences, zero and unstable modes can be handled easily and without loss of numerical accuracy. Moreover the algorithm is so fast that the whole calculation presented here can be performed on a standard PC with a 486 processor within one hour for one value of α.

This work was sponcored by Alexander von Humboldt Foundation and supported, in its initial stage, by Byelorussian Foundation for Fundamental Research, grant No Ф2-23.

References

[1] J. S. Langer, Ann.Phys. (N.Y.) **41** (1967) 108

[2] J. S. Langer, Ann.Phys. (N.Y.) **54** (1969) 258

[3] S. Coleman, Phys. Rev. **D15** (1977) 2929

[4] C. G. Callan and S. Coleman. Phys. Rev. **D16** (1977) 1762

[5] J. Baacke and V. G. Kiselev, Phys. Rev. **D48** (1993) 5648

[6] M. Dine, R. G. Leigh, P. Huet A. Linde, D. Linde, Phys.Rev. **D46** (1992) 550; Phys. Lett. **B283** (1992) 319

[7] V. G. Kiselev and K. G. Selivanov, Pis'ma Zh. Eksp. Teor. Fiz. **39** (1984) 72 [Sov. Phys. JTEP Lett. **39** (1984) 85]; V. G. Kiselev and K. G. Selivanov, Yad. Phys. **43** (1986) 239 [Sov. J. Nucl. Phys. **43** (1986) 153]; K. G. Selivanov, Z. Exp. Teor. Fiz. **94** (1988) 57 [Sov. Phys. JETP **67** (1988) 1548

[8] R. F. Dashen, B. Hasslacher, and A. Neveu, Phys.Rev. **D10** (1974) 4114; I. M. Gel'fand and A. M. Yaglom, J. Math. Phys. **1** (1960) 48; R. H. Cameron and W. T. Martin, Bull. Am. Math. Soc. **51** (1945) 73; J. H. van Vleck, Proc. Nat. Acad. Sci. **14** (1928) 178

[9] S. Coleman, 'The Uses of Instantons' in "The Whys of Subnuclear Physics", A. Zichichi ed., Plenum Press, New York 1979

[10] J. Baacke, Acta Phys. Pol. **B22** (1991) 127 and references therein

HYDRODYNAMICS OF BUBBLE GROWTH

M. Laine

Department of Theoretical Physics,
P.O.Box 9, 00014 University of Helsinki, Finland

INTRODUCTION

An important detail of the first-order electroweak phase transition is the growth velocity of bubbles of the low-temperature phase. Such bubbles are formed inside the supercooled high-temperature phase. Recently, several estimates of the bubble wall velocity have been given [1]-[3], based on considerations of the microscopic physics at the phase transition front. The purpose of this note is to point out that not only the microscopic physics at the phase transition front, but also the macroscopic hydrodynamical features of the growth process [4]-[7] must be understood if one is to calculate the bubble wall velocity. Especially, we stress that the temperature changes across the phase transition front.

RESULTS

The mean free path of particles at the EW phase transition is $\lambda_{\rm mfp} \approx 5 - 10\ T_c^{-1}$ [1, 2]. This is much shorter than the size of the growing bubbles but only a bit shorter than the width d of the bubble surface, $d \approx 15\ T_c^{-1}$. Therefore, at large scales there is local thermodynamical equilibrium, but at the bubble surface no equilibrium exists: entropy is produced. This suggests that in calculating the bubble wall velocity, one should combine hydrodynamics with some consistent way of accounting for the entropy production. In practice, however, it seems that some authors more or less neglect the hydrodynamics part and concentrate only on what is essentially an equation for the entropy production. In particular, we think that it may be dangerous to assume that temperature does not change across the phase transition surface.

To see what can go wrong, we review what can be said from hydrodynamics alone [4, 5]. Energy-momentum conservation $\partial_\mu T^{\mu\nu} = 0$ and *boundary conditions* imply that there are two types of solutions, deflagrations and detonations. Deflagrations are the more usual solutions. In a deflagration bubble, there is first a shock front which heats the matter, and then the phase transition front follows in which the pressure *decreases*.

Now consider calculating the bubble wall velocity. Let us assume that the temperature changes very little over the phase transition front. Then the pressure would be larger at the low-T phase (see Fig. 1), producing a force outwards. It seems that if one could just calculate the velocity-dependent friction force produced by the collisions of particles with the phase transition front, the problem would be solved. In hydrodynamics, however, the situation is qualitatively different: as mentioned above, the pressure difference is of *the opposite sign* than we expected from the intuitive assumption!

Based on the above argument, we believe that it is important to develop a method of taking consistently into account both the microscopic and the hydrodynamical features of bubble growth. A model which tries to achieve this has been described in Ref. [6]. This model consists of equations for energy-momentum conservation and entropy production. The equation for the entropy production is parametrized with a

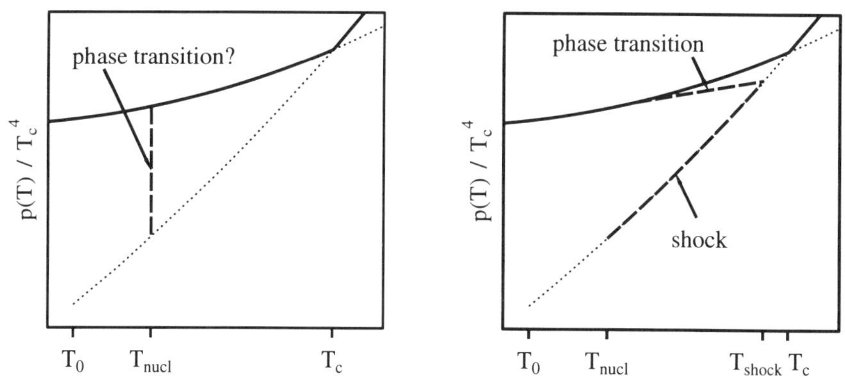

Figure 1. On the left hand side, the naive expectation of the pressure and temperature differences across the phase transition front is shown. On the right hand side is the true situation in a hydrodynamical deflagration solution.

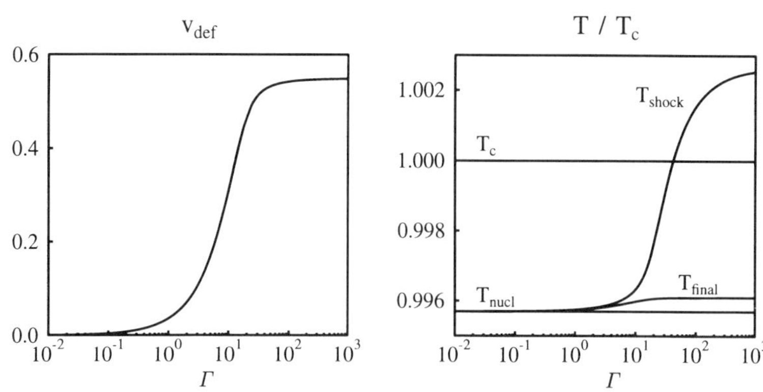

Figure 2. The deflagration front velocity and the temperatures as a function of Γ for one set of parameters. $T_{\rm nucl}$ is the nucleation temperature, $T_{\rm shock}$ is the temperature of the high-T phase after it has been heated by the shock front, and $T_{\rm final}$ is the temperature of the low-T phase behind the phase transition front.

constant Γ, which is in principal calculable from the field theory. Large Γ implies small entropy production and large velocity. An estimate for Γ in the case of the electroweak phase transition was obtained by comparing the results of Refs. [2, 6], yielding $\Gamma \approx 30\, T_c^{-1}$.

In Fig. 2, some results of the above-mentioned model are shown. It is very explicitly seen that the temperature changes significantly both across the shock front and across the phase transition front. In conclusion, the hydrodynamical features of the growth process seem to be important.

REFERENCES

[1] M. Dine, R. G. Leigh, P. Huet, A. Linde and D. Linde, Phys. Rev. D46 (1992) 550

[2] B. H. Liu, L. McLerran and N. Turok, Phys. Rev. D46 (1992) 2668

[3] S. Yu. Khlebnikov, Phys. Rev. D46 (1992) R3223

[4] M. Gyulassy, K. Kajantie, H. Kurki-Suonio and L. McLerran, Nucl. Phys. B237 (1984) 477

[5] H. Kurki-Suonio, Nucl. Phys. B255 (1985) 231

[6] J. Ignatius, K. Kajantie, H. Kurki-Suonio, and M. Laine, Phys. Rev. D49 (1994), in press

[7] M. Laine, Phys. Rev. D49 (1994), in press

QUANTUM SCATTERING AND CLASSICAL SOLUTIONS*

Stephen D.H. Hsu

Lyman Laboratory of Physics
Harvard University
Cambridge, MA 02138

INTRODUCTION

The behavior of quantum scattering amplitudes at large center of mass energies, $E \sim E_* \equiv m/g^2$, is an open question currently under investigation. Here, m is a mass scale characterizing the particle excitations of the theory, and g is a small coupling constant controlling the nonlinear interactions. It has been speculated that multiparticle scattering from initial states with only a small number of particles could be unsuppressed at these energies. A second related issue is whether anomalous processes, such as baryon number violation in electroweak theory where $E_* \simeq \mathcal{O}(M_w/\alpha_w) \sim 10\ TeV$, are unsuppressed at energies of the same order [1].

Standard perturbation theory is clearly inadequate to address either of these issues. Recent investigations have focused on expansions about nonperturbative solutions of classical field equations, such as instantons [1, 2]. However, none of the previously proposed solutions provides a stationary point for a controlled semiclassical expansion at the energies of interest, $E \sim E_*$ [3]. In general, the semiclassical expansion about the proposed instanton-like saddlepoints is controlled only at low energies and provides little or no information at $E \sim E_*$.

In this talk, I will discuss a method for finding classical solutions which will prove suitable saddlepoints of scattering amplitudes even at $E \sim E_*$. The equations and boundary conditions which must be satisfied are dramatically simplified compared to a previous approach to the same problem by Mattis, McLerran and Yaffe [3]. In particular, the resulting equations and boundary conditions for the classical saddlepoint are no longer of the integro-differential type. Rather, the equations of motion are source-free,

$$\frac{\delta S}{\delta \phi(x)} = 0$$

and in certain cases the boundary conditions are of the initial value type, which can be integrated forward in time using standard numerical algorithms. This is also a departure from the approach of Khlebnikov, Rubakov and Tinyakov [2, 4], who derive similar source-free equations, but for an initial *coherent* state, rather than a wave-packet type state which is relevant for high energy collisions.

*Talk presented at Sintra '94 workshop on Electroweak Physics and the Early Universe.

In light of this new formalism, I will discuss the implications of some recent numerical simulations in classical field theory for high energy two-particle scattering. In particular, in 4-dimensional pure SU(2) Yang-Mills theory, Gong, Matinyan, Müller and Trayanov [5] have found an instability of high frequency standing waves of pure glue to decay to low frequency modes. When interpreted within our formalism, this result leads to the speculation that multiparticle scattering in $SU(2)$ gauge theories may be unsuppressed (at least exponentially) at high energies. A generalization of our method to complex trajectories which allows their application to essentially any wave packet - coherent state scattering amplitude. I finally discuss how our formulation relates to some previous work by other authors on high energy multiparticle and anomalous processes in electroweak theory [3, 4].

Before launching into the formalism, let me make a few remarks on the method and the results we can expect. We are interested in quantum scattering amplitudes (or S-matrix elements) between arbitrary initial and final states $|i\rangle$ and $|f\rangle$. In most of what follows in the talk I will be specifically interested in states that are relevant to accelerator experiments (i.e. wave packet states), but here I will be completely general.

Now, imagine thinking about the scattering process in terms of paths inside the functional integral. The quantum fields do whatever they want, weighted by the appropriate action $exp[iS]$, but also by the overlap of the asymptotic part of the path with the initial and final states. Therefore we expect that the amplitude must be expressible in the form:

$$\langle f|S|i\rangle \sim \int d\phi_i \; d\phi_f \; D\phi \; \Psi_i[\phi(T_i)] \; \Psi_f[\phi(T_f)] \; e^{iS[\phi]}, \tag{1}$$

where I have explicitly written the measure of path integration to include fluctuations of the fields in the asymptotic past and future. The wave-functionals $\Psi_{i,f}$ measure the overlap of the initial and final states with position (or field operator) eigenstates ϕ at asymptotic times. Below I will give an actual derivation of (1) along with the explicit form of the wavefunctionals in the case that the initial and final states are either wave-packets or coherent states. (Recall that a plane wave is a special type of wave packet that is narrowly peaked in momentum space, so we can work with plane waves as well if we wish.)

Now, the usual way to compute (1) at weak coupling is by expanding the functional integral about either the vacuum - which generates standard perturbation theory - or about a more exotic classical path such as the instanton, but which also happens to obey vacuum boundary conditions. (In the case of the instanton (1) is first continued to Euclidean space, where finite action solutions must satisfy vacuum boundary conditions. The results of expanding the path integral are then analytically continued back term by term. More on this later.) Clearly, for some initial and final states it is not a big loss to evaluate the overlaps $\Psi_{i,f}[\phi]$ on the asymptotic vacuum. We know that in most instances standard perturbation theory works quite well. We also know that instanton methods give a controlled way of computing tunneling phenomena - at least at low energies.

However, to get a better approximation to the above path integral, one could try to extremize the entire integrand, which will yield some nontrivial boundary conditions on the classical solutions, as determined by the specific form of the wavefunctionals, or in other words the initial and final states of interest.

FORMALISM

As I mentioned above, we wish to calculate the S-matrix element between an initial two particle state and an arbitrary final state †. The result will be a boundary value problem which determines the classical field which is the stationary point of the S-matrix element.

†The authors of [3] extremize momentum-space Greens functions. See Section 6.

To this end, we express the kernel of the S-matrix in a basis of coherent states, first used in this context in [4]. The initial and final states are defined by sets of complex variables $\mathbf{a} \equiv \{a_{\mathbf{k}}\}$, $\mathbf{b}^* \equiv \{b^*_{\mathbf{k}}\}$, respectively. A coherent state $|a_{\mathbf{k}}\rangle$ is an eigenstate of the annihilation operator $\hat{a}_{\mathbf{k}}$: $\hat{a}_{\mathbf{k}}|a_{\mathbf{k}}\rangle = a_{\mathbf{k}}|a_{\mathbf{k}}\rangle$. The relevance of coherent states for the scattering problem stems from the fact that upon differentiation they generate all momentum eigenstates (see eq.11 below).

For simplicity, we illustrate the case of a single real massive scalar field with self-coupling g^2 in 3+1 dimensions. The results can be generalized to the case of gauge fields. First, we express the transition amplitude from an initial coherent state $|\mathbf{a}\rangle$ at time T_i to a final coherent state $|\mathbf{b}^*\rangle$ at time T_f, in terms of the path integral by inserting a complete set of eigenstates of the field operator on the initial and final time slices (see figure 1)

$$\langle \mathbf{b}^* | U | \mathbf{a} \rangle = \int d\phi_i \, d\phi_f \, \langle \mathbf{b}^* | \phi_f \rangle \, \langle \phi_f | U | \phi_i \rangle \, \langle \phi_i | \mathbf{a} \rangle \, . \tag{2}$$

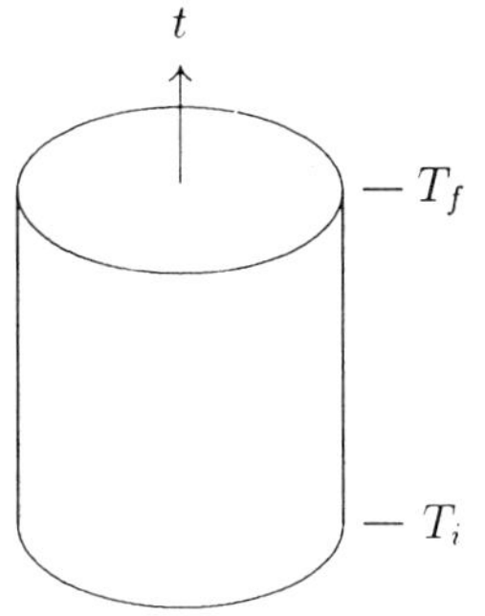

Figure 1. Spacetime, with two slices.

U is the evolution operator between time T_i and T_f. A "position" eigenstate of the field operator ϕ is denoted $|\phi\rangle$ and $\phi_{i,f} = \phi(T_{i,f})$. Then, from (2), we obtain the S-matrix kernel in a compact form in terms of path integrals

$$\langle \mathbf{b}^* | S | \mathbf{a} \rangle \equiv S[\mathbf{b}^*, \mathbf{a}] = \lim_{T_i, T_f \to \mp\infty} \int d\phi_f \, d\phi_i \, e^{B_f} \, e^{B_i} \int_{\phi_i}^{\phi_f} D\phi \, e^{iS[\phi]} \, , \tag{3}$$

where $S[\phi]$ is the action functional. The path integral appearing here is over fields obeying the boundary conditions $\phi(T_{i,f}) = \phi_{i,f}$. The functional B_f is

$$\begin{aligned} B_f[\mathbf{b}^*, \phi_f] &= -\frac{1}{2}\int d^3k \; b^*_{\mathbf{k}} b^*_{-\mathbf{k}} \; e^{2i\omega_k T_f} - \frac{1}{2}\int d^3k \; \omega_{\mathbf{k}} \; \phi_f(\vec{k}) \; \phi_f(-\vec{k}) \\ &+ \int d^3k \; \sqrt{2\omega_{\mathbf{k}}} \; e^{i\omega_k T_f} \; b^*_{\mathbf{k}} \; \phi_f(-\vec{k}) \, , \end{aligned} \tag{4}$$

in terms of which the wave functional of the final coherent state is

$$\langle \mathbf{b}^* | \phi_f \rangle \equiv \exp(B_f[\mathbf{b}^*, \phi_f]) \, . \tag{5}$$

Similarly, the functional B_i is

$$\begin{aligned} B_i[\mathbf{a}, \phi_i] &= -\frac{1}{2}\int d^3k \; a_{\mathbf{k}} a_{-\mathbf{k}} e^{-2i\omega_k T_i} - \frac{1}{2}\int d^3k \; \omega_{\mathbf{k}} \phi_i(\vec{k}) \phi_i(-\vec{k}) \\ &+ \int d^3k \; \sqrt{2\omega_{\mathbf{k}}} \; e^{-i\omega_k T_i} \; a_{\mathbf{k}} \; \phi_i(\vec{k}) \end{aligned} \tag{6}$$

is related to the wave functional of the initial coherent state

$$\langle \phi_i \,|\, \mathbf{a} \rangle \;\equiv\; \exp\left(B_i \left[\mathbf{a}, \phi_i\right] \right) . \tag{7}$$

Here the 3-dimensional Fourier transform is defined

$$\phi_{i,f}(\vec{k}) \;=\; \int \frac{d^3x}{(2\pi)^{3/2}}\; e^{i\vec{k}\cdot\vec{x}}\; \phi(T_{i,f}, \vec{x}) \;, \tag{8}$$

and is related to the residue of the 4-dimensional Fourier transform when the field reduces to a plane wave superposition as $|\,T_{i,f}\,| \to \infty$ [4].

The kernel (3) is a generating functional for S-matrix elements between any initial and final N particle states, by functional differentiation with respect to arbitrary $a_{\mathbf{k}}$ and $b^*_{\mathbf{k}}$. We now use this fact to construct a kernel for scattering from initial two particle states. We define an initial two particle (wave packet) state at $t = T_i$

$$|\,\vec{p}, -\vec{p}\,\rangle \;\equiv\; \int d^3k\; \alpha_R(\vec{k})\; \hat{a}^\dagger_{\mathbf{k}} \int d^3k'\; \alpha_L(\vec{k}')\; \hat{a}^\dagger_{\mathbf{k}'}\; |\,0\,\rangle \;, \tag{9}$$

where $\hat{a}^\dagger_{\mathbf{k}}$ is a creation operator, and $\alpha_{R,L}(\vec{k})$ are arbitrary smearing functions of $\vec{k}$, localized around some reference momenta $\vec{p}$ and $-\vec{p}$ respectively. The wave packets are normalized so that [‡]

$$\int d^3k\; |\,\alpha_{R,L}(k)\,|^2 \;=\; 1 \,. \tag{10}$$

The regime relevent to the high energy multiparticle scattering problem is where, in the limit $g^2 \;\to\; 0$, the wave packet is localized around momenta $|p| \;\sim\; m/g^2$ with a characteristic width, $\Delta p \sim m$.

This state can be generated by functional differentiation of the coherent state $|\mathbf{a}\rangle$ with respect to $a_{\mathbf{k}}$

$$|\vec{p}, -\vec{p}\rangle \;=\; \int d^3\vec{k}\; d^3\vec{k}'\; \alpha_R(\vec{k})\; \alpha_L(\vec{k}')\; \frac{\delta}{\delta a_{\mathbf{k}}}\; \frac{\delta}{\delta a_{\mathbf{k}'}}\; |\,\mathbf{a}\,\rangle \bigg|_{\,\mathbf{a}\,=\,0} . \tag{11}$$

So, differentiating under the functional integral, the S-matrix element between the two particle state (11) and any final state $|\{b^*_{\mathbf{k}}\}\rangle$ involves the following functional at $t = T_i$

$$\frac{\delta}{\delta a_{\mathbf{k}}}\frac{\delta}{\delta a_{\mathbf{k}'}} \exp\left(B_i \left[\mathbf{a}, \phi_i\right] \right) \bigg|_{\,a=0} \;= \tag{12}$$
$$2\; \sqrt{\omega_{\mathbf{k}}\omega_{\mathbf{k}'}}\; \phi_i(\vec{k})\; \phi_i(\vec{k}')\; e^{-i(\omega_{\mathbf{k}}+\omega_{\mathbf{k}'})T_i}\; \exp\left(B_i \left[0, \phi_i\right] \right) ,$$

after dropping a term which vanishes in the limit $T_i \;\to\; -\infty$. The last factor here is simply the normalization of the initial position eigenstate

$$\exp\left(-\frac{1}{2} \int d^3k\; \omega_k\; \phi_i(\vec{k})\; \phi_i(-\vec{k}) \right) . \tag{13}$$

We combine this with the smearing functions and finally obtain an S-matrix kernel for the scattering of two wave packets into arbitrary final states,

$$S\left[\mathbf{b}^*, 2\right] \;=\; \lim_{T_i, T_f \to \mp\infty} \int d\phi_f\, d\phi_i\; \alpha_R \cdot \phi_i\; \alpha_L \cdot \phi_i\; e^{B_f\left[b, \phi_f\right] \,+\, B_i\left[0, \phi_i\right]} \int_{\phi_i}^{\phi_f} D\phi\; e^{iS\,[\phi]} \;, \tag{14}$$

where we have denoted the initial state (11) by "2". Here we have used the following compact notation for the initial state factors

$$\alpha \cdot \phi_i \;\equiv\; \int d^3k\; \sqrt{2\omega_k}\; \alpha(k)\; \phi_i(k)\; e^{-i\omega_k T_i} \,. \tag{15}$$

---

‡We use the non-relativistically normalized commutator $\left[\, \hat{a}_{\mathbf{k}'} \,,\, \hat{a}^\dagger_{\mathbf{k}} \,\right] \;=\; \delta^3\left(\vec{k} - \vec{k}'\right)$.

We now derive the boundary value problem obeyed by the classical field in the stationary phase approximation of (14). The stationary phase approximation of ordinary integrals suggests that the initial state factors be included in the extremization of the path integral §. These factors will certainly impact the stationary phase in the kinematic regime relevant to the high energy scattering problem, where the initial wave packets are peaked around $|p| \sim m/g^2$, as expansions around the instanton indicate [4]. More generally though, we will show in Section 4 that this procedure provides a consistent weak coupling expansion at *any* fixed center of mass energy.

We include initial state factors in a straightforward manner by first exponentiating the initial state factors into an "effective action", so that

$$S\,[\mathbf{b}^*, 2] \;=\; \lim_{T_i, T_f \to \mp\infty} \int d\phi_f \, d\phi_i \, D\phi \; e^{\Gamma} \;, \tag{16}$$

where the effective action Γ is

$$\Gamma\,[\,\phi\,] \;=\; \ln \alpha_R \cdot \phi_i \; \alpha_L \cdot \phi_i \;+\; B_i\,[0, \phi_i] \;+\; iS\,[\,\phi\,] \;+\; B_f\,[b^*, \phi_f] \;, \tag{17}$$

after dropping a term which vanishes as $T_i \to -\infty$.

We can now derive the boundary value problem by varying the effective action. Varying the entire exponent Γ with respect to $\phi(x)$ for $T_i < t < T_f$ gives the source-free equations of motion

$$\frac{\delta S}{\delta \phi(x)} \;=\; 0 \;. \tag{18}$$

Varying the entire exponent with respect to $\phi_i(k)$, gives

$$i\,\dot{\phi}_i(\vec{k}) \;+\; \omega_k \phi_i(\vec{k}) \;=\; \sqrt{2\,\omega_{\mathbf{k}}} \left(\frac{\alpha_R(\vec{k})}{\alpha_R \cdot \phi_i} \;+\; \frac{\alpha_L(\vec{k})}{\alpha_L \cdot \phi_i} \right) e^{-i\omega_k T_i} \;. \tag{19}$$

The first term on the left hand side comes from a surface term in the action S. The other terms come from variation of the wave functional at $t = T_i$. This boundary condition involves both the positive and negative frequency parts of the field, unlike the boundary condition which arises for an initial coherent state $|\mathbf{a}\rangle$, which depends on the negative frequency component of ϕ only [2].

The boundary condition (19) at the initial time slice is rather complicated. However, it can be simplified since a real field ϕ may be written in the asymptotic region $t = T_i \to -\infty$ as a plane wave superposition

$$\phi_i(\vec{k}) \;=\; \frac{1}{\sqrt{2\,\omega_{\mathbf{k}}}} \left(u_{\mathbf{k}} \, e^{-i\omega_k T_i} \;+\; u^*_{-\mathbf{k}} \, e^{i\omega_k T_i} \right) \;. \tag{20}$$

Equation (19) then reduces to the requirement

$$u_{\mathbf{k}} \;=\; \frac{\alpha_R(\vec{k}) \;+\; \alpha_L(\vec{k})}{\left(1 \;+\; \int d^3k \; \alpha_R(\vec{k}) \; \alpha_L(\vec{k})\right)^{1/2}} \;, \tag{21}$$

using the normalization (10). This solution is consistent with physical intuition, the classical field reducing to the initial particles at early times. The overlap of the left- and right- moving wave packets in the denominator is very small for narrow high energy wave packets.

If the field is real, then the negative frequency part equals the complex conjugate of the positive frequency part, and the field ϕ_i is determined as in (20) and (21). The

§The Stirling approximation to the Gamma function is a classic example.

real initial condition can be integrated forward, and uniquely determines the final field ϕ_f. We say more about the case of *complex* stationary points of real fields in Section 5.

It remains to be seen in what sense it is a good approximation to include the initial state prefactors in the stationary phase calculation. We will demonstrate in Section 4 that the stationary phase solution which results from this approach provides a *controlled* weak coupling expansion of the scattering amplitude.

Finally, varying with respect to $\phi_f(k)$ gives

$$-\, i\,\dot{\phi}_f(k) \;+\; \omega_{\mathbf{k}}\phi_f(k) \;=\; \sqrt{2\,\omega_k}\; b^*_{-\mathbf{k}}\; e^{i\omega_k T_f}\;. \tag{22}$$

Again with a free field form as in (20), this boundary condition may be re-expressed as

$$\phi_f(\vec{k}) \;=\; \frac{1}{\sqrt{2\,\omega_{\mathbf{k}}}}\; \left(b_{\mathbf{k}}\; e^{-i\omega_k T_f} \;+\; b^*_{-\mathbf{k}}\; e^{i\omega_k T_f} \right)\;. \tag{23}$$

We see that the precise scattering amplitude which is extremized is only determined at the end of the calculation, by the asymptotic form of the classical solution in the far future, $T_f \rightarrow \infty$. For the initial conditions given by (19), the scattering amplitude corresponds to a transition from wave packet states to a final coherent state described by $|b^*\rangle$ from (22). Note that neither a wave packet state ¶ nor a coherent state are eigenstates of the Hamiltonian, so that neither have a definite energy.

When the classical field satisfies the equation of motion (18) and the boundary conditions (19) and (22), there are no initial and final state corrections to leading order in the semiclassical expansion. This is not the case for the instanton, which satisfies the equation of motion but not the correct boundary conditions. Instead, there are linear terms in the fluctuation about the instanton and these generate initial and final state corrections.

SEARCH FOR SOLUTIONS

The results of the previous section establish an important fact:

> The classical field obtained by starting with two incident wave packets and evolving them forward using (18) is the dominant contribution to *some* two particle scattering amplitude, in the semiclassical approximation.

In the case of a purely Minkowskian classical solution, the result for the corresponding scattering amplitude will *not* contain an exponential suppression. Thus, if a real time classical trajectory can be found which connects wave packet initial conditions to an "interesting" final state, for example a multiparticle or fermion number violating state, our results imply that the corresponding scattering amplitude is unsuppressed. Classical solutions which either exist in complex time or are themselves complex are relevant to classically forbidden transitions and will be discussed in Section 5.

The boundary conditions given by (20) and (21) are amenable to straightforward numerical integration. The boundary conditions specify $\phi(x, T_i)$ and $\dot{\phi}(x, T_i)$, which are sufficient to construct $\phi(x, t)$ for all subsequent $t > T_i$, given a discretization of the equations of motion (18). In principle, it is possible to generate an infinite number of classical trajectories, each relevant to a particular initial wave packet scattering amplitude.

Of course, it is not clear *a priori* that the final state which results from the classical evolution will be one which is interesting to the problem of multiparticle production or anomalous baryon number violation. However, the above observation summarizes the relevance for scattering amplitudes of computations by Rajagopal and Turok [6], and also

¶Except in the limit of $\alpha_R(\vec{k}) \sim \delta^3(\vec{k} - \vec{p})$, where it reduces to a plane wave.

Goldberg, Nash and Vaughn [7]. Rajagopal and Turok studied the classical scattering of wave packets in the Abelian Higgs model, while Goldberg et al. studied the classical dynamics of ϕ^4 theory. Neither group found that energy was readily transferred from high frequency to low frequency modes, a signal which would indicate the production of a final state with many (soft) particles. Therefore, the scattering amplitudes which are dominated by their classical trajectories are probably not relevant to multiparticle production or fermion number violation.

On the other hand, the initial indications on the behavior of non-Abelian classical trajectories seem more promising. Some recent results on the classical behavior of pure $SU(2)$ Yang-Mills theory by Gong, Matinyan, Müller and Trayanov [5] may provide insight into solutions of the boundary value problem presented above. These authors have considered the classical stability of a stationary mono-color wave in Yang-Mills theory (in $A_0^c = 0$ gauge):

$$A_i^c(x,t) \;=\; \delta_{i3}\,\delta_{c3}\,A\,\cos k_0 x\,\cos\omega_0 t, \tag{24}$$

where c is a color and i is a spatial index [||]. Small amplitude variations of the field in directions of different color are found to lead to an instablility with long wavelength.

The Hamiltonian evolution of a perturbed standing wave is equivalent to the evolution of the initial conditions relevant to the scattering of two perturbed plane waves in $SU(2)$ gauge theory. The discovery of an instability of the initial configuration to decay into long wavelength modes implies the existence of a classical trajectory connecting initial high energy plane waves to long wavelength modes in the final state. It therefore suggests that the corresponding $2 \to$ *many* gluon scattering amplitude may be unsuppressed ! The relevance of weak coupling calculations to pure gauge theory is unclear, due to the asymptotic freedom of the theory. However, it is possible that arguments similar to those from deep inelastic scattering may be used to justify the semiclassical result. In the limit that all of the energy scales (including the energies of the many "soft" outgoing gluons) are large compared to the intrinsic mass scale of the theory $\Lambda_{SU(2)}$, it seems plausible that the relevant running coupling constant is small.

It is also important to determine whether the instability persists for wave packet initial conditions, as the plane wave limit is never achieved at an actual accelerator. Since wave packets are localized in space, two packets have only a finite amount of time to interact before passing completely through each other. This is in contrast to plane waves, which exist everywhere in spacetime. Since the plane waves of Müller et al. [5] have finite, nonzero amplitude, they also contain the equivalent of an infinite number of particles. It is quite possible that the behavior of wave packets representing a finite number of particles is very different.

Another class of trajectories in pure $SU(2)$ Yang-Mills theory are those found by Farhi et al. [9]. These solutions in the spherical ansatz correspond to spherical shells of glue that originate at infinity, collapse inward, and bounce back to infinity, leaving behind some fractional topological charge. There are some open questions concerning such solutions - in particular, their stability to perturbations which are outside the spherical ansatz, and their relevance to fermion number violation. However, in the absence of fermions, they can be interpreted within our formalism as extremizing scattering amplitudes between spherically symmetric initial and final quantum states.

Of course, our eventual goal is to address the electroweak theory, where the Higgs mechanism cuts off the infrared growth of the coupling constant. It would be extremely interesting if simulations of wave packet scattering could be performed in a spontaneously broken $SU(2)$ gauge theory.

Clearly, there is much numerical work to be done to address these issues.

THE EXPANSION

[||] NB - A *travelling* wave is stable in Yang-Mills theory [8].

Let us now construct the perturbative expansion around the classical solution of the BVP. We will demonstrate that corrections to the leading contribution are systematically suppressed in powers of the small coupling constant for any fixed energy. The resulting small coupling expansion differs from the usual one (e.g.– instanton) and the subtleties involved will be discussed at the end of this section.

As usual, we proceed by expanding the quantum field ϕ around a classical background field ϕ^c, which is a solution to the classical boundary value problem derived in Section 2.

$$\phi(x) \;=\; \phi^c(x;\mu) \;+\; \nu(x)\;. \tag{25}$$

The classical solution depends in general on a collection of collective coordinates denoted here by μ, corresponding to the invariances of the *scattering amplitude* [3, 10]. We expect invariances of translation and scale size to be broken by the choice of $\alpha_{R,L}(\vec{k})$; however, μ may include gauge coordinates if we are working in a gauge theory. Equation (25) involves an expansion on the initial and final time slices, which we can express in momentum space as

$$\phi_{i,f}(\vec{k}) \;=\; \phi^c_{i,f}(\vec{k};\mu) \;+\; \nu_{i,f}(\vec{k}) \;\;\text{for}\;\; t = T_{i,f}\;, \tag{26}$$

where $\phi^c_{i,f}(\vec{k})$ is known explicitly for any given initial and final wave packets, from (20) and (23)

Note that the classical solution is of order $\mathcal{O}(g^0)$, since it is non-trivial in the limit as $g \to 0$. In fact, when g vanishes, the background field is simply the initial condition propagated forward in time with the free Hamiltonian,

$$\phi^c(t,\vec{k}) \;=\; \frac{1}{\sqrt{2\omega_{\mathbf{k}}}}\left(u_{\mathbf{k}}\,e^{-i\omega_k t} \;+\; u^*_{-\mathbf{k}}\,e^{i\omega_k t}\right)\;, \tag{27}$$

where $u_{\mathbf{k}}$ is as in (21). In this limit, the two wave packets simply pass through each other. For non-vanishing g, the classical field ϕ^c will have some complicated dependence on g, reflecting the non-linearity of the theory and the external boundary conditions.

The kernel of the S-matrix (14) is composed of three functional integrals which we write as

$$S\,[\mathbf{b}^*,2] \;=\; \int d\phi_f\; d\phi_i\; \alpha_R\cdot\phi_i\;\alpha_L\cdot\phi_i\; e^{B_f[b^*,\phi_f] \,+\, B_i[0,\phi_i]}\int_{\phi_i}^{\phi_f} D\phi\; e^{iS\,[\phi]} \tag{28}$$

where the limit $|T_{i,f}| \to \infty$ is understood. Now we formally expand each of the terms in (28) using (25). The action becomes

$$S\,[\phi^c+\nu] \;=\; S\,[\phi^c] \;+\; \int d^3x\;\nu(x)\;\dot{\phi}^c(x)\Big|_{T_i}^{T_f} \;+\; S_2\,[\nu] \;+\; S_{int}\,[\nu]\;, \tag{29}$$

after integrating by parts and retaining the surface terms in the time direction, and using the equations of motion (18). All of the non-quadratic dependence is contained in the interaction terms $S_{int} \simeq \mathcal{O}(\nu^3,\nu^4)$ which are suppressed by powers of g^2. Here S_2 is the part of the action quadratic in the fluctuation field, which can be expressed as

$$S_2\,[\nu] \;=\; -\frac{1}{2}\int d^3x\;\nu(x)\;\dot{\nu}(x)\Big|_{T_i}^{T_f} \;-\; \frac{1}{2}\int d^4x\;\nu(x)\;\Delta\,[x]\;\nu(x)\;, \tag{30}$$

in terms of the operator Δ of quadratic fluctuations in the ϕ^c background,

$$\delta^4(x-y)\;\Delta\,[x] \;\equiv\; \frac{\delta S}{\delta\phi(x)\;\delta\phi(y)}\Big|_{\phi^c} \;=\; \delta^4(x-y)\left\{\,\partial_\mu^2 \;+\; V''(\phi^c)\right\}\;. \tag{31}$$

Expanding the boundary functionals gives

$$B_f\left[\mathbf{b}^*,\phi^c_f+\nu_f\right] \;=\; -i\int d^3k\;\nu_f(k)\;\dot{\phi}^c_f(-k) \;+\; B_f\left[\mathbf{b}^*,\phi^c_f\right] \;+\; B_f\,[0,\nu_f]\;, \tag{32}$$

after using the boundary condition (22), and

$$B_i\left[0, \phi_i^c + \nu_i\right] = \tag{33}$$
$$i\int d^3k\ \nu_i(k)\ \dot{\phi}_i^c(-k)\ +\ B_i\left[0, \phi_i^c\right]\ +\ B_i\left[0, \nu_i\right]\ -\ \frac{\alpha_R\cdot\nu_i}{\alpha_R\cdot\phi_i^c}\ -\ \frac{\alpha_L\cdot\nu_i}{\alpha_L\cdot\phi_i^c}\ ,$$

after using the boundary condition (19). The $\dot{\phi}_{i,f}$ terms in (32) and (33) cancel with the boundary terms in the action (29). Note the appearance of additional terms linear in ν_i in (33), which arise because we are expanding around a stationary point of the *effective* action (17), instead of the action. Note also that the terms independent of $\nu_{i,f}$ are the overlaps of the classical field with the initial and final states, respectively.

Now, substituting (29), (32) and (33) in (28), we obtain

$$S\left[\mathbf{b}^*, 2\right]\ =\ \int d\mu\ e^{B_f\left[\mathbf{b}^*, \phi_f^c\right]\ +\ B_i\left[0, \phi_i^c\right]\ +\ iS\left[\phi^c\right]}\ \mathcal{Z}\left[\phi^c\right]\ \left[1 + \mathcal{O}(g^2)\right] \tag{34}$$

where the quadratic integral over fluctuations is contained in

$$\mathcal{Z}\left[\phi^c\right]\ \equiv \tag{35}$$
$$\int d\nu_f\ d\nu_i\ D\nu\ \left\{1\ +\ \alpha_R\cdot\nu_i\right\}\ \left\{1\ +\ \alpha_L\cdot\nu_i\right\}\ e^{B_f\left[0, \nu_f\right] + B_i\left[0, \nu_i\right] - \alpha_R\cdot\nu_i - \alpha_L\cdot\nu_i + iS_2\left[\nu\right]}\ ,$$

after using $\alpha_{R,L}\cdot\phi_i^c = 1$ in the initial state factors and in the exponent. An integration over collective coordinates has been factored out of (34) and the remaining functional integral (35) involves only fluctuations orthogonal to any zero modes of the classical background.

The $\mathcal{O}(g^2)$ corrections result from the interactions in S_{int}. Expanding $\exp(iS_{int}[\nu])$ inside (28) yields the leading term plus corrections which have the form of vacuum to vacuum loops. The vertices used to construct these loops each carry a suppression of g^2. The propagators inside these loops are the usual Feynman propagators evaluated in the classical background ϕ^c **. In the instanton case [2, 11], dangerous initial state corrections arose from the residues of propagators in the instanton background, which displayed an energy dependence $\sim\ g^2 s$, where s is the center of mass energy of the collision and could be parametrically of order $1/g^4$. Note that it is only the *residue* of the instanton propagator that is ill-behaved and not the propagator itself. Since the corrections in our case are proportional to loop integrals rather than the residues of individual propagators [2, 11], we expect them to remain subleading even at high energies.

The Gaussian path integral appearing in (35) can now be done exactly. First factor out the quadratic prefactor by defining a more general functional integral,

$$\mathcal{Z}\left[j\right]\ \equiv\ \int d\nu_f\ d\nu_i\ D\nu\ e^{B_f\left[0, \nu_f\right]\ +\ B_i\left[0, \nu_i\right]\ +\ iS_2\left[\nu\right]\ -\ \int d^3k\ j(k)\,\nu_i(-k)}\ , \tag{36}$$

in terms of which (35) may be written

$$\mathcal{Z}\left[\phi^c\right]\ =\ \left\{1\ +\ \alpha_R\cdot\frac{\delta}{\delta j}\right\}\ \left\{1\ +\ \alpha_L\cdot\frac{\delta}{\delta j}\right\}\ \mathcal{Z}\left[j\right]\Big|_{j=j_*}\ , \tag{37}$$

where we will set the arbitrary current $j(k)$ equal to

$$j_*(k)\ =\ \sqrt{2\,\omega_{\mathbf{k}}}\ \left(\alpha_R(k)\ +\ \alpha_L(k)\right)\ e^{-i\omega_{\mathbf{k}}T_i}\ , \tag{38}$$

after the functional differentiation. The stationary point of (36) is a "classical" fluctuation field ν^c satisfying

$$\Delta\left[x\right]\ \nu^c\left(x\right)\ =\ 0\ , \tag{39}$$

**It is worth remembering that since we are working directly in Minkowski space (instead of beginning in Euclidean space and analytically continuing back), the Feynman boundary conditions on Greens functions are only obtained through some form of regularization, such as an $i\epsilon$ prescription.

with boundary conditions

$$i\,\dot{\nu}_i^c(k) \;+\; \omega_{\mathbf{k}}\nu_i^c(k) \;=\; j(k) \quad \text{and} \quad -\,i\,\dot{\nu}_f^c(k) \;+\; \omega_{\mathbf{k}}\nu_f^c(k) \;=\; 0\,. \tag{40}$$

The solution of this linear homogeneous boundary value problem can be expressed in terms of a Greens function

$$\nu(x) \;=\; \int d^3q\; G(\,\vec{x},\vec{q};t\,)\, j(\vec{q})\;, \tag{41}$$

where G obeys

$$\Delta\,[x]\;\; G(\,\vec{x},\vec{x}';t\,) \;=\; 0\;, \tag{42}$$

with boundary conditions in terms of its Fourier transform

$$i\,\dot{G}\left(\vec{k},\vec{q};T_i\right) \;+\; \omega_{\mathbf{k}}\, G\left(\vec{k},\vec{q};T_i\right) \;=\; \delta^3\left(\vec{k}-\vec{q}\right) \tag{43}$$

and

$$-\,i\,\dot{G}\left(\vec{k},\vec{q};T_f\right) \;+\; \omega_{\mathbf{k}}\, G\left(\vec{k},\vec{q};T_f\right) \;=\; 0\,. \tag{44}$$

Using the above expressions, we obtain a compact expression for (36) entirely in terms of this Greens function, evaluated at $t = T_i$, and the known initial state source.

$$\mathcal{Z}\,[j] \;=\; \det{}^{-1/2}\,[\Delta]\;\exp\;\left(\frac{1}{2}\int d^3k\; d^3q\; j(\vec{k})\; G\left(\vec{k},\vec{q};T_i\right)\;\; j(\vec{q})\right)\,. \tag{45}$$

The exponent appearing here is due entirely to the boundary functional B_i in (36). All other terms vanish on the stationary phase solution (41). With this result, we obtain the Gaussian integral (37) by functional differentiation and setting $j = j_*$

$$\mathcal{Z}\,[\phi^c] \;=\; \det{}^{-1/2}\,[\Delta]\;\,. \tag{46}$$

Several terms in (37) have vanished in the limit of $T_i \;\to\; \infty$, most notably initial state factors. This is because both $j_*(k)$ and our inner product (15) contain rapidly oscillating factors $e^{-i\omega_k T_i}$. This rapid oscillation allows the application of the Riemann-Lebesgue lemma in the limit $T_i \;\to\; -\infty$, which guarantees that for any function $f(k)$ whose Fourier transform exists,

$$\lim_{T_i \to -\infty} \int dk\;\, f(k)\; e^{-i\omega_k T_i} \;=\; 0. \tag{47}$$

Here the functional determinant $\det^{-1/2}[\Delta]$ is actually the product of three determinants: the standard 4-dimensional determinant of the operator Δ satisfying Feynman boundary conditions, and two 3-dimensional determinants representing the edge fluctuations $\nu(\vec{x}, T_{i,f})$. The latter determinants can be expressed in terms of a homogeneous Greens function similar to G described above.

We have demonstrated that the kernel of the S-matrix (28) is obtained to leading order in g^2 from the action of the classical solution and the determinant of quadratic fluctuations around it.

$$S\,[\mathbf{b}^*, 2] \;=\; \int d\mu\;\, e^{B_f[\mathbf{b}^*,\phi_f^c] \;+\; B_i[0,\phi_i^c] \;+\; iS\,[\,\phi^c\,]}\;\det{}^{-1/2}\,[\Delta]\;\;\left[1+\mathcal{O}(g^2)\right]\,. \tag{48}$$

The classical factors here are known functions, depending only on the initial and final boundary values of the classical field, (20) and (23). Thus, we may further reduce (48) to

$$S\,[\mathbf{b}^*, 2] \;=\; e^{-1}\int d\mu\;\, e^{\tilde{N}/2 \;+\; iS\,[\,\phi^c\,]}\;\det{}^{-1/2}\,[\Delta]\;\;\left[1+\mathcal{O}(g^2)\right]\,. \tag{49}$$

where $\bar{N} \equiv \int d^3k\ b^*_{\mathbf{k}} b_{\mathbf{k}}$ is the average number of particles in the final coherent state, which depends implicitly on the classical field. The origin of this factor is the use of coherent states which are not normalized. The standard normalization for coherent states which we use here is

$$\langle \mathbf{b}^* | \mathbf{b}^* \rangle = e^{\bar{N}} \ . \tag{50}$$

The correctly normalized S-matrix element would not include this factor, leaving the classical action and the determinant.

There are *no* additional tree-level radiative corrections to (49), as there are in an expansion around a stationary point which does not obey the correct boundary conditions (c.f.– the instanton [2, 4]). This is a direct result of the fact that our classical solution obeys both the correct equation of motion and the correct boundary conditions.

It is worth noting that our semiclassical expansion differs from the usual one in a crucial way. Our classical background depends on the coupling constant itself. In the usual instanton calculation, the field equations and boundary conditions can be made independent of g by rescaling $\phi \to \bar{\phi} = g\phi$. In our case, since the boundary conditions are fixed by the wave packet shapes $\alpha_{R,L}(\vec{k})$, the equations of motion for ϕ contain a dependence on g which cannot be scaled away. It is clear that a change in g will lead to a different nonlinear evolution of ϕ and hence a *different* classical trajectory. In the limit of strictly zero coupling constant, the evolution of the initial wave packet is clearly trivial, and we get only a contribution to the "diagonal" part of the S-matrix (i.e.– no scattering). Therefore, in order to obtain a nontrivial result, we have to take g large enough to find an interesting classical trajectory, and yet small enough to justify the semiclassical g^2 expansion derived above. We have not demonstrated that the two requirements can be satisfied simultaneously in any theory. This can only be checked *a posteriori* by computing the size of the subleading $\mathcal{O}(g^2)$ corrections in the background of a proposed trajectory. Of course, our experience with perturbation theory leads us to believe that for $g^2 \ll 1$ the corrections will be small. Thus, a classical trajectory found using weakly coupled equations of motion is likely to provide a controlled approximation.

COMPLEX TRAJECTORIES

In the preceding discussion, we have focused on classical trajectories dependent on a real time variable. These are relevant to classically allowed amplitudes, such as those for multiparticle scattering or transitions at energies well above an energy barrier in configuration space. As we noted previously, the existence of a real time trajectory implies an unsuppressed scattering amplitude for the corresponding transition. However, there are many instances in which classically disallowed transitions are of interest. In particular, baryon number violation in the electroweak theory at energies less than the sphaleron energy $E_s \sim M_w/\alpha_w \sim 10\ TeV$ is classically forbidden and therefore cannot be described by the same type of real time trajectories. The application of our formalism to such cases therefore requires some generalization, which we will describe in this section.

Consider the initial value problem given by (18) and (19). Since both the initial field values $\phi_i(x)$ and their time derivatives $\dot{\phi}_i(x)$ are given, the classical field has a well defined energy. Since real time trajectories conserve energy, it is clear that a suitable trajectory will not exist for some choices of initial and final conditions. In particular, in the electroweak theory for $E_{cl} < E_s$, there are no real time trajectories which connect initial and final conditions with different topological charge.

In order to find such trajectories, previous authors [12, 13, 14] have proposed complex *time* paths, which involve alternately Minkowskian, Euclidean and Minkowskian paths joined together at two times, say 0 and iT as in figure 2. In this approach, $t = 0$ and $t = iT$ must be turning points ($\dot{\phi}(\vec{x}, T_i) = 0$ for all $\vec{x}$) of the field equations with given initial conditions. As is familiar from one dimensional quantum mechanics [13], complex time contours describe an incident configuration which reaches a turning point, and then tunnels under the barrier in Euclidean time, returning to real time at the second turning point upon reaching the other side of the barrier. The field $\phi(x)$ can be taken to be real on the entire contour because of the existence of the two turning points [††]. The action is then purely imaginary on the Euclidean part of the contour and yields the familiar exponential tunneling suppression.

$$|\langle f \,|\, i \rangle|^2 \;\sim\; e^{-2\,\mathrm{Im}\, S} \,. \tag{51}$$

As a matter of practice, one might try to find a suitable complex time trajectory by integrating our initial conditions (21) forward in real time in hopes that a turning point is encountered. Then, the time variable should be Wick rotated $t \to ix_4$ and the turning point configuration integrated forward in Euclidean time until another turning point is encountered. Finally, the contour should be rotated back to real time and the second turning point integrated forward until it achieves its asymptotic plane wave state. The final state $|b^*\rangle$ can be read off from (23).

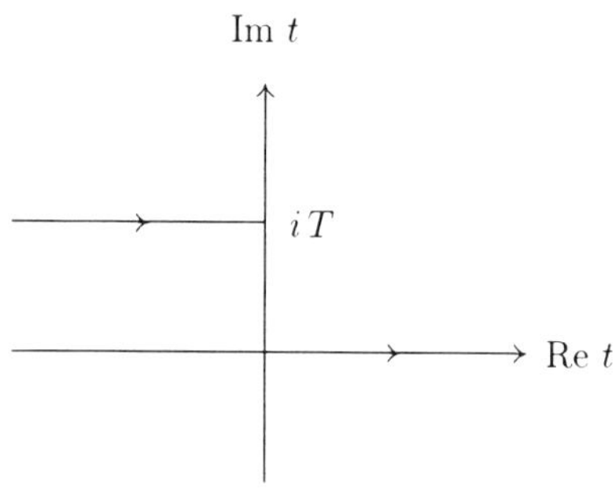

Figure 2. Complex time contour

However, in the most general case, there is no guarantee of encountering turning points. Here we propose a more general method for extremizing S-matrix elements for which there is not a corresponding real, Minkowskian trajectory. Rather than complexifying time, we instead search for a general *complex* saddlepoint configuration satisfying (18) and (19). Our motivation stems from the usual method of steepest descents applied to a real integral of the form

$$I(h) \;=\; \int_{x_a}^{x_b} dx\; e^{\frac{1}{h} f(x)}. \tag{52}$$

In general, the asymptotic series in h for this integral can be obtained by expanding about a saddlepoint of the function $f(x) : f'(x)|_{x^*} = 0$. However, the saddlepoint x^* is often complex, and to apply the method, one must first deform the original real integral into the complex plane (see figure 3).

In our case, we first think of the Feynman Path Integral as the product of a large number of regular integrals by discretizing spacetime:

$$\int D\phi\; e^{iS[\phi]} \;\;\to\;\; \int \prod_x d\Phi_x\; e^{iS(\Phi_x, \Phi_{x\pm 1})}. \tag{53}$$

For a real field $\phi(x)$ each integral in the product is a real integral like (52). We generalize the method of steepest descents by allowing the variable of each integration, Φ_x, to

[††]A fact guaranteed in one dimensional QM, but not in higher dimensions or in field theory [14].

become complex. There should be no obstruction to this generalization as the function $S(\Phi_x, \Phi_{x\pm1})$ is analytic in the variables Φ_x. This procedure leads us to search for a complex trajectory $\{\Phi_x\}$ (or in the continuum, $\phi^c(x)$) satisfying (18), (19) and (22). The integral over small fluctuations about $\phi^c(x)$ must be performed along the path of steepest descent in configuration space. In other words, along small $\nu(x)$ fluctuations which keep the imaginary part of $iS\left[\phi^c(x) + \nu(x)\right]$ constant.

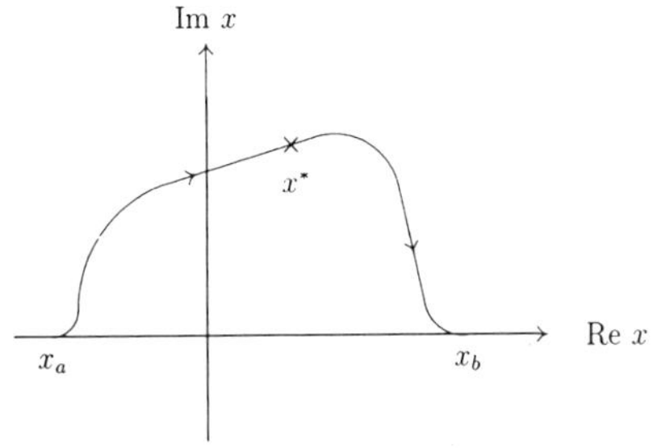

Figure 3. Contour in complex plane, deformed through saddlepoint x^*

For a complex field, there is no longer a relation between the positive and negative energy Fourier components. Without loss of generality, we can write

$$\phi_i(\vec{p}) = \frac{1}{\sqrt{2\omega_{\mathbf{p}}}} \left(u_{\mathbf{p}}\, e^{-i\omega_p T_i} + v_{\mathbf{p}}\, e^{i\omega_p T_i} \right) , \tag{54}$$

where $u_{\mathbf{p}}$, $v_{\mathbf{p}}$ are independent complex functions. The initial boundary condition (19) is satisfied by

$$u_{\mathbf{p}} = \frac{\alpha_R(\vec{p})}{\int d^3k\ \alpha_R(\vec{k})\ v_{\mathbf{k}}} + \frac{\alpha_L(\vec{p})}{\int d^3k\ \alpha_L(\vec{k})\ v_{\mathbf{k}}} , \tag{55}$$

where $v_{\mathbf{k}}$ is arbitrary. The initial configurations which satisfy (54), (55) have negative frequency modes which "resemble" the wave packet $\alpha_L + \alpha_R$ (up to complex multiplicative factors) and positive frequency modes which are arbitrary.

It is then clear that there are an infinite number of complex trajectories which can result from evolving (54) using (18). The final quantum state $|\mathbf{b}^*\rangle$ which results from a given trajectory depends only on the positive frequency component of the trajectory in the far future (see Eq. (22)). The multiplicity of trajectories allows a very large set of scattering amplitudes to be addressed within our formalism. In fact, it seems in general there always exists a complex trajectory which satisfies (18), (19) and (22) simultaneously for any $\alpha(k)$ and $\mathbf{b}^*$. This is because the initial and final conditions (19) and (22) each correspond to conditions on a combination of $\phi(x)|_{T_i}$ and $\dot{\phi}(x)|_{T_i}$, and $\phi(x)|_{T_f}$ and $\dot{\phi}(x)|_{T_f}$, respectively (mixed Dirichlet-Neumann boundary conditions). Since the equations of motion (18) are second order in time, this mixed set of boundary conditions should be enough to specify a unique solution.

Since the trajectories found here are complex, the action $iS[\phi^c]$ may contain a real suppression factor. This is to be expected, as some of the trajectories will correspond to classically disallowed tunnelling transitions. Thus, although potentially *any* S-matrix element can be approximated in this way, the result for particular trajectories may be exponentially small.

RELATION TO PREVIOUS WORK

Here we comment on the relationship between our approach and some previous work by other authors on high energy multi-particle and anomalous processes in electroweak theory. First, we will recall the expansion around the constrained instanton in electroweak theory [2, 4]. This approach neglects effects of initial and final states on the saddlepoint and results in a low energy ($E \ll E_* \simeq M_w/\alpha_w$) approximation to the total two-particle cross-section. Then, we remark on a previous approach to account for the impact of initial and final states on the saddlepoint [3] and comment on the formal resemblance which it bears to our approach.

A few of the properties of the electroweak instanton will make clear the relationship to our work: (*i*) The electroweak instanton is a solution of the Euclidean equations of motion. So, it will be necessary to rotate our real time formalism to Euclidean space in the analysis above, $t \to ix_4$. We would then be considering a saddlepoint approximation to an on-shell truncated Euclidean Greens function. (*ii*) The electroweak instanton has finite Euclidean action, $S = 8\pi^2/g^2$, and therefore satisfies *vacuum* boundary conditions. So, it is a suitable saddlepoint point for a Greens function with external fields only insofar as the effect of initial and final states on the saddlepoint are neglected (dropping the right hand side of (19) and (22)).

Since the instanton does not obey the correct boundary conditions to be a saddlepoint of a scattering amplitude [2], linear terms in the fluctuation expansion do not cancel, and the expansion of the scattering amplitude entails corrections which are formally large $O(1/g^2)$ in the exponent. Fortunately, these corrections are under control and calculable for sufficiently small energies, $E \ll E_*$ [4]. The effect of the final states can be taken into account, in a perturbative expansion in powers of $x \equiv E/E_* \ll 1$ [4]. For instance, the total inclusive cross section in the one-instanton sector is given by the well-known result

$$\sigma_{tot}(E) \sim \exp\left\{ \frac{16\,\pi^2}{g^2} \left(-1 + c_1 x^{4/3} + c_2 x^2 + \mathcal{O}\left(x^{8/3}\right)\right)\right\}, \tag{56}$$

in the limit $E/E_* = fixed$ and $g \to 0$. Here $\sim$ implies that only the exponential behavior of the cross-section is shown. The first term here is just twice the instanton action, the 'tHooft suppression for vacuum tunnelling, while the next term indicates the exponential growth of the cross-section at low energies [1]. The higher order terms are determined from all tree-level corrections to the many soft ($E \simeq M_w$) final state particles, and the first few coefficients c_i are known. Thus, provided the energy is sufficiently low $E \ll E_*$, tree-level corrections to the final state can be described *semiclassically*.

It has not been similarly demonstrated that the effect of *initial* state corrections, involving many loops, can be described semiclassically. However, some results indicate that these corrections may run counter to naive intuition and the initial two particle state may be described semiclassically as well [11, 3, 2]. As we noted in Section 4, initial state corrections in the instanton expansion appear as factors of the residue of the propagator in the instanton background. The residue of the propagator has hard high energy behavior ($\sim g^2 s$) where s is the center of mass energy [11], so that initial state corrections can contribute at the *semiclassical* level when s is of order $1/g^4$. Mueller [11] has shown that they contribute first at order $(E/E_*)^{10/3}$ to (56).

By contrast, in our formalism the classical background field is constructed to solve the correct multi-particle boundary value problem derived in Section 2. So, there are no additional tree-level corrections in the limit $E = fixed$ while $g \to 0$. All corrections are formally suppressed by powers of g, as shown in Section 4. The size of subleading corrections to our result depend on the properties of the Feynman propagator in the background of our proposed classical trajectory, ϕ^c. It would therefore be very interesting to understand the high energy behavior of the propagator in our background field. Also, for the purposes of direct comparison to (56), it would be interesting to obtain its behavior in the same limit, where $E/E_* = fixed$ while $g \to 0$. We have reason to expect that the propagator behaves less severely than the instanton propagator in this limit, due to the lack of translational zero modes of the background field [3].

A previous approach to account for the impact of initial and final states on the saddlepoint [3] bears a resemblance to ours but results in a different classical equation. We discuss it here for the purposes of comparison. These authors extremized Minkowskian n-point Greens functions of electroweak gauge fields

$$G^{(n)}(p_1, p_2, \ldots) = \int \mathcal{D}A \; A^{\mu_1 a_1}(p_1) \cdots A^{\mu_n a_n}(p_n) \; e^{iS[A]/g^2} \; . \tag{57}$$

It is not clear to us whether the Greens function is the proper quantity to extremize to yield information about the corresponding scattering amplitude (and ultimately the cross-section). It may be that extremizing the entire Greens function, rather than its LSZ projection, is too strict a requirement.

An equation of motion for the saddlepoint of (57) can be derived in a manner with formal similarity to our own. To do so, first express the n fields as an exponential, using $A = \exp \ln A$, and then extremize the exponent. This yields [3]

$$\frac{\delta S}{\delta A^{\nu b}(x)} = i \sum_i \delta^{\mu_i \nu} \; \delta^{a_i b} \; \frac{e^{ip_i \cdot x}}{A^{\mu_i a_i}(p_i)} \; . \tag{58}$$

This is a non-linear integro-differential field equation, depending on both the field $A(x)$ and its Fourier transform $A(p)$. Not much is known about equations of this type, and despite much effort (58) defies solution [15]. It appears more difficult to solve in practice than the boundary value problem we have derived (18), (20) and (22).

Before concluding I would like to point out a difference between studying scattering via the Minkowskian and Euclidean functional integral. The Euclidean counterparts to our solutions (i.e. their analytic continuations to imaginary time $t \to ix_4$) are very badly behaved asymptotically. That is, the 'wrong frequency' components required by the prescence of particles in the initial and final states (see equations (19) and (22)) imply an exponential blow-up of the solution at large imaginary times. In other words, our BC's are incompatible with the usual Feynman BC's (only positive frequencies in the future, and negative in the past) which arise automatically from requiring finite action in Euclidean space.

SUMMARY

In this lecture I have outlined a connection between certain classical solutions and corresponding quantum S-matrix elements in a weakly coupled field theory. The connection is made via a semiclassical approximation which appears to be controlled at small coupling, regardless of the scattering energy. This is in contrast to methods which involve expansions about instanton-like trajectories (i.e.– satisfying vacuum boundary conditions), which have been shown to break down at nonperturbative energy scales $E \sim m/g^2$. Real Minkowskian trajectories can be found by the straightforward time integration of a well-defined initial value problem which is determined by the inital quantum state. Complex Minkowskian trajectories require the implementation of mixed boundary conditions at both initial and final times. This procedure, while more difficult in practice, can be guaranteed in principle to describe any wave packet to coherent state transition specified by the complex functions $a(k), b_k^*$. The formalism described here can thus be used to compute intrinsically nonperturbative scattering amplitudes in a variety of field theories.

Some examples discussed here are the problem of baryon number violation in the electroweak theory and multiparticle scattering in gauge and scalar theories. The discovery of nontrivial or "interesting" real classical trajectories in these theories can now be related to unsuppressed scattering amplitues which are potentially observable experimentally. Alternatively, nontrivial complex trajectories, of which there are an infinite number, allow the calculation of a much larger class of scattering amplitudes, including some that are classically forbidden and therefore involve tunneling. We hope that our results will stimulate future numerical work in this area, particularly the search for nontrivial trajectories.

Acknowledgements

This talk is based on research [16] performed in collaboration with Thomas Gould and Erich Poppitz, both of The Johns Hopkins University. SDH acknowledges support from the National Science Foundation under grant NSF-PHY-87-14654, the state of Texas under grant TNRLC-RGFY-106, the Harvard Society of Fellows and an SSC Fellowship.

REFERENCES

1. A. Ringwald, High Energy Breakdown of Perturbation Theory in the Electroweak Instanton Sector, *Nucl. Phys.* B330:1 (1990) ; O. Espinosa, High Energy Behavior of Baryon and Lepton Number Violating Scattering Amplitudes and Breakdown of Unitarity in the Standard Model, *Nucl. Phys.* B343:310 (1990) ; For a recent review, see M.P. Mattis, The Riddle of High Energy Baryon Number Violation, *Phys. Rep.* 214:159 (1992) .
2. P.G. Tinyakov, Instanton-like Transitions in High Energy Collisions, *Int. J. Mod. Phys.* A8 (1993) 1823.
3. M.P. Mattis, L. McLerran, and L.G. Yaffe, High Energy Anomalous Scattering: Is it Semiclassical ?, *Phys. Rev.* D45:4294 (1992) .
4. S.Yu. Khlebnikov, V.A. Rubakov, and P.G. Tinyakov, Instanton Induced Cross Sections Below the Sphaleron, *Nucl. Phys.* B350:441 (1991) .
5. C. Gong, S.G. Matinyan, B. Müller, and A. Trayanov, Manifestation of Infrared Instabilities in High Energy Processes in Nonabelian Gauge Theories, *Phys. Rev.* D49:607 (1994) .
6. K. Rajagopal and N. Turok, Classical High Energy Scattering in the Abelian Higgs Model, *Nucl. Phys.* B375:299 (1992) .
7. H. Goldberg, D. Nash, and M.T. Vaughn, Classical Lambda Φ^4 Theory in 3+1 Dimensions, *Phys. Rev.* D46:2585 (1992) .
8. S. Coleman, Nonabelian Plane Waves, *Phys. Lett.* 70B:59 (1977) ;
   G. Passarino, Yang-Mills Theories in the Presence of Classical Plane Wave Fields: Stability Properties, *Phys. Lett.* 176B:135 (1986) .
9. E. Farhi, V. Khoze, and R. Singleton, Minkowski Space Nonabelian Classical Solutions With Non-integer Winding Number Change, *Phys. Rev.* D47:5551 (1993) ;
   E. Farhi, V. Khoze, K. Rajagopal, and R. Singleton, Spherical Shells of Classical Gauge Field and Their Topological Charge as a Perturbative Expansion, MIT preprint CTP-2287, hep-ph/9403285.
10. T.M. Gould and S.D.H. Hsu, Spacetime Symmetries and Semiclassical Amplitudes, Harvard preprint HUTP-93-A035, hep-ph/9311291.
11. A. Mueller, First Quantum Corrections to Gluon-Gluon Collisions in the One Instanton Sector, *Nucl. Phys.* B348:310 (1991) ; Leading Power Corrections to the Semiclassical Approximation for Gauge Meson Collisions in the One Instanton Sector, *ibid.* B353:44 (1991) .
12. S.Yu. Khlebnikov, V.A. Rubakov, and P.G. Tinyakov, Periodic Instantons and Scattering Amplitudes, *Nucl. Phys.* B367:334 (1991) .
13. W.H. Miller, Classical Limit Quantum Mechanics and the Theory of Molecular Collisions, *Adv. Chem. Phys.* **3** (1975) 3;
    D. McLaughlin, Complex Time, Contour Independent Path Integrals and Barrier Penetration, *J. Math. Phys.* **13** (1972) 1099;
    E.M. Ilgenfritz and H. Perlt, Complex Time Path for Tunneling at Intermediate Energy, *J. Phys. A* **25** (1992) 5729.
14. T.M. Gould and E.R. Poppitz, *Phys. Lett.* 312B:299 (1993) ; Semiclassical Scattering in Yang-Mills Theory, Johns Hopkins preprint JHU-TIPAC-930012, *Nucl. Phys.* B, in press.

15. L. McLerran, private communication;
B+L Non-Conservation as a Semiclassical Process, Minnesota preprint, TPI-MINN-93/51-T, hep-ph/9311239.
16. T.M. Gould, S.D.H. Hsu, and E.R. Poppitz, Quantum Scattering From Classical Field Theory, JHU-TIPAC-940003, hep-ph/9403353.

SPHERICAL SHELLS OF CLASSICAL GAUGE FIELD AND THEIR TOPOLOGICAL CHARGE AS A PERTURBATIVE EXPANSION

Krishna Rajagopal

Lyman Laboratory of Physics
Harvard University
Cambridge, MA 02138

Fermion number is not conserved in the standard model[1], yet reliable techniques for calculating fermion number violating rates in high energy collisions have not been fully developed. At very low energy where the processes are best described as tunnelling events, the calculational methods make use of solutions to the Euclidean field equations — instantons. At energies comparable to but below the sphaleron barrier, Euclidean methods[2] and other methods in which part of the calculation is done in Euclidean space[3] have also been applied. However, at energies well above the tunnelling barrier, it may be more appropriate to work directly in Minkowski space.[4, 5] To this end, E. Farhi, V. V. Khoze, R. Singleton, and I are investigating solutions to the Minkowski space classical equations of motion. In the work I sketch here,[6] we study $SU(2)$ gauge theory without the Higgs field. Working in the (spatial) spherical ansatz,[7] we explore solutions which have the property that in the far past they describe freely propagating incoming shells of energy. We do this by developing a perturbative expansion in the coupling g which can be used to systematically solve the equations of motion once arbitrary initial profiles have been specified. We discuss the topological charge of these solutions, which we also develop in a power series expansion in g. We show that the topological charge is nonzero at order g^5, by doing an explicit analytic calculation of the order g^5 contribution to Q for specific initial pulses. We then describe the associated anomalous fermion production in the presence of these solutions, and speculate on the physical implications of our results.

SPHERICAL ANSATZ SOLUTIONS

Recently, it was shown that in the spherical ansatz[7] the classical equations of motion of $SU(2)$ gauge theory can be reduced to two coupled nonlinear partial differential equations of motion for two gauge invariant functions of $r = |\mathbf{x}|$ and t.[5]

The spherical ansatz is given by expressing the gauge field A_μ in terms of four functions a_0 , a_1 , α and γ of r and t:

$$\begin{aligned} A_0(\mathbf{x},t) &= \frac{1}{2}a_0(r,t)\sigma\cdot\hat{\mathbf{x}} \\ A_i(\mathbf{x},t) &= \frac{1}{2}[a_1(r,t)\sigma\cdot\hat{\mathbf{x}}\hat{x}_i + \frac{\alpha(r,t)}{r}(\sigma_i - \sigma\cdot\hat{\mathbf{x}}\hat{x}_i) + \frac{\gamma(r,t)}{r}\epsilon_{ijk}\hat{x}_j\sigma_k] , \end{aligned} \tag{1}$$

where $\hat{\mathbf{x}}$ is the unit three-vector in the radial direction. The action

$$S = -\frac{1}{2}\int d^4x\, \mathrm{Tr}(F_{\mu\nu}F^{\mu\nu}) , \tag{2}$$

Electroweak Physics and the Early Universe, Edited by
Romao and Freire, Plenum Press, New York, 1994

in the spherical ansatz takes the form of an Abelian Higgs model in curved space,[7]

$$S = 4\pi \int dt \int_0^\infty dr \left[-\frac{1}{4} r^2 f_{\mu\nu} f^{\mu\nu} - (D_\mu \chi)^* D^\mu \chi - \frac{g^2}{2r^2} \left(|\chi|^2 - \frac{1}{g^2} \right)^2 \right] , \tag{3}$$

where $f_{\mu\nu} = \partial_\mu a_\nu - \partial_\nu a_\mu$ is the field strength, $\chi = \alpha + i(\gamma - 1/g)$ is a complex scalar and $D_\mu \chi = (\partial_\mu - i g a_\mu)\chi$ is the covariant derivative. The indices are raised and lowered with the $1+1$ dimensional metric $ds^2 = -dt^2 + dr^2$.

It is convenient to define the gauge invariant field ψ through $r^2 f_{\mu\nu} = -2\epsilon_{\mu\nu}\psi$, and to write the complex field χ in polar form $\chi(r,t) = -i\rho(r,t)\exp[ig\varphi(r,t)]$. Finally, because $\rho = 1/g$ in the vacuum, it is convenient to introduce a shifted field δ given by $2g\delta = 1 - g^2\rho^2$. The equations of motion obtained from the action (2) in the ansatz (1), when written using the variables just defined, become the equations of motion

$$\begin{aligned} \partial_\mu \partial^\mu \psi - \frac{2}{r^2}\psi &= -2g\partial_\mu \left(\frac{\delta\, \partial^\mu \psi}{1 - 2g\delta} \right) \\ \partial_\mu \partial^\mu \delta - \frac{2}{r^2}\delta &= g\frac{\partial_\mu \psi\, \partial^\mu \psi}{1 - 2g\delta} - g\frac{\partial_\mu \delta\, \partial^\mu \delta}{1 - 2g\delta} - \frac{4g}{r^2}\delta^2 \end{aligned} \tag{4}$$

for the gauge invariant variables δ and ψ. Once a gauge is chosen then the gauge-variant variables φ and a_μ can be determined. Equations (4) can be solved after specifying initial value data, that is the values of ψ , $\dot{\psi}$, δ and $\dot{\delta}$ at some fixed time. Any initial value data expressed in terms of ψ , $\dot{\psi}$, δ and $\dot{\delta}$ will be consistent with the Gauss's law constraint.

The equations (4) have a wide class of finite energy solutions all of which have certain general features in common. At early times they depict a thin spherical shell of energy imploding towards the origin at near the speed of light. We see this by noting that if $\psi(r,t) = \psi_p(t+r)$ and $\delta(r,t) = \delta_p(t+r)$ then equations (4) with the $1/r^2$ terms neglected are satisfied. If $\psi_p(u)$ and $\delta_p(u)$ are close to zero except for u in a region of size Δ around $u = 0$, then the solution describes an incoming shell of width Δ moving undistorted along $r = -t$. This description remains valid for all $t \ll -\Delta$. At intermediate times the region around the origin is energetically excited. At late times, the $1/r^2$ terms can again be neglected, $\psi(r,t) = \psi_f(t-r)$ and $\delta(r,t) = \delta_f(t-r)$, and we have a shell of energy expanding outward, asymptotically approaching the speed of light.

We are particularly interested in solutions which look free in the far past. This is because our ultimate goal is to connect the classical solutions to the quantum mechanical description of the scattering of quanta which in the far past are well separated and non-interacting. For a classical solution to be said to linearize in the far past we require that there exists a gauge in which the commutator term in $F_{\mu\nu}$ can be neglected in the far past. Such a gauge can be found.[6] We also want A_ν in this gauge to be independent of g in the far past. This means that the initial value data can be specified without reference to g. We show[6] how to specify $\psi_p(u)$ and $\delta_p(u)$ so that this criterion is met.

We now expand the fields ψ and δ in the coupling g,

$$\psi = \psi_0 + g\psi_1 + g^2\psi_2 + \dots , \qquad \delta = \delta_0 + g\delta_1 + g^2\delta_2 + \dots , \tag{5}$$

change variables to $u = t + r$ and $v = t - r$, and find that the equations (4) become

$$\begin{aligned} \left[\partial_u \partial_v + \frac{2}{(u-v)^2} \right] \psi_0 &= I_0^\psi , & \left[\partial_u \partial_v + \frac{2}{(u-v)^2} \right] \delta_0 &= I_0^\delta , \\ &\vdots & & \\ \left[\partial_u \partial_v + \frac{2}{(u-v)^2} \right] \psi_n &= I_n^\psi , & \left[\partial_u \partial_v + \frac{2}{(u-v)^2} \right] \delta_n &= I_n^\delta , \end{aligned} \tag{6}$$

where $I_0^\psi = I_0^\delta = 0$ and I_n^ψ and I_n^δ depend only on $\psi_0, \dots \psi_{n-1}$ and $\delta_0, \dots \delta_{n-1}$. For given initial pulses $\psi_p(u)$ and $\delta_p(u)$, equations (6) can be solved order by order in g. [6]

If we focus on the region near the origin we see that before the shell reaches it is a vacuum configuration. The energy shell excites this region and then leaves so that at late times the region is once again a vacuum configuration. However, the late time vacuum structure of this region need not coincide with its early time structure, and therefore the solutions we have constructed can have nonzero topological charge Q. We define Q as the $R \to \infty$ limit of the topological charge inside a sphere of radius R:

$$Q = \lim_{R\to\infty} \frac{g^2}{32\pi^2} \int_{-\infty}^{\infty} dt \int_R d^3x\, \epsilon^{\mu\nu\alpha\beta} \mathrm{Tr}(F_{\mu\nu} F_{\alpha\beta}) \; . \tag{7}$$

For the spherical ansatz solutions which we consider, the $R \to \infty$ limit in (7) is well defined, and yields a generically nonzero, noninteger, g-dependent value for Q:[5, 6]

$$Q = \frac{g^2}{\pi} \int_{-\infty}^{\infty} dv \left\{ \frac{\delta_f(v)\, \partial_v \psi_f(v)}{1 - 2g\delta_f(v)} \right\} - \frac{g^2}{\pi} \int_{-\infty}^{\infty} du \left\{ \frac{\delta_p(v)\, \partial_u \psi_p(u)}{1 - 2g\delta_p(u)} \right\} \; . \tag{8}$$

Note that topological charge develops only to the extent that the outgoing profiles differ from the incoming profiles.

The topological charge Q can take any value, not just integer values.[5] The usual argument that leads to integer topological charge assumes that the space-time region where the energy density is nonzero is bounded. For solutions to the equations of motion, energy is conserved and the energy computed on *any* equal time surface is non-vanishing. It is therefore impossible to surround the space-time region of nonzero energy density with a three sphere. We have no reason to expect and we do not find integer Q.

We can develop (8) perturbatively in g using the perturbative expansion for the solutions $\psi(u, v)$ and $\delta(u, v)$. We prove that in the spherical ansatz Q is zero up to order g^3. For every choice of initial profiles for which we have done the calculation, we find that the order g^4 contribution to Q vanishes. We have evaluated the fifth order topological charge for several initial profiles. In every case, we have found a nonzero result. As an example, for the initial pulses

$$\psi_p(u) = \frac{u^2 - 1}{(1 + u^2)^2} \qquad \delta_p(u) = -g\frac{(u^2 - 1)^2}{2(1 + u^2)^4} \tag{9}$$

for which A_μ is in fact g-independent in the far past, we find

$$Q = -\frac{15\pi^2}{1024} \frac{g^5}{\pi} + \text{higher order in } g \; . \tag{10}$$

We have demonstrated that spherical solutions to the Minkowski space classical equations of motion for pure $SU(2)$ which propagate freely in the far past can have nonzero topological charge. The power series expansion of Q in g can be nonzero at finite order — in the examples we have considered, at order g^5.

The solutions we have constructed have the property that the topological charge inside a sphere of radius R, approaches a fixed non-integer value as R gets large. Suppose that we have a quantized chiral fermi field coupled to this classical non-Abelian background. Without introducing a Higgs field, we cannot give the fermion a gauge invariant mass, so we take the fermion to be massless. Due to the anomaly, the fermion number current is not conserved:

$$\partial_\mu \hat{J}^\mu = \frac{g^2}{32\pi^2} \epsilon^{\mu\nu\alpha\beta} \mathrm{Tr}(F_{\mu\nu} F_{\alpha\beta}) \; . \tag{11}$$

If we couple quantized chiral fermions to classical gauge field solutions with nonzero Q, anomalous fermion number violation will occur. We construct operators[6] from integrals of the fermion number density and fermion flux which have finite fluctuations, and whose measurement is therefore well-defined. Using these operators, we show that in a series of measurements with the same background gauge field and the same quantum mechanical fermion state, the average number of fermions anomalously produced is given by the topological charge Q. For the solutions we have discussed, the topological charge vanishes as a power of g. This suggests that anomalous fermion number violation in high energy scattering processes may occur at finite order in g.

We cannot yet apply our results directly to scattering of W bosons at energies much higher than the sphaleron energy because important open questions remain. We have done a classical calculation in the spherical ansatz for an $SU(2)$ gauge theory without a Higgs field. Let us consider each of these idealizations in turn.

First, we have treated the gauge field classically, whereas we are ultimately interested in quantum scattering processes with a two-particle initial state. We need to relate our classical results to quantum mechanical scattering. One possible approach is to relate small amplitude classical solutions to quantum mechanical coherent states with small mean particle number. In this context, it is interesting that the g-expansion we have done is equivalent to an expansion in the amplitude of the initial pulse. Therefore, Q at order g^5 is proportional to the fifth power of the amplitude of the initial pulse. Also, Q is nonzero for g-independent initial conditions, which suggests that the number of particles in the initial state is g-independent, and need not be large at small g. This suggests that there may be scattering processes with few particles in the initial state in which Q is not exponentially suppressed. Another approach might be to relate the lowest order term in a semi-classical expansion of the S-matrix for a quantum scattering process to the solutions of classical equations of motion for a specified initial configuration. Some progress in this direction was reported at this conference by Gould and Hsu.

Second, we have worked within the spherical ansatz. This is a reasonable starting point since it makes the calculation tractable, and typically the lowest partial waves dominate scattering processes. An important open question here is that of stability — we have not checked that the solutions we discuss are stable to non-spherical perturbations.

Finally, we have worked in $SU(2)$ gauge theory without a Higgs field. It remains to be determined[8] how the inclusion of the Higgs field affects our results. It is possible that in scattering at very high energies the W mass and sphaleron barrier are irrelevant and our results are a good approximation to results in the full theory. However, including the Higgs field changes the theory in qualitative ways. The theory is no longer scale invariant. The behavior we describe at early and late times — pulses propagating without distortion — will not occur once the Higgs field is included. Thus, while the results reported here are likely applicable in the full theory only in the limit of infinite energy scattering, it remains to be seen[8] how they are modified at high but finite energy.

Acknowledgments

This work was done with E. Farhi of MIT, V. V. Khoze of Stanford University and R. Singleton, Jr. of Boston University. K.R. acknowledges the support of the Harvard Society of Fellows, the Milton Fund, and the N.S.F. under grant PHY-92-18167.

REFERENCES

1. G. 't Hooft, *Phys. Rev.* D14:3432 (1976) ; *Phys. Rev.* D18:2199 (1978) .
2. A. Ringwald, *Nucl. Phys.* B330:1 (1990) ; O. Espinosa, *Nucl. Phys.* B343:310 (1990) .
3. V. A. Rubakov, D. T. Son and P. G. Tinyakov, *Phys. Lett.* B287:342 (1992) .
4. N. H. Christ, *Phys. Rev.* D21:1591 (1980) .
5. E. Farhi, V.V. Khoze and R. Singleton, Jr., *Phys. Rev.* D47:5551 (1993) .
6. E. Farhi, V. V. Khoze, K. Rajagopal, and R. Singleton, Jr., Spherical Shells of Classical Gauge Field and Their Topological Charge as a Perturbative Expansion, CTP#2287, hep-ph/9403285.
7. E. Witten, *Phys. Rev. Lett.* 38:121 (1977) ; B. Ratra and L.G. Yaffe, *Phys. Lett.* B205:57 (1988) .
8. E. Farhi, K. Rajagopal, and R. Singleton, Jr., in preparation.

IN SEARCH OF CLASSICAL TRAJECTORIES

Thomas M. Gould

Department of Physics and Astronomy
The Johns Hopkins University
Baltimore, MD 21218

INTRODUCTION

In this short talk, I will emphasize some of the key points of a new Minkowski space formulation of nonperturbative contributions to quantum scattering [1], described at length by Hsu elsewhere in this volume. I will especially comment on some of the practical considerations of computing classical trajectories.

Our formulation [1] is based on a stationary phase approximation to a scattering amplitude involving an initial two-particle state. A real or complex classsical stationary trajectory is determined by an initial value or boundary value problem, respectively. The work establishes the role of Minkowski trajectories for quantum scattering amplitudes in weakly coupled field theories, and outlines a procedure for finding such trajectories, in principle via computation. Phenomena of particular interest in this regard involve final states with (i) large numbers of particles ($N \sim 1/g^2$ where g is the coupling constant of the theory), and/or (ii) anomalous violation of global quantum numbers, like fermion number violation in the electroweak theory.

Standard perturbative methods cannot describe these phenomenon. Current non-perturbative Euclidean space methods break down at high energies ($E \sim M_w/g^2$ in electroweak theory) because they do not account for the non-vacuum boundary conditions appropriate for a scattering problem. Our approach is to find the stationary point of a scattering amplitude, to include the effects of initial and final states on the stationary trajectory, and then to assess in what regime the expansion around this classical trajectory is controlled.

The formalism was illustrated for the simple case of a real scalar field [?]. A distinction is then made between real and complex stationary trajectories of the scattering amplitude of the real scalar field. I emphasize this here because the computation of the two types of trajectories could in practice be quite different.

REAL TRAJECTORIES

A given initial state on a time slice $t = T_i$ implies initial values for ϕ and $\dot{\phi}$, which are sufficient to determine a real trajectory uniquely [1]. The initial condition expressed

Electroweak Physics and the Early Universe, Edited by
Romao and Freire, Plenum Press, New York, 1994

in terms of Fourier components of the field is

$$\phi_i(\vec{k}) \;=\; \frac{1}{\sqrt{2\,\omega_{\mathbf{k}}}} \left(u_{\mathbf{k}}\; e^{-i\omega_k T_i} \;+\; u^*_{-\mathbf{k}}\; e^{i\omega_k T_i} \right) \,, \tag{1}$$

where the negative frequency component is

$$u_{\mathbf{k}} \;=\; \frac{\alpha_R(\vec{k}) \;+\; \alpha_L(\vec{k})}{\left(1 \;+\; \int d^3k\; \alpha_R(\vec{k})\; \alpha_L(\vec{k})\right)^{1/2}} \,, \tag{2}$$

with the (normalized) right- and left-moving wave packets $\alpha_R(\vec{k})$ and $\alpha_L(\vec{k})$ of the two-particle initial state. One assumes here that the field behaves freely for times $t < T_i$, so that the field and its time derivative are related in the obvious way. The field evolves forward in time from (1) according to the source-free equations of motion

$$\frac{\delta S}{\delta \phi(x)} \;=\; 0 \,. \tag{3}$$

So, a real trajectory is determined by an initial value problem. A real trajectory connects every initial wave packet state with a unique final state. The corresponding scattering amplitude is then not exponentially suppressed, since its action is real. But, the final state may or may not be interesting from the point of view of (i) and (ii) above. So, one must search the space of fields for interesting trajectories, by varying the initial wave packets. One might hope to find evidence of instabilities for the generation of long wavelength modes from the initial short wavelength modes, the signal of many (soft) particles in the final state.

I stress that a negative result to the search does not rule out the existence of unsuppressed amplitudes involving interesting final states. There is no one-to-one correspondence between real trajectories and scattering amplitudes. (more on this below) There may well be amplitudes not dominated in any way by real classical trajectories !

Hints for Classical Trajectories

Several previous investigations into the classical behavior of a variety of field theories provide some hints about the behavior of their real trajectories. I refer you to the discussion and references contained in Gould et al. [1] Here I would like to mention one most promising situation: evidence for an instability in Yang-Mills theory. Müller et al. [2] have considered the classical stability of a stationary mono-color wave in Yang-Mills theory (in $A^c_0 = 0$ gauge):

$$A^c_i(x,t) \;=\; \delta_{i3}\,\delta_{c3}\; A\; \cos k_0 x\; \cos\omega_0 t, \tag{4}$$

where c is a color and i is a spatial index. Small amplitude variations of the field in directions of different ($c \neq 3$) color are found to lead to an instability with long wavelength. This implies the existence of a classical trajectory connecting initial high energy plane waves to long wavelength modes in the final state. It therefore suggests that the corresponding $2 \rightarrow$ *many* gluon scattering amplitude may be unsuppressed !

For the purpose of the scattering problem, it would be interesting to see whether the instability in Yang-Mills theory persists for wave packets of finite spatial extent. As the width of a wave packet in momentum space decreases, approaching the plane wave limit (4) in which the instability has been observed, its amplitude decreases also, with the normalization held fixed [1]. However, since the instability persists for arbitrarily small amplitude [?], one may hope that it still produces many long wavelength modes.

In the context of electroweak theory, it will also be necessary to consider how the instability is modified in the presence of an explicit symmetry breaking scale. Some benchmark figures for a study of electroweak theory might be the following. One could consider initial wave packets with average momenta

$$k_{avg} \sim M_w/g^2 \ ,$$

and width

$$\Delta k \sim M_w \ .$$

Such wave packets overlap for a very short time for weak coupling, $\sim g^2/M_w$, during which an instability must generate long wavelength amplitudes. I emphasize though that a quantitative understanding of these estimates should be obtainable from direct computation of classical trajectories in a chosen field theory.

COMPLEX TRAJECTORIES

Complex trajectories of the real field arise from an analytic continuation of the path integral describing the scattering amplitude [1]. Complex trajectories can in principle connect a given initial state with *any* final state, due to the additional degrees of freedom. They include the case of classically forbidden transitions, in the case where initial and final states are based on different vacua separated by an energy barrier in configuration space. This is the situation relevant for processes with anomalous fermion number violation in electroweak theory and center of mass energies below the sphaleron energy, $\sim M_w/g^2$. Complex trajectories will have complex actions, so the corresponding amplitudes can in general have some exponential suppression.

The field is initially free, so it has a form similar to (1)

$$\phi_i(\vec{p}) = \frac{1}{\sqrt{2\omega_{\mathbf{p}}}} \left(u_{\mathbf{p}} \, e^{-i\omega_p T_i} + v_{\mathbf{p}} \, e^{i\omega_p T_i} \right) , \tag{5}$$

but where $u_{\mathbf{p}}$ and $v_{\mathbf{p}}$ are now independent complex functions. The initial boundary condition is satisfied by

$$u_{\mathbf{p}} = \frac{\alpha_R(\vec{p})}{\int d^3k \; \alpha_R(\vec{k}) \; v_{\mathbf{k}}} + \frac{\alpha_L(\vec{p})}{\int d^3k \; \alpha_L(\vec{k}) \; v_{\mathbf{k}}} , \tag{6}$$

where $v_{\mathbf{k}}$ is arbitrary. The initial configurations which satisfy (5), (6) have negative frequency modes which "resemble" the wave packet $\alpha_L + \alpha_R$ (up to complex multiplicative factors) and positive frequency modes which are arbitrary.

An additional degree of freedom is present, which must be determined by the additional information in the final state. The final state enforces a condition on the positive frequency part of the complex trajectory at final time T_f. The negative frequency part of the field remains independent and undetermined. The boundary condition takes the form

$$\phi_f(\vec{k}) = \frac{1}{\sqrt{2\,\omega_{\mathbf{k}}}} \left(c_{\mathbf{k}} \, e^{-i\omega_k T_f} + b^*_{-\mathbf{k}} \, e^{i\omega_k T_f} \right) , \tag{7}$$

for a final state $|\mathbf{b}^*\rangle$. Finally, the field evolves again with the source-free equations of motion, (3).

So, a complex trajectory is determined by a boundary value problem, with boundary conditions given by the initial and final states. One can consider solving this problem with a search procedure similar to the case of a real trajectory. A given $u_{\mathbf{p}}$ specified by the initial state gives rise to any number of final states, by varying $v_{\mathbf{k}}$. Alternatively, a direct solution of the boundary value problem for given initial and final states may be possible.

A WINDOW OF INTEREST

Once an interesting classical trajectory is found, one must turn to the problem of the quantum corrections to determine the regime in which the semiclassical expansion is controlled. I would like to emphasize in this regard how our semiclassical expansion differs subtly from the more familiar instanton expansion. In the familiar semiclassical expansion around an instanton, the vacuum boundary conditions of the classical field allow us to scale the field by the coupling constant, both in the boundary conditions and in the action

$$S[\phi] = \frac{1}{g^2}\,\hat{S}[g\phi]$$

so that $\hat{S}$ is independent of the coupling g. Then, the expansion is good, and the instanton contribution dominates, for sufficiently small coupling g, where the semiclassical exponent is large.

In our formulation, we account for the non-vacuum boundary conditions relevant to a scattering amplitude. As such, the boundary conditions are fixed by physical initial and/or final states, and the scaling above cannot be made. Now, we must require the coupling to be sufficiently large to produce an instability for the production of long wavelength modes from short wavelength modes. Meanwhile, our experience with ordinary perturbation theory and the semiclassical expansion leads us to expect that the the coupling must be sufficiently small to control the expansion. Thus, there must be a *window* in the coupling in order that our method be both interesting and controlled. While we have not proven that a window exists in any field theory, the means by which to do so are clear. The upper bound on the coupling can be determined by computations of classical solutions to the field equations, of the type outlined above. The lower bound on the coupling will have to come from computations of the quantum corrections, as discussed in Gould et al. [1]

COMMENTS

Current methods are inadequate to describe nonperturbative contributions to scattering amplitudes at very high energies. An outstanding example is the unsolved problem of the rate of fermion number violation in high energy collisions ($E \sim M_w/g^2$) in electroweak theory. I will close by emphasizing that *any* solution of the classical problems which I have outlined here corresponds to the stationary point of *some* scattering amplitude, and therefore yields *some* nontrivial information about the nonlinear aspects of quantum field theory, entirely inaccessible to perturbation theory. While much is known about semiclassical calculations in the vacuum sectors of field theories, our work points towards the need for a study of a much wider class of classical solutions, involving multi-particle boundary conditions.

Acknowledgements

The author acknowledges the support of the National Science Foundation under grant NSF-PHY-90-96198.

REFERENCES

1. T.M. Gould, S.D.H. Hsu, and E.R. Poppitz, Quantum Scattering From Classical Field Theory, JHU-TIPAC-940003, hep-ph/9403353.
2. C. Gong, S.G. Matinyan, B. Müller, and A. Trayanov, Manifestation of Infrared Instabilities in High Energy Processes in Nonabelian Gauge Theories, *Phys. Rev.* D49:607 (1994) .

POSSIBLE INSTABILITY OF THE EFFECTIVE POTENTIAL IN A SCALAR MODEL DUE TO EXTRA DIMENSIONS

E. Elizalde [1], K. Kirsten [2] and Yu. Kubyshin [3]

Universitat de Barcelona, Av. Diagonal 647
Departament d'E.C.M., 08028 Barcelona, Spain

1. We present a brief discussion of the effective potential in models with an infinite number of fields with a certain, increasing spectrum of masses. Such kind of models naturally appears in Kaluza-Klein type theories where the infinite set of fields is the Kaluza-Klein tower of modes[1]. Also similar systems might appear in certain models of composite particles with the infinite set of fields consisting of a ground state and its excitations.

As a concrete model we consider the $\lambda\phi^4$-theory on the six-dimensional spacetime $M^4 \times T^2$, where $T^2 = S^1 \times S^1$ is the two-dimensional torus with equal radii L. Instead of choosing L to be a constant of the order of the Planck length, as it is given by most of the spontaneous compactification solutions, we consider it as a parameter which takes values in a wide range. The interest in such an analysis is motivated by some cosmological scenarios[2], in which L varies significantly, as well as by some arguments coming from superstrings[3] (see also a discussion in ref. 4). The latter suggest that L^{-1} is of the order of $M_{SUSY} \sim (1 \div 10)$ TeV. Thus the model might be relevant for the description of the Higgs sector in certain Kaluza-Klein extentions of the Standard Model.

The action of the model is given by

$$S = \int_{M^4} d^4x \int_{T^2} d^2y \left[\frac{1}{2} \left(\partial \hat{\phi}(x,y) \right)^2 - \frac{m^2}{2} \hat{\phi}^2(x,y) - \frac{\hat{\lambda}}{4!} \hat{\phi}^4(x,y) \right], \qquad (1)$$

where $0 \leq y^1, y^2 < 2\pi$ and $\hat{\phi}(x,y)$ is periodic in y^1, y^2. Substituting the Fourier expansion

$$\hat{\phi}(x,y) = \sum_{N=-\infty}^{\infty} \frac{\phi_N(x)}{2\pi L} \exp\left(\frac{i}{L}(n_1 y^1 + n_2 y^2) \right), \quad N = (n_1,\ n_2),$$

[1]E-mail address: eli@ebubecml.bitnet, eli@zeta.ecm.ub.es

[2]Alexander von Humboldt fellow. E-mail address: kirsten@ebubecml.bitnet

[3]On leave of absence from Nuclear Physics Institute, Moscow State University, 117234 Moscow, Russia. E-mail address: kubyshin@ebubecml.bitnet

into the action (1) we get the effective theory in four dimensions with an infinite number of fields ϕ_N:

$$S = \int_{M^4} d^4x\,(\mathcal{L}_{M^4} + \Delta\mathcal{L}), \tag{2}$$

$$\mathcal{L}_{M^4} = \frac{1}{2}(\partial_\mu\phi) + \frac{m^2}{2}\phi^2 - \frac{\lambda}{4!}\phi^4, \tag{3}$$

$$\begin{aligned}\Delta\mathcal{L} = & \frac{1}{2}\sum_N{}' \left(|\partial_\mu\phi_N|^2 - M_N^2|\phi_N|^2\right) - \frac{\lambda}{4}\phi^2\sum_N{}'|\phi_N|^2 \\ & -\frac{\lambda}{6}\phi\sum_{N,M}{}'\phi_N\phi_M\phi_{-N-M} - \frac{\lambda}{4!}\sum_{N,M,K}{}'\phi_N\phi_M\phi_K\phi_{-N-M-K},\end{aligned} \tag{4}$$

where the prime means that the terms with $N = M = K = (0,0)$ are ommitted from the summation, the masses M_N are given by the formula

$$M_N^2 = m^2 + (n_1^2 + n_2^2)/L \tag{5}$$

and $\phi(x)$ stands for the zero mode $\phi_{00}(x)$.

2. The 1-loop effective potential $V_{M^4\times T^2}(\phi)$ of the field $\phi(x)$ in the model (2)-(4) is given by (see refs. 5 and 6 for the general theory)

$$V_{M^4\times T^2}(\phi) = V_{M^4}(\phi) + \Delta V^{1-loop}(\phi) + \frac{1}{6!}h_B\phi^6, \tag{6}$$

where

$$V_{M^4}(\phi) = \frac{m_B^2}{2}\phi^2 + \frac{\lambda_B}{4!}\phi^4 + V_{M^4}^{1-loop}(\phi) \tag{7}$$

is the unrenormalized Coleman-Weinberg potential for the $\lambda\phi^4$ model in four dimensions[5] and ΔV^{1-loop} is given by the sum over the non-zero values of $N = (n_1, n_2)$ of the sum of contributions of all 1-loop diagrams where the field ϕ_N propagates along the loop. The sum over N creates additional divergences. This makes the effective theory (2)-(4) non-renormalizable, hence the ϕ^6-counterterm in eq. (6).
We choose the renormalization conditions as follows:

$$V_{M^4\times T^2}(0) = V_{M^4}(0) = 0, \tag{8}$$

$$\frac{\partial^2 V_{M^4\times T^2}(\phi)}{\partial\phi^2}|_{\phi=0} = \frac{\partial^2 V_{M^4}(\phi)}{\partial\phi^2}|_{\phi=0} = m^2, \tag{9}$$

$$\frac{\partial^4 V_{M^4\times T^2}(\phi)}{\partial\phi^4}|_{\phi=\mu} = \frac{\partial^4 V_{M^4}(\phi)}{\partial\phi^4}|_{\phi=\mu} = \lambda, \tag{10}$$

$$\frac{\partial^6 V_{M^4\times T^2}(\phi)}{\partial\phi^6}|_{\phi=\mu} = \frac{\partial^6 V_{M^4}(\phi)}{\partial\phi^6}|_{\phi=\mu} + h, \tag{11}$$

where μ is a renormalization point, m and λ are the renormalized mass and the quartic coupling constant, correspondingly, and h is a constant of the ϕ^6-vertex, which in principle appears due to the non-renormalizability of the model. The qualitative behaviour of the potential for small and for large values of $\phi(x)$ does not depend on h.
The second equations in (8)-(10) define the part $V_{M^4}(\phi)$ to be just the standard renormalized Coleman-Weinberg potential in four dimensions[5]. To explain the meaning of the other conditions we consider the case when $m \neq 0$ and take $\mu = 0$. After renormalization of the effective potential, the function $v_{M^4\times T^2}(x,\nu) \equiv \lambda V_{M^4\times T^2}/m^4$ can be written as a function of two dimensionless parameters: $x = \lambda\phi^2/(2m^2)$ and $\nu = (mL)^2$. It is easy to show that $v_{M^4\times T^2}(x,\nu) - v_{M^4}(x) \sim \mathcal{O}(x^4\nu)$, where $v_{M^4}(x) = \lambda V_{M^4}(\phi)/m^4$. This

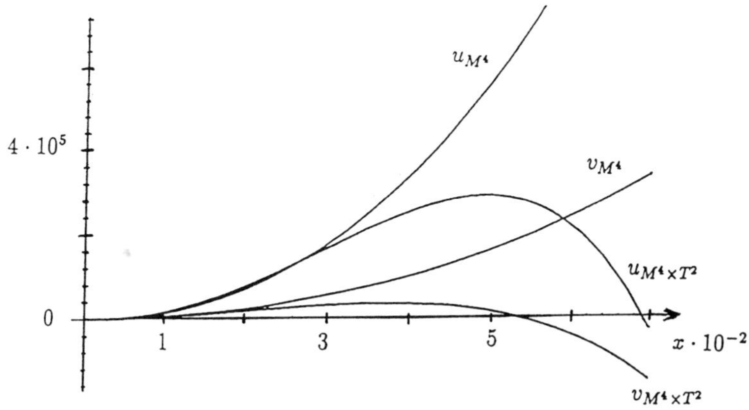

Figure 1. Plots of functions v_E and u_E for $E = M^4$ and $E = M^4 \times T^2$.

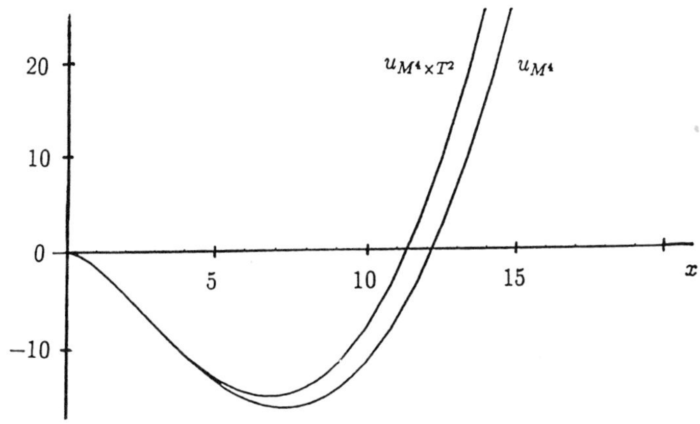

Figure 2. Plots of functions u_{M^4} and $u_{M^4 \times T^2}$ for $x \leq 20$.

result means that when $x^4\nu^2 \sim \lambda\phi^2 L/m \ll 1$ the extra dimensions do not reveal themselves and the vacuum states of the complete theory (2)-(4) and of the four-dimensional model (2) are the same. The condition above is satisfied when the field is small enough or when the size of the extra dimensions is sufficiently small, or both. Thus, the renormalization conditions (8)-(11) provide the physically correct correspondence between these two theories.

The functions $v_{M^4 \times T^2}(x, \nu)$ for $\nu = 10^{-2}$ and $v_{M^4}(x)$ are plotted in Fig. 1 (here and in the following we choose $\lambda = (32\pi^2/5)$ and $h = 0$). For other values of ν the shape of the potential remains the same with increasing scale along the x-axis. In the case $m = 0$, because of infrared divergences in the 1-loop diagrams contributing to the effective potential, the renormalization point μ must be kept non-zero. The plots of the functions

$$u_{M^4}(x) \equiv 24V_{M^4}(\phi)/(\lambda\mu^4) \quad \text{and} \quad u_{M^4 \times T^2}(x, t) \equiv 24V_{M^4 \times T^2}(\phi)/(\lambda\mu^4)$$

for $t = (\lambda/2)(\mu L)^2 = 10^{-2}$ are presented in Fig. 1 and Fig. 2, where now $x = \phi^2/\mu^2$. As before, for small x these two functions have the same behaviour. In particular they possess a non-trivial minimum which, for $V_{M^4}(\phi)$, is the well known result of Coleman and Weinberg[5].

We see that, though the renormalized effective potential for weak fields has the same behaviour as the one for the $\lambda\phi^4$ theory in four dimensions, the effect of two compact extra dimensions is rather drastic. For strong fields, $V_{M^4 \times T^2}(\phi)$ is unbounded from below, and thus the minimum $\phi = 0$ in the case $m \neq 0$ or the non-trivial minimum in the case $m = 0$ (see Fig. 2) are metastable. The lifetime of these states will be estimated elsewhere.

Of course, all these non-trivial effects take place in the interval of values of ϕ where the 2-loop correction seems to be important. Analysis of this correction needs more computation and is beyond the scope of the present contribution. However, we believe that the 1-loop result provides some hints of what might happen in effective theories appearing in the framework of the Kaluza-Klein approach.

3. We would like to add that another effect due to extra dimensions have been studied recently in ref. 4. It has been shown there that the total cross section of scattering of two light particles (zero modes) in the theory (1) on the space $M^4 \times T^2$ or $M^4 \times S^2$ differs significantly from the total cross section of the same process in the $\lambda\phi^4$ theory in four dimensions. Though, as it is natural to expect, the difference decreases when the energy $\sqrt{s} \to 0$, it is quite noticeable far below the threshold of the first heavy particle (the mode with $n_1^2 + n_2^2 = 1$, see eq. (5)). This is a possible manifestation of the multidimensionality of the original model, which is of physical interest and could in principle be measured in high energy experiments.

ACKNOWLEDGEMENTS: KK and YK would like to thank the members of the Department ECM, Barcelona University, for the kind hospitality. This work has been supported by the Alexander von Humboldt Foundation (Germany), by DGICYT (Spain) and by CIRIT (Generalitat de Catalunya).

REFERENCES

1. M.J. Duff, B.E.W. Nilsson and C.N. Pope, Kaluza-Klein Supergravity, *Phys. Rep.* C130:1 (1986).
2. S.K. Blau, E.I. Guendelman, A. Taormina, and L.C.R. Wijewardhana, On the stability of toroidally compact Kaluza-Klein theories, *Phys. Lett.* B144:30(1984).
3. L.J. Dixon, V.S. Kaplunovsky and J. Louis, Moduli dependence of string loop corrections to gauge coupling constants, *Nucl. Phys.* B355:649(1991).
   I. Antoniadis, A possible new dimension at a few TeV, *Phys. Lett.* B246:377(1990).
4. A. Demichev, Yu. Kubyshin and J.I. Perez Cadenas, Manifestations of space-time multidimensionality in scattering of scalar particles, *Phys. Lett.* B323:139(1984).
   E. Elizalde and Yu. Kubyshin, Possible evidences of Kaluza-Klein particles in a scalar model with spherical compactification, UB-ECM-PF 94/4 (hep-th/9402086).
5. S. Coleman and E. Weinberg, Radiative corrections as the origin of spontaneous symmetry breaking, *Phys. Rev.* D7:1888(1973).
6. S.W. Hawking, Zeta function regularization of path integrals in curved spacetime, *Commun. Math. Phys.* 110:439(1987).
   R. Critchley and J.S. Dowker, Effective Lagrangian and energy-momentum tensor in de Sitter space, *Phys. Rev.* D13:3224(1976).
   E. Elizalde and K. Kirsten, Topological symmetry breaking in self-interacting theories on toroidal spacetime, *J. Math. Phys.* 35:1260(1994).

GRAVITINOS IN THE EARLY UNIVERSE[1]

Willy Fischler

Theory Group
Department of Physics
University of Texas
Austin, TX 78712

ABSTRACT

The production of gravitinos is much larger at high temperature than it is in a high energy zero temperature environment.

In this talk I will make two essential assumptions. The first one is that whatever the ultimate theory of nature is, it leaves below the Planck scale M_p, $N = 1$ supergravity. The second assumption is that the superpartners of known particles have masses of $0\,(G_F^{-1/2})$ so that there is no fine tuning for the Higgs sector.

The existence of gravitinos in the spectrum has important model independent implications. Depending on the gravitino mass $m_{3/2}$ there are two quite different alternatives. Either $m_{3/2} < 1$ keV, then supersymmetry has to be broken at a scale smaller than $10^5 - 10^6$ GeV, and the feeddown of this breaking to the visible sector cannot be gravitational. The alternative $m_{3/2} > 1$ keV implies that the maximum temperature of the universe T_{max} (or reheating temperature if there is a period of inflation) does not exceed $10^5 - 10^6$ GeV. This bound on T_{max}, in the case of unstable gravitinos, comes from requiring that the abundance of gravitinos be such as not to affect the relative abundance of light nuclei.[2] For stable gravitinos[3] one requires their present energy density to be smaller than the critical energy density and obtains a bound $T_{max} \lesssim 10^5 - 10^6$ GeV.

The production rate of gravitinos depends on their coupling to the other degrees of freedom. Both the spin 3/2 and spin 1/2 components of the gravitinos couple in a universal way. The spin 3/2, Ψ_μ^α couples with gravitational strength $\frac{1}{M_p}$ to the Noether supercurrent S_μ^α. The spin 1/2 component, Ψ_α being the would be goldstino couples derivatively to the supercurrent

$$\frac{1}{F}\,\partial_\mu \Psi_\alpha\, S^{\mu\alpha}$$

where F is the scale of supersymmetry breaking. The gravitino acquires a mass by the superhiggs mechanism, $m_{3/2} = \frac{F}{M_p}$.

We will show that for temperatures T bigger than the weak scale $G_F^{-1/2}$ but smaller

Electroweak Physics and the Early Universe, Edited by
Romao and Freire, Plenum Press, New York, 1994

than $\sqrt{F}$ the coupling of the spin 1/2 component of the gravitino is the dominant one for estimating the abundance of gravitinos. For the sake of clarity let us first consider the production of gravitinos at zero temperature for energies E above $G_F^{-1/2}$ but smaller than $\sqrt{F}$. To focus consider the process gluon + gluon $\rightarrow$ gluino + gravitino (see Figure 1).

$$V^b + V^c \rightarrow \lambda^a + \psi_S$$

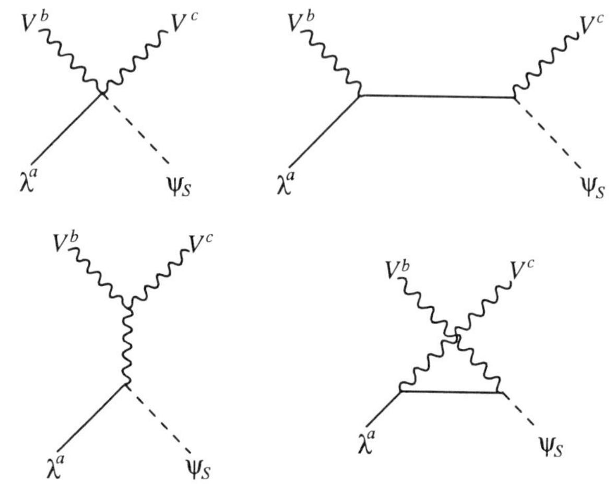

Figure 1. Feynman diagram involved in the calculation for the cross section of gluon + gluon $\rightarrow$ gluino + gravitino.

The spin 1/2 component dominates because it couples with a strength $\frac{E}{F}$ (E is the energy of the goldstino) as compared to $\frac{1}{M_p}$ for the spin 3/2 component. Naively the cross section for such a process can be expected to be of $0\,(E^2/F^2)$. This however is not correct, a careful calculation of the various contributions give a cross section of $0\left(\frac{(\Delta M)}{F^2}\right)$ where ΔM for this example is the gluino mass and would in general be the splitting in a supermultiplet. The reason for this behavior of the cross section is supersymmetry. Indeed at energies E much larger than ΔM one can neglect to first order ΔM, in that case the members of a supermultiplet are degenerate and there is no other source of supersymmetry breaking in diagrams, hence the goldstino decouples. This ensures that the leading contribution to the amplitude is proportional to $\frac{\Delta M}{F}$. This is the clue for understanding why the production of gravitinos at finite temperature does not saturate for $T > \Delta M$ unlike the case for zero temperature. Indeed recall that fermions can have at most one quantum per state whereas no such limitation exists for bosons. This implies that at finite temperature where bosonic states with momenta $|\bar{k}| < T$ have large occupation, supersymmetry is "broken". Therefore in a thermal environment the argument based on supersymmetry that ensured the saturation of the amplitude is now invalid. One then obtains a cross section for the production of gravitinos where the dependence on temperature is as follows:

$$\sigma \sim T^2/F^2\,.$$

This in turn leads to a production rate

$$\Gamma \sim T^5/F^2$$

which in the relevant temperature range

$$\sqrt{G_F^{-1/2}} < T < \sqrt{F}$$

is much larger than previous estimates. One can then use this rate to estimate the abundance of gravitinos. This in turn leads to the bounds on T_{max}. For gravitinos of mass smaller then 1 keV their abundance is such that at the best they could provide the dark matter for 1 keV or be insignificant for lower masses.

In conclusion for the case where one has an ultimate temperature of $0\,(10^5 - 10^6$ GeV) one obviously needs a low temperature mechanism for baryogenesis. Also such a low reheating temperature in the case of inflationary cosmology requires an extreme weak coupling between the inflationary sector and the low energy visible degrees of freedom. Indeed the cosmological constant Λ must have been of $O\,(10^{17}$ GeV) during inflation to account for $\frac{\delta_\rho}{\rho} \sim 0\,(10^{-5})$ and hence $\frac{T_{max}}{\Lambda} \sim 0\,(10^{-12})$. On the other hand for $m_{3/2} < 1$ keV one has low energy supersymmetry breaking and the emergence of new degrees of freedom at energy of $0\,(10^6$ GeV) or even lower.

ACKNOWLEDGEMENTS

I would like to thank the organizers of this workshop for their hospitality and for providing a stimulating environment. Research supported in part by the Robert A. Welch Foundation and NSF Grant PHY 9009850.

REFERENCES

1. W. Fischler, Gravitinos and a low ultimate temperature for the early universe, HEP-TH 9404044, preprint to be published in Phys. Lett. B.

2. H. Pagels, T.R. Primack, *Phys. Rev. Lett.* **48,** 223 (1982) ;
M.Y. Khlopov, A.D. Linde, *Phys. Lett.* **B138,** 265 (1984) ;
T. Ellis, J.E. Kim, D.V. Nanopoulos, *Phys. Lett.* **B145,** 181 (1984) ;
T Moroi, H. Murayama, M. Yamaguchi, *Phys. Lett.* **B303,** 289 (1993) ;
V.S. Berezinski, *Phys. Lett.* **B261,** 71 (1991) ;
M. Kawasaki, T. Moroi, HEP-PH 9403364, preprint.

3. H. Pagels, T.R. Primack, *Phys. Rev. Lett.* **48,** 223 (1982) ;
M.Y. Khlopov, A.D. Linde, *Phys. Lett.* **B138,** 265 (1984) ;
T. Ellis, J.E. Kim, D.V. Nanopoulos, *Phys. Lett.* **B145,** 181 (1984) .

ON THE PROBLEM OF THE MAGNETIC MASS

O. Philipsen

Theory Group, DESY
Notkestr. 85
D-22603 Hamburg, Germany

1. INTRODUCTION

A complete perturbative description of non-abelian field theories at finite temperature is prohibited due to the occurrence of infrared divergencies for massless fields and the corresponding breakdown of perturbation theory. In particular, this renders impossible a discussion of the electroweak phase transition in the immediate vicinity of the critical temperature, as well as an evaluation of the effective potential in the symmetric phase. A potential cure for this problem is the dynamical generation of a "magnetic mass", which has to be of the form $\sim g^2T$, but whose coefficient cannot be calculated perturbatively, since it receives contributions from all orders of perturbation theory.[1]

Recently an estimate of this coefficient was attempted[2] by studying a set of gap equations for the full propagators of the theory. In this approach one rearranges the perturbation series by adding and subtracting mass counterterms to the action,

$$S = S_0 + \delta S_m + S_{int} - \delta S_m \ , \tag{1}$$

and expanding in the modified interaction term $S_{int} - \delta S_m$ with propagators derived from $S_0 + \delta S_m$. If the up to now arbitrary counterterms δS_m are chosen to represent the radiative corrections to the masses of the corresponding fields, one obtains propagators containing full masses $m^2 = m_0^2 + \delta m^2$, which are to be used in the evaluation of loop diagrams in the modified perturbation theory. The counterterm $-\delta S_m$ in the new interaction term ensures that the higher order contributions, which have been resummed into the propagator, do not contribute twice when calculations in higher orders are done. The value of the mass corrections δm^2 can be determined implicitly and self-consistently by calculating the self-energies of the fields in the modified perturbative scheme using full propagators. This corresponds to an infinite iteration of one-loop mass corrections and leads to a gap equation for the full masses m.[3] In the case of the SU(2)-Higgs model the authors of Refs. 2 considered the gap equation

$$m_T^2 = \Pi_T(p_0 = 0, p \to 0) \tag{2}$$

for the transverse gauge boson mass m_T in the symmetric phase (i. e. for a tree level mass $m_{T0} = 0$), and they found in Landau gauge

$$m_T^2 = \frac{m_T}{3\pi} g^2 T \ , \tag{3}$$

yielding a nonvanishing magnetic mass $m_T = g^2 T/(3\pi)$.

Most equilibrium properties of field theories at very high temperature can be investigated by studying the corresponding dimensionally reduced field theory.[4] It is therefore instructive to examine, if the dynamical generation of a mass for the spatial components of the vector fields can also be understood in the three-dimensional theory.[5]

2. PURE GAUGE THEORY IN 3 DIMENSIONS

Consider a pure gauge theory in three euclidean dimensions, where the gauge fields are known to remain massless at every finite order perturbation theory. In order to study non-perturbative dynamical mass generation a mass counterterm $A^a_\mu A^a_\mu m^2/2$ is added and subtracted to the action following the rationale explained in the last section. After covariant gauge fixing and including the ghost term the Lagrangian density relevant for studying the one loop self-energy of the gauge fields is given by

$$\mathcal{L} = \frac{1}{4} F^a_{\mu\nu} F^a_{\mu\nu} + \frac{m^2}{2} A^a_\mu A^a_\mu + \frac{1}{2\xi} (\partial_\mu A^a_\mu)^2 - \bar{c}_a \partial_\mu \left[\delta_{ab} \partial_\mu + g \varepsilon_{abc} A^c_\mu \right] c_b \ . \tag{4}$$

In three dimensions there are only two linearly independent tensors with two Lorentz indices which can be chosen to be the transverse and longitudinal projectors,

$$\begin{aligned} P_{L\mu\nu} &= \frac{p_\mu p_\nu}{p^2} \ , \\ P_{T\mu\nu} &= \delta_{\mu\nu} - \frac{p_\mu p_\nu}{p^2} \ . \end{aligned} \tag{5}$$

The propagator and the self-energy can thus be decomposed,

$$\begin{aligned} D^{ab}_{\mu\nu}(p) &= \delta_{ab} [D_T(p) P_{T\mu\nu} + D_L(p) P_{L\mu\nu}] \ , \\ \Pi^{ab}_{\mu\nu}(p) &= \delta_{ab} [\Pi_T(p) P_{T\mu\nu} + \Pi_L(p) P_{L\mu\nu}] \ , \end{aligned} \tag{6}$$

yielding the full propagator

$$\begin{aligned} \tilde{D}^{ab}_{\mu\nu}(p) &= \sum_{n=0}^{\infty} D(p) [\Pi(p) D(p)]^n \\ &= \delta_{ab} \left[\frac{P_{T\mu\nu}}{p^2 - \Pi_T(p)} + \frac{P_{L\mu\nu}}{p^2 - \Pi_L(p)} \right] \ . \end{aligned} \tag{7}$$

In order to find the magnetic mass the transverse self-energy Π_T has to be calculated with the propagators following from (4). If the magnetic mass determined in this way is to have a physical meaning instead of being just a technical quantity appearing in calculations, it has to be independent of the gauge fixing parameter. We therefore do the calculation in a general covariant gauge in order to check the gauge dependence explicitly. Thus, we study the gap equation

$$m^2 = -\Pi_T(p \to 0, m, \xi) \ , \tag{8}$$

which is shown diagrammatically in Fig. 1. Removing linearly divergent terms by means of dimensional regularization we obtain

$$m^2 = -g^2 \frac{m}{\pi} \left[-\frac{5}{6} + \frac{1}{3} \xi^{3/2} + \frac{3 - 5\xi + 2\xi^{5/2}}{6(1-\xi)} \right] \ . \tag{9}$$

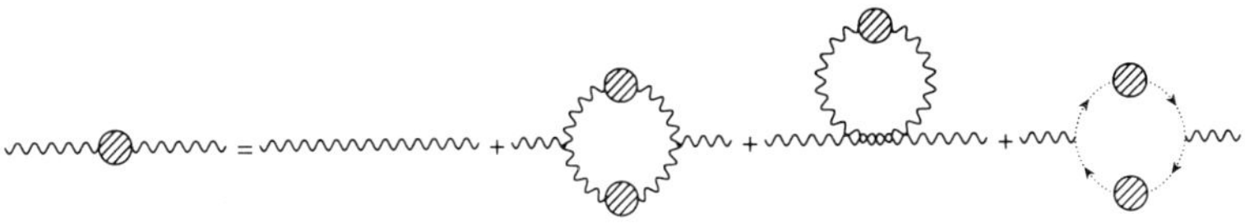

Figure 1. Gap equation for the self-consistent determination of the full mass correction for the propagator. The blobs on the lines indicate full propagators.

In Landau gauge, $\xi = 0$, and with the replacement $g^2 \to g^2 T$ this reduces to equation (3). For four-dimensional theories at $T \neq 0$ we can therefore conclude that the generation of a magnetic mass as described by gap equations at one-loop order is a feature of the Matsubara zero modes.

However, the mass term generated in this way is gauge dependent. In a study of the Debye mass it has been pointed out by Rebhan[6] that, in order to obtain gauge invariant results, one has to define the mass by the position of the pole in the propagator rather than by the zero momentum limit of the self-energy. Adopting this prescription one has to consider the gap equation

$$m^2 = -\Pi_T(p^2 = -m^2, m, \xi) \, , \tag{10}$$

instead of (8). However, the corresponding expression for the transverse self-energy still depends on ξ.

3. MASSIVE YANG-MILLS FIELDS IN 3 DIMENSIONS

A gauge invariant way to implement mass terms for non-abelian gauge fields is provided by the mechanism of spontaneous symmetry breaking. Consider the gauge theory supplemented by a scalar field,

$$\mathcal{L} = \tfrac{1}{4} F^a_{\mu\nu} F^a_{\mu\nu} + \mathrm{Tr}(D_\mu \Phi)^\dagger (D_\mu \Phi) + \mu^2 \mathrm{Tr}(\phi^\dagger \phi) + \tfrac{\lambda}{6} \left[\mathrm{Tr}(\phi^\dagger \phi)^2 \right] \, , \tag{11}$$

where Φ is a complex 2×2 matrix field transforming as the (1/2,1/2) representation of SU(2)$\times$SU(2). This field develops a nonvanishing vacuum expectation value $\langle \Phi \rangle = v$, and can be parametrized by four real fields,

$$\Phi = \tfrac{1}{2}(\sigma + i\tau^a \pi^a) \, . \tag{12}$$

Inserting this into (11) leads to the σ-model. Upon imposing the condition $\sigma^2 + (\pi^a)^2 = v^2$ and sending $\lambda, \mu \to \infty$ with v fixed the σ field decouples and one obtains the non-linear σ-model,

$$\mathcal{L} = \tfrac{1}{4} F^a_{\mu\nu} F^a_{\mu\nu} + \tfrac{g^2 v^2}{8} A^a_\mu A^a_\mu + \tfrac{1}{2}(\partial_\mu \pi^a)^2 + \tfrac{gv}{2}(\partial_\mu \pi^a) A^a_\mu - \tfrac{g}{2}\varepsilon_{abc} \partial_\mu \pi^a A^b_\mu \pi^c$$
$$+ O(\pi^4, \pi^3 A, \ldots) \, . \tag{13}$$

where the gauge field has acquired a mass $m_0 = gv/2$. We now perform the same procedure for this model as in the last section, the difference being that in this case we start out with a tree level mass $m_0 \neq 0$, which has to be sent to zero at the end of the calculation. To this Lagrangian we add the mass correction for the gauge fields, the R_ξ gauge fixing term

$$\mathcal{L}_{gf} = \tfrac{1}{2\xi}(\partial_\mu A^a_\mu + m\xi \pi^a)^2 \, ,$$

and the corresponding ghost term to arrive at

$$\mathcal{L} = \frac{1}{4}F^a_{\mu\nu}F^a_{\mu\nu} + \frac{m^2}{2}A^a_\mu A^a_\mu + \frac{1}{2\xi}(\partial_\mu A^a_\mu)^2 + \frac{1}{2}(\partial_\mu \pi^a)^2 + \frac{1}{2}m^2\xi\pi^a\pi^a - \frac{g}{2}\varepsilon_{abc}\partial_\mu\pi^a A^b_\mu\pi^c$$
$$- \bar{c}_a\left[\partial_\mu\left(\delta_{ab}\partial_\mu + g\varepsilon_{abc}A^c_\mu\right) + \frac{g^2 v}{4}\xi\varepsilon_{abc}\pi^c - m^2\xi\delta_{ab}\right]c_b\,. \quad (14)$$

The Goldstone fields π^a are unphysical and can be gauged away by going to the unitary gauge, in which (14) reduces to a theory describing a massive vector field,

$$\mathcal{L}_{unitary} = \frac{1}{4}F^a_{\mu\nu}F^a_{\mu\nu} + \frac{m^2}{2}A^a_\mu A^a_\mu\,. \quad (15)$$

The self-energy of the gauge fields now gets an additional contribution from the Goldstone loop, Fig. 2.

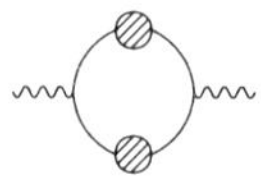

Figure 2. Contribution of the Goldstone modes to the gap equation for the model (14).

We have derived the gap equations (8) and (10) for this model. In the first case we recover equation (9) with an additional term $\sim \sqrt{\xi}$ from the Goldstone and ghost diagrams, leaving the expression gauge dependent. However, for the gap equation (10) we find in the limit $m_0 = 0$

$$m^2 = g^2\frac{m}{\pi}\left[\frac{63}{64}\ln 3 - \frac{3}{16}\right]\,. \quad (16)$$

This gap equation is manifestly gauge invariant and has, besides $m = 0$, the non-trivial solution

$$m = 0.28g^2\,. \quad (17)$$

In conclusion it was found that dynamical generation of a transverse gauge field mass as it can be investigated by the study of gap equations occurs in purely three-dimensional theories and hence, in the case of four-dimensional theories at $T \neq 0$, can be attributed to the static modes. We found a gauge invariant pole of the resummed gauge field propagator of the non-linear σ-model in the limit of vanishing tree level mass. If one interprets this model as the high temperature limit of four-dimensional SU(2) pure gauge theory, the value (17) for the magnetic mass is consistent with the results of other non-perturbative studies.[7] In order to understand the precise meaning of the above results for physical, four-dimensional theories further studies are required. An investigation of the connection to the linear σ-model as well as a more systematic discussion of renormalization in the approach presented here is currently in progress.[5]

REFERENCES

1. A. Linde, *Phys. Lett.* 93B:327 (1980);
   D. J. Gross, R. D. Pisarski and L. G. Yaffe, *Rev. Mod. Phys.* 53:43 (1981).
2. W. Buchmüller, Z. Fodor, T. Helbig and D. Walliser, *DESY Preprint* DESY-93-021 (1993);
   J.R. Espinosa, M. Quiros and F. Zwirner, *Phys. Lett.* 314B:206 (1993).
3. L. Dolan and R. Jackiw, *Phys. Rev.* D9:3320 (1974).
4. T. Appelquist and R. Pisarski, *Phys. Rev.* D23:2305 (1981).
5. W. Buchmüller and O. Philipsen, *in preparation.*
6. A. K. Rebhan, *Phys. Rev.* D48:3967 (1993).
7. A. Billoire, G. Lazarides and Q. Shafi, *Phys. Lett.* 103B:450 (1981);
   T. S. Biró and B. Müller, *Nucl. Phys.* A561:477 (1993).

NEXT-TO-LEADING ORDER DEBYE-SCREENING IN SPONTANEOUSLY BROKEN GAUGE THEORIES

Anton K. Rebhan

Theory Group, DESY
Notkestr. 85
D-22603 Hamburg, Germany

1. SUMMARY

The main rôle of the Debye screening mass in the perturbative treatment of the electro-weak phase transition is the reduction of the cubic term that determines the strength of a first-order transition. In this note I point out that the standard definition of the Debye mass is unphysical. Its next-to-leading order corrections in resummed perturbation theory are gauge dependent generally in nonabelian gauge theories, and even in Abelian theories when in the Higgs phase. A gauge independent definition can be obtained from a gap equation for the propagator rather than the self-energy, which turns out to be perturbatively under control in the Higgs phase, but sensitive to the nonperturbative magnetic mass scale in the symmetric phase of nonabelian theories.

2. ABELIAN HIGGS MODEL

In the Abelian Higgs model with Lagrangian

$$\mathcal{L} = -\tfrac{1}{4}F^2 + |D_\mu\Phi|^2 + \lambda v^2|\Phi|^2 - \lambda|\Phi|^4, \qquad \sqrt{2}\Phi = \varphi + i\chi \tag{1}$$

and R_ξ-gauge fixing term $\mathcal{L}_{g.f.} = -\frac{1}{2\xi}(\partial A - \frac{\xi}{2}e\varphi\chi)^2$, the leading-order results for the various masses at high temperature read

$$m_L^2 = \tfrac{1}{3}e^2T^2 + m^2, \qquad m_T^2 = m^2 \equiv e^2\varphi^2, \tag{2}$$

$$m_\varphi^2 = 3\lambda\varphi^2 - \lambda v^2 + (\tfrac{e^2}{4} + \tfrac{\lambda}{3})T^2, \tag{3}$$

$$m_\chi^2 = \lambda\varphi^2 - \lambda v^2 + (\tfrac{e^2}{4} + \tfrac{\lambda}{3})T^2 + \xi m^2, \tag{4}$$

where m_L and m_T are the longitudinal (Debye) and the transverse (magnetic) mass of the photon propagator.[1]

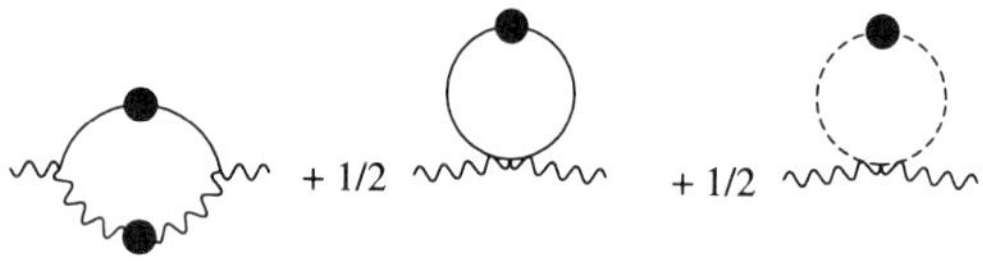

Figure 1. Dressed one-loop diagrams for Π_{00} in the Abelian Higgs model.

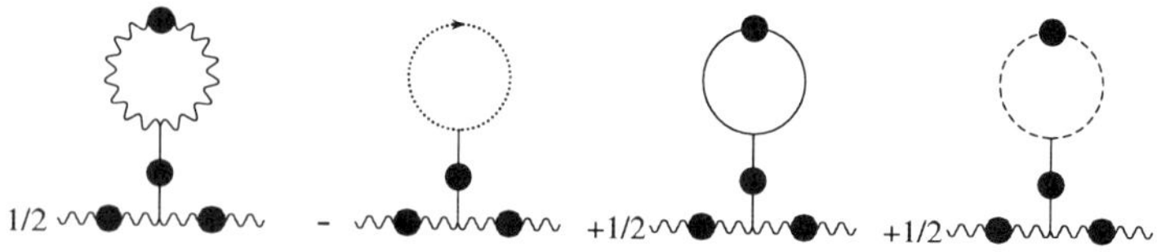

Figure 2. Additional dressed one-loop corrections to the longitudinal gauge boson propagator. Wavy, dotted, full, and dashed lines correspond to gauge bosons, Faddeev-Popov ghosts, Higgs and Goldstone particles, resp.; a blob on these lines marks one-loop dressed propagators.

The next-to-leading order result for $\Pi_{00}(0)$, which is usually taken as the definition of the Debye mass squared, is given by the dressed one-loop diagrams of Fig. 1,[1]

$$\Pi_{00}(k_0=0, k \to 0) = \tfrac{1}{3}e^2T^2 + m^2 - \tfrac{e^2T}{4\pi}\left(\tfrac{4m^2}{m_L+m_\varphi} + m_\varphi + m_\chi\right). \tag{5}$$

Because all higher-order calculation to date have been performed in the Landau gauge, it seems to have gone unnoticed that this definition of the Debye mass is gauge dependent through its dependence on the Goldstone boson mass (4), so that it cannot be the correct one.

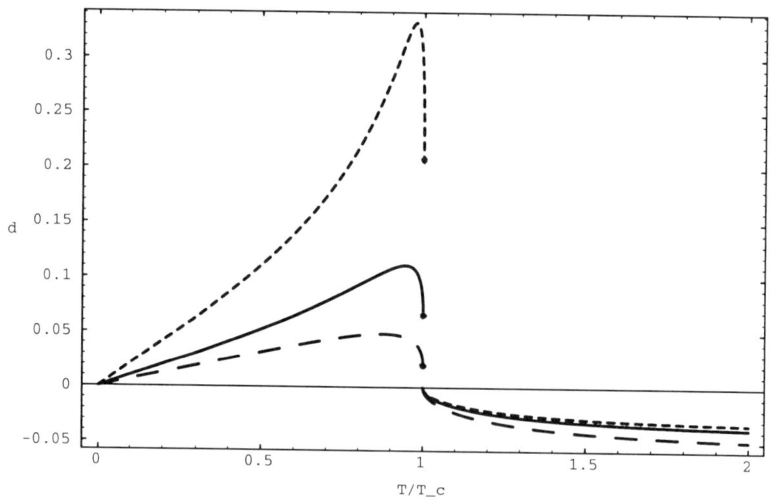

Figure 3. $d \equiv \delta m_L/m_L$ over T/T_c in the Abelian Higgs model for $e = 0.3$. The full line is for $\lambda = e^3$; the short-dashed one for $\lambda = e^4$, where the high-T expansion ceases to apply; the long-dashed one for $\lambda = e^2$, where perturbation theory breaks down.

Defining instead[2] the correction to the leading-order Debye mass (2) through the position of the pole of the longitudinal gauge boson propagator at imaginary wave vector $k^2 = -m_L^2$,

$$\delta m_L^2 = \delta\mathcal{D}_{00}^{-1}(k^2 \to -m_L^2), \tag{6}$$

one indeed is led to a different result. Firstly, the momentum-dependence of the self-energy diagrams has to be taken into account, which is non-trivial in the first diagram of Fig. 1, and secondly, there are additional next-to-leading order corrections in the Higgs phase coming from the reducible diagrams of Fig. 2, which account for the corrections to $V'(\varphi)$. Together this yields

$$\begin{aligned}\delta m_L^2 = -\frac{e^2T}{4\pi}\Big\{ & 2\frac{m^2}{m_L}\ln\frac{2m_L+m_\varphi}{m_\varphi} + m_\varphi \\ & + m_\chi\Big(1-\frac{2\lambda m^2}{e^2 m_\varphi^2}\Big) - \frac{6\lambda m^2}{e^2 m_\varphi} - \frac{2m^2}{m_\varphi^2}(2m+m_L)\Big\}.\end{aligned} \tag{7}$$

Evaluated at the minimum of $V(\varphi)$, the term involving the Goldstone boson mass m_χ becomes gauge independent. In the symmetric phase one has $m_\chi = m_\varphi$, whereas in the Higgs phase the coefficient of the then gauge dependent m_χ vanishes.

In Fig. 3, $\delta m_L/m_L$ is given as a function of T/T_c for $e = 0.3$ and some values of λ. Remarkably, this correction term is discontinuous at T_c even when the phase transition itself were second order. However, perturbation theory breaks down for T very close to T_c.

3. SU(2) HIGGS MODEL

In the nonabelian case, there is a more complicated gauge dependence introduced additionally by the diagrams of Fig. 4. In $\Pi_{00}(0)$ they contribute

$$\frac{g^2T}{\pi}\Big[-m_T + \xi\frac{m_L^2}{m_L+\sqrt{\xi}m_T}\Big] \tag{8}$$

in the case of the SU(2) Higgs model. The last term did not show up in previous analyses because of their restriction to Landau gauge[3].

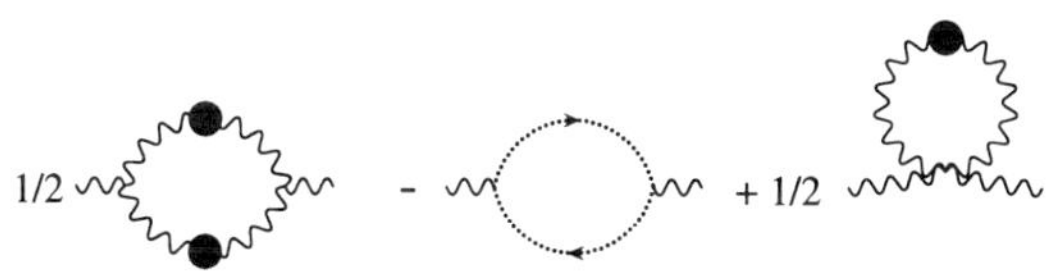

Figure 4. Additional dressed one-loop contributions to Π_{00} in the nonabelian case.

Evaluating these diagrams at general k however yields

$$\begin{aligned}&\frac{g^2T}{\pi}\Big[-\tfrac{1}{2}m_T - \tfrac{1}{2}m_L + \tfrac{1}{k}\big(m_L^2 - \tfrac{1}{2}m_T^2 - k^2\big)\arctan\frac{k}{m_L+m_T} \\ &+ \big(k^2+m_L^2\big)\Big\{\frac{k^2+m_L^2}{m_T^2 k}\Big(\arctan\frac{k}{m_L+\sqrt{\xi}m_T} - \arctan\frac{k}{m_L+m_T}\Big) + (\sqrt{\xi}-1)\frac{1}{m_T}\Big\}\Big],\end{aligned} \tag{9}$$

and the gauge parameter is seen to drop out only at $k^2 = -m_L^2$ as prescribed by the definition (6) for the Debye mass.

Together with the other contributions which are analogous to the Abelian case, the next-to-leading order correction to the SU(2) Debye mass reads

$$\begin{aligned}\delta m_L^2 = \frac{g^2T}{\pi}\Big[& -\tfrac{1}{2}m_L - \tfrac{1}{2}m_T + \Big(m_L - \frac{m_T^2}{4m_L}\Big)\ln\frac{2m_L+m_T}{m_T} \\ & - \frac{m^2}{8m_L}\ln\frac{2m_L+m_\varphi}{m_\varphi} + \tfrac{1}{8}m_\varphi + \frac{3g^2}{64\lambda}(2m_T+m_L)\Big]\end{aligned} \tag{10}$$

in the Higgs phase. In the symmetric phase, however, one encounters a logarithmic singularity due to the vanishing of m_T in perturbation theory. Assuming a nonvanishing magnetic mass $m_T \ll m_T$, one is led to

$$\delta m_L^2 = \frac{g^2 T}{\pi}\left[m_L\left(\ln\frac{2m_L}{m_T} - \frac{1}{2}\right) - \frac{1}{4}m_\varphi\right]. \tag{11}$$

The sensitivity of (11) to the magnetic mass scale in the symmetric phase means however that perturbation theory is no longer under control. Only the coefficient in front of $\ln(2m_L/m_T) \sim \ln(1/g)$ is reliably calculable. The full result (11) only holds for the (rather crude) assumption of a simple mass term in the transverse propagator.

In Fig. 5, $\delta m_L/m_L$ is plotted for $g = 0.66$, $m_H = m_W$ and $m_{top} = 2m_W$. For T very close to T_c, where a (finite) discontinuity in δm_L arises, perturbation theory breaks down (as in the Abelian case) because of the vanishing of the Higgs mass. In the symmetric phase a value $m_T = 0.28g^2T$ for the hypothetical magnetic screening mass has been adopted, which is consistent with some lattice simulations[4], a more recent semiclassical result[5], and also with the result presented at this Workshop by Philipsen[6].

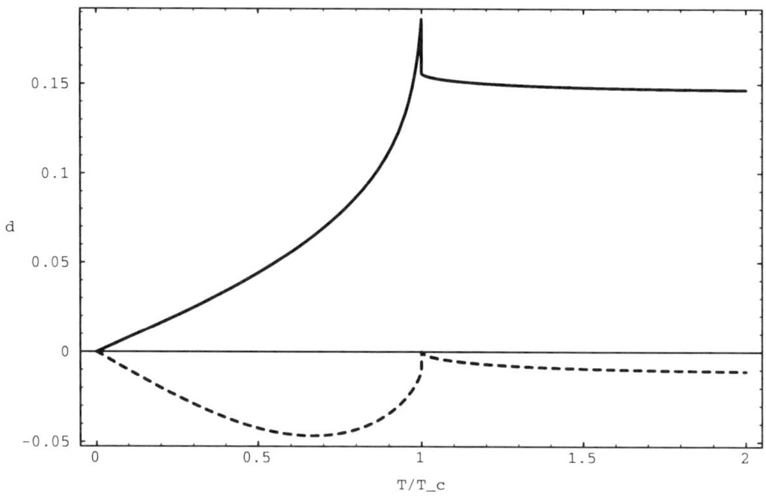

Figure 5. $d \equiv \delta m_L/m_L$ over T/T_c in the SU(2) Higgs model for $g = 0.66$ and $m_H = m_W = m_{top}/2$. The result in the symmetric phase ($T > T_c$) depends on the magnetic screening mass, which has been chosen as $0.28g^2T$. The dashed line gives the "off-pole" result in Landau gauge[3].

REFERENCES

1. W. Buchmüller, T. Helbig and D. Walliser, *Nucl. Phys.* B407:387 (1993).
2. A. K. Rebhan, *Phys. Rev.* D48:3967 (1993).
3. P. Arnold and O. Espinosa, *Phys. Rev.* D47:3546 (1993);
   W. Buchmüller, Z. Fodor, T. Helbig and D. Walliser, DESY report 93-021 (1993);
   J. R. Espinosa, M. Quirós and F. Zwirner, *Phys. Lett.* B314:206 (1993).
4. A. Billoire, G. Lazarides and Q. Shafi, *Phs. Lett.* 103B:450 (1981);
   T. A. DeGrand and D. Toussaint, *Phys. Rev.* D25:526 (1982).
5. T. S. Biró and B. Müller, *Nucl. Phys.* A561:477 (1993).
6. O. Philipsen, this volume.

HIERARCHY OF EFFECTIVE MODELS

A. Jakovác

Department of Atomic Physics
Eötvös University, Budapest, Hungary

INTRODUCTION

It is well known, that because of various infrared problems in a system with spontaneously broken symmetry at finite temperature a simple one-loop calculation is not satisfactory e.g. for calculating the effective potential. That means that at least some fields or components need more precise treatment. This suggests a hierarchy of the fields with respect their relevance in the transition range (long range behaviour). By gradually integrating out the less important fields we get a hierarchy of effective models of the remaining degrees of freedom [1].

The organizing principle of this hierarchy is the mass (correlation length) of the fields. This assumption can be based on general experience, but it is very similar to the RG techniques, where again we concentrate on the lightest components, the more massive ones just modify the interaction of them. With this gradual elimination of massive modes a relative simple one loop integration might be enough at each step to give a more precise result.

The hierarchy of course is not given a priori, in general a self-consistent treatment is needed. However we expect – if perturbation theory is valid – that well separated tree-level mass scales remain well separeted also after the quantization. In this sense at sufficiently high temperature the nonstatic Matsubara modes having a tree-level mass $m_{n\neq 0} = 2\pi T n$ will be at the bottom of this hierarchy; they can be integrated out first giving an effective 3D theory of the static fields: this is the dimensional reduction [2]. However, it could be just the first step: if there exist further mass scales, further integrations can be performed. For example the A_0^a static fields can be integrated out safely in the case of the SU(2) Higgs model [1].

After each integration an infinite number of terms (even in the Gaussian case) is generated to the effective action. In order to keep control on the theory we have to truncate it at some stage. If there exists a large mass scale (e.g. T is large), the operators suppressed by inverse powers of it are usually dropped. Clearly, the consistency check includes also the checking quantitatively the effect of the omission of these pieces of the effective action.

In the rest of this contribution I would like to examine the dimensional reduction of the SU(2) Higgs model, generating the effective potential up to 6 dimensional operators. I shall find the region where the higher dimensional terms can be omitted giving the commonly used [3,4] effective theory which contains only 4 dimensional renormalizable operators.

DIMENSIONAL REDUCTION OF THE SU(2) HIGGS MODEL

I would like to go through the steps of the dimensional reduction including the generation of the higher dimensional operators in the effective potential. The Lagrangian of the theory:

$$\mathcal{L} = \frac{1}{4} F^a_{\mu\nu} F^a_{\mu\nu} + \frac{1}{2}(D_\mu \Phi)^\dagger (D_\mu \Phi) + \frac{1}{2} m^2 \Phi^\dagger \Phi + \frac{\lambda}{24}(\Phi^\dagger \Phi)^2 \tag{1}$$

We choose a general background (A^a_i, A^a_0, Φ; the nonstatic fields are a^a_i, a^a_0, ϕ) what can be held constant for the calculation of the effective potential. Then we fix the thermal axial gauge $a^a_0 = 0$. This is a full gauge fixing for nonstatic modes, but leaves the gauge symmetry of the 3D theory. It makes possible to choose $A^a_i = 0$, the A^a_i dependence can be restored by minimal coupling principle (setting all wave function renormalization constant to be 1). Then we can integrate out the nonstatic fields at one loop level:

$$V^{3D}_{eff} = V^{(0)} + \frac{1}{2} \int_p' \ln \det K \tag{2}$$

where K is the fluctuation matrix, $\int_p' = T \sum\limits_{n \neq 0} \int \frac{d^3 p}{(2\pi)^3}$. Hence

$$\ln \det K = \ln \text{polynomial}\,(p^2, \omega_n^2; \lambda, g; A^a_0, \Phi) \tag{3}$$

and because there are no IR divergencies ($n \neq 0$!) the integral can be expanded with respect to the fields and mass around 0. By dimensional reasons this is at once a $1/T$ expansion. So the higher dimensional operators are suppressed at high temperature; the question is, how relevant are they at the critical temperature T_c. The last step is the renormalization. It can be done with zero temperature counterterms, but the renormalization prescription might also depend on T (e.g. through renormalization scale). After the renormalization the theory is totally fixed: we did not get the freedom of choosing the counterterms of the 3D theory. They will arise in the reduction as T-dependent divergencies. The action can be trivially completed with the assumption $Z = 1$ for all wave function renormalization constants. However keeping a general background with the help of the gauge invariance one can find the Z factors as well [5].

Choosing the simplest renormalization prescription

$$\left.\frac{\partial^2 U_{T-indep.}}{\partial \Phi^2}\right|_{\Phi=0} = m_R^2 \qquad \left.\frac{\partial^4 U_{T-indep.}}{\partial \Phi^4}\right|_{\Phi=0} = \lambda_R \qquad \left.\frac{\partial^4 U_{T-indep.}}{\partial A^{a\,2}_0 \partial \Phi^2}\right|_{\Phi=0} = \frac{1}{2} g_R \tag{4}$$

we obtain the following result:

$$\mathcal{L}^{3D}_{SU(2)} = \frac{1}{4}F^a_{ij}F^a_{ij} + \frac{1}{2}(D_iA_0)^a(D_iA_0)^a + \frac{1}{2}\left(m^2 + \Delta m^2 + \delta_\Phi m^2 + \frac{T^2}{12}\left(\frac{9}{4}g^2 + \lambda\right)\right.$$
$$\left. - \frac{\Lambda T}{2\pi^2}\left(\frac{9}{4}g^2 + \lambda\right)\right)\Phi^\dagger\Phi + \frac{1}{2}\left(\frac{5}{6}g^2T^2 + \frac{m^2}{8\pi^2} + \delta_A m^2 - \frac{\Lambda T}{2\pi^2}5g^2\right)A_0^aA_0^a$$
$$+ \frac{1}{24}(\lambda + \Delta\lambda + \delta\lambda)(\Phi^\dagger\Phi)^2 + \frac{1}{8}(g^2 + \delta g^2)\Phi^\dagger\Phi A_0^aA_0^a + \frac{17g^4}{192\pi^2}(A_0^aA_0^a)^2$$
$$+ \frac{1}{T^2}(a(\Phi^\dagger\Phi)^3 + b(\Phi^\dagger\Phi)^2A_0^aA_0^a + c\Phi^\dagger\Phi(A_0^aA_0^a)^2 + d(A_0^aA_0^a)^3) + \ldots \tag{5}$$

Here the A_0^a fields become scalars in the adjoint representation, we get the usual thermal masses for A_0^a and Φ, and we get the 3D counterterms as well.

The Δ terms refer to the corrections which would come by choosing another renormalization prescription, δ terms refer to the corrections from the expansion of the logarithm (3) with respect to m^2/T^2. I have written out the 6 dimensional operators as well.

How important are these corrections?

If we choose another, physically more meaningful renormalization condition, e.g.

$$\left.\frac{\partial U_{T-indep.}}{\partial\Phi}\right|_{\Phi=v_0} = 0 \qquad \left.\frac{\partial^2 U_{T-indep.}}{\partial\Phi^2}\right|_{\Phi=v_0} = -2m_R^2 \tag{6}$$

Linde-type one ($v_0 = -6m^2/\lambda$ the classical vacuum expectation value for the Higgs field), which allows us to fix the parameters in the perturbative region and to find a smooth $T \to 0$ connection, we get for instance:

$$\frac{\Delta m^2}{m^2} = -\frac{1}{128\pi^2}\left(\frac{27g^4}{\lambda} + 20\lambda + 45g^2\right) - \frac{1}{32\pi^2}\left(\left(\frac{9}{2}g^2 + \lambda\right)\ln\frac{3g^2v_0^2}{4T^2} + \lambda\ln\frac{\lambda v_0^2}{3T^2}\right) \tag{7}$$

Because of the logarithm this is important when T is much smaller or much larger than v_0.

A typical example of the δ-type corrections is:

$$\frac{\delta_\Phi m^2}{m^2} = \frac{m^2}{T^2}\frac{\zeta(3)}{128\pi^4}\left(\lambda - \frac{9}{4}g^2\right) \tag{8}$$

This is important at low temperature as compared to m, however its actual relevance is still suppressed by the small numerical coefficients.

Finally I quote some example of the coefficients of the 6 dimensional operators: $d = 0$ i.e. there is no direct induced $(A_0^aA_0^a)^3$ term; the coefficient of $(\Phi^\dagger\Phi)^3$:

$$a = \frac{\zeta(3)}{1024\pi^4}\left(\frac{3}{16}g^6 - \frac{3}{8}\lambda g^4 - \frac{1}{4}\lambda^2g^2 + \frac{5}{27}\lambda^3\right), \tag{9}$$

it is the most relevant for the effective potential. These terms are important at low temperature or at large Φ values (i.e. in case of strong first order phase transition), but here again the small numerical factors suppress the actual relevance of them.

A physically appealing quantitative answer can be given to the question of relevance of the corrections after a one loop integration of the 3D action. The results are

Table 1. The effect of the corrections in the 1-loop treatment of the phase transition

m_H		T_c/v_0	Φ_c/v_0	σ/T_c^3
20 GeV	a	.3165	1.5756	0.804
	b	.4427	3.3253	2.317
	c	.4351	3.1877	2.219
35 GeV	a	.4080	0.5857	$3.25 \cdot 10^{-2}$
	b	.3946	0.6743	$5.06 \cdot 10^{-2}$
	c	.3947	0.6740	$5.05 \cdot 10^{-2}$
80 GeV	a	.7837	0.2110	$4.88 \cdot 10^{-4}$
	b	.7725	0.2109	$5.06 \cdot 10^{-4}$
	c	.7724	0.2110	$5.05 \cdot 10^{-4}$

a. truncation at 4 dim terms, $\mu = T$ renorm. scale
b. truncation at 4 dim terms, Linde renormalization
c. truncation at 6 dim terms, Linde renormalization

illustrated in the Table. We show in units of v_0 the transition temperature (T_c), the order parameter discontinuity (Φ_c) and the surface tension calculated in the thin wall approximation at T_c. Beyond the global tendency for the weakening phase transition with rising Higgs mass we find the difference between the different parametrisations (a, b, c) vanish as m_H becomes larger: at $m_H = 20GeV$ the discrepancy is more than 100% in the discontinuity, at $m_H = 80GeV$ the difference is almost 0. We can say, that above 35 GeV Higgs mass all the corrections are negligible, and we can work with the simplest action. Below 20 GeV Higgs mass the corrections are of the same order as the leading terms, so the dimensional reduction cannot be applied.

SUMMARY

The consistency of each step in the hierarchical integration of the more massive fields implies the stability of the mass hierarchy and a check of the irrelevance of the omitted operators in the effective action. In the case of the dimensional reduction the first condition is fulfilled as the static Matsubara modes get a mass $\sim gT$. The answer to the second question at one loop level is, that in the experimentally allowed regime of the Higgs boson mass ($m_H > 50GeV$) all the corrections (renormalization corrections, corrections of higher dimensional operators) are negligible, so we can work with the simplest action (5) without Δ and δ terms. The consistency at higher loops is a task for future investigations.

REFERENCES

[1] A. Jakovác, K. Kajantie and A. Patkós, HU-TFT-94-1 to appear in Phys.Rev. D.
[2] T. Appelquist and R. Pisarski, Phys.Rev. **23** (1981) 2305.
[3] K. Kajantie, K. Rummukainen and M. Shaposnikov, Nucl.Phys. **B407** (1993) 356.
[4] P. Arnold and L.G. Yaffe, Phys.Rev. **D49** (1994) 3003.
[5] A. Jakovác and A. Patkós work in progress.

DERIVATIVE CORRECTIONS IN A SCALAR THEORY

David E. Brahm

Physics Dept., Carnegie Mellon University
Pittsburgh, PA 15213 (USA)

THE DECAY RATE OF THE FALSE VACUUM

A scalar field theory whose potential V has two local minima may tunnel out of the false vacuum (ϕ_f) by the nucleation and subsequent growth of bubbles of true vacuum (ϕ_t). The nucleation rate per unit volume at temperature T is, to 1-loop order,[1,2]

$$\frac{\Gamma}{\mathcal{V}} = \frac{T^4}{2\pi} \left(\frac{E_c}{2\pi T}\right)^{3/2} \frac{|\omega_-|/2T}{\sin(|\omega_-|/2T)} \, e^{-F_c/T} \,. \tag{1}$$

E_c is the classical energy of the critical bubble, a static and spherically symmetric field configuration $\bar{\phi}(r)$, which extremizes the classical action[3] subject to periodic boundary conditions in Euclidean time.

F_c is the critical bubble free energy, defined at 1 loop by

$$F_c = E_c + \sum_j{}' \left[\frac{\omega_j - \omega_j^0}{2} + T \ln\left(\frac{1 - e^{-\omega_j/T}}{1 - e^{-\omega_j^0/T}}\right)\right] + F^{\rm ct} \,. \tag{2}$$

The sum is over eigenvalues of the operator $[-\nabla^2 + V'']$, evaluated about the bubble (ω_j) or about the vacuum (ω_j^0). The prime on the sum indicates omission of the bubble's breathing mode $\omega_-^2 < 0$ and three translational modes, which show up instead in the prefactor of eq. (1). The fact that the sum in eq. (2) contains 4 more vacuum eigenvalues than bubble eigenvalues is essentially responsible for the T^4 in the prefactor of eq. (1). Counterterms $F^{\rm ct}$ will not be discussed here.

EFFECTIVE POTENTIAL APPROXIMATION

In place of the difficult sum of eq. (2), one often uses the effective potential approximation to the critical bubble free energy,

$$F_c^{pot} = \int d^3x \left[\tfrac{1}{2}(\partial\phi)^2 + V_{\rm eff}(\phi)\right] = E_c + \int d^3x \left[V_1(\phi) + V_T(\phi)\right] \qquad (3)$$

where (after adding the counterterms)

$$V_1(\phi) = \frac{1}{64\pi^2}\left\{m^4 \ln\left(\tfrac{m^2}{\mu^2}\right) - \tfrac{3}{2}m^4 + 2m^2\mu^2 - \tfrac{1}{2}\mu^4\right\}, \qquad (4)$$

$$V_T(\phi) = \frac{T^4}{2\pi^2}\, I(m/T), \qquad I(y) \equiv \int_0^\infty dx\, x^2 \ln\left(1 - e^{-\sqrt{x^2+y^2}}\right). \qquad (5)$$

Here $m^2 = V''$ evaluated on the relevant configuration, and $\mu^2 = V''(\phi_f)$. The expansion of $I(y)$ for real $y < 2\pi$ is[4,5]

$$I(y) = \frac{-\pi^4}{45} + \frac{\pi^2}{12}y^2 - \frac{\pi}{6}y^3 - \frac{y^4}{32}\left[\ln y^2 - c_3 + \sum_{k=1}^{\infty}\frac{4(2k)!\,\zeta(2k+1)}{k!(k+2)!}\left(\frac{-y^2}{16\pi^2}\right)^k\right] \qquad (6)$$

with $c_3 = \frac{3}{2} + 2\ln(4\pi) - 2\gamma \approx 5.4076$.

THE SIZE OF DERIVATIVE CORRECTIONS

The effective potential is the free energy density for a homogeneous background, so integrating it over a non-homogeneous configuration such as the bubble introduces errors, which we call "derivative corrections". Also, no omission of breathing and translational modes is attempted in the effective potential approximation. Finally, the effective potential as defined above is complex, so we must choose a prescription for deriving a real decay rate.

By numerically comparing the true critical bubble free energy, eq. (2), to the effective potential approximation, eq. (3), we[2] derived the following false vacuum decay rate to replace eq. (1):

$$\frac{\Gamma}{\mathcal{V}} = X\,\frac{\mu^4}{2\pi}\left(\frac{E_c}{2\pi T}\right)^{3/2}\frac{|\omega_-|/2T}{\sin(|\omega_-|/2T)}\; e^{-{\rm Re}\{F_c^{pot}\}/T}, \qquad (7)$$

Three points are worth mentioning:

- The real part of F_c^{pot} appears in the exponent. We tried other, fancier ways of eliminating the imaginary contributions, but none did better than this.
- The scale of the prefactor is now μ^4 [where $\mu^2 = V''(\phi_f)$] instead of T^4.
- The derivative corrections appear in X, a dimensionless number in the prefactor. For thick-walled bubbles we examined, X was between 10^{-1} and 10^{-2}. However, a thin-wall analysis indicated $X > 1$.

These constitute the main conclusions of our work.

THINGS GET WEIRD

We thought we would plug in numbers for the Standard Model (taking m_h = 70 GeV, m_t = 150 GeV). To do so, we first integrated out the non-Higgs particles at the Electroweak Phase Transition critical temperature, and treated the resulting effective potential as if it were the bare potential V in the above analysis. Derivative-type errors arising from this procedure can be substantial[6], but we ignored them. The free energy we then calculated (in the thin-wall limit) turned out negative!

We believe this result is erroneous, and results from our use of the classical extremum $\bar{\phi}(r)$ as our starting point. An analogous and informative problem is the determination of the critical temperature of a scalar theory from the vanishing of the domain wall free energy, a problem discussed in 1986[7] but never resolved.

DOMAIN WALLS AT A GLANCE

A single-component scalar theory has a (renormalized) potential

$$V = \tfrac{\lambda}{4}\phi^4 - \tfrac{\mu^2}{2}\phi^2 \,, \tag{8}$$

with $\lambda \ll 1$. It undergoes a second-order phase transition at a critical temperature

$$T_c = \tfrac{2\mu}{\sqrt{\lambda}}[A + \mathcal{O}(\lambda)] \,. \tag{9}$$

What is A?[8]

A domain wall may be constructed which obeys the classical equation of motion, and its free energy may be calculated[9]. Setting this $F^{wall} = 0$ yields $A = \sqrt{\frac{2}{3}}$.

The "effective mass" approach, setting $V''_{\text{eff}}(0) = 0$, yields $A = 1$. Equivalently, a domain wall can be constructed that interpolates between the minima of V_{eff}, and its free energy approximated by integrating V_{eff} plus a canonical kinetic term over the wall. This free energy vanishes at $A = 1$.

Significantly, the "domain wall free energies" calculated these two different ways agree at low temperature[7]. But near the critical temperature, for $\sqrt{\frac{2}{3}}\frac{2\mu}{\sqrt{\lambda}} < T < \frac{2\mu}{\sqrt{\lambda}}$, the system appears to still have two degenerate minima, but the classical domain wall has negative free energy. We believe this is analogous to our Standard Model bubble result.

CONCLUSIONS

Two methods at our disposal for calculating the critical bubble free energy are:

1) Using the classical bubble solution, add the classical energy to the "ln det" term, eq. (2). The domain wall results indicate that this method leads to errors at higher temperatures.
2) Using the extremum of the effective potential, integrate the canonical kinetic term plus the effective potential, eq. (3). This method misses derivative corrections.

Using Method (1), we found that integrating the real part of V_{eff} was a satisfactory way of dealing with the complexity issue, that the prefactor contains μ^4, and that derivative corrections introduce a dimensionless number in the prefactor of order 10^{-1} or 10^{-2}. These results are only strictly true at low temperatures, since Method (1) is only reliable there, but should be indicative of the effects of derivative corrections.

REFERENCES

1. S. Coleman, *Phys. Rev.* **D15**:2929 (1977), **16**:1248(E) (1977);
   C.G. Callan & S. Coleman, *Phys. Rev.* **D16**:1762 (1977);
   S. Coleman, "The Uses of Instantons", *Proc. 1977 Int. School of Subnuclear Physics, Ettore Majorana*, ed. A. Zichichi (Plenum, New York, 1979); reprinted in *Aspects of Symmetry* (Cambridge University Press, 1985);
   I. Affleck, *Phys. Rev. Lett.* **46**:388 (1981);
   J.W. Cahn & J.E. Hilliard, *J. Chem. Phys.* **31**:688 (1959);
   J.S. Langer, *Ann. Phys.* **41**:108 (1967); *ibid.* **54**:258 (1969);
   M.B. Voloshin, I.Y. Kobzarev & L.B. Okun', *Yad. Fiz.* **20**:1229 (1974) [*Sov. J. Nucl. Phys.* **20**:644 (1975)];
   P.H. Frampton, *Phys. Rev.* **D15**:2922 (1977);
   A.D. Linde, *Phys. Lett.* **70B**:306 (1977); *ibid.* **100B**:37 (1981);
   Nucl. Phys. **B216**:421 (1983); *ibid.* **B223**:544 (1983)(E);
   O.J.P. Éboli & G.C. Marques, *Rev. Bras. Fis.* **16**:147 (1986).
2. D. Brahm & C.L.Y. Lee, *Phys. Rev.* **D49**:4094 (1994);
   C.L.Y. Lee, *Phys. Rev.* **D49**:4101 (1994).
3. M. Gleiser, G.C. Marques & R.O. Ramos, *Phys. Rev.* **D48**:1571 (1993).
4. L. Dolan & R. Jackiw, *Phys. Rev.* **D9**:3320 (1974).
5. H. Haber & H.A. Weldon, *J. Math. Phys.***23**, 1852 (1982).
6. E. Weinberg, *Phys. Rev.* **D47**:4614 (1993); and these Proceedings.
7. C. Aragão de Carvalho, O.J.P. Éboli, & G.C. Marques, *Phys. Rev.* **D34**:659 (1986);
   H. Aoyama & H.R. Quinn, *Phys. Rev.* **D34**:662 (1986).
8. D. Brahm, *The Critical Temperature of* $\lambda\phi^4$ *Theory*, CMU-HEP94-14 (in progress).
9. I. Ventura, *Phys. Rev.* **B24**:2812 (1981);
   C. Aragão de Carvalho *et.al.*, *Phys. Rev.* **D31**:1411 (1985);
   C. Aragão de Carvalho *et.al.*, *Nucl. Phys.* **B265**:45 (1986);
   D. Bazeia *et.al.*, *Z. Phys.* **C46**:457 (1990).

INDEX